Dinosaur
Facts and Figures
The Theropods
and Other Dinosauriformes

Dinosaur
Facts and Figures
The Theropods
and Other Dinosauriformes

Rubén Molina-Pérez and Asier Larramendi
Illustrated by Andrey Atuchin and Sante Mazzei

Translated by David Connolly and Gonzalo Ángel Ramírez Cruz

Princeton University Press
Princeton and Oxford

Published by Princeton University Press
41 William Street, Princeton, New Jersey 08540

press.princeton.edu

LCCN 2018961537
ISBN 978-0-691-18031-1

Editorial: Robert Kirk and Kristen Zodrow
Production Editorial: Kathleen Cioffi
Jacket Design: Lorraine Doneker
Production: Steven Sears
Publicity: Sara Henning-Stout
Copyeditor: Charles J. Hagner

Printed on acid-free paper. ∞
Printed in China
10 9 8 7 6 5 4 3 2

Prologue

Which dinosaur was the largest? What about the smallest? The oldest? These are questions that most children and quite a few adults have asked themselves. Answering such questions has never been easy, and doing so nowadays is even more troublesome. The science of paleontology is evolving at unprecedented speed. Constant discoveries are making many of our previously held beliefs outdated. Preventing one's knowledge from becoming obsolete requires constant attention. New technology and the vast number of studies published each year suggest that a real revolution in dinosaur paleontology (dinosaurology) is occurring—so much so, that buying a book on dinosaurs published in 1995 is as useful to us today as having a look at a catalog of mobile phones from the year 2005.

Under these circumstances, we found ourselves forced to be continually aware of everything being published about these wonderful animals. After years of consulting thousands of scientific articles, visiting museum collections, seeing in many blogs, forums, and social networks very interesting debates about which dinosaur was the largest or smallest, finding that inaccuracies plague different printed and virtual encyclopedias (although it would only be fair to note that they improve daily), observing that very few scientific papers answer these questions, and watching as even the famous *Guinness World Records* removed its section on dinosaurs a few years ago, for not meeting the demands of truthfulness, we decided to work on a book on dinosaur records that had never been done before, by gathering records up to the year 2016.

At first, we considered including every dinosaur group in one book. But we soon realized that such a book would have to be close to 900 pages long. So in the end, we focused on records related to theropods and dinosauromorphs, leaving sauropodomorphs and ornithischians for a future project.

This book is our attempt to respond in the most comprehensive manner possible to any record related to theropods. In addition, each record and all of the data presented have bibliographic references so that the reader can corroborate or explore the information. This makes the book especially reliable and verifiable.

On the other hand, the reader will be able to enjoy plentiful tables containing information that before publication of this book was scattered in many scientific articles. For instance, we present a list of the different bite forces of numerous living and extinct animals. And not only have we collected data already published, but we have done many calculations ourselves, such as the size estimates of theropods. We even propose new methods for estimating the speed of bipedal dinosaurs.

All this information and much more is distributed across eight sections. Most of them are illustrated by fantastic reconstructions based on the latest discoveries. In fact, as mentioned at the outset, paleontology is advancing so fast that we had to redo some dinosaurs, such as *Spinosaurus* and *Deinocheirus*, while writing.

Returning to the chapters of the book, the classification shown in the first chapter is organized into arbitrary groups that are based on the most prominent clades. They collect a good number of specimens, which makes comparison easier. The following two sections have a format similar to the first: Largest and smallest dinosaurs are shown by taxonomic group, Mesozoic epoch, and geographic zone, and an extensive list of pertinent chronological records is also included. In these first three sections, each dinosaur record is depicted alongside a human being and measuring bars for scale, all of which have been conscientiously and rigorously resized so the reader gets the most accurate idea possible of the size of these animals. This aspect has been given special attention because, in the vast majority of books, the scales that compare dinosaurs with humans leave much to be desired.

Subsequently, we move on to chapter 4 ("Prehistoric Puzzle"), which presents all of the dinosaur bone groups with their respective records, such as the largest and smallest femur found, etc. As additional support, a diagram of the *Tyrannosaurus* skeleton aids in visualizing the location of each bone. This section is followed by "Theropod Life," which not only contains a wide range of records related to the biology of theropods but also includes very interesting information about their maturation and diet. The listing of all the first-discovered theropod eggs can also be found in this chapter.

The book's next chapter deals with footprints. Their silhouettes are displayed along with several records from these epochs. "Chronicle and Dinomania" follows, in which a huge compilation of records through history can be found, some dating as far back as 10,000 BC! The types of records vary from artistic and literary to cinematographic and musical.

The last section is the first formally published complete listing of all theropod species up to 2016. This list may be of special interest to all dinosaur specialists. We are convinced that many will discover dinosaurs they were not even aware of.

The book ends with a glossary of terms, an index of species, and a bibliography that—due to its great length—is stored in the cloud at {http://www.eofauna.com/book/en/theropoda_records_refB.pdf}. Also available is an online appendix that presents statistics.

Asier Larramendi and Rubén Molina-Pérez, May 2016

Contents

Comparing species

Mesozoic calendar

The world of dinosaurs

Prehistoric puzzle

Theropod life

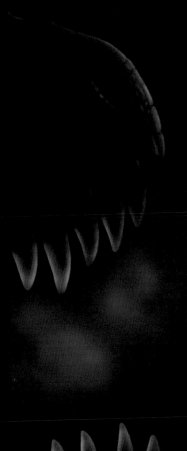

Testimony in stone

Chronicle and dinomania

Theropod list

How to use this book

This book begins with an introduction that explains how each dinosaur's size was calculated and how its appearance was reconstructed. Also provided is an overview that aids in understanding what dinosaurs are. It should be noted that each of the species records included in this book are the largest or smallest *specimens* discovered, as size within a species can vary considerably.

Eight chapters follow the introduction: "Comparing Species," "Mesozoic Calendar," "The World of Dinosaurs," "Prehistoric Puzzle," "Theropod Life," "Testimony in Stone," "Chronicle and Dinomania," and "Theropod List." All sections show the records and information corresponding to these chapters. It should be noted that the identification of footprints is part of an unpublished study (Molina-Pérez et al., manuscript in preparation). The last chapter presents a complete list of all discovered theropods up to this edition's publication. The book ends with a glossary, a taxonomic index, a bibliography, and an appendix.

Animals frequently used for comparison in this book

Animal		Size	Weight
Bee hummingbird		5 cm long	2 g
Sparrow		15 cm long	30 g
City pigeon		35 cm long	300 g
Crow		60 cm long	1.2 kg
Ferret		50 cm long	1.5 kg
House cat		45 cm long (not including tail)	4.5 kg
Iberian lynx		90 cm long (not including tail)	15 kg
Emperor penguin		1.2 m high	30 kg
German shepherd		60 cm high at shoulder	35 kg
Human being		1.8 m high	75 kg
Ostrich		2.7 m high	120 kg
Lion		1.05 m high at shoulder	180 kg
Brown bear		1.25 m high at shoulder	400 kg
Fighting bull		1.4 m high at shoulder	600 kg
Saltwater crocodile		5 m long	600 kg
Great white shark		5 m long	1000 kg
Giraffe		5.3 m high	1.2 t
Hippopotamus		1.5 m high at shoulder	1.5 t
White rhino		1.75 m high at shoulder	2.5 t
Asian elephant		2.75 m high at shoulder	4 t
African elephant		3.2 m high at shoulder	6 t
Killer whale		8 m long	7 t

Other objects used for comparison

Object		Size	Weight
Car		4 m long	-
City bus		12 m long and 3 m high	-

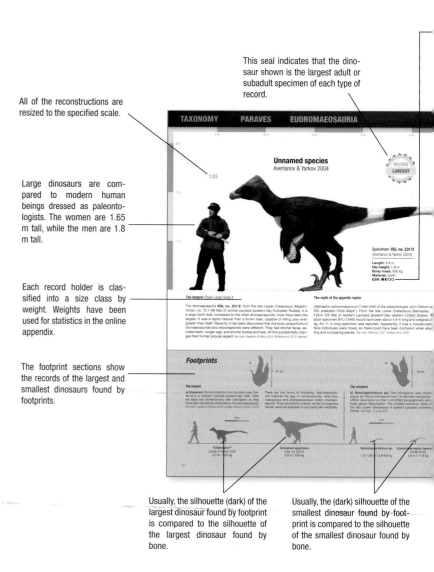

This seal indicates that the dinosaur shown is the largest adult or subadult specimen of each type of record.

All of the reconstructions are resized to the specified scale.

Large dinosaurs are compared to modern human beings dressed as paleontologists. The women are 1.65 m tall, while the men are 1.8 m tall.

Each record holder is classified into a size class by weight. Weights have been used for statistics in the online appendix.

The footprint sections show the records of the largest and smallest dinosaurs found by footprints.

Usually, the silhouette (dark) of the largest dinosaur found by footprint is compared to the silhouette of the largest dinosaur found by bone.

Usually, the (dark) silhouette of the smallest dinosaur found by footprint is compared to the silhouette of the smallest dinosaur found by bone.

Dinosaur lengths in this book

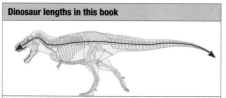

Length ◄───►
Approximate length of the live animal from the tip of the snout to the tip of the tail, measured along the **vertebral centra.**

Scientific books and publications often do not specify how they obtained the length cited. Among vertebrates, length is usually measured as the distance from the tip of the snout or nose to the tip of the tail, along the dorsal side of the animal (along the back or following the spine). This means that animals with very prominent vertebrae (e.g. *Spinosaurus* or *Deinocheirus*) have an exaggerated length, which is not useful for comparing them to animals with less prominent vertebrae. For that reason, the lengths displayed in this book reflect the distance from the tip of the snout to the end of the tail, measured along the **vertebral centra** (not including feathers).

Abbreviations used in this book

Ma	millions of years
mm	millimeters
cm	centimeters
m	meters
km	kilometers
km²	square kilometers
g	grams
kg	kilograms
t	metric ton (1000 kg)
~	approximately
cc	cubic centimeters
N	North
S	South
°C	degrees centigrade
CO₂	Carbon dioxide

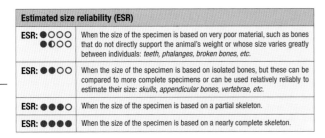

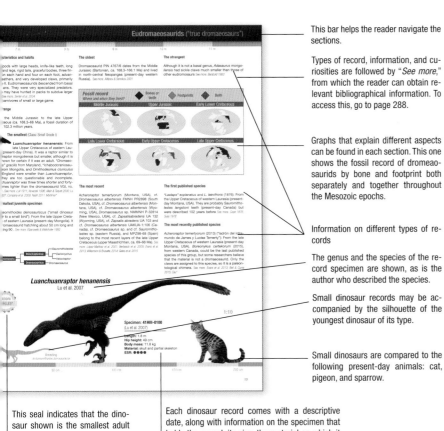

Eudromaeosaurids ("true dromaeosaurs")

Luanchuanraptor henanensis
Lu et al. 2007

Specimen: 41HIII-0100
(Lu et al. 2007)
Length: 1.8 m
Hip height: 49 cm
Body mass: 11.8 kg
Material: skull and partial skeleton
ESR: ●●●●

1:10

This bar helps the reader navigate the sections.

Types of record, information, and curiosities are followed by "*See more*," from which the reader can obtain relevant bibliographical information. To access this, go to page 288.

Graphs that explain different aspects can be found in each section. This one shows the fossil record of dromeaosaurids by bone and footprint both separately and together throughout the Mesozoic epochs.

Information on different types of records

The genus and the species of the record specimen are shown, as is the author who described the species.

Small dinosaur records may be accompanied by the silhouette of the youngest dinosaur of its type.

Small dinosaurs are compared to the following present-day animals: cat, pigeon, and sparrow.

This seal indicates that the dinosaur shown is the smallest adult or subadult specimen of each type of record.

Each dinosaur record comes with a descriptive date, along with information on the specimen that holds the record, its size, the material on which it is based, and the ESR.

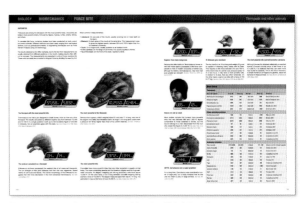

The pages with lists of records in small type are records and curiosities ordered chronologically.

Each chapter contains numerous tables, in which interesting comparative information is summarized. This table shows a list of current and extinct animals with their bite force

Record samples (animal, footprint, or egg) are grouped by their size class. This helps classify them and generate statistics.

SIZE CLASSES (by weight)

MINUSCULE		TINY	
Grade I	0-3 g	Grade I	+12-24 g
Grade II	+3-6 g	Grade II	+24-50 g
Grade III	+6-12 g	Grade III	+50-100 g
DWARF		**VERY SMALL**	
Grade I	+100-200 g	Grade I	+0.8-1.5 kg
Grade II	+200-400 g	Grade II	+1.5-3 kg
Grade III	+400-800 g	Grade III	+3-6 kg
SMALL		**MEDIUM**	
Grade I	+6-12 kg	Grade I	+50-100 kg
Grade II	+12-25 kg	Grade II	+100-200 kg
Grade III	+25-50 kg	Grade III	+200-400 kg
LARGE		**VERY LARGE**	
Grade I	+400-800 kg	Grade I	+3-6 t
Grade II	+0.8-1.5 t	Grade II	+6-12 t
Grade III	+1.5-3 t	Grade III	+12-25 t
GIANT			
	Grade I		+25-50 t
	Grade II		+50-100 t
	Grade III		+100 t

SIZE CLASSES (by footprint length)

MINUSCULE	0-15 mm
TINY	+15-30 mm
VERY SMALL	+30-60 mm
SMALL	+60-120 mm
MEDIUM	+120-250 mm
LARGE	+250-500 mm
VERY LARGE	+50-100 cm
GIANT	+100-200 cm

SIZE CLASSES (by egg weight)

VERY SMALL		SMALL	
Grade I	0-0.25 g	Grade I	+1-2 g
Grade II	+0.25-0.5 g	Grade II	+2-4 g
Grade III	+0.5-1 g	Grade III	+4-8 g
MEDIUM		**LARGE**	
Grade I	+8-15 g	Grade I	+60-120 g
Grade II	+15-30 g	Grade II	+120-240 g
Grade III	+30-60 g	Grade III	+240-500 g
VERY LARGE		**GIANT**	
Grade I	+0.5-1 kg	Grade I	+4-8 kg
Grade II	+1-2 kg	Grade II	+8-16 kg
Grade III	+2-4 kg	Grade III	+16 kg

Methodology and calculations

Size, shape, and appearance of dinosaurs

The different sizes, shapes, and external appearances of the dinosaurs displayed in this book were calculated by different methods, depending on each case. Primarily, we relied on rigorous skeletal reconstructions carried out by today's most respected specialists: Gregory Paul, Scott Hartman, Jaime A. Headden, and Ville Sinkkonen et al. Reconstructions made by the book's authors are also used (figure 1).

The more completely a dinosaur has been preserved, the more easily and accurately its size and appearance in life can be represented. Unfortunately, many dinosaurs are known by only very fragmentary skeletons and sometimes by a single element that may be incomplete, worn, or deformed (bone, tooth, osteoderm, etc.). In these cases, we deduced the proportions and external appearance of the theropods by making phylogenetic, geographical, morphometric, and anatomical comparisons with the closest and best-preserved relatives. We tried to be as precise and conservative as possible when calculating the size of a dinosaur from a single element, reducing the estimate to its minimum. For instance, when we had an isolated tooth, we estimated the dinosaur's size by comparing it with the largest and most robust tooth of another specimen that is phylogenetically close and whose skeleton is more complete. It is assumed that the isolated tooth was the largest of the specimen's teeth, so we present the most conservative estimate possible.

How much does a theropod dinosaur weigh?

Different methods have been proposed throughout history to calculate the mass of dinosaurs. These systems are principally based on two basic forms: allometric and volumetric methods.

Allometric methods are based on the mathematical relationship between measurements of different elements of a body (generally bones) and the body mass of an animal. To do this, a large number of current animals are measured and after translating the results into a graph a regression line is obtained that translates into a mathematical formula that is then applied to the extinct forms. The main problem with this method is that many of the dinosaurs had proportions and sizes very different from today's animals. This often leads to exaggerated results. Moreover, the obtained estimates are frequently too broad, since the different bone elements of the same individual can give very different and imprecise figures—e.g., a dinosaur weighing between 7 and 15 metric tons (7.7–16.5 short tons).

Volumetric methods are based on physical or digital models of the animals to be estimated. The results of this system depend on the accuracy of the reconstructions. For a model to be rigorous, it must be based on scientifically described material (osteological measurements, photos, etc.) and extensive knowledge in comparative anatomy. Once the volume of the model has been determined, the animal's volume in natural size is obtained by multiplying the result by the scale factor cubed. Finally, the mass is obtained by multiplying the volume by the density that the animal is estimated to have had in life. This method's main problem is that in many cases the extinct animal is not preserved sufficiently well for an adequate three-dimensional model.

Today, volumetric methods are becoming more common when estimating the mass of extinct vertebrates. The first to use this method made physical models to scale that were submerged in water tanks, using the displaced water to calculate the volume. Nowadays, this method has been modernized, and the models are usually digital.

For this book, we have decided on the volumetric method known as *graphic double integration* (GDI) because it is a simple, fast, and very effective way to obtain the mass of dinosaurs. The GDI method was created by Jerison (1973) to estimate the volumes of endocranial models from dorsal and lateral views. Thereafter, different authors began to apply this method to whole animals. In GDI, the body or parts of the body are treated as elliptical cylinders. This requires two perspectives of the model to be studied (lateral and dorsal). The model is divided into slices, and its volume is calculated using the radiuses or diameters of both perspectives. The figures obtained from each slice are added up to obtain the final result using the following equation:

$$V = \pi (r^1)(r^2)(L)$$

Although volumetric methods have been refined over time, some results pertain-

ing to dinosaurs are still controversial. As an example, the volume of the *Tyrannosaurus rex* specimen famously known as "Sue" (FMNH PR2081) has been estimated. Studies published in recent years estimate volumes ranging between 7 and 10 m³ (247.2 and 353.1 ft³). The result obtained from the rigorous skeletal reconstruction (figure 1) and the application of the GDI method is 8.7 m³ (307.2 ft³).

Once a dinosaur's volume (or volumes, as it is often more practical to divide the body into sections) is known, it can be multiplied by the density or average specific gravity of the animal's body (or that of the sections) to obtain the dinosaur's mass. Furthermore, it should be noted that feathers account for about 6% of the total weight of heavily feathered birds. This percentage ought to be added to the body mass of the most heavily feathered theropods.

Specific gravity

In this case, density or specific gravity (SG) is the comparison of the body density of an animal with that of water as a reference, which is equal to 1.0.

It should be made clear that the SG of any living or extinct animal varies depending on the air within the respiratory system. It would be ideal to calculate the SG on a live and relaxed animal (normal breathing). This is the most common state during the animal's daily life (walking, sleeping, feeding, etc.). In these circumstances, the lungs are not full of air. This is generally not taken into account when calculating the SGs of animals, leading to quite conservative results.

As dinosaurs are an extinct group, the SGs applied to the volume obtained from their volumetric reconstructions will always be a rough estimate.

To better understand the SG that is applied to theropods, it would be good to provide examples from different kinds of live animals, since there are large differences between groups. Among today's animals, only birds exhibit ample pneumatization in the postcranial skeleton and a respiratory system constituted of air sacs. These characteristics are shared by saurischians (sauropodomorphs and theropods), but not by ornithischians. The latter would be more comparable with present mammals and reptiles, so the applied SG is quite varied among dinosaurs.

It would be instructive to determine the density or SG of the human being. To do so, we must first know the different lung volumes of an adult person:

Tidal volume (TV): Volume of air displaced with each normal breath, which is about 500 mL (17.6 oz) or 7 mL/kg (0.1 fl oz/lb) of body mass for a human adult.

Inspiratory reserve volume (IRV): Maximum volume of air that can be inspired over the normal tidal volume; this is usually about 3,300 mL (111.6 fl oz).

Expiratory reserve volume (ERV): Maximum additional amount of air that can be exhaled by forced exhalation, after normal expiration; this is generally around 1,000 mL (33.8 fl oz).

Residual volume (RV): Volume of air remaining in the lungs after forced expiration; on average, this is about 1,200 mL (40.6 fl oz).

The total lung capacity (TLC) would be the sum of the four constants mentioned above, resulting in 6 L of air (202.8 fl oz), representing around 8% of the total volume of the person. This percentage (8%–10%) is common for all terrestrial mammals, and a TV of about 10% of the TLC is typical in terrestrial mammals. Aquatic mammals may have a TV in excess of 75% of their TLC.

In a relaxed adult, the air within his or her lungs is 2.2–2.7 L (74.4–91.3 oz), RV (1.2 L) + ERV (1 L) ± (TV 0.5 L). This is 37%–45% of the adult's TLC and represents about 3.5% of the body's total volume. When a person is relaxed in fresh water, the SG of which equals 1.0, he or she barely floats. However, most people sink when they expel the TV and a small part of the ERV, which represent about 1% of their total volume. This tells us that the average SG of a human being is 0.99, practically the same as that of fresh water. An SG of about 1.0 is common for the vast majority of terrestrial mammals.

As far as the archosaurs are concerned, with the exception of birds, they have a lung capacity considerably lower than that of mammals. Several studies have calculated the SGs of different reptiles today. Colbert (1962) calculated SGs as low as 0.89 for alligators (*Alligator mississippiensis)* and 0.81 for Gila monsters (*Heloderma suspectum).* In contrast, Cott (1961) calculated an SG of 1.08 for

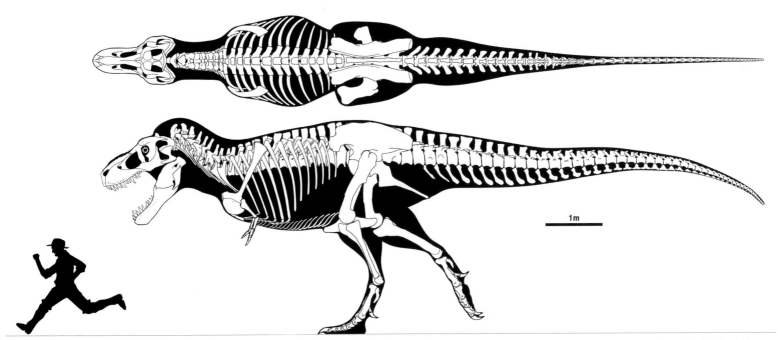

Figure 1. Rigorous reconstruction of the skeleton of Tyrannosaurus FMNH PR2081, popularly known as "Sue." The total volume estimated by GDI is 8,700 L (2,298.3 US gal), which means a body mass of about 8,265 kg (18,221.2 lb) after applying a SG of 0.95.

the Nile crocodile *(Crocodylus niloticus)* after analyzing nine individuals. These variations can be explained by several reasons: The low density for the alligator may be due to measuring a young individual weighing only 280 g (0.6 lb). This would mean that its bones would have been less compact than those of an adult and its muscle and bone ratio would likely have been considerably lower, so it should not be taken as representative data. Cott (1961) obtained results on dead individuals, which means the specimen's lungs would have been completely empty of air. It is also possible that some of the crocodiles presented gastroliths in the stomach, which would have produced higher results. The observation of live alligators and crocodiles, including the youngest individuals and animals that had not ingested gastroliths, shows that they have the ability to rest submerged at the bottom of rivers.

The SG of relaxed alligators and their close relatives should, therefore, be around 1.00–1.05.

On the other hand, birds are a present-day group that is most similar to the majority of the theropods. These dinosaurs may have the most complicated SG to estimate, as birds unlike mammals have air sacs in their respiratory system as well as lungs. They also have pneumatic bone in their postcranial skeleton. These factors lead to birds' very varied and relatively low SGs. To further complicate matters, birds are usually covered by dense plumage, which increases their total volume. Most birds, except several families that comprise mostly land birds, have a gland known as the uropygial gland. Waxes are secreted from this gland and, through preening, impregnate the plumage. This creates a water-repellent coating that increases the buoyancy of birds. This is why it's important to exclude feathers when calculating the SG. However, the weight of their plumage is later added to obtain the animal's total mass.

Going back to the lung capacity of birds, two different concepts are often confused or mixed up. The lungs and air sacs of birds can occupy 10%–20% of the total body volume (King 1966). This does not necessarily mean that the total capacity of the respiratory system is 10%–20% of the total volume of the body, since the lungs and air sacs have their own mass (parabronchi, soft tissues, fluids, blood, etc.). Moreover, many studies do not generally specify if they are filled with air to maximum capacity.

Specific gravities as diverse as 0.73, 0.8, 0.9, and up to 0.937 have been presented for birds.

Hazlehurst and Rayne (1992) obtained an average SG of 0.73 with a sample of twenty-five birds from twelve different species. In order to calculate this mean SG, the study was performed on dead birds with their respiratory system artificially inflated. The results obtained are unreliable, since bird lungs and air sacs are not completely filled with air in a state of normal breathing.
It is possible that the total lung capacity of the lightest flying birds with an over-

developed sternum, which gives them more space for the respiratory system, can reach almost 20% of the total body volume. What is more, another small percentage of air could be in their pneumatic bones or their extraskeletal diverticulum. In the case of these less dense birds, their SG would be around 0.85 in a relaxed state.

The respiratory system of the ostrich *(Struthio camelus)* has a total capacity that is considerably lower than that of the lightest flying birds. This is because one of the main functions of air sacs is to ventilate the lungs due to the high metabolic rate required for flight. The ostrich, owing to its large size and inability to fly, has a slower metabolic rate and does not need such a large respiratory system. As a result, it is foreseeable that its SG would be significantly higher than that of flying birds.

The total capacity of the respiratory system (including air sacs) of an adult ostrich that weighs 100 kg (220.4 lb) is around 13.5% of the animal's total volume. The TV represents 10% of the total respiratory system (Schmidt-Nielsen et al. 1969), which is comparable to some terrestrial mammals. The respiratory system of a relaxed ostrich is not completely full. Moreover, the muscular system, solid bones, and most body tissues have a density greater than 1.0, so a minimum SG of 0.9 and probably closer to 0.95 is expected for this bird. Furthermore, if one observes great terrestrial birds like emus and ostriches swimming, it seems obvious that they have an SG superior to that of flying birds. When one of these birds swims, only the head, neck, and a small part of the back are visible, which represent about 10%–15% of the total body volume (figure 2). In addition, one must take into account that, by swimming, the body is propelled toward the surface, so in a relaxed state the body would sink more. Therefore, an SG around 0.9 or higher can be expected for today's large land birds. These estimates are similar to the results of Welty (1962) and Alexander (1983).

The larger an animal, the lower its metabolic rate. The reduction of lung capacity in larger animals is also common in the animal kingdom. This is evident in both birds and mammals. Members of the porpoise family (Phocoenidae), the smallest of today's cetaceans, are comparable in size to humans. The ratio between their lung capacity and body mass is about 8%, which is comparable to ours, but in large whales, the proportion drops to 1.6%–2.7%.

This leads us to believe that large theropods must have had a respiratory system that was considerably smaller than that of present-day terrestrial and flying birds and, therefore, a greater body density. In addition, theropods—with the precursors of avian air sacs—may have had a less-developed respiratory system.

Finally, it can be said that animals with semi-aquatic habits develop certain adaptations for diving and swimming. For example, the bones of semi-aquatic animals are often considerably denser than those of purely terrestrial animals. This property helps them increase their body density, allowing greater control over

Methodology and calculations

buoyancy. The SGs of this type of animal are usually above 1.0. A recent study of *Spinosaurus* clearly indicates that this theropod was fully adapted to aquatic life. This is corroborated by its external anatomy and its long bones, which are about 30%–40% denser than those of any other theropod. It is possible that the SG of the *Spinosaurus* was comparable to that of the current crocodiles.

In summary, the most derivative and smaller avian theropods would be the least dense dinosaurs, with SGs of around 0.85. In general, as the body size of the theropods increased, their SGs would tend to grow, as their respiratory system would be reduced, just as in today's birds. Their skeletons would have tended to be more robust and heavier. Theropods with semi-aquatic habits would be the densest.

See more: *Scholander 1940; Cott 1961; Colbert 1962; Welty 1962; King 1966; Ganong 1969; Jerison 1973; Dorst 1974; Schmidt-Nielsen 1969, 1984; Lockley 1981; Alexander 1983; Perry 1988; Hazlehurst & Rayne 1992; Hurlburt 1999; Wartzok 2002; Henderson 2004; Wedel 2004, 2005; O'Connor & Claessens 2006; Wartzok 2002; Vickers-Rich 2004; Sellers et al. 2012; Yates et al. 2012; Ibrahim et al. 2014; Larramendi 2016*

Dinosaur	SG	Condition
Theropoda 1	0.85	More derived and a body mass of less than 30 kg (66.1 lb)
Theropoda 2	0.9	Body mass: 30–500 kg (66.1–1,102.3 lb)
Theropoda 3	0.925	Body mass: +500–1,000 kg (1,102.3–2,204.6 lb)
Theropoda 4	0.95	Basal or of a mass exceeding 1,000 kg (2,204.6 lb)
Theropoda 5	1.0 -1.05	Aquatic or semi-aquatic

Table of specific gravities. Densities proposed and applied in this book to the different theropods due to their body size, evolutionary and adaptive states.

Scales, filaments, or feathers?

There are doubts about what kind of integument each species of dinosaur would have had, since specimens have been preserved with the impressions or presence of complex scales, osteoderms, filaments, and feathers. For a long time, feathers were considered to be unique to birds, but we now know that feathers were present not only in theropods close to birds but even in dinosaurs very distant from them, such as some ornithischians *(Kulindadromeus, Psittacosaurus,* and *Tianyulong)*. Furthermore, it is possible that the filaments of pterosaurs had the same origin, since these—in conjunction with dinosaurs—come from avemetatarsalians, though there are some doubts.

We now know that some dinosaurs, like birds, had scales and feathers on different parts of the body, but unless there is evidence, there is no reason to reconstruct a dinosaur with scales or filaments. However, it is necessary to consider that feathers with rachises and barbs are present from ornithomimosaurs to modern birds. It should be noted that contour feathers with rachises and barbs are found only in *Archeopteryx* and in the other avialae. Other theropods, even in those closest to birds, were covered by monofilaments. It is a common mistake to illustrate theropods such as oviraptorosaurs and paraves covered in feathers around the body.

The color of dinosaurs

It was long thought that knowing the color of dinosaurs was practically impossible, and it mostly still is. The only exceptions are samples whose fossils retained the structure of their melanosomes. In this book, species' color has been applied as accurately as possible when it is known. In other cases, colors are deduced from the animals' ways of life, size, climate, or customs so that they match the model in life. For instance, in the 1990s, some unpublished sketches experimented with the possible color of various dinosaurs, and several of these experiments are applied in some of this book's illustrations. In the case of three "Coelurosaurs" (figure 3), a reddish hue with white stripes on the tail was fortunately seen in *Sinosauropteryx prima* in 2010. Rather than mere chance, this was because the patterns were based on the Malay civet *(Viverra tangalunga)*, small Indian civet *(Viverricula indica)*, and coati *(Nasua nasua)*. See more: *Longrich 2002; Zhang et al. 2010*

Figure 2. Reliable recreation of a swimming emu. As can be seen, most of the body is submerged (about 85%). In a state of relaxation, it would sink even more, as fluttering pushes the body upward. This is proof that large terrestrial birds are remarkably dense, with SGs quite likely higher than 0.9.

Figura 3. Reconstruction of "coelurosaurs" made in 1998 by Rubén Molina (background by Marlene Moreno). Twelve years after this illustration was made, it was found that *Sinosauropteryx prima* had similar colors. (See page 41.)

Definitions

Paleontology

Paleontology is a discipline of biology and geology that helps us to understand organisms from the past, their way of life, their evolution, their relation to current species, and their distribution in the world, among other aspects. The most emblematic animals that are studied by paleontology are ammonites, trilobites, mammoths, and dinosaurs.

Dinosaurs

Almost everyone is familiar with dinosaurs, animals we usually see as gigantic reptiles similar to the dragons and feathered serpents that appear in different cultures, but dinosaurs were real, and part of their lineage (birds) lives on.

The origin of dinosaurs: avemetatarsalians

Avemetatarsalians were very agile archosaurs. Under their bodies were long, thin legs that allowed them to assume a completely erect posture, something that differs from the extended legs of current reptiles. Filaments may have been present on avemetatarsalians, as we see in their descendants: the pterosaurs and dinosaurs. Early avemetatarsalians existed only during the Triassic. See more: *Witton 2013*

Halfway dinosaurs: dinosauromorphs

Dinosauromorphs were agile predators or specialized bipedal and four-legged herbivores. Except for their direct descendants, the dinosaurs, they disappeared at the end of the Triassic. One of the evolutionary novelties that stands out is the reduction of metatarsals I and V, giving rise to tridactyl and tetradactyl treads. See more: *Sereno 1991; Sereno & Arcucci 1994; Benton 1999, 2004; Dzik 2003; Ezcurra 2006; Brusatte et al. 2010; Nesbitt 2011*

The famous dinosaurs

The Dinosauria clade presents unique characteristics that gave them an advantage over other contemporary archosaurs. These properties allowed them to dominate the terrestrial environment for millions of years. Some of their features can still be seen in modern birds.

Their skulls have wide openings in the temporal region (supratemporal foramen) that allowed large muscles and thus a stronger bite force. In the area of the orbits, the jugal bone bifurcates and articulates with the quadratojugal. Its function is unknown.

Their neck vertebrae have elevations or extra processes (epipophyses) that allow a better grip of the neck muscles, improving their movement. The arm bone (humerus) has a large deltopectoral ridge, which gave dinosaurs greater musculature and greater strength for catching prey against their bodies.

Their hips have a perforated acetabulum, giving them the possibility of having a vertical posture.

The femur (thigh bone) has a small, asymmetrical, rearward projection called the fourth trochanter. This structure was what allowed the insertion of the caudofemoral muscle. This muscle was the prime locomotive force for walking and running.

At the ankle (calcaneus bone), the reduction of the joint with the fibula is remarkable because neither the fibula nor the calcaneus is important for muscle insertions. In modern birds, the bones are either fused or absent. See more: *Glut 1997; Cadbury 2002; Brusatte 2012*

Knowing the primitive (basal) saurischians

The saurischians ("reptile-hipped") form the group of dinosaurs that contains the theropods (mostly carnivores), birds, and sauropodomorphs (the long-necked giants). Saurischians were characterized by a long neck and a palm with a robust and clawed thumb. They appeared in the Middle Triassic and are the only group of dinosaurs that still exists today.

The primitive saurischians were bipedal dinosaurs that can't be categorized as theropods and sauropodomorphs. They were carnivores or omnivores, with four- to five-digit hands and tetradactyl feet. They existed only in the Triassic.

The impressive theropods

Theropods ("beast-footed") are the famous terrestrial predatory dinosaurs of the Mesozoic, although there were also several omnivorous and herbivorous species. Now we know that theropods came in a great diversity of sizes, from the small bee hummingbird *(Mellisuga helenae)*, weighing about 2 g, to large theropods such as the *Giganotosaurus* and *Tyrannosaurus*, which weighed about 8.5 t.

They were characterized by having semirigid tails and developed air sacs. See more: *Gauthier & Padian 1989; Grady et al. 2014*

The successful birds

The term *bird* is ambiguous at present, since it was originally coined to designate the group of feathered and biped vertebrates with beaks and wings. Scientific advances, however, have shown that most of these anatomical details are also present in dinosaurs, so the concept fades. Therefore, we use the term bird for the avian theropods (clade Avialae).

We will use the term to refer to species from *Archeopteryx* to modern birds, taking into account that they have bodies covered with feathers, instead of filaments, as in the case of the paravians or troodontids. Some specializations of avian theropods are nonserrated teeth with bulbous roots, a humerus and an ulna that are longer than the femur, and an elongated ilium toward the posterior. During their evolution, modern birds developed a short tail and fused bones, such as the pygostyle, notarium, synsacrum, and skull. In addition, they lost their teeth, developed large air sacs, pneumatic bones, and a large sternum with substantial keels to accommodate large flight muscles. At present, there are 10,157 species of modern birds. (An additional 153 have become extinct in recent times). See more: *Huxley 1868, 1870a, 1870b; Ostrom 1969, 1973; Bakker & Galton 1974; Clemens 2011; Foth et al. 2014*

The most robust dinosauromorph
Teyuwasu barbarenai
Specimen: JVP 16:728
Whereas dinosauromorphs generally have very thin bones, Teyuwasu is very massive, even when compared with primitive dinosaurs. See more: *Huene 1938, 1942; Kischlat 1999; Ezcurra 2012*

Classification

Evolutionary history of Mesozoic theropod dinosaurs

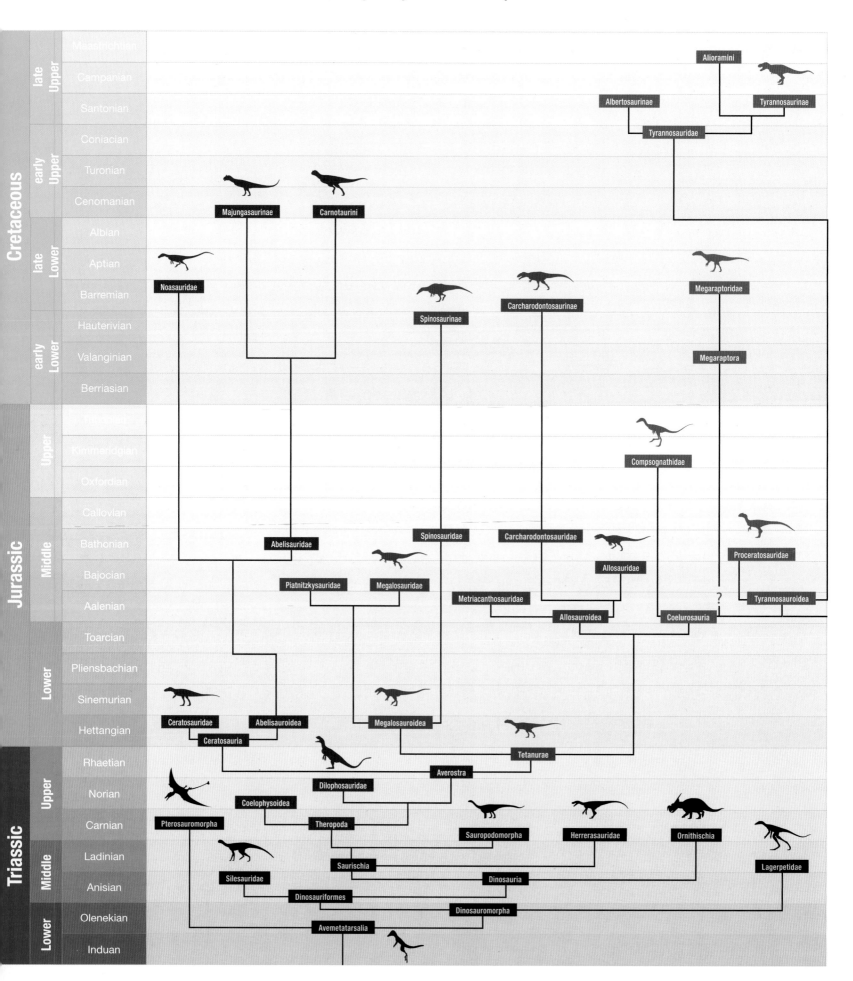

This cladogram represents the most outstanding clades of dinosauromorphs, basal saurischians, theropods, and Mesozoic birds. Also indicated is the period in which the oldest known specimen of each taxonomic group is found.

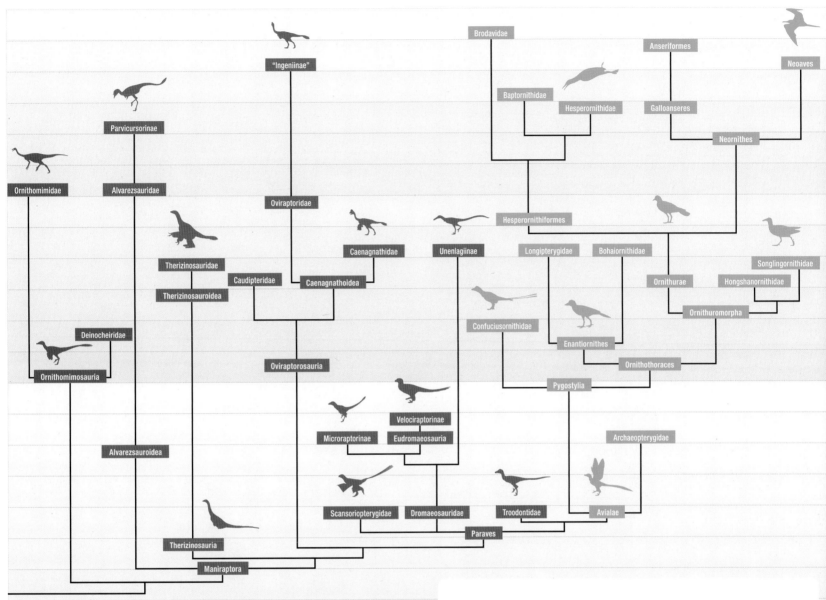

Linnaean system

The Linnaean classification system preceded the cladistic system. It was created by Carl Nilsson Linnaeus in 1758. In this method, a large taxon contains smaller ones, and these, in turn, include others successively, as if they were an index for classifying documents. Without exception, every taxon is named and rigorously organized within a group.

This system worked for many years, but new phylogenetic knowledge contributed by genetics, paleontology, and anatomy have caused it to be replaced by cladistics, which is more precise. Cladistics contains many of the contributions of the Linnaean system, such as binomial nomenclature and some taxa such as families. In specific cases, the Linnaean hierarchy is useful—for example, when presenting a certain group of living beings to a non-professional audience for didactic purposes. It is also useful when studying specialized disciplines, such as ornithology (the study of living dinosaurs).
See more: *Lucas 2007; Martyniuk 2012*

Cladistic system

Cladistics (also known as phylogenetic systematics) is used to understand the evolutionary relationships of dinosaurs and the groups that compose them. It rigorously defines the shared derived similarities (synapomorphies). The more information we integrate into the analysis, the greater becomes its capacity to model the real phylogeny. Unfortunately, cladistics is not exempt from problems, as many dinosaurs were frequently kept incomplete, in an immature state, or mixed with other individuals and organisms. Consequently, not all genres can be securely placed, so it is often better to leave them uncatalogued until more information is obtained. In the case of present-day birds, we have additional information, beyond being able to study their complete anatomy, such as their DNA sequence and biochemical data. Finally, all the data obtained can be represented graphically in a phylogenetic tree, where nodes represent the existence of a common ancestor.

Geological time is shown from bottom to top or from right to left, often with information from the paleontological record. See more: *Lucas 2007*

Comparing species
Dinosaur taxonomy

Records: The largest and smallest by taxonomic group

It is necessary to classify organisms to understand their phylogenetic position and their place in the tree of life, especially when those that make up a particular group turn out to be numerous. With the passage of time, the traditional classification that divided tetrapod vertebrates into amphibians, reptiles, birds, and mammals has become obsolete because it is superficial and unrealistic. The "reptiles" turned out to be a paraphyletic group (it includes their common ancestor but not all its descendants), which is why it has disappeared. At present, since birds are the direct descendants of dinosaurs and are anatomically similar, they are included within the clade Dinosauria.

Nowadays, dinosaurs are defined as a group of avemetatarsalian archosaurs that have a perforated acetabulum (that is, the hollow between the ilium, pubis, and ischium that allows the union with the femur and thus keeps the legs under the trunk), a large ridge in the humerus for the insertion of strong muscles, and a large supratemporal foramen that gives dinosaur's bite more power.

1 m   2 m   3 m

RECORD
LARGEST

Unnamed species
Peecook et al. 2013

1 m

1:15

Specimen: **uncataloged**
(Peecook et al. 2013)

Length: 3 m
Hip height: 92 cm
Body mass: 83 kg
Material: femur
ESR: ●●○○

The largest (Class: Medium Grade I)

This enormous silesaurid was as heavy as an adult person. It lived in the Middle Triassic (Anisian, ca. 247.2–242 Ma) in south-central Pangea (present-day Zambia). It was herbivorous and could move on two or four legs. It was larger than any dinosauromorph, but the land was dominated by herbivorous archosaurs and carnivores that were even larger but less swift. Having lived at the same time as *Lutungutali sitwensis*, it could be a large individual of that species. See more: *Peecook et al. 2013*

Not as big but more robust

Teyuwasu barberenai ("paleontologist Mario Costa Barberena's great lizard"): It was probably a silesaurid or primitive dinosaur of the Upper Triassic (present-day Brazil). Unlike other dinosauromorphs, it is very robust and was even confused with remains of the archosaur *Hoplitosuchus*. It was 2.3 m long and weighed 75 kg. It is unknown if it was an adult or not. See more: *Huene 1938; Kischlat 1999; Ezcurra 2012*

Footprints

a)

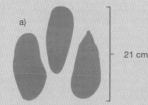

21 cm

b)

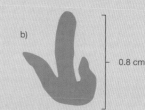

0.8 cm

The largest

a) Unnamed: Triassic footprints of north-central Pangea (present-day France) were attributed to probable dinosaurs. Their shape coincides with the footprints of dinosauromorphs similar to derived silesaurids. The owner of these footprints would have been even larger than the largest specimen we know from bones. See more: *Demathieu 1970*

Numerous tracks named *Atreipus, Banisterobates, Paratrisauropus, Prorotodactylus,* and *Rotodactylus* were present only in the Triassic. They belong to bipedal and quadruped avemetatarsalians and primitive dinosauromorphs.

The smallest

b) Atreipus-Grallator: Tiny footprints from the Middle Triassic were located in south-central Pangea (present-day Morocco). They were left by dinosauromorphs similar to *Saltopus*, since they display a mesaxonic and symmetrical morphology unlike the oldest species. Because of their extremely small sizes, only 0.8–1.3 cm, it is likely that they belong to newborn or young individuals. See more: *Klein et al. 2011*

2 m

20 cm

Footprint 21.6E
Demathieu 1970
3.5 m / 120 kg

Unnamed species
Peecook et al. 2013
3 m / 83 kg

Atreipus-Grallator
CDUE 260
11 cm / 2 g

Scleromochlus taylori
BMNH R3556
18.5 cm / 18 g

4 m   5 m   6 m   7 m

Avemetatarsalian characteristics and habits

Small to large heads, knife-shaped teeth in carnivores and conical teeth in herbivores, short to medium necks, with five fingers on each hand and five on each foot, light or robust bodies, and medium to long tails. See more: *Langer et al. 2013*

Diet: small game carnivores, omnivores, and/or selective herbivores.

Time range

From the Lower to Upper Triassic (ca. 251.2–201.3 Ma), a duration of 49.9 million years in the fossil record.

The smallest (Class: Tiny Grade I)

Scleromochlus taylori ("William Taylor's hard fulcrum") lived during the Upper Triassic (Carnian, ca. 237–227 Ma) in north-central Pangea (present-day Scotland). It is the most primitive genus known, since it is close to the evolutionary line that gave rise to dinosaurs and pterosaurs. It was so tiny that a sparrow has almost twice its mass. It had a short body with legs that were better adapted to jumping than running. Among the dinosauromorphs, the smallest were *Agnosphitys cromhallensis*, which were 35 cm long and weighed 45 g, and *Saltopus elginensis*, 50 cm and 110 g. It is not known if they were adults. *Scleromochlus taylori* was 16 times shorter and 4,600 times lighter than the silesaurid from Zambia. See more: *Woodward 1907; Fraser et al. 2002; Benton & Walker 2011*

The smallest juvenile specimens

The specimens of *Scleromochlus taylori* BMNH R3146B and BMNH 3146B were 85% the size of the type specimen BMNH R3556. These two individuals barely reach a length of 15 cm and a weight of 11 g. See more: *Benton 1999*

The oldest

Lutungutali sitwensis, *Asilisaurus kongwe*, the silesaurid NHMUK R16303, and others have been located in the Middle Triassic (Anisian, ca. 247.2–242 Ma) in south-central Pangea (present-day Zambia and Tanzania). The ichnological record suggests that they existed at the end of the Lower Triassic (Olenekian, ca. 251.2–247.2 Ma) in north-central and northwestern Pangea (present-day Germany, Poland, and Arizona), represented by the ichnogenera *Rotodactylus* and *Prorotodactylus*. See more: *Nesbitt et al. 2010; Peecook et al. 2013; Barrett et al. 2014*

The first published species

The first avemetatarsalian to be described was *Scleromochlus taylori* (1907), which was considered a possible dinosaur. The first dinosauromorph was discovered in the same area three years later: *Saltopus elginensis* ("jumping feet of the city of Elgin"), which was considered a small dinosaur for about one hundred years. See more: *Woodward 1907; Huene 1910; Benton & Walker 2000*

The most recent

Eucoelophysis sp. from northwestern Pangea (present-day New Mexico) and the *Coelurosaurichnus sassendorfensis* footprint, from north-central Pangea (present-day Germany), date from the late Upper Triassic (Rhaetian, ca. 208.5–201.3 Ma). The prints cf. *Rotodactylus* from the early Upper Cretaceous of Egypt and cf. *Atreipus* of the late Upper Cretaceous of Peru are questionable assignments and may be impressions of other types of quadruped animals, as basal dinosauromorphs became extinct at the end of the Triassic. See more: *Kuhn 1958; Demathieu & Wycisk 1990; Leonardi 1994; Noblet et al. 1995; Rinehart et al. 2009*

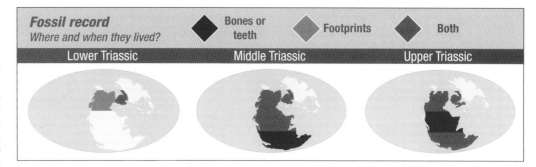

Fossil record
Where and when they lived?   ◆ Bones or teeth   ◆ Footprints   ◆ Both

| Lower Triassic | Middle Triassic | Upper Triassic |

The most recently published species

Lutungutali sitwensis (2013) ("high hip of the valley of Sitwe"): From the Upper Triassic of south-central Pangea (present-day Zambia). It was previously known as "N'tawere form." See more: *Peecook et al. 2013*

The strangest

Teyuwasu barberenai was an enigmatic avemetatarsalian. Its body was too robust compared to any dinosauromorph or basal dinosaur. Its appearance was similar to that of some bipedal sauropodomorphs of the time. See more: *Huene 1938; Kischlat 1999, 2000; Ezcurra 2012*

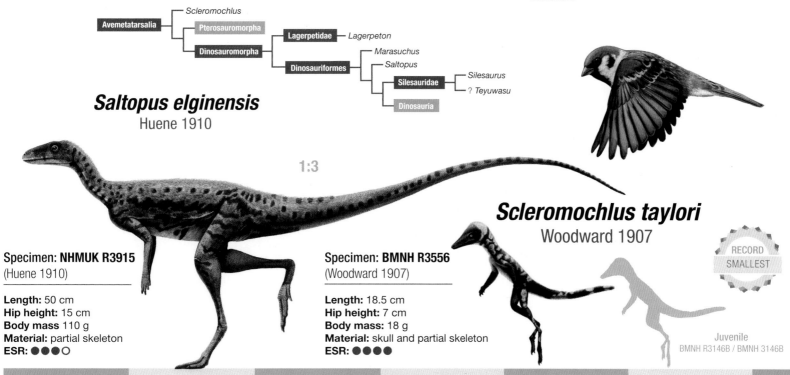

Avemetatarsalia — Scleromochlus
— Pterosauromorpha
— Dinosauromorpha — Lagerpetidae — Lagerpeton
— Dinosauriformes — Marasuchus
— Saltopus
— Silesauridae — Silesaurus
— ? Teyuwasu
— Dinosauria

Saltopus elginensis
Huene 1910

1:3

Scleromochlus taylori
Woodward 1907

RECORD SMALLEST

Specimen: NHMUK R3915
(Huene 1910)

Length: 50 cm
Hip height: 15 cm
Body mass: 110 g
Material: partial skeleton
ESR: ●●●○

Specimen: BMNH R3556
(Woodward 1907)

Length: 18.5 cm
Hip height: 7 cm
Body mass: 18 g
Material: skull and partial skeleton
ESR: ●●●●

Juvenile
BMNH R3146B / BMNH 3146B

10 cm   20 cm   30 cm   40 cm   50 cm   60 cm

RECORD
LARGEST

Frenguellisaurus ischigualastensis
Novas 1986

1:20

Specimen: PVSJ 53
(Novas 1986)
___

Length: 5.3 m
Hip height: 1.55 m
Body mass: 360 kg
Material: skull, incomplete bones, and vertebrae
ESR: ●●●○

The largest (Class: Medium Grade III)

Frenguellisaurus ischigualastensis ("Joaquin Frenguelli lizard of the Ischigualasto Formation"): From the Upper Triassic (lower Norian, ca. 227–218 Ma) of southeastern Pangea (present-day Argentina). It was one of the largest carnivorous dinosaurs of the Triassic, almost as heavy as a brown bear. Despite its size, it was not the dominant predator of its time, as some primitive archosaurs, such as *Saurosuchus galilei*, were larger. *Frenguellisaurus* may have been an adult *Herrerasaurus ischigualastensis*, but the former is more recent than the latter. See more: *Reig 1959; Novas 1986*

A false contender

Ratrichema pellati lived during the Upper Triassic of north-central Pangea (present-day France). It was believed that it may have been a dinosaur, but a more recent revision classifies it as an ichthyosaur, an aquatic animal similar in appearance to dolphins. The genus is based on a very large vertebral arch that, if it had been a saurischian, its size would have been even larger than that of *Frenguellisaurus*, which was approximately 6.6 m long and weighed 695 kg. See more: *McGowan & Motani 2003; Fischer & Goolaerts 2013*

Footprints

a)

21 cm

b)

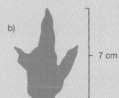

7 cm

The largest

a) Unnamed: Footprints located in the Upper Triassic of northeastern Pangea (present-day Thailand) have a very long hallux (toe I), a typical trait of primitive saurischians. Due to their fingers' proportions, they could be related to *Herrerasaurus*. See more: *Le Loeuff et al. 2009*

Some footprints that are found only from the Middle to Upper Triassic known as *Cridotrisauropus* and *Qemetrisauropus* had primitive morphology that matches the footprints of *Guaibasaurus* and other primitive saurischians.

The smallest

b) Grallator pisanus: A little-known footprint and might not be a valid name. It has been identified as a possible ornithischian dinosaur footprint. It dates from the Upper Triassic and lived in north-central Pangea (present-day Italy). See more: *Bianucci & Landini 2005*

2 m

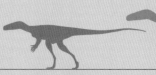

30 cm

Unnamed footprint	**Frenguellisaurus ischigualastensis**	**Grallator pisanus**	**cf. Agnosphitys cromhallensis**
Le Loeuff et al. 2007	PVSJ 53	Bianucci & Landini 2005	VMNH 1751
4.9 m / 295 kg	5.3 m / 360 kg	80 cm / 660 g	1 m / 1.3 kg

Characteristics and habits

Large-headed dinosaurs, knife-shaped teeth, relatively short necks, five fingers on each hand and four on each foot, semirigid bodies, and long tails. Some had a double-hinged jaw that allowed greater flexibility in order to withstand stronger tensions when biting large animals.

The basal saurischians descend from the primitive dinosauromorphs and precede theropods and sauropodomorphs. They, together with ornithischians, comprise the dinosaur group. See more: *Novas 1996; Langer 2004; Sereno 2007*

Diet: small or large game carnivores.

Time range

Possibly present from the Middle to Upper Triassic (ca. 247.2–201.3 Ma). They existed for about 45.6 million years.

The smallest (Class: Very small Grade I)

The VMNH 1751 specimen was mistakenly identified as a large individual of **Agnosphitys cromhallensis** ("of unknown lineage from Cromhall"). Only 1 m long and weighing 1.3 kg, it is the smallest of the basal saurischians, but it is unknown if it was a juvenile or an adult. It lived during the Upper Triassic (Rhaetian, ca. 208.5–201.3 Ma) in the north-central portion of Pangea (present-day England). It was 5.3 times shorter and 280 times lighter than *Frenguellisaurus ischigualastensis*. See more: *Fraas 1913; Sereno & Wild 1992; Fraser et al. 2002; Langer et al. 2013*

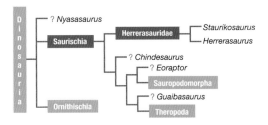

The smallest juvenile specimen

The incomplete femur ZPAL V.39/47 (present-day Poland) probably belonged to a young herrerasaurid. It was found associated with other remains of older individuals that might have been 2.4 m long and weighed 34 kg, while the smallest was about 1.35 m long and weighed 6 kg. See more: *Niedzwiedzki et al. 2014*

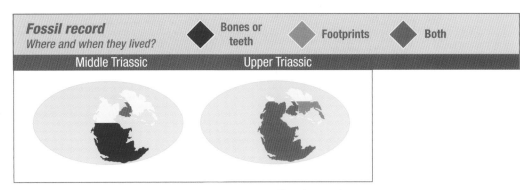

Fossil record
Where and when they lived?   ◆ Bones or teeth   ◆ Footprints   ◆ Both

Middle Triassic   Upper Triassic

The oldest

There are numerous basal saurischians from the early Upper Triassic (Carnian, ca. 237–227 Ma). Among them are *Staurikosaurus pricei, Herrerasaurus ischigualastensis, Alwalkeria maleriensis,* and *Eoraptor lunensis* from the southern area of western Pangea (present-day India and Argentina). Other remains known as *Nyasasaurus parringtoni* from the Middle Triassic (Anisian, ca. 247.2–242 Ma) from the south-central zone of Pangea (present-day Tanzania) may have belonged to very primitive saurischians. Unfortunately, it can't be identified correctly because of how incomplete it is. The *Grallator* sp. footprint (present-day France) may be the oldest that has been attributed to a dinosaur, since it dates from the Middle Triassic (Anisian-Ladinian, ca. 242 Ma). Another even older footprint (present-day Czech Republic) is likely to be dinosauromorph. See more: *Montenat 1968; Zajíc 1998; Nesbitt et al. 2013, Sereno et al. 2013, Langer 2014; Madzia 2014*

The most recent

A maxilla with teeth belonging to *Agnosphitys cromhallensis* from the late Upper Triassic (Rhaetian, ca. 208.5–201.3 Ma) of north-central Pangea (present-day England). It could be similar to *Guaibasaurus*, while the type material was of a dinosauromorph. See more: *Fraser 2002; Langer et al. 2013*

The strangest

Eoraptor lunensis ("dawn thief from the Valley of the Moon"): From the Upper Triassic of southwestern Pangea (present-day Argentina). It had very large eyes and rudimentary teeth on the palate in the same fashion as the theropod *Eodromaeus*. See more: *Martínez et al. 2011; Sereno et al. 2013*

The first published species

"Thecodontosaurus" primus (1905): From the Middle Triassic of north-central Pangea (present-day Poland). It may have belonged to a dinosaur or a primitive archosaur. We are also unsure of *"Zanclodon" silesiacus* (1910), a tooth similar to those found in theropods. See more: *Huene 1905; Jaekel 1910; Carrano et al. 2012*

The most recently published species

Nyasasaurus parringtoni ("lizard of the Lake Nyaza of Francis Rex Parrington"): From the Middle Triassic of south-central Pangea (present-day Tanzania). It was officially presented in 2013, although it had held the name "Nyasasaurus cromptoni" from a thesis for forty-six years. See more: *Charig 1967; Harland et al. 1967; Nesbitt et al. 2013*

cf. *"Agnosphitys cromhallensis"*
Fraser et al. 2002

Juvenile
ZPAL V.39/47

RECORD
SMALLEST

1:6

Specimen: **VMNH 1751**
(Fraser et al. 2002)

Length: 1 m
Hip height: 30 cm
Body mass: 1.3 kg
Material: maxilla with teeth
ESR: ●●○○

"Sinosaurus shawanensis"
Young in Anonymous 1979

RECORD LARGEST

1:35

Specimen: IVPP V31
(Young 1948)

Length: 9.2 m
Hip height: 2.55 m
Body mass: 1.7 t
Material: thoracic vertebra
ESR: ●●○○

The largest (Class: Large Grade III)

A large vertebra from the Lower Jurassic (Hettangian, ca. 201.3–199.3 Ma) of northeastern Pangea (present-day China) was found among remains of sauropodomorphs and other mixed theropods. It may belong to a large specimen of *"Dilophosaurus" sinensis*, weighing more than a current hippopotamus. It was a dangerous predator, as it was not only big for its time but also quite agile, due to its slender build. The name **"Sinosaurus shawanensis"** appears on a list of fauna somewhere but was not formally named. See more: *Young 1948; Yang 1979*

The second largest

"Lufeng beds Ceratosaurus" is a huge talus-calcaneus from the Lower Jurassic (Sinemurian, ca. 199.3–190.8 Ma) of northeastern Pangea (present-day China). It could be an adult *Sinosaurus triassicus*, which was 7.5 m long and weighed 820 kg. See more: *Young 1951; Welles & Long 1974*

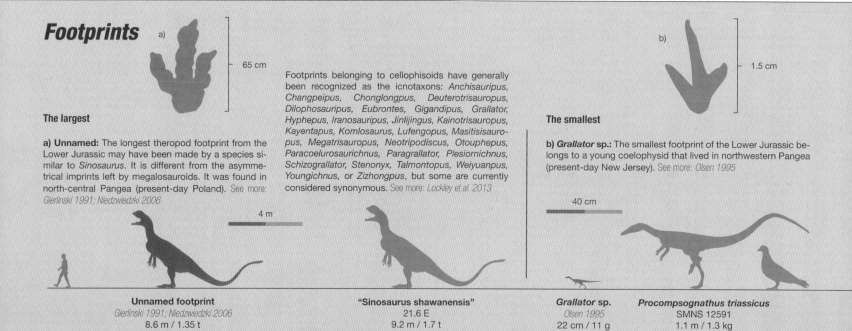

Footprints

a)
— 65 cm

The largest

a) **Unnamed:** The longest theropod footprint from the Lower Jurassic may have been made by a species similar to *Sinosaurus*. It is different from the asymmetrical imprints left by megalosauroids. It was found in north-central Pangea (present-day Poland). See more: *Gierlinski 1991; Niedzwiedzki 2006*

Footprints belonging to cellophisoids have generally been recognized as the icnotaxons: *Anchisauripus, Changpeipus, Chonglongpus, Deuterotrisauropus, Dilophosauripus, Eubrontes, Gigandipus, Grallator, Hyphepus, Iranosauripus, Jinlijingus, Kainotrisauropus, Kayentapus, Komlosaurus, Lufengopus, Masitisisauropus, Megatrisauropus, Neotripodiscus, Otouphepus, Paracoelurosaurichnus, Paragrallator, Plesiornichnus, Schizograllator, Stenonyx, Talmontopus, Weiyuanpus, Youngichnus,* or *Zizhongpus*, but some are currently considered synonymous. See more: *Lockley et al. 2013*

b)
— 1.5 cm

The smallest

b) *Grallator* sp.: The smallest footprint of the Lower Jurassic belongs to a young coelophysid that lived in northwestern Pangea (present-day New Jersey). See more: *Olsen 1995*

40 cm

4 m

Unnamed footprint	**"Sinosaurus shawanensis"**	***Grallator* sp.**	***Procompsognathus triassicus***
Gierlinski 1991; Niedzwiedzki 2006	21.6 E	*Olsen 1995*	SMNS 12591
8.6 m / 1.35 t	9.2 m / 1.7 t	22 cm / 11 g	1.1 m / 1.3 kg

10 m   12 m   14 m   16 m

Characteristics and habits

Theropods with long or short heads, with or without crests, knife-shaped teeth, long necks, four fingers on each hand and foot, graceful bodies, and very long tails. Basal theropods descend from primitive saurischians and precede ceratosaurs and megalosauroids. The most successful group was the coelophysoids, which (apart from their very diverse teeth and ornamentation) have very similar forms. See more: *Carrano & Sampson 2004*

Diet: small game carnivores and/or possible piscivores.

Time range

They are known to have existed for about 82 million years, from the Upper Triassic to the Middle Jurassic (ca. 227–145 Ma).

The smallest Class: Very small Grade I

Procompsognathus triassicus ("before elegant Triassic jaw"): From the Upper Triassic (Norian, ca. 227–208.5 Ma) of north-central Pangea (present-day Germany). It had a weight comparable to that of a ferret, although it is unknown whether it was an adult. Some specialists suspect that the skeleton's skull belongs to an archosaur similar to *Saltoposuchus connectens*. It was about 9 times shorter and 1,300 times lighter than "Sinosaurus shawanensis." See more: *Fraas 1913; Sereno & Wild 1992*

The smallest juvenile specimen

The metatarsus TTU P 9201 belonged to a coelophysoid that was assigned along with other remains of various animals to *Protoavis texensis*. It might be an unidentified coelophysoid offspring. If we were to compare it with *Coelophysis* juveniles, it would have been about 43 cm long and weighed 110 g. See more: *Chatterjee 1991; Hutchinson 2001; Nesbitt et al. 2005; Rinehart et al. 2009*

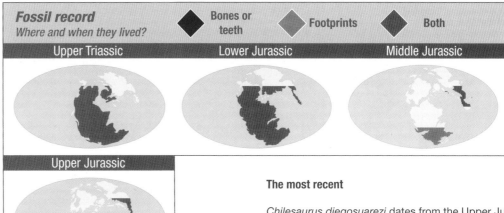

Fossil record
Where and when they lived?   ◆ Bones or teeth   ◆ Footprints   ◆ Both

Upper Triassic   Lower Jurassic   Middle Jurassic

Upper Jurassic

The oldest

Lepidus praecisio and the remains of a young specimen of an indeterminate coelophysoid that belonged to the *Protoavis texensis* chimera date from the Upper Triassic (lower Norian, ca. 227–217 Ma) in northwestern Pangea (present-day western United States). Theropod footprints (present-day Argentina) that had been dated to the Middle Triassic belong to the Upper Triassic (lower Carnian, 237–232 Ma). See more: *Hutchinson 2001; Nesbitt et al. 2005; Marsicano et al. 2007, 2015; Nesbitt & Ezcurra 2015*

The strangest

Chilesaurus diegosuarezi ("Chilean lizard of Diego Suárez"): From the Upper Jurassic of southwestern Neopangea (present-day Chile). It has a strange combination of traits, such as being a herbivore with very wide feet. Although it is very complete, it is very difficult to classify, as it fits in different phylogenetic positions. See more: *Salgado et al. 2008; Novas et al. 2015; Mortimer**

The most recent

Chilesaurus diegosuarezi dates from the Upper Jurassic (Tithonian, ca. 152.1–145 Ma). *Monolophosaurus jiangi* and Coelophysoid SGP 2000/2 lived in the Middle Jurassic (Callovian, ca.166.1–163.5 Ma) of eastern Paleoasia (present-day China). See more: *Zhao & Currie 1994; Maisch et al. 2001; Maisch & Matzke 2003; Novas et al. 2015*

The first published species

Coelophysis bauri and *Longosaurus logicollis* (formerly *Coelurus bauri* and *C. longicollis* in 1887), from the Upper Triassic of northwestern Pangea (present-day New Mexico), were the first published reports. "*Tanystropheus*" *willistoni* appeared in another publication that same year. *Arctosaurus osborni* precedes them by twelve years, but it isn't known if these thoracic vertebrae belong to a theropod, archosaur, or trilophosaur. See more: *Adams 1875; Cope 1887a, 1887b; Steel 1970; Nesbitt et al. 2007; Russell 1989*

The most recently published species

Lepidus praecisio (2015) ("fascinating fragments"): From the Upper Triassic of northwestern Pangea (present-day Texas). It is one of the oldest theropods in North America. See more: *Sterling, Nesbitt & Ezcurra 2015*

```
         ┌ Daemonosaurus
    ┌────┤ Coelophysoidea ─┬─ Coelophysis
T   │    └                 └─ Segisaurus
h   │  ── Liliensternus
e   │    ┌ Dilophosauridae ── Dilophosaurus
r ──┤    │
o   │    │           ┌─ Ceratosauria
p   │    └ Averostra ─┤      ┌─ ? Chilesaurus
o   │                 └ Tetanurae ┬─ Monolophosaurus
d   │                             └─ Megalosauroidea
a
```

RECORD SMALLEST

Procompsognathus triassicus
Fraas 1913

Breeding
TTU P 9201

1:6

Specimen: SMNS 12591
(Fraas 1912)

Length: 1 m
Hip height: 28 cm
Body mass: 1.3 kg
Material: skull and partial skeleton
ESR: ●●●●

20 cm   40 cm   60 cm   80 cm   100 cm   120 cm

Spinostropheus gauthieri
Sereno et al. 2004

1:30

RECORD
LARGEST

2 m

Specimen: MNHN coll.
(Lapparent 1960)

Length: 8.5 m
Hip height: 2.4 m
Body mass: 600 kg
Material: Ulna
ESR: ●●○○

The largest (Class: Large Grade I)

Spinostropheus gauthieri ("vertebral spine of Francois Gautier"): from south-central Neopangea (present-day Niger). The best specimen mentioned is a forearm bone from the Middle Jurassic that was never illustrated and is now lost. If the identification was correct, the theropod would have weighed as much as a grizzly bear and a lion combined. See more: *Stromer 1934; Sereno et al. 1996; Wilson et al. 2003; Mortimer**

Longer but lighter

A tibia known as cf. *Elaphrosaurus* sp. from the Early Upper Cretaceous in the central Gondwana zone (present-day Morocco), it may belong to this type of theropod or to another group. As a basal avetheropod, it would have been 6.7 m long and weighed 470 kg. See more: *Lavocat 1954*

Footprints

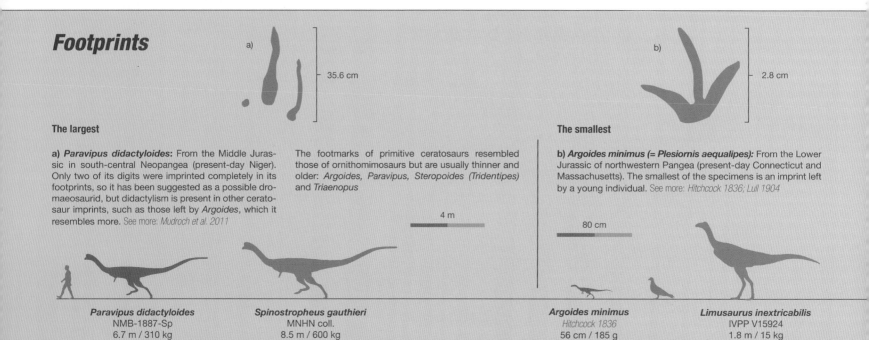

a)

35.6 cm

b)

2.8 cm

The largest

a) *Paravipus didactyloides*: From the Middle Jurassic in south-central Neopangea (present-day Niger). Only two of its digits were imprinted completely in its footprints, so it has been suggested as a possible dromaeosaurid, but didactylism is present in other ceratosaur imprints, such as those left by *Argoides*, which it resembles more. See more: *Mudroch et al. 2011*

The footmarks of primitive ceratosaurs resembled those of ornithomimosaurs but are usually thinner and older: *Argoides, Paravipus, Steropoides (Tridentipes)* and *Triaenopus*

The smallest

b) *Argoides minimus* (= *Plesiornis aequalipes*): From the Lower Jurassic of northwestern Pangea (present-day Connecticut and Massachusetts). The smallest of the specimens is an imprint left by a young individual. See more: *Hitchcock 1836; Lull 1904*

4 m

80 cm

Paravipus didactyloides
NMB-1887-Sp
6.7 m / 310 kg

Spinostropheus gauthieri
MNHN coll.
8.5 m / 600 kg

Argoides minimus
Hitchcock 1836
56 cm / 185 g

Limusaurus inextricabilis
IVPP V15924
1.8 m / 15 kg

8 m   10 m   12 m   14 m

Characteristics and habits

Theropods with small heads, tiny or nonexistent teeth, long necks, four fingers on each hand and foot, underdeveloped arms and hands, long legs, and very light bodies. The first averostra were small unspecialized carnivores that descended from coelophysoids and gave rise to ceratosaurids and megalosauroids. Elaphrosaurus and its close relatives moved away from this line, developing adaptations similar to those of ornithomimosaurs and some terrestrial birds. The popular name "Elaphrosaurus" has not been accepted by the scientific community. See more: *Paul 1988, 2010*
Diet: small game carnivores, omnivores and/or herbivores.

Time range

Averostra definitely lived during the Lower Jurassic, although some very incomplete remains and some traces suggest that they may have appeared in the Upper Triassic. They disappeared in the early Upper Cretaceous (ca. 208.5–93.9 Ma). They are present in the fossil record for approximately 114.6 million years.

The smallest (Class: Small Grade II)

Limusaurus inextricabilis ("lizard trapped in the mud"): From the Upper Jurassic (Oxfordian, ca. 163.5–157.3 Ma) of eastern Paleoasia (present-day China). It was the size of an Iberian lynx. Most of the specimens that have been found so far are of subadults or juveniles. It was about five times shorter and forty times lighter than *Spinostropheus gauthieri*. See more: *Xu et al. 2009*

The smallest juvenile specimen

The offspring of *Limusaurus inextricabilis* (IVPP V15303) is the smallest specimen of the group. It was about 50 cm long and weighed 340 g. See more: *Eberth 2010*

The oldest

"Newtonsaurus" cambrensis ("E.T. Newton lizard of Mount Cambré"): It is the oldest known specimen, dating from the early Triassic (Rhaetian, ca. 208.5–201.3 Ma). It lived in north-central Pangea (present-day England). Footprints of *Steropoides ingens* (formerly known as *Ornithichnites* or *Tridentipes*) from central northwestern Pangea (present-day New Jersey) are equally ancient and perhaps belonged to primitive averostra. See more: *Hitchcock 1836, 1889; Carrano, Benson & Sampson 2012*

The most recent

Deltadromeus agilis and cf. *"Elaphrosaurus"* sp. lived during the early Upper Cretaceous (Cenomanian, ca. 100.5–93.9 Ma) in central Gondwana (present-day Morocco and Sudan, respectively). Some tracks attributed to ornithomimosaurs from a relatively near region (present-day Israel) and of similar age may have belonged to this type of theropod. See more: *Werner 1993; Lavocat 1954; Avnimelech 1966*

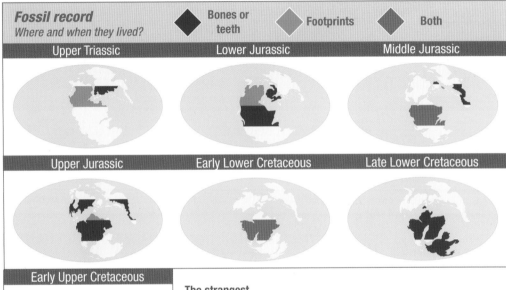

Fossil record
Where and when they lived?

◆ Bones or teeth   ◆ Footprints   ◆ Both

Upper Triassic	Lower Jurassic	Middle Jurassic
Upper Jurassic	Early Lower Cretaceous	Late Lower Cretaceous
Early Upper Cretaceous		

The strangest

Limusaurus inextricabilis is different from other known averostra because it lacks teeth. Its general appearance is very similar to that of nonflying birds. The ischial bone was very long, so it probably had a large stomach, as seen in herbivores. See more: *Xu et al. 2009*

The first published species

Elaphrosaurus bambergi (1920) From the Upper Jurassic of south-central Neopangea (present-day Tanzania). It is one of the first theropods described in Africa. HMN Gr.S. 38-44 is a very complete specimen that was considered an ornithomimosaur for decades due to its graceful body and its ability to walk quickly. See more: *Janench 1920*

The most recently published species

Camarillasaurus cirugedae (2014) ("Camarillas lizard of Pedro Cirugeda Buj"): From the Upper Cretaceous of central Laurasia (present-day Spain). It was formally presented in 2014, although it was digitally available to the public while it was being studied (2012). See more: *Sánchez-Hernández & Benton 2014*

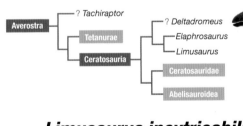

? Tachiraptor
Averostra
Tetanurae
Ceratosauria
? Deltadromeus
Elaphrosaurus
Limusaurus
Ceratosauridae
Abelisauroidea

Limusaurus inextricabilis
Xu et al. 2009

Juvenile
IVPP V15303

RECORD
SMALLEST

1:12

Specimen: IVPP V15924
(Xu et al. 2009)

Length: 1,8 m
Hip height: 70 cm
Body mass: 15 kg
Material: partial skeleton
ESR: ●●●●

50 cm   100 cm   150 cm   200 cm   250 cm

Ceratosaurus sp.
Raath & McIntosh 1987

RECORD
LARGEST

2 m

1:30

Specimen: QG 65
(Raath & McIntosh 1987)

Length: 8.4 m
Hip height: 2.55 m
Body mass: 1.9 t
Material: femur
ESR: ●●○○

The largest (Class: Large Grade III)

***Ceratosaurus* sp.** ("horned lizard"): It is a recognized genus found in northwestern, north-central, and possibly south-central Neopangea (present-day western United States, Portugal, Switzerland, Tanzania, and Zimbabwe). The largest individual dates from the Upper Jurassic (Tithonian, ca. 152.1–145 Ma) of Zimbabwe. It weighed as much as two great white sharks and was longer than an orca. Despite its large size, other contemporary carnivores were larger, so it may have specialized in hunting particular prey. Because of its flattened tail, its wide teeth, and fish remains found nearby, some authors speculate that it may have consumed aquatic prey. The largest specimen might be a large *Ceratosaurus roechlingi*, but, as of now, it can't be confirmed. See more: *Raath & McIntosh 1987; Bakker 2004*

A possible contender

Ceratosaurus sp.: From northwestern Neogondwana (present-day western United States). At 7.5 m in length and probably weighing 1.35 t, it was smaller than the one located in Zimbabwe, but it had not yet reached maturity. *"Megalosaurus" ingens* was interpreted as a gigantic ceratosaurid, but now we know it was in fact a very ancient carcharodontosaurid. See more: *Janensch 1920; Paul 1988; Britt et al. 2000; Rauhut 2011*

Footprints

a)

53 cm

b)

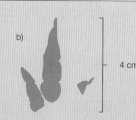

4 cm

The largest

a) ***Carmelopodus* sp.:** Different from megalosauroids, these were large carnivores that were present in the Lower Jurassic of south-central Pangea (present-day Morocco). They left both direct evidence (fossilized bones) and indirect evidence (imprints). See more: *Ishigaki 1988*

There are no known theropods with legs that match those of *Carmelopodus*, but the feet of the ceratosaurids have not been preserved. Their kinship is suggested because of the similarities in their footprints with abelisaurids, and they were contemporaries.

The smallest

b) ***Carmelopodus untermannorum*:** From the Middle Jurassic of northwestern Neopangea (present-day Utah). This ichnogenus left prints up to 20 cm long, although the smaller ones were up to one-fifth that size. See more: *Lockley et al. 1988, 1998*

4 m

3 m

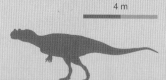

***Carmelopodus* sp.**
Ishigaki 1988
8 m / 1.65 t

***Ceratosaurus* sp.**
QG 65
8.4 m / 1.9 t

Carmelopodus untermannorum
Lockley et al. 1988
68 cm / 1 kg

Sarcosaurus woodi
BMNH R4840/1
3.35 m / 71 kg

Characteristics and habits

Large-headed theropods with or without nasal crests, large flat teeth in the shape of knives, short necks, four fingers on each hand and foot, graceful or robust bodies. Ceratosaurids were carnivores that descended from the averostra and gave rise to the abelisauroids. Theropod predators with very extravagant bony crests do not seem to have been as reckless as those with less vulnerable structures. See more: *Bakker 2004*
Diet: small game carnivores and/or piscivores

Time range

From the Lower Jurassic to the late Lower Cretaceous (ca. 201.3–113 Ma). They appear in the fossil record for approximately 88.3 million years.

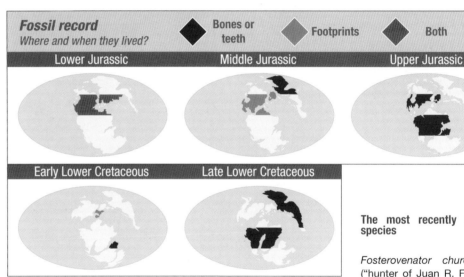

The smallest (Class: Medium Grade I)

Sarcosaurus woodi ("scavenging wood lizard"): From the Lower Jurassic (Sinemurian, ca. 199.3–190.8 Ma) of north-central Pangea (present-day England). It was an adult 2.5 times shorter and 27 times lighter than *Ceratosaurus* sp. Approximately 1.25 m long and weighing 4.2 kg, IGM 6625 from the Lower Jurassic (present-day Mexico) was even smaller. Unfortunately, it is not known if it was an adult. See more: *Andrews 1921; Munter 1999*

The smallest juvenile specimen

Lukousaurus yini ("lug of the Lugou bridge of Yin") was barely longer than 1 m and weighed 2 kg. It must have been very young, since the largest specimen of the same species (V263) was 4 m long and weighed 144 kg. Some authors consider it a theropod related to ceratosaurs, while others suspect that it was a crocodile. See more: *Young 1948; Knoll 2012*

The oldest

Sarcosaurus andrewsi ("Andrew's scavenger lizard"): From the Lower Jurassic (Hettangian, ca. 201.3–199.3 Ma) of north-central Pangea (present-day England). It might have been a primitive ceratosaurid. The *Anticheiropus hamatus* and *A. pilulatus* imprints, which are equally old, had fingers whose proportions were very similar to those of *Carmelopodus* and may have belonged to ceratosaurids. See more: *Huene 1932; Hitchcock 1865*

The strangest

Berberosaurus liassicus ("lizard of the Berber tribe"), from the Lower Jurassic (present-day Morocco), could be a very ancient abelisauroid, but the shape of its teeth suggests that it was a ceratosaurid. See more: *Allain et al. 2007; Hendrickx & Mateus 2014*

The most recent

Genyodectes serus ("recent biting jaws"): From the late Lower Cretaceous (Aptian, ca. 125–113 Ma) of southwestern Gondwana (present-day Argentina). It was characterized by having four teeth on each premaxillary bone, while *Ceratosaurus* had only three. The specimen TNM 03041 (present-day Tanzania) and *Ceratosaurus* sp. (present-day Uzbekistan), both from the late Lower Cretaceous (Albian, ca. 113–100.5 Ma), may have been ceratosaurids. See more: *Woodward, 1901; Ryan 1997; Rauhut 2004; O'Connor et al. 2006*

The first published species

Ceratosaurus meriani (known as Megalosaurus in 1870): From the Upper Jurassic of north-central Neopangea (present-day Switzerland). It is based solely on a premaxillary tooth. See more: *Greppin 1870*

Fossil record *Where and when they lived?*	◆ Bones or teeth	◇ Footprints	◆ Both

Lower Jurassic   **Middle Jurassic**   **Upper Jurassic**

Early Lower Cretaceous   **Late Lower Cretaceous**

The most recently published species

Fosterovenator churei (2014) ("hunter of Juan R. Foster and Daniel J. Chure"): From the Upper Jurassic of northwestern Neopangea (present-day Wyoming). It has been described only from young specimens that could be ceratosaurids. See more: *Dalman 2014*

	Ceratosauridae	— Berberosaurus
Ceratosauria		— Ceratosaurus
	Abelisauroidea	

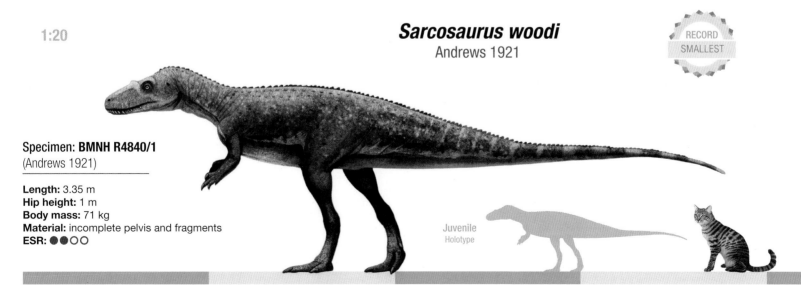

1:20

Sarcosaurus woodi
Andrews 1921

RECORD SMALLEST

Specimen: BMNH R4840/1
(Andrews 1921)

Length: 3.35 m
Hip height: 1 m
Body mass: 71 kg
Material: incomplete pelvis and fragments
ESR: ●●○○

Juvenile
Holotype

4 m

Dryptosauroides grandis
Huene & Matley 1933

RECORD
LARGEST

2 m

1:40

Specimen: GSI Collection
(Huene & Matley 1933)

Length: 10 m
Hip height: 2.8 m
Body mass: 1.5 t
Material: caudal vertebra
ESR: ●●○○

The largest (Class: Large Grade II)

Dryptosauroides grandis ("resembling a large lacerating lizard"): It lived in the Upper Cretaceous (upper Maastrichtian, ca. 72.1–66 Ma) of Hindustan (present-day India). It was an abelisauroid as heavy as a hippopotamus and almost as long as a city bus. Due to its light build, it is possible that it captured small prey or took advantage of the carrion abandoned by the dominant hunters of its time, the abelisaurids. See more: *Huene & Matley 1933; Novas, Agnolin & Bandyopadhyay 2004; Mortimer**

Another contender

Austrocheirus isasii ("southern hands of Marcelo Pablo Isasi"): From the late Upper Cretaceous of western Gondwana (present-day Argentina). A young adult of a very large abelisauroid. It isn't known if it was a noasaurid or a member of another unknown group, but its arms were not atrophied, as they are in abelisaurids. It was about 9.3 m long and weighed about 930 kg. See more: *Ezcurra et al. 2010*

Footprints

a)

49 cm

b)

3.5 cm

c)

4.4 cm

The largest

a) Morphotype B: An unusual footprint with a very elongated central finger compared to the lateral ones. Similar to the impressions left by ornithomimosaurs and noasaurids. Both were very fast animals. Because it dates from the Middle Jurassic and lived in north-central Neopangea (present-day Portugal), it is more likely that it was an abelisauroid. See more: *Frazão et al. 2010*

It has been proposed that some fingerprints with a very developed middle finger could have been made by primitive abelisauroids, such as *Deferrariischnium*, *Sarmientichnus* and *Zhengichnus*. See more: *Sacchi et al. 2008*

The smallest

b and c) *Sarmientichnus scagliai (Casamiquelichnus navesorum)*: Monodactyl (only one imprinted digit) or tridactyl footprints from the Middle Jurassic of south-central Neopangea (present-day Argentina). Because of their small size, they are considered to have been made by young individuals. See more: *Casamiquela 1964; Coria & Paulina Carabajal 2004*

4 m

1 m

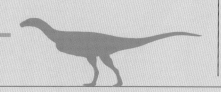

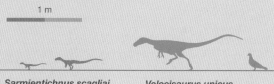

Unnamed footprint
Morphotype B
5.4 m / 200 kg

Dryptosauroides grandis
GSI Col.
10 m / 1.5 t

Sarmientichnus scagliai
Casamiquela 1964
37-65 cm / 285-495 g

Velocisaurus unicus
MUCPv 41
1.5 m / 4.7 kg

0 m   12 m   14 m   16 m   18 m

Characteristics and habits

Theropods with small heads, large anterior teeth, short necks, four fingers on each hand and foot, and very light bodies. They were less specialized carnivores, unlike those with forward-facing teeth. They descended from ceratosaurids and gave rise to abelisaurids. Noasaurids are the best-known members of this group.
Diet: small game carnivores and/or piscivores.

Time range

From the Middle Jurassic to the late Upper Cretaceous (ca. 201.3–66 Ma), a duration of approximately 135.3 million years in the fossil record.

The smallest (Class: Very small Grade III)

Velocisaurus unicus ("single speedy lizard"): From the late Upper Cretaceous of western Gondwana (present-day Argentina). It was smaller than other noasaurids, but as is the case for the majority of them, its degree of development isn't known. *Velocisaurus*'s weight was similar to that of a domestic cat. It was almost 7 times shorter and about 350 times lighter than *Dryptosauroides grandis*. See more: *Bonaparte 1991; Sampson et al. 2001*

The smallest juvenile specimen

In New Zealand, pelvic remains were found of a diminutive juvenile that must have been only 20 cm long and weighed about 5.3 g. See more: *Molnar et al. 2006*

The oldest

Berberosaurus liassicus ("lizard of the Berber tribe"), from the Lower Jurassic (Pliensbachian-Toarcian, ca. 182.7 Ma) of present-day Morocco, could be a primitive abelisauroid, although some doubt it. *Zhengichnus jinningensis* is a footprint that, like that of the noasaurids, has a significantly more developed middle digit. Perhaps it could have been made by an abelisauroid from the Lower Jurassic (Hettangian, ca. 201.3–199.3 Ma) of eastern Pangea. See more: *Zhen et al. 1985; Allain et al. 2007; Sacchi et al. 2008; Rauhut 2012*

The most recent

Coeluroides largus, Compsosuchus solus, Laevisuchus indicus, Jubbulpuria tenuis, Ornithomimoides barasimlensis, O. mobilis and *Dryptosauroides grandis* are abelisauroids that were described from very incomplete remains. They were contemporaries, since they come from the Lameta Formation (upper Maastrichtian, ca. 69–66 Ma) in Hindustan (present-day India). It is thought that many of them may belong to only one or two species of noasaurid. See more: *Huene & Matley 1933; Novas et al. 2004*

The strangest

Masiakasaurus knopfleri ("vicious lizard of the musician Mark Knopfler"): From the late Upper Cretaceous of central Gondwana (present-day Madagascar). Unlike in other theropods, its front teeth project outward and downward, meaning, probably, that its diet was very specialized. See more: *Sampson et al. 2001*

The first published species

Betasuchus bredai (known as "Megalosaurus" in 1883) ("Crocodile Beta of Gijsbertus Samuel Jacob van Breda"): From the late Upper Cretaceous of central Laurasia (present-day Holland). It has been identified as "Ornithomimidorum" by mistake. This term refers to its possible affinity with ornithomimosaurs, but it is not really a genus. See more: *Seeley 1883*

The most recently published species

Dahalokely tokana (2013) ("little solitary thief"): From the early Upper Cretaceous of central Gondwana (present-day Madagascar). It lived around the time that Madagascar became an island. Some authors consider it a primitive abelisaurid. See more: *Farke & Sertich 2009, 2013*

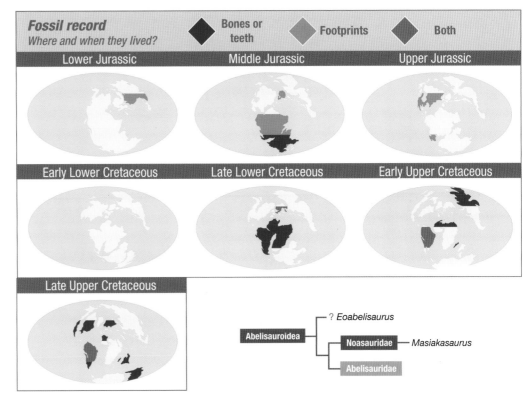

Fossil record
Where and when they lived?

◆ Bones or teeth   ◆ Footprints   ◆ Both

Lower Jurassic | Middle Jurassic | Upper Jurassic

Early Lower Cretaceous | Late Lower Cretaceous | Early Upper Cretaceous

Late Upper Cretaceous

Abelisauroidea
— ? *Eoabelisaurus*
— **Noasauridae** — *Masiakasaurus*
— **Abelisauridae**

RECORD
SMALLEST

Velocisaurus unicus
Bonaparte 1991

Specimen: MUCPv 41
(Bonaparte 1991)

Length: 1.5 m
Hip height: 41 cm
Body mass: 4.7 kg
Material: tibia, astragalus, and incomplete foot
ESR: ●●○○

Neonate

1:10

50 cm   100 cm   150 cm   200 cm

4 m

Unnamed
Candeiro 2014

RECORD
LARGEST

2 m

1:35

Specimen: URC R 44
(Candeiro 2014)

Length: 11.3 m
Hip height: 2.8 m
Body mass: 3.7 t
Material: premaxillary
ESR: ●◑○○○

The largest (Class: Very large Grade I)

URC R 44 is a large abelisaurid from the early Upper Cretaceous (Cenomanian, ca. 100.5–93.9 Ma) of northwestern Gondwana (present-day Brazil). It is known only by a premaxilla that belonged to a carnivore that was as heavy as two hippos and as long as two saltwater crocodiles lined up. Some robust abelisaurids had curved teeth that were suited more to holding prey than to slicing meat. In addition, they had a flexible jaw in order to avoid fractures. These adaptations were useful for catching and killing small or young sauropods. See more: *Sampson & Witmer 2007; Candeiro 2014*

Other contenders

Rajasaurus narmadensis ("royal lizard from the Narmada River"): From the Upper Cretaceous (upper Maastrichtian, ca. 69–66 Ma) of Hindustan (present-day India). The largest specimen of this species was based on some remains that were assigned to the armed dinosaur *Lametasaurus indicus*, which was actually made up of a mixture of crocodile remains, titanosaur sauropods, and *Rajasaurus* bones. It was about 10 m long and weighed 3 t. Another huge abelisaurid, from the late Upper Cretaceous of central Gondwana (present-day Kenya), was reported to be 11 m long but has not been formally described yet. See more: *Matley 1923; Wilson et al. 2003; Sertich et al. 2013*

Footprints

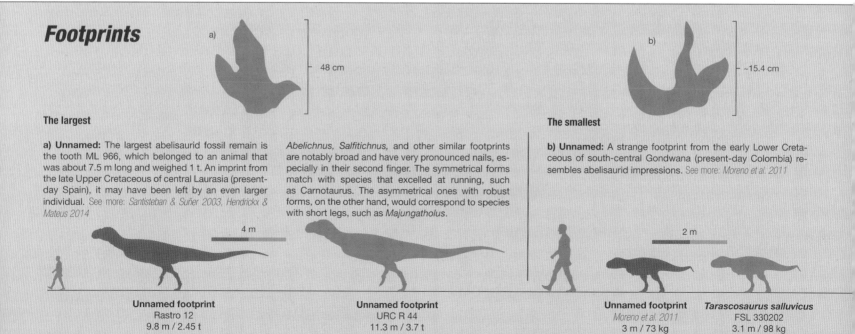

a)

48 cm

b)

~15.4 cm

The largest

a) Unnamed: The largest abelisaurid fossil remain is the tooth ML 966, which belonged to an animal that was about 7.5 m long and weighed 1 t. An imprint from the late Upper Cretaceous of central Laurasia (present-day Spain), it may have been left by an even larger individual. See more: *Santisteban & Suñer 2003, Hendrickx & Mateus 2014*

Abelichnus, Salfitichnus, and other similar footprints are notably broad and have very pronounced nails, especially in their second finger. The symmetrical forms match with species that excelled at running, such as Carnotaurus. The asymmetrical ones with robust forms, on the other hand, would correspond to species with short legs, such as *Majungatholus*.

The smallest

b) Unnamed: A strange footprint from the early Lower Cretaceous of south-central Gondwana (present-day Colombia) resembles abelisaurid impressions. See more: *Moreno et al. 2011*

4 m

2 m

Unnamed footprint
Rastro 12
9.8 m / 2.45 t

Unnamed footprint
URC R 44
11.3 m / 3.7 t

Unnamed footprint
Moreno et al. 2011
3 m / 73 kg

Tarascosaurus salluvicus
FSL 330202
3.1 m / 98 kg

10 m   12 m   14 m   16 m

Characteristics and habits

Theropods with short heads, small, thick, knife-shaped teeth, short necks, four digits on each hand and foot, short or long legs, and heavy bodies. They have epidermal scales. Abelisaurids were carnivores that descended from the abelisauroids. Highly specialized hunters, some had very wide jaws, while others could run at very high speeds. See more: *Mazzetta & Farina 1999*

Diet: small or large prey carnivores.

Time range

From the Middle Jurassic to the Late Upper Cretaceous (ca. 168.3–66 Ma), a duration of approximately 102.3 million years in the fossil record.

The smallest (Class: Medium Grade I)

The age of numerous tiny abelisaurids is unknown, so there is doubt about which is the smallest. *"Carnosaurus"* sp. from Argentina, *"Megalosaurus"* sp. from Brazil, *"Massospondylus" rawesi*, and *"Megalosaurus"* sp. from India are based on teeth, possibly from juveniles. The smallest of the adult specimens is ***Tarascosaurus salluvicus*** ("Tarasque lizard"), from the late Upper Cretaceous of central Laurasia (present-day France). It was 3.5 times shorter and about 40 times lighter than URC R 44. See more: *Le Loeuff & Buffetaut 1991*

The smallest juvenile specimen

CMN 50382, a complete femur, is from the late Lower Cretaceous of Morocco. It belonged to a small carnivore that was 1 m long and weighed 3.4 kg. See more: *Russell 1996*

The oldest

MSNM V5800 dates from the Middle Jurassic (Bathonian, ca. 168.3–166.1 Ma) of present-day Madagascar. *Eoabelisaurus mefi* is even older (Aalenian, ca. 174.1–170.3 Ma), from modern Argentina. It is possible that it was an abelisauroid. See more: *Maganuco, et al. 2005; Pol & Rauhut 2012; Novas et al. 2013*

The most recent

Indosaurus raptorius, Indosuchus matleyi, "Massospondylus" rawesi, Orthogoniosaurus matleyi, Rahiolisaurus gujaratensis and *Rajasaurus narmadensis.* They lived during the Upper Cretaceous (upper Maastrichtian, ca. 69–66 Ma) in Hindustan (present-day India). It is likely that some are synonymous. Cf. *Majungatholus crenatissimus* from Madagascar and FGGUB R.351 from Romania are from the same period, but the latter was a hadrosaurid. See more: *Piveteau 1926; Huene & Matley 1933; Csiki & Grigorescu 1998; Novas et al. 2004; Kessler et al. 2005*

The strangest

Carnotaurus sastrei ("carnivorous bull of the Sastre family"): From the late Upper Cretaceous of western Gondwana (present-day Argentina). It stands out because of its two small horns, one above each eye. Its face was extremely short and proportioned very differently than other abelisaurids. It had very long legs and forward-facing eyes, suggesting that it specialized in fast, small prey. See more: *Bonaparte et al. 1990; Mazzetta & Farina 1999*

The first published species

"Massospondylus" rawesi (1890): From the late Upper Cretaceous of Hindustan (present-day India). It was thought to be a sauropodomorph, but it was too recent; sauropodomorphs became extinct in the middle of the Jurassic. Later it was identified as an abelisaurid and has even been assigned to the genus *"Orthogoniosaurus" rawesi*. See more: *Lydekker 1890*

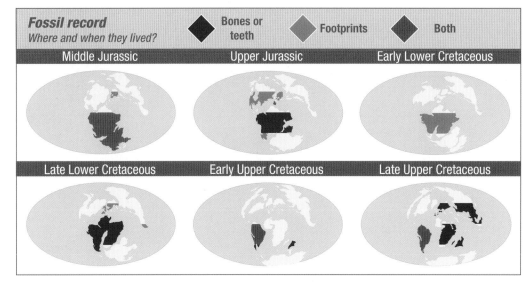

Fossil record
Where and when they lived?

◆ Bones or teeth   ◆ Footprints   ◆ Both

Middle Jurassic | Upper Jurassic | Early Lower Cretaceous

Late Lower Cretaceous | Early Upper Cretaceous | Late Upper Cretaceous

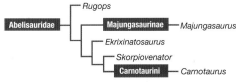

Rugops

Abelisauridae — Majungasaurinae — Majungasaurus
— Ekrixinatosaurus
— Skorpiovenator
— Carnotaurini — Carnotaurus

The most recently published species

Arcovenator escotae (2013) ("arc hunter of Escota"): From the late Upper Cretaceous of central Laurasia (present-day France). At 7.2 m in length and weighing 950 kg, it was one of the largest abelisaurids in Europe. See more: *Hendrickx & Mateus 2014; Tortosa et al. 2013*

Tarascosaurus salluvicus
Le Loeuff & Buffetaut 1991

RECORD SMALLEST

1:15

Specimen: FSL 330202
(Le Loeuff & Buffetaut 1991)

Length: 3.1 m
Hip height: 98 cm
Body mass: 90 kg
Material: thoracic vertebrae
ESR: ●●○○

Juvenile
CMN 50382

1 m   2 m   3 m

Edmarka rex
Bakker et al. 1992

1:40

4 m

2 m

RECORD
LARGEST

Specimen: CPS 1010
(Bakker et al. 1992)

Length: 12 m
Hip height: 3.1 m
Body mass: 4.2 t
Material: incomplete pubis
ESR: ●●○○

The largest (Class: Very large Grade I)

Edmarka rex ("King of Bill Edmark") lived in the Upper Jurassic (Kimmeridgian, ca. 157.3–152.1 Ma) of northwestern Neopangea (present-day western United States). A huge theropod, it was as long as a city bus and heavier than an Asian elephant. It was one of the largest predators of its time and area. It was equal in size with the enormous *Saurophaganax* allosauroid, which was its contemporary. It is possible that both giants specialized in different types of prey, but it is also very likely that they competed for some food sources. It is suspected that *Edmarka* was a large specimen of *Torvosaurus tanneri*, but a detailed analysis has not been carried out yet. See more: *Bakker et al. 1992; Carrano et al. 2012*

A triple tie

The largest specimens of *Torvosaurus tanneri* are BYUVP 2003 and BYUVP 4882, *Torvosaurus gurneyi*, and cf. *Torvosaurus* sp. (present-day western United States, Portugal, and Tanzania, respectively). They were very close in size to *Edmarka rex*. See more: *Janensch 1925; Siegwarth et al. 1996; Hendrickx & Mateus 2014*

Footprints

a)

82 cm

b)

~24.2 cm

The largest

a) Unnamed: The largest footprints of the Upper Jurassic were found in south-central Neopangea (present-day Morocco). They reached sizes of up to 90 cm, although they include the metatarsal. Another footprint that is shorter but lacks the metatarsal marking must have been a gigantic megalosaurus greater than *Edmarka rex*. See more: *Boutakiout et al. 2008, 2009*

The footprints of the great theropods of the Upper Jurassic are divided into two large groups: asymmetrical, which coincide with the finger proportions of the megalosauroids *Changpeipus, Gabirutosaurichnus, Megalosauripus, Samandrinda, Turkmenosaurus,* and *Tuojiangpus*; and symmetrical, which belong to allosauroids.

The smallest

b) Unnamed: It is difficult to differentiate the footprints of small nonspecialized theropods. Thanks to the proportions of the fingers, the asymmetrical shape, and the age of this specimen, however, we can deduce that it may have belonged to a megalosauroid (present-day Italy). See more: *Leonardi & Mietto 2000*

2 m

4 m

Unnamed footprint
161GR1.1
12.7 m / 5.1 t

Edmarka rex
CPS 1010
12 m / 4.2 t

Unnamed footprint
Leonardi & Mietto 2000
3 m / 70 kg

Magnosaurus nethercombensis
OUM J12143
4.5 m / 220 kg

Characteristics and habits

Theropods with small to large heads, crests in the most primitive forms, knife-shaped teeth (ziphodonts), short to long necks, two to four fingers on each hand and four on each foot, and semirobust to robust bodies. Basal megalosauroids descended from the averostra and preceded spinosaurids and allosauroids. There were carnivorous forms that became dominant in their environment.
Diet: carnivores of small or large game.

Time range

From the Lower to Upper Jurassic (ca. 201.3–145 Ma). They appear in the fossil record for approximately 56.3 million years. This would increase to 115 million years if MLP 89-XIII-1-1 is an authentic megalosauroid. See more: *Molnar et al. 1986; Carrano et al. 2012*

The smallest (Class: Medium Grade II)

Magnosaurus nethercombensis ("big lizard of Nethercombe"): From the Middle Jurassic (Bajocian, ca. 170.3–168.3 Ma) of north-central of Neopangea (present-day England). Despite its name, it was the smallest adult basal megalosauroid, comparable to a lion. *Marshosaurus bicentessimus* was of similar size. *Magnosaurus nethercombensis* was somewhat shorter but more robust, 4.4 m long and weighing 225 kg, but it is unknown if it was an adult. Both are 2.7 times shorter and 19 times lighter than *Edmarka rex*. See more: *Madsen 1976; Benson et al. 2014*

The smallest juvenile specimen

Some embryos of *Torvosaurus* sp., from Portugal, are the smallest specimens of this group. The maxilla with teeth ML1188 may be the smallest. It belonged to a small youngling that was 20 cm long and weighed about 30 g. See more: *Araújo et al. 2013*

The oldest

Despite its large size, "Saltriosaurus" is the oldest representative of this group. It dates from the Lower Jurassic (Sinemurian, ca. 199.3–190.8 Ma) and lived in south-central and north-central Neopangea (present-day India and Italy, respectively). Footprints from the north-central area (present-day Italy), with fingers arranged asymmetrically as in megalosauroids, date from the Lower Jurassic (Hettangian, ca. 201.3–199.3 Ma). See more: *Leonardi & Mietto 2000; Dalla Vecchia 2001; Carrano et al. 2012*

The most recent

"*Allosaurus*" *tendagurensis* and cf. "*Torvosaurus*" sp.: These are the most recent. They date from the Upper Jurassic (Upper Tithonian, ca. 148.5–145.5 Ma). An enigmatic fossil is MLP 89-XIII-1-1 (present-day Antarctica), which might be a megalosauroid. It is quite recent; almost 60 million years separate it from any other specimen in the group (Coniacian, ca. 89.8–86.3 Ma). See more: *Janensch 1925; Molnar et al. 1996; Carrano et al. 2012*

The most mysterious

Teinurosaurus sauvagei ("extended lizard of the paleontologist Henri Émile Sauvage"): A very large posterior caudal vertebra from the Upper Jurassic of north-central Pangea (present-day France). Its owner must have been enormous, but it has not been possible to deduce its affinity with certainty. Because of its size and location, it may have been a megalosaurid. See more: *Sauvage 1897*

The first published species

Megalosaurus ("big lizard"): From the Middle Jurassic of north-central Neopangea (present-day England). The name was mentioned for the first time in 1822, although the official description was published two years later. The name "Scrotum humanum," which appeared in 1763 and was later considered to be remains of Megalosaurus, was not conceived to be assigned to a species. The piece can not be assigned to any particular genus since there were several large theropods where it was found, and it is lost. See more: *Parkinson 1822; Buckland 1824; Mortimer**

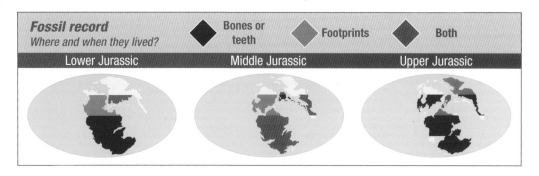

Fossil record
Where and when they lived?

| | Bones or teeth | Footprints | Both |

Lower Jurassic   Middle Jurassic   Upper Jurassic

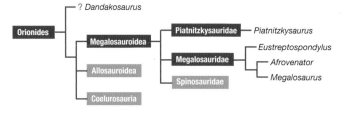

? *Dandakosaurus*

Orionides

Megalosauroidea — Piatnitzkysauridae — *Piatnitzkysaurus*

Megalosauridae — *Eustreptospondylus*

Afrovenator

Allosauroidea

Spinosauridae — *Megalosaurus*

Coelurosauria

The most recently published species

Torvosaurus gurneyi ("wild paleoartist lizard James Gurney"): From the Upper Jurassic of north-central Neopangea (present-day Portugal). It was named in 2014, although it was already known as *Torvosaurus* sp. fourteen years before. See more: *Mateus & Telles-Antunes 2000; Hendrickx & Mateus 2014; Novas et al. 2015*

RECORD SMALLEST

Magnosaurus nethercombensis
Huene 1932

1:30

Specimen: OUM J12143
(Huene 1923)

Length: 4.5 m
Hip height: 1.25 m
Body mass: 220 kg
Material: jaw and partial skeleton
ESR: ●●●○

Neonate
ML 1188

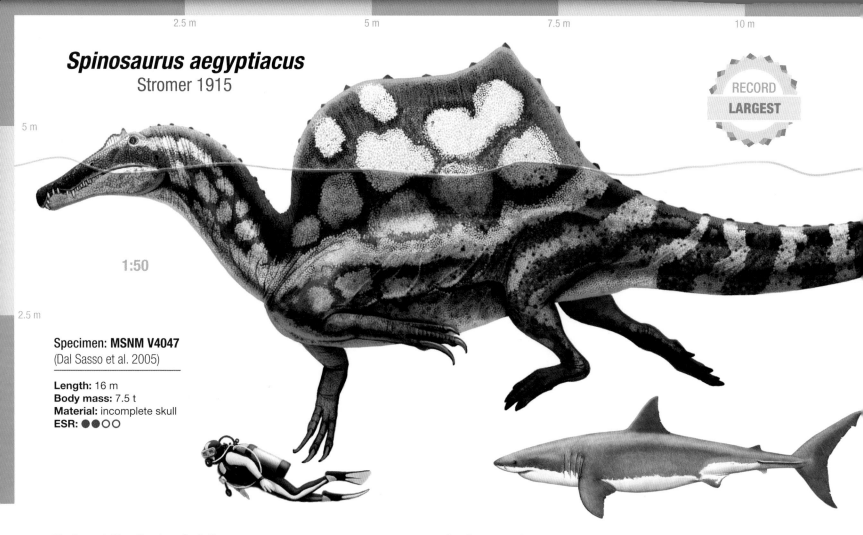

Spinosaurus aegyptiacus
Stromer 1915

RECORD
LARGEST

5 m

1:50

2.5 m

Specimen: MSNM V4047
(Dal Sasso et al. 2005)

Length: 16 m
Body mass: 7.5 t
Material: incomplete skull
ESR: ●●○○

The largest (Class: Very large Grade II)

Spinosaurus aegyptiacus ("lizard with thorns of Egypt"): It lived during the early Upper Cretaceous (Cenomanian, ca. 100.5–93.9 Ma) in north-central Gondwana (present-day Algeria, Egypt, and Morocco). It was a gigantic theropod and the longest of them all. It was as long as a bus and car lined up and weighed as much as an African elephant and a hippopotamus together. The traditional appearance of this dinosaur changed due to a new specimen that showed its true proportions: It had very short hind legs and a very long body. See more: *Dal Sasso et al. 2005; Ibrahim et al. 2014*

Another contender

In 1996, the species *Spinosaurus marocannus* (present-day Egypt, Morocco, and Niger) was described. It was thought to be even greater than *Spinosaurus aegyptiacus*, but it was discovered over time that it was similar to *Sigilmassasaurus*. Its estimated length was about 14.4 m; its weight was 6.5 t. See more: *Russell 1996; Evers et al. 2015*

Footprints

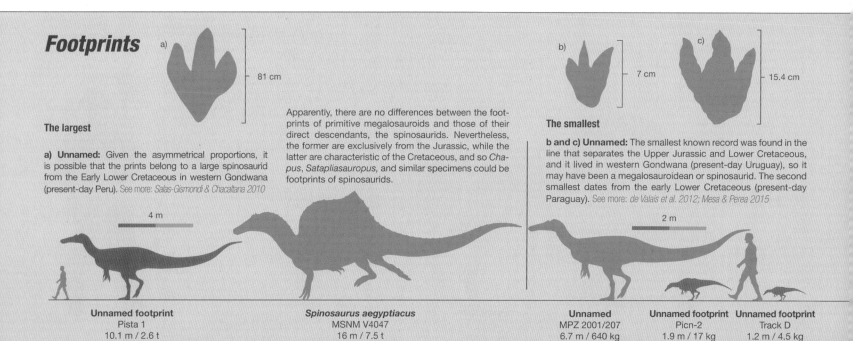

a)

81 cm

b)

7 cm

c)

15.4 cm

The largest

a) Unnamed: Given the asymmetrical proportions, it is possible that the prints belong to a large spinosaurid from the Early Lower Cretaceous in western Gondwana (present-day Peru). See more: *Salas-Gismondi & Chacaltana 2010*

Apparently, there are no differences between the footprints of primitive megalosauroids and those of their direct descendants, the spinosaurids. Nevertheless, the former are exclusively from the Jurassic, while the latter are characteristic of the Cretaceous, and so *Chapus, Satapliasauropus,* and similar specimens could be footprints of spinosaurids.

The smallest

b and c) Unnamed: The smallest known record was found in the line that separates the Upper Jurassic and Lower Cretaceous, and it lived in western Gondwana (present-day Uruguay), so it may have been a megalosauroidean or spinosaurid. The second smallest dates from the early Lower Cretaceous (present-day Paraguay). See more: *de Valais et al. 2012; Mesa & Perea 2015*

4 m

2 m

Unnamed footprint
Pista 1
10.1 m / 2.6 t

Spinosaurus aegyptiacus
MSNM V4047
16 m / 7.5 t

Unnamed
MPZ 2001/207
6.7 m / 640 kg

Unnamed footprint
Picn-2
1.9 m / 17 kg

Unnamed footprint
Track D
1.2 m / 4.5 kg

2.5 m   15 m   17.5 m   20 m   22.5 m

Characteristics and habits

Theropods with long heads, conical teeth, long necks, three fingers on each hand and four on each foot, large nails, and high neural spines. Bipedal or quadrupedal, some have a central crest.

Spinosaurids descended from basal megalosauroids.

Adapted to semi-aquatic or aquatic life, they possessed faces similar to crocodiles. Most had high or very high vertebral spines whose function is not entirely clear. It's suspected that they were used as a storehouse for fat and water or perhaps sexual exhibition. See more: *Bailey 1997; Rayfield et al. 2007; Amiot et al. 2010; Ibrahim et al. 2014*

Diet: carnivores of small game and/or piscivores.

Time range

From the Middle Jurassic to the late Upper Cretaceous (ca. 168.3–83.6 Ma). They lived for approximately 84.7 million years.

The smallest (Class: Large Grade I)

There are numerous remains of small spinosaurids, especially loose teeth, but it is not known how spinosaurids matured. The smallest adult is MPZ 2001/207 (present-day Spain), which was very similar to *Baryonyx walkeri*. It dates from the early Lower Cretaceous. It was about the size of a saltwater crocodile, almost 2.4 times shorter and almost 12 times lighter than *Spinosaurus aegyptiacus*. *Siamosaurus fusuiensis* (formerly *Sinopliosaurus*) and *S. suteethorni* (of present-day China and Thailand, respectively) may have been smaller, but it is difficult to prove this in the absence of more complete remains. See more: *Hou, Yeh & Zhao 1975; Buffetaut & Ingavat 1986; Canudo & Ruiz-Omeñaca 2003*

The smallest young specimen

"Weenyonyx" is an informal name given to a small claw. It could have belonged to a young *Baryonyx walkeri*, which was 1.6 m long and weighed 10 kg. See more: *Martill & Naish 2001*

The oldest

TP4-2, from the Middle Jurassic (Bathonian, ca. 168.3–166.1 Ma) of south-central Neopangea (present-day Niger), seems to have belonged to a primitive spinosaurid. Due to the high spines of the archosaur *Hypselorhachis mirabilis* from the Middle Triassic (present-day Tanzania), it was proposed as an ancestral spinosaurid. See more: *Charig 1967; Moody & Naish 2010; Serrano-Martínez et al. 2014*

The most recent

The unnamed XMDFEC V0010, from east Laurasia (present-day China), lived during the beginning of the late Upper Cretaceous (Santonian, ca. 86.3–83.6 Ma). See more: *Hone, Xu & Wang 2010*

The strangest

Ichthyovenator laosensis ("fish hunter of Laos"), from the late Lower Cretaceous of Cimmeria (present-day Laos), and *Spinosaurus aegyptiacus* were the strangest spinosaurids. *Ichthyovenator* had very wide vertebral spines. It also had an uncommon shape: at the hips was a very pronounced indentation that separated the dorsal sail into two parts. *Spinosaurus*, on the other hand, was so adapted to aquatic life that its hind legs were very short. It occupied the niche of gigantic crocodiles. See more: *Allain et al. 2012, 2014; Ibrahim et al. 2014*

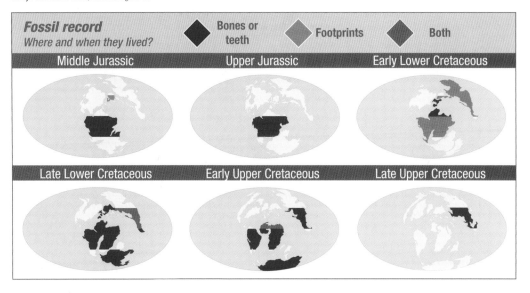

Fossil record *Where and when they lived?*	Bones or teeth	Footprints	Both
Middle Jurassic	Upper Jurassic	Early Lower Cretaceous	
Late Lower Cretaceous	Early Upper Cretaceous	Late Upper Cretaceous	

The first published species

Suchosaurus cultridens (1841) ("crocodile lizard with sharp teeth"): From the late Lower Cretaceous of central Laurasia (present-day England). It was considered a crocodile for many years due to its conical teeth. A later study revealed it to be a spinosaurid dinosaur. See more: *Owen 1841; Buffetaut 2007*

The most recently published species

Ichthyovenator laosensis (2012) ("fish hunter of Laos"): From the Lower Cretaceous of Cimmeria (present-day Laos). It has a mixture of characteristics found in *Baryonyx* and spinosaurids. See more: *Allain et al. 2012, 2013*

Spinosauridae — Baryonyx
— Spinosaurinae — Ichthyovenator
— Spinosaurus

Unnamed
Canudo & Ruiz-Omeñaca 2003

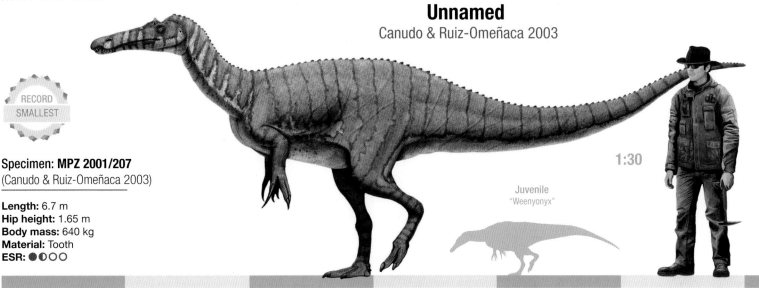

RECORD SMALLEST

Specimen: MPZ 2001/207
(Canudo & Ruiz-Omeñaca 2003)

Length: 6.7 m
Hip height: 1.65 m
Body mass: 640 kg
Material: Tooth
ESR: ●○○○

1:30

Juvenile "Weenyonyx"

1 m   2 m   3 m   4 m   5 m   6 m

Saurophaganax maximus
Stovall 1950

RECORD
LARGEST

4 m

2 m

1:40

Specimen: OMNH 1935
(Chure 1995)

Length: 12 m
Hip height: 3.25 m
Body mass: 4.5 t
Material: humerus
ESR: ●●○○

The largest (Class: Very large Grade I)

Saurophaganax maximus ("supreme among murderous lizards") dates from the Upper Jurassic (Kimmeridgian, ca. 157.3–152.1 Ma) and lived in northwestern Neopangea (present-day Oklahoma, USA). It was one of the largest land predators of its time, weighing more than an Asian elephant. Its neck was strong, and its jaws could open very wide, allowing it to tear large pieces of flesh from its victims, weakening them. In order to subdue some giants, it needed to hunt in groups. See more: *Bakker 1988; Paul 1988; Chure 1995*

Other contenders

Epanterias amplexus, from the Upper Jurassic (present-day Colorado, USA), is another large allosaurid. It is unknown whether it was a large *Allosaurus fragilis* or a subadult of *Saurophaganax maximus*. It reached a length of 10.4 m and a weight of 2.9 t. *Sinraptor dongi* and the metriacanthosaur IVPP V 15310 from the Upper Jurassic (present-day China) were longer and heavier than *Epanterias*. They were about 11.5 m long and weighed 3.9 t. See more: *Cope 1878; Williamson & Chure 1996; Xing & Clark 2008; Paul & Carpenter 2010; Mortimer**

Footprints

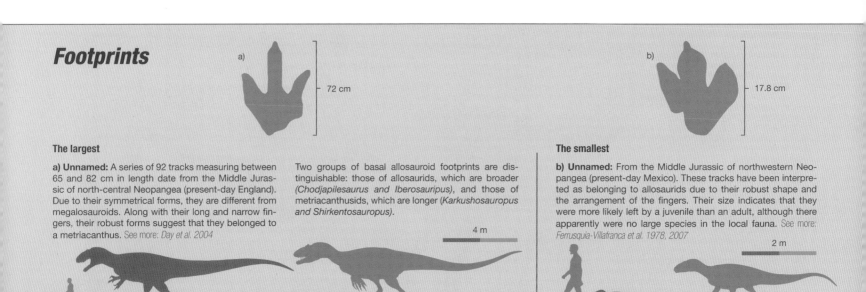

a)

72 cm

b)

17.8 cm

The largest

a) Unnamed: A series of 92 tracks measuring between 65 and 82 cm in length date from the Middle Jurassic of north-central Neopangea (present-day England). Due to their symmetrical forms, they are different from megalosauroids. Along with their long and narrow fingers, their robust forms suggest that they belonged to a metriacanthus. See more: *Day et al. 2004*

Two groups of basal allosauroid footprints are distinguishable: those of allosaurids, which are broader (*Chodjapilesaurus* and *Iberosauripus*), and those of metriacanthusids, which are longer (*Karkushosauropus* and *Shirkentosauropus*).

4 m

The smallest

b) Unnamed: From the Middle Jurassic of northwestern Neopangea (present-day Mexico). These tracks have been interpreted as belonging to allosaurids due to their robust shape and the arrangement of the fingers. Their size indicates that they were more likely left by a juvenile than an adult, although apparently there were no large species in the local fauna. See more: *Ferrusquia-Villafranca et al. 1978, 2007*

2 m

Unnamed footprint	*Saurophaganax maximus*	Unnamed footprint	*Xuanhanosaurus qilixiaensis*
T80	OMNH 1935	Morfotipo C - IGCU-2532	IVPP V6729
11.9 m / 3.2 t	12 m / 4.5 t	2 m / 36 kg	4.8m / 265 kg

10 m   12 m   14 m   16 m   18 m

Characteristics and habits

Theropods with large or medium heads, knife-like teeth, short necks, three or four fingers on each hand and four on each foot, and moderately robust bodies. Allosauroids descended from basal megalosauroids and preceded carcharodontosaurids and coelurosaurs. They were secondary or dominant hunters. Allosaurids and metriacanthosaurids propagated during the Jurassic, while other forms similar to *Erectopus* were common in the Cretaceous.

See more: *Brusatte & Sereno 2007; Carrano et al. 2012*

Diet: carnivores of small or large game.

Time range

From the Middle Jurassic to the early Upper Cretaceous (ca. 174.1–89.8 Ma), a duration of about 84.3 million years in the fossil record.

The smallest (Class: Medium Grade III)

Xuanhanosaurus qilixiaensis ("lizard from Xuanhan County and Qilixia City"): From the Middle Jurassic (Bajocian, ca. 170.3–168.3 Ma) of eastern Paleoasia (present-day China). It was almost as long as a saltwater crocodile and weighed as much as a lion and an adult man combined. As in various allosauroids, its development stage is not known. It is estimated to be 2.5 times shorter and 17 times lighter than *Saurophaganax amplexus*. See more: *Dong 1984*

The smallest juvenile specimen

Several teeth of young allosaurids were located in Portugal. The tiniest of all is IPFUB GUI Th 4 a, which was only about 55 cm long and weighed 130 g. See more: *Rauhut & Fecner 2005*

The oldest

Shidaisaurus jinae from the Middle Jurassic (Aalenian, ca. 174.1–170.3 Ma) of eastern Paleoasia (present-day China) was a metriacanthus, the most basal group of allosauroids, suggesting that they originated on this continent. See more: *Wu et al. 2009*

The most recent

The MCF-PVPH 320 theropod comes from the early Lower Cretaceous (Turonian, ca. 93.9–89.8 Ma) of southwestern Gondwana (present-day Argentina). This specimen comprises an incomplete frontal with affinity with the syraptorids. Another remain that is just as old was located in Syria and has been considered similar to *Carcharodontosaurus* or *Erectopus*. See more: *Hoojier 1968, Paulina-Carabajal & Coria 2015*

The strangest

Yangchuanosaurus shangyouensis CV 00215) is a subadult whose skull is 9% shorter than its femur. In the adult specimen CV 00216, however, the skull is 15% longer than the leg bone. This indicates that body proportions changed drastically as *Yangchuanosaurus* developed, and more remarkably than in its relatives. See more: *Dong et al. 1978, 1983*

The first published species

Antrodemus valens (1870) ("cavernous and healthy body"): From the Upper Jurassic of northwestern Neopangea (present-day Colorado, USA). It is an incomplete caudal vertebra, possibly of a species already known but not identified with certainty. It is often attributed to *Allosaurus fragilis*. See more: *Leidy 1870*

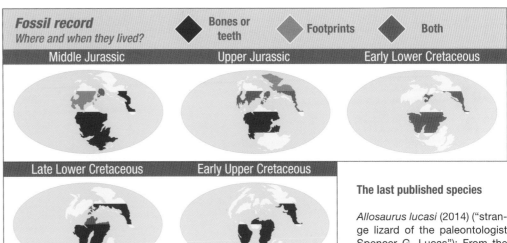

Fossil record
Where and when they lived?

◆ Bones or teeth   ◇ Footprints   ◆ Both

Middle Jurassic   Upper Jurassic   Early Lower Cretaceous

Late Lower Cretaceous   Early Upper Cretaceous

The last published species

Allosaurus lucasi (2014) ("strange lizard of the paleontologist Spencer G. Lucas"): From the Upper Jurassic of northwestern Neopangea (present-day Colorado, USA). Some authors do not accept the species, which they consider an example of *Allosaurus fragilis*, but it is an adult and shorter than the other species. See more: *Dalman 2014*

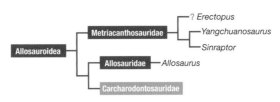

Allosauroidea
- Metriacanthosauridae
  - ? *Erectopus*
  - *Yangchuanosaurus*
  - *Sinraptor*
- Allosauridae — *Allosaurus*
- Carcharodontosauridae

Xuanhanosaurus qilixiaensis
Dong 1984

RECORD SMALLEST

1:30

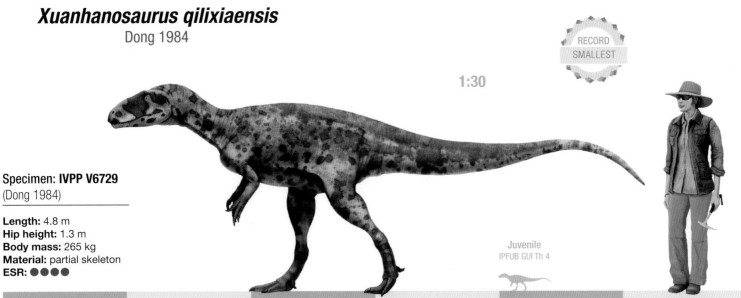

Specimen: IVPP V6729
(Dong 1984)

Length: 4.8 m
Hip height: 1.3 m
Body mass: 265 kg
Material: partial skeleton
ESR: ●●●●

Juvenile
IPFUB GUI Th 4

1 m   2 m   3 m   4 m   5 m   6 m

2 m   4 m   6 m   8 m

Giganotosaurus carolinii
Coria & Salgado 2000

RECORD
LARGEST

4 m

2 m

1:40

Specimen: MUCPv-95
(Calvo & Coria 2000)

Length: 13.2 m
Hip height: 3.85 m
Body mass: 8.5 t
Material: incomplete dentary with teeth
ESR: ●●○○

The largest (Class: Very large Grade II)

Giganotosaurus carolinii ("southern giant lizard of Rubén Carolini"): It lived during the late Upper Cretaceous (lower Cenomanian, ca. 100.5–93.9 Ma) in southwestern Gondwana (present-day Argentina). It was one of the largest predators of all time, as long as an urban bus, and weighing as much as an African elephant and a white rhinoceros together. The largest specimen is based on a very incomplete tooth that turns out to be 6.5% larger than that of the MUCPv-Ch1 type specimen. This individual would be as heavy as the largest specimen of *Tyrannosaurus rex*. *Giganotosaurus* was longer than *Tyrannosaurus*, but its body was less sturdy. See more: *Calvo & Coria 2000; Paul 2010; Hartman**

Other gigantic species

There are several carcharodontosaurids that reached or exceeded 7 t. Among them are *Tyrannotitan chubutensis* (12.5 m and 7 t) from Argentina, MB R2352 (12.5 m and 7 t) from South Africa, *Mapusaurus roseae* (12.6 m and 7.6 t) from Argentina, *Carcharodontosaurus* sp. (12.7 m and 7.8 t) from Niger, and *Carcharodontosaurus saharicus* (12.7 m and 7.8 t) from Morocco. See more: *Lapparent 1960; Sereno et al. 1996; Novas et al. 2005; Coria & Currie 2006; Galton & Molnar 2011*

Footprints

a)

— 78.2 cm

b)

— 28.5 cm

The largest

a) Unnamed: One of the largest theropods that left its imprints was a carcharodontosaurid from the early Upper Cretaceous of southwestern Gondwana (present-day Brazil). See more: *Carvalho 2001*

The footprints known as *Buckeburgichnus* and others that are similar may have belonged to carcharodontosaurids. They differ from those of *Megalosauripus* because they are symmetrical.

The smallest

b) Unnamed: Identifying traces of nonspecialized theropods is especially problematic when the specimens are small, because the ignorable margin is much larger. A young carcharodontosaurid could have produced this footprint because it is robust, symmetrical, and from the early Lower Cretaceous of western Gondwana (present-day Brazil). See more: *Leonardi 1980*

4 m

3 m

Unnamed footprint	*Giganotosaurus carolinii*	Unnamed footprint	*Concavenator concorvatus*
SLPG - D	MUCPv-95	SOPI 1	MCCM-LH 6666
10.6 m / 5 t	13.2 m / 8.5 t	4.2 m / 230 kg	5.2 m / 400 kg

10 m   12 m   14 m   16 m   18 m

Characteristics and habits

Theropods with large heads, knife-like teeth, short necks, short arms with three fingers on each hand and four on each foot, and semirugged or robust bodies. The carcharodontosaurids descended from basal allosauroids. They were secondary or dominant hunters. Many of the largest known theropods belonged to this family. See more: *Carrano et al. 2012*
Diet: carnivores of small or large game.

Time range

From the Upper Jurassic to the late Upper Cretaceous (ca. 168.5–66 Ma), a duration of approximately 102.3 million years in the fossil record.

The smallest (Class: Large Grade I)

Concavenator corcovatus ("hunchback hunter"): From the late Lower Cretaceous (Barremian, ca. 129.4–125 Ma) of north-central Asia (present-day Spain). It was as heavy as a great brown bear and longer than a saltwater crocodile. The type specimen is an adult almost 2.5 times shorter and 21 times lighter than *Giganotosaurus carolinii*. See more: *Ortega et al. 2010*

The smallest juvenile specimen

MACN-PV RN 1086 is known only from an incomplete tooth that belonged to a small carcharodontosaurid from Argentina. The animal in life would have had a minimum length of only about 2.6 m and a weight of 64 kg. See more: *Martinelli & Forasiepi 2004*

The oldest

It is believed that TP4-6, from the Middle Jurassic (Bathonian, ca. 168.3–166.1 Ma) area of south-central Neopangea (present-day Niger), is the oldest carcharodontosaurid. See more: *Serrano-Martínez, Ortega & Knoll 2013*

The most recent

Carcharodontosaurids were affected by the extinction that occurred between the late and early Upper Cretaceous, but one group survived in northern South America and perhaps in Europe. It is believed that the most recent specimen is the one classified as UFRJ-DG 379-Rd, since it dates from the early Upper Cretaceous (upper Maastrichtian, ca. 69–66 Ma). It lived in the southwestern zone of Gondwana (present-day Brazil). See more: *Candeiro et al. 2006*

The strangest

In the sacrum, *Concavenator corcorvatus* had highly developed vertebral spines that might have been used in visual exhibitions, to store fat, or as a thermal regulator. A close relative, *Becklespinax altispinax*, had a similar backbone, but the material is not complete. See more: *Owen 1855; Bailey 1997; Ortega et al. 2010*

The first published species

"*Megalosaurus*" *ingens* (1920): From the Upper Jurassic of south-central Neopangea (present-day Tanzania). It has been called "*Ceratosaurus*" *ingens* because it was speculated to be a giant ceratosaurid. Now it is known that it was a basal carcharodontosaurid. It was a contemporary of *Veterupristisaurus milneri*, but they can't be compared, since they are known by different materials. See more: *Janensch 1920; Rauhut 2011*

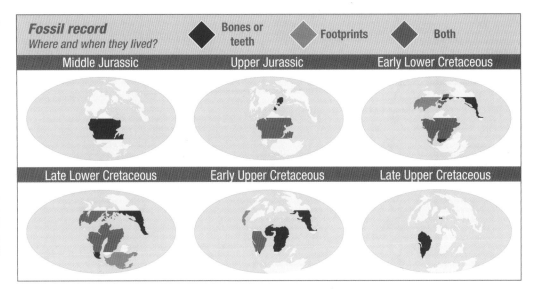

Fossil record
Where and when they lived?

◆ Bones or teeth   ◆ Footprints   ◆ Both

| Middle Jurassic | Upper Jurassic | Early Lower Cretaceous |
| Late Lower Cretaceous | Early Upper Cretaceous | Late Upper Cretaceous |

Carcharodontosauridae
├─ Neovenator
├─ Acrocanthosaurus
└─ Carcharodontosaurinae ── Carcharodontosaurus

The most recently published species

Datanglong guangxiensis (2014) ("dragon from the Datang locality of the Guangxi Province"): From the late Lower Cretaceous of eastern Laurasia (present-day China). It is an adult specimen of medium size, but NHMG 10858 was found in the same locality. It probably belongs to this species and must have been larger. See more: *Mo et al. 2014a, 2014b*

Concavenator corcovatus
Ortega et al. 2010

RECORD SMALLEST

1:30

Juvenile
MAC-PV RN 1086

Specimen: MCCM-LH 6666
(Ortega et al. 2010)

Length: 5.2 m
Hip height: 1.9 m
Body mass: 400 kg
Material: partial skeleton
ESR: ●●●●○

1 m   2 m   3 m   4 m   5 m   6 m

Siamotyrannus isanensis
Buffetaut et al 1996

RECORD **LARGEST**

2 m

1:30

Specimen: PW9-1
(Buffetaut et al 1996)

Length: 10 m
Hip height: 2.5 m
Body mass: 1.75 t
Material: partial skeleton
ESR: ●●●○

The largest (Class: Large Grade III)

Siamotyrannus isanensis ("Siam tyrant of northeast Thailand"): Lived during the late Lower Cretaceous (Barremian, ca. 129.4–125 Ma) in Cimmeria (present-day Thailand). It was a theropod as heavy as four brown bears and as long as two saltwater crocodiles. It was considered a probable tyrannosaurid or allosauroid, but it has recently been suggested that it might be a great primitive coelurosaur. See more: *Buffetaut et al. 1996; Carrano et al. 2012; Buffetaut & Suteethorn 2012; Samathi 2015*

The second largest

Basal coelurosaurs were mostly small, except for some of the more primitive species. *Lourinhanosaurus antunesi* ("lizard of the Lourinhã Formation of the paleontologist Miguel Telles Antunes"), from the Middle Jurassic (Kimmeridgian, ca. 157.3–152.1 Ma) of north-central Neopangea (present-day Portugal), was 5.2 m long and weighed approximately 300 kg. It was twice as short and six times lighter than *Siamotyrannus*. See more: *Mateus 1998; Carrano et al. 2012*

Footprints

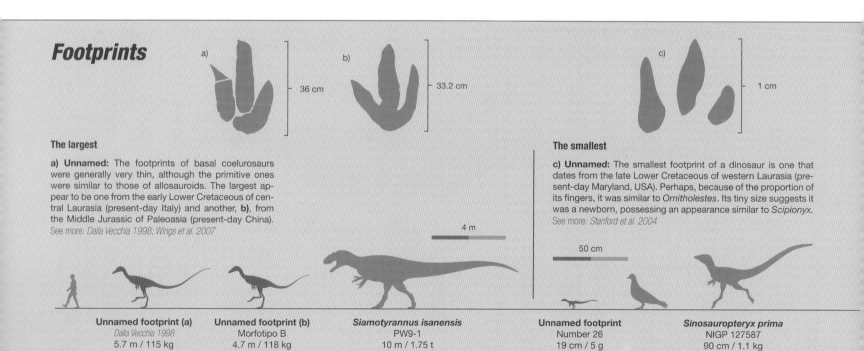

a)   36 cm

b)   33.2 cm

c)   1 cm

The largest

a) Unnamed: The footprints of basal coelurosaurs were generally very thin, although the primitive ones were similar to those of allosauroids. The largest appear to be one from the early Lower Cretaceous of central Laurasia (present-day Italy) and another, **b),** from the Middle Jurassic of Paleoasia (present-day China). See more: *Dalla Vecchia 1998; Wings et al. 2007*

The smallest

c) Unnamed: The smallest footprint of a dinosaur is one that dates from the late Lower Cretaceous of western Laurasia (present-day Maryland, USA). Perhaps, because of the proportion of its fingers, it was similar to *Ornitholestes*. Its tiny size suggests it was a newborn, possessing an appearance similar to *Scipionyx*. See more: *Stanford et al. 2004*

4 m

50 cm

Unnamed footprint (a)	Unnamed footprint (b)	*Siamotyrannus isanensis*	Unnamed footprint	*Sinosauropteryx prima*
Dalla Vecchia 1998	Morfotipo B	PW9-1	Number 26	NIGP 127587
5.7 m / 115 kg	4.7 m / 118 kg	10 m / 1.75 t	19 cm / 5 g	90 cm / 1.1 kg

8 m   10 m   12 m   14 m

Characteristics and habits

Theropods with small to medium heads, large brains, small to medium knife-like teeth, medium to long arms and necks, long legs and tails, graceful bodies, and three fingers on each hand and four on each foot. Coelurosaurs descended from basal allosauroids. Most of them were small, fast hunters. See more: Senter 2007

Diet: carnivores from small to large game and/or scavengers.

The most recent

Basal coelurosaurs declined halfway through the Cretaceous, but some managed to survive until the late Upper Cretaceous (Coniacian, ca. 89.8–86.3 Ma). An example of this was the coelurosaur MAU-PV-PH-447/1 from southwestern Gondwana (present-day Argentina). It could have been similar to *Bicentenaria argentina* but was more recent. See more: *Canudo et al. 2009*

The strangest

Ornitholestes hermanni ("Adam Hermann's bird thief"), from the Upper Jurassic of northwestern Neopangea (present-day Wyoming, USA), was a strange coelurosaur that differed from others. Its legs were less adapted to running, and it had premaxillary teeth that were more robust, a common feature of tyrannosauroids. Some authors place it with coelurosaur maniraptors. See more: *Osborn 1903; Brusatte 2013*

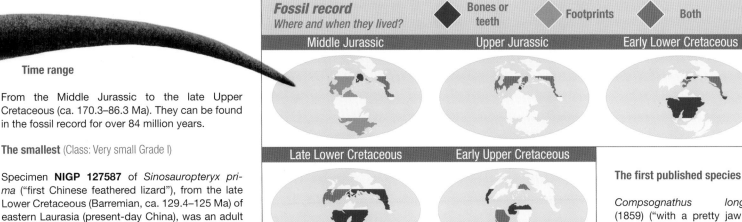

Fossil record *Where and when they lived?* — Bones or teeth ◆   Footprints ◆   Both ◆

Middle Jurassic   Upper Jurassic   Early Lower Cretaceous

Late Lower Cretaceous   Early Upper Cretaceous

Time range

From the Middle Jurassic to the late Upper Cretaceous (ca. 170.3–86.3 Ma). They can be found in the fossil record for over 84 million years.

The smallest (Class: Very small Grade I)

Specimen **NIGP 127587** of *Sinosauropteryx prima* ("first Chinese feathered lizard"), from the late Lower Cretaceous (Barremian, ca. 129.4–125 Ma) of eastern Laurasia (present-day China), was an adult female; we know because it had two eggs inside. It was the size of a crow, almost 11 times shorter and 1,600 times lighter than *Siamotyrannus isanensis*. See more: *Chen et al. 1998*

The smallest juvenile specimen

Nearly 300 osseous elements of *Lourinhanosaurus antunesi* embryos are known. No more than 47 cm long and weighing 70 g, the specimen ML 565 was perhaps the smallest of these. See more: *Martinelli & Forasiepi 2004*

The oldest

Gasosaurus constructus ("lizard found during the construction of a gas mine"), from the Middle Jurassic (Bajocian, ca. 170.3–168.3 Ma) of eastern Paleoasia (present-day China), was considered an allosaurid. New studies identify it as a coelurosaur. See more: *Dong & Tang 1985; Holtz 2000*

The first published species

Compsognathus longipes (1859) ("with a pretty jaw and long feet"), from the Upper Jurassic of north-central Neopangea (present-day Germany), was the first articulated dinosaur. Among the only missing pieces are the back of the tail and some fingers. It was considered to be didactyl at the time. Back then, dinosaurs were thought to have had very grotesque forms; since this one was as nice as a bird, it was not considered a dinosaur until 1896. See more: *Wagner 1859; Marsh 1896*

The most recently published species

Aorun zhaoi (2013): From the Middle Jurassic of eastern Paleoasia (present-day China). Its name is based on the Chinese Mandarin language: *Ao Run* is "the dragon king of the western sea from the *Journey to the West* epic." See more: *Choiniere et al. 2013*

Coelurosauria
- ? *Lourinhanosaurus*
- **Tyrannosauroidea**
- ? **Megaraptora**
- ? *Coelurus*
- ? *Ornitholestes*
- **Compsognathidae** — *Compsognathus*
- **Ornithomimosauria**
- **Maniraptora**

RECORD SMALLEST

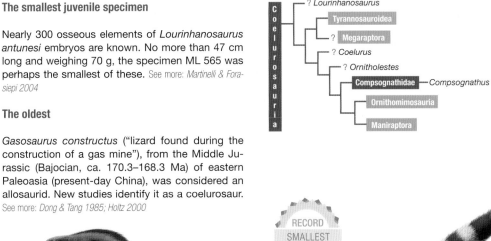

Sinosauropteryx prima
Ji & Ji 1996

1:5

Specimen: NIGP 127587
(Chen et al. 1998)

Length: 90 cm
Hip height: 33 cm
Body mass: 1.1 kg
Material: partial skeleton
ESR: ●●●●

Juvenile
ML 565

25 cm   50 cm   75 cm   100 cm

Unnamed
Mo & Xu 2015

RECORD
LARGEST

4 m

2 m

1:40

Specimen: NHMG 8500
(Mo & Xu 2015)

Length: 12.9 m
Hip height: 3.6 m
Body mass: 5.2 t
Material: tooth
ESR: ●●○○

The largest (Class: Very large Grade I)

NHMG 8500, from the late Upper Cretaceous of eastern Laurasia (present-day China), is known from a huge tooth as large as that of the largest carcharodontosaurids or tyrannosaurids, although its morphology indicates that it belonged to a theropod other than these. Huge megaraptors coexisted with *Tarbosaurus bataar*, but this was larger. It was so immense that it would be necessary to combine an Asian elephant and a giraffe to match its weight. It coexisted with large-clawed theropods, heavy titanosaurs, hadrosaurs, ankylosaurids, and dangerous ceratopsids that could have been its prey. See more: *Mo & Xu 2015; Mortimer**

The closest in size

Bahariasaurus ingens from the late Upper Cretaceous of central Gondwana (present-day Egypt) is the remains of a very large megaraptor that reached a length of 12.2 m and a weight of 4.6 t. *Chilantaisaurus tashuikouensis* and *Siats meekerorum* (present-day China and Utah, USA, respectively) are smaller, only 11.7 m long and weighing 4.1 t. The latter could have been even larger, since it was a subadult specimen. "Osteoporosia gigantea" is the informal name of a supposed gigantic megaraptorid that was 15 m long, but it seems to have been a significantly smaller carcharodontosaurid. See more: *Stromer 1934; Singer*; Mortimer**

Footprints

a)

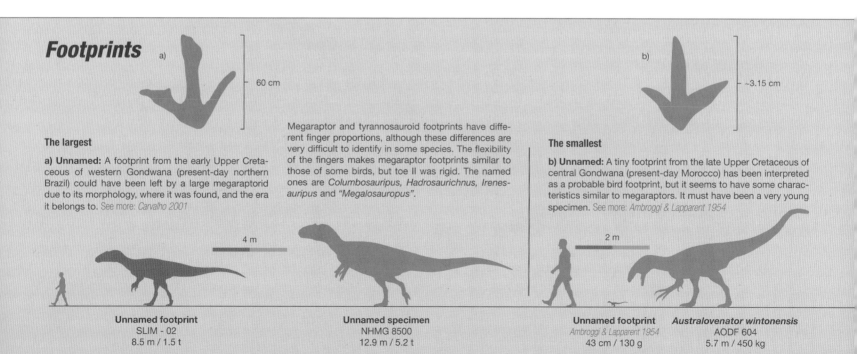

60 cm

~3.15 cm

b)

The largest

a) Unnamed: A footprint from the early Upper Cretaceous of western Gondwana (present-day northern Brazil) could have been left by a large megaraptorid due to its morphology, where it was found, and the era it belongs to. See more: *Carvalho 2001*

Megaraptor and tyrannosauroid footprints have different finger proportions, although these differences are very difficult to identify in some species. The flexibility of the fingers makes megaraptor footprints similar to those of some birds, but toe II was rigid. The named ones are *Columbosauripus*, *Hadrosaurichnus*, *Irenesauripus* and *"Megalosauropus"*.

The smallest

b) Unnamed: A tiny footprint from the late Upper Cretaceous of central Gondwana (present-day Morocco) has been interpreted as a probable bird footprint, but it seems to have some characteristics similar to megaraptors. It must have been a very young specimen. See more: *Ambroggi & Lapparent 1954*

4 m

2 m

Unnamed footprint
SLIM - 02
8.5 m / 1.5 t

Unnamed specimen
NHMG 8500
12.9 m / 5.2 t

Unnamed footprint
Ambroggi & Lapparent 1954
43 cm / 130 g

Australovenator wintonensis
AODF 604
5.7 m / 450 kg

Characteristics and habits

Theropods with elongated heads, large knife-like teeth, fused nostrils, short or medium arms and necks, hands with large, flat fingernails on the thumbs, semirigid bodies, and three fingers on each hand and four on each foot. Megaraptors, like tyrannosauroids, descended from the primitive coelurosaur. In general, they were small, fast hunters, but some were very large. They were the dominant predators in Gondwana and part of Laurasia. See more: *Novas et al 2013; Yun et al. 2015*

Diet: small or large game carnivores

Time range

From the early Lower Cretaceous to the late Upper Cretaceous (ca. 129.4–66 Ma). They lasted approximately 63.4 million years and were present throughout the Cretaceous.

The smallest (Class: Medium Grade II)

There are several specimens of small megaraptors in Australia, but it is unknown if they were adults or juveniles. ***"Megaraptor" sp.***, from the late Lower Cretaceous (Aptian, ca. 125–113 Ma), was the smallest individual and was older than both *Rapator ornitholestoides* and *Australovenator wintonensis*. It was comparable in size to a lion. It is based on an ulna, and it is believed that it could represent a new genus. It is three times shorter and thirty times lighter than the unnamed NHMG 8500. See more: *Chen et al. 1998*

The smallest juvenile specimens

Several juvenile specimens of *Fukuiraptor kitadaniensis*, from the early Lower Cretaceous (present-day Japan), are known. The smallest was specimen FPDM-V980805018, which was perhaps 76 cm long and weighed 1.9 kg. Numerous remains of teeth or vertebrae are just as old but in the south (present-day Australia). The smallest of all could be NMV P221205, known by a tiny tooth that belonged to an individual that could have been 56 cm long and weighed 450 g. See more: *Currie & Azuma 2006; Benson et al. 2012*

The strangest

On each thumb, *Megaraptor namunhuaiquii* had two huge, very noticeable nails that were up to 34 cm along the curvature, the largest known in a terrestrial predator. See more: *Novas 1998*

The most recent

NS CD 583 from Chatham Island, New Zealand, belongs to the Upper Cretaceous (Upper Maastrichtian, ca. 69–66 Ma). Some authors consider it a primitive coelurosaur, which means it may have been a megaraptor. See more: *Stilwell et al. 2006*

The oldest

"Allosaurus" sibiricus was found in the Tignin Formation (also known as Turgin or Zugmar). Its precise age is unknown. It is believed that it belonged to the early Lower Cretaceous (Berriasian-Hauterivian, ca. 129.4 Ma). It lived in eastern Laurasia (present-day eastern Russia). See more: *Huene 1926; Benton & Spencer 1995*

The first published species

"Megalosaurus" sp. (1913): From the late Upper Cretaceous of western Gondwana (present-day northern Brazil). Perhaps it was a megaraptor because its teeth were similar to those of a tyrannosaurid *(Dryptosaurus aquilungus)*, an uncommon group in South America. *Rapator ornitholestoides* is the first genus of this group to be created, since other species had been assigned to *"Allosaurus"* or *"Megalosaurus"*. See more: *Pacheco 1913; Huene 1932*

The most recently published species

Siats meekerorum (2013) ("Siats, John Caldwell Meeker, Withrow Meeker, and Lis Meeker"): From the early Upper Cretaceous of western Laurasia (present-day Utah, USA). It was described as a possible neovenatorid allosauroid, but it is now known that it was a basal megaraptor. See more: *Zanno & Makovicky 2013*

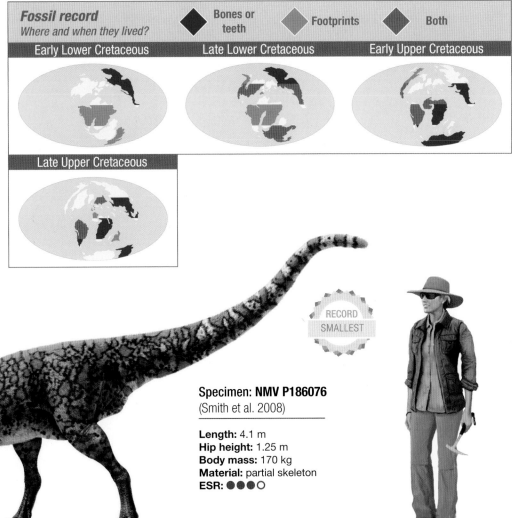

Fossil record
Where and when they lived? ◆ Bones or teeth ◆ Footprints ◆ Both

Early Lower Cretaceous   Late Lower Cretaceous   Early Upper Cretaceous

Late Upper Cretaceous

```
                ? Bahariasaurus
               ┌ Fukuiraptor
  ┌ Megaraptora ┤
  │             │              ┌ Australovenator
  │             └ Megaraptoridae ┤
  │                            └ Megaraptor
```

"Megaraptor" sp.
Smith et al. 2008

1:25

Juvenile
FPDM-V980805018

Specimen: NMV P186076
(Smith et al. 2008)

Length: 4.1 m
Hip height: 1.25 m
Body mass: 170 kg
Material: partial skeleton
ESR: ●●●○

RECORD
SMALLEST

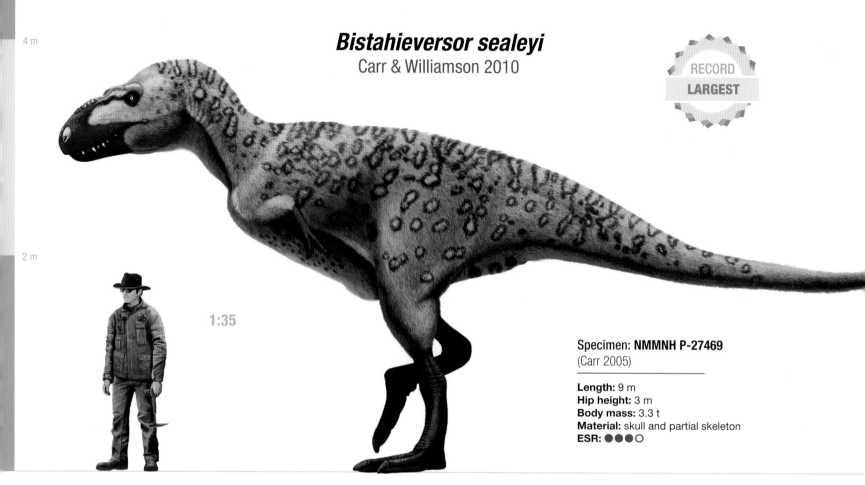

Bistahieversor sealeyi
Carr & Williamson 2010

RECORD
LARGEST

4 m

2 m

1:35

Specimen: NMMNH P-27469
(Carr 2005)

Length: 9 m
Hip height: 3 m
Body mass: 3.3 t
Material: skull and partial skeleton
ESR: ●●●●○

The largest (Class: Very large Grade I)

Bistahieversor sealeyi ("Paul Sealey's destroyer of Bistahi"): It lived during the late Upper Cretaceous (Campanian, ca. 78–72.1 Ma) in western Laurasia (present-day New Mexico, USA). It was a sturdy carnivore with an appearance similar to tyrannosaurids, leading it to be known previously as *Aublysodon* cf. *mirandus* or *Daspletosaurus* sp. At the end of the Cretaceous in western North America (paleocontinent Laramidia), tyrannosauroid derivatives were replaced by tyrannosaurids. In the east (Appalachia), where there was no exchange of fauna with Asia, the dominant predators were tyrannosaurids similar to *Bistahieversor*. See more: *Carr 2005; Lehman & Carpenter 1990; Carr and Williamson 2000; Carr & Williamson 2010*

Longer but less heavy

cf. *Prodeinodon mongoliensis* ("previous to Deinodon of Mongolia"): From the late Lower Cretaceous (Valanginian, ca. 139.8–132.9 Ma) area of eastern Laurasia (present-day Mongolia). The incomplete remains of a maxillary tooth and a tibia and a fibula that are about 1 m long are attributed to *Prodeinodon mongoliensis*, although these remains can not be compared with this species. It may have been a predator whose appearance was similar to *Yutyrannus huali*, so perhaps its length, 9.8 m, was longer than that of *Bistahieversor*, but its body was lighter, weighing close to 2.3 t. *Labocania anomala* would be heavier at 2.6 t but shorter, only 8.2 m long. See more: *Osborn 1924; Bohlin 1953; Molnar 1974; Xu et al. 2012*

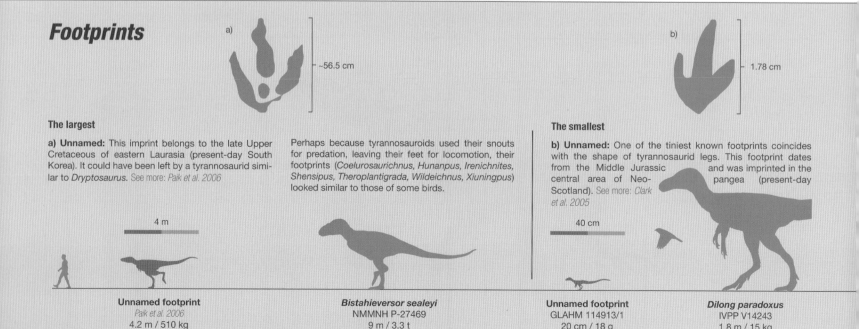

Footprints

a)   ~56.5 cm

b)   1.78 cm

The largest

a) Unnamed: This imprint belongs to the late Upper Cretaceous of eastern Laurasia (present-day South Korea). It could have been left by a tyrannosaurid similar to *Dryptosaurus*. See more: *Paik et al. 2006*

Perhaps because tyrannosauroids used their snouts for predation, leaving their feet for locomotion, their footprints (*Coelurosaurichnus, Hunanpus, Irenichnites, Shensipus, Theroplantigrada, Wildeichnus, Xiuningpus*) looked similar to those of some birds.

The smallest

b) Unnamed: One of the tiniest known footprints coincides with the shape of tyrannosaurid legs. This footprint dates from the Middle Jurassic   and was imprinted in the central area of Neo-   pangea (present-day Scotland). See more: *Clark et al. 2005*

4 m

Unnamed footprint
Paik et al. 2006
4.2 m / 510 kg

Bistahieversor sealeyi
NMMNH P-27469
9 m / 3.3 t

40 cm

Unnamed footprint
GLAHM 114913/1
20 cm / 18 g

Dilong paradoxus
IVPP V14243
1.8 m / 15 kg

10 m   12 m   14 m   16 m

Characteristics and habits

Theropods with medium heads, large knife-like teeth, fused nostrils, medium arms and necks, long tails, graceful bodies, and three fingers on each hand and four on each foot. Basal tyrannosauroids descended from primitive coelurosaurids and preceded the tyrannosaurids. There were small, light, and fast forms in addition to other large and more robust forms See more: *Senter 2007*

Diet: carnivores of small to large game

Time range

From the Middle Jurassic to the late Upper Cretaceous (ca. 167.7–66 Ma). They appear in the fossil record for about 100 million years.

The smallest (Class: Small Grade II)

Dilong paradoxus ("paradoxical emperor dragon"): From the late Lower Cretaceous (Barremian, ca. 129.4–125 Ma) of eastern Laurasia (present-day China). In Chinese mythology, the Dilong or Ti-Lung are land dragons that rule over rivers and streams. The largest individual was a subadult, with a size comparable to that of a lynx. It was 5.5 times shorter than cf. *Prodeinodon mongoliensis* and 220 times lighter than *Bistahieversor sealeyi*. See more: *Chen et al. 1998*

The smallest juvenile specimen

Nuthetes sp. (CHEm03.537) of the Lower Cretaceous of present-day France is a tiny tooth of a basal tyrannosauroid. Its owner must have been 85 cm long and weighed 1.5 kg. See more: *Mazin et al. 2006*

The oldest

Kileskus aristotocus and *Proceratosaurus bradleyi* date from the Middle Jurassic (Bathonian, ca. 168.3–166.1 Ma). They lived in eastern Paleoasia and north-central Neopangea (present-day Russia and England, respectively). See more: *Woodward 1910; Averianov et al. 2010*

The most recent

Dryptosaurus aquilunguis ("lacerating eagle-clawed lizard"): From the late Upper Cretaceous (middle Maastrichtian, ca. 69 Ma) of eastern Laurasia (present-day New Jersey, USA). It was the first theropod to be reconstructed in a bipedal posture. See more: *Cope 1866*

The strangest

The external aspect of *Guanlong wucaii* is strange because of its elongated nose and highly developed bony crest. It seems that this would be a common trait among proceratosaurids, but it is difficult to identify other members of the family. On the surface, *Bagaraatan ostromi* had a very typical skeleton, but it stands out due to its many fused leg bones. *Labocania anomala* has some characteristics similar to carcharodontosaurids, but new material that has not been formally described yet reveals that it was a tyrannosauroid derivative or a basal tyrannosaurid See more: *Molnar 1974; Osmolska 1996 ; Xu et al. 2006; Loewen et al. 2013; Ramírez-Velasco et al. 2015*

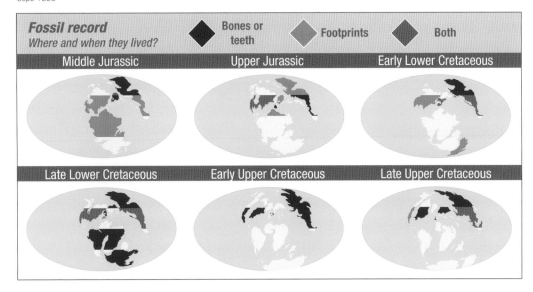

Fossil record Where and when they lived?	Bones or teeth	Footprints	Both
Middle Jurassic	Upper Jurassic		Early Lower Cretaceous
Late Lower Cretaceous	Early Upper Cretaceous		Late Upper Cretaceous

The first published species

Laelaps aquilunguis (1866) ("storm-wind with eagle claws"): From the Upper Cretaceous in the western Laurasia zone (present-day New Jersey, USA). The name changed to *Dryptosaurus* owing to the fact that thirty years previously, an *Acarus* had already been given the name. See more: *Koch 1836; Cope 1866; Marsh 1877*

The most recently published species

The species *Xiongguanlong baimoensis* was published in March 2010. In June of that year, *Bistahieversor sealeyi* and *Kileskus aristotocus* were described. The name "Bistahieversor sealeyi" is from a 2008 thesis. In 2013, some teeth attributed to cf. *Iliosuchus incognitus* were reported in Germany, but they are of an indeterminate proceratosaurid. See more: *Carr 2005; Li et al. 2010; Averianov et al. 2010; Carr & Williamson 2010; Larser & Heimhofer 2013*

Dilong paradoxus
Xu et al. 2004

1:10

RECORD SMALLEST

Specimen: IVPP V14243
(Xu et al. 2004)

Length: 1.8 m
Hip height: 55 cm
Body mass: 15 kg
Material: partial skeleton
ESR: ●●●○

Juvenile
CHEm03.537

50 cm   100 cm   150 cm   200 cm

2 m   4 m   6 m   8 m

4 m

2 m

Tyrannosaurus rex
Osborn 1905

RECORD
LARGEST

1:40

Specimen: UCMP 137538
(Longrich et al. 2010)

Length: 12.3 m
Hip height: 3.75 m
Body mass: 8.5 t
Material: pedal phalanx
ESR: ●○○○

The largest (Class: Very large Grade II)

Carpenter 2000; Brochu 2003; Happ 2008; Longrich et al. 2010; Mortimer 2015

Tyrannosaurus rex ("lizard king of tyrants"): The largest individual may be represented by the **UCMP 137538** specimen, a phalanx of the toe IV-2. The piece is slightly larger than that of the famous individual "Sue" FMNH PR2081 (the largest and most complete skeleton found so far). If this specimen (UCMP 137538) had body proportions similar to those of "Sue," it could easily have been as heavy as the largest *Giganotosaurus carolinii* (heavier than two Asian elephants). "Celeste" MOR 1126 is another very large specimen, but, unfortunately, the description has yet to be published. As far as their way of life is concerned, there is evidence that tyrannosaurs were hunters of ankylosaurids, hadrosaurs, and ceratopsids, as some show marks of nonfatal attacks. See more: *Tumanova et al. 1998;*

Other contenders

Tarbosaurus bataar and *Zhuchengtyrannus magnus* were previously thought to be similar in size to *T. rex*. In reality, these species were of lighter constitution and possessed a more elongated skull. Other large species come from *Tyrannosaurus* rex specimens: *Dyamosaurus imperiosus,* "Tyrannosaurus gigantus," "T. imperator," and "T. stanwinstonorum." See more: *Maleev 1955; Molnar 1991; Currie, Hurum & Sabath 2003*

Footprints

a)

─ 86 cm

b)

─ 36.7 cm

The largest

a) *Tyrannosauripus pillmorei:* From the late Upper Cretaceous of western Laurasia (present-day New Mexico, USA). It could belong to *Tyrannosaurus rex* due to its location and age. Another footprint, called *Tyrannosauropus petersoni*, turned out to be of a hadrosaurid dinosaur. See more: *Haubold 1971; Lockley & Hunt 1994*

Tyrannosaurids existed in the late Upper Cretaceous. They lived in western North America (Laramidia) and in western and eastern Asia (Paleoasia). Depending on whether they were adults or juveniles, their tracks were either thin or very sturdy (*Bellatoripes* and *Tyrannosauripus*).

The smallest

b) Unnamed: A footprint from the late Upper Cretaceous of eastern Laurasia (present-day Mongolia) has more primitive anatomical features than those of *Tarbosaurus*. Due to its shape, it could have been imprinted by a tyrannosaurid similar to *Alioramus* See more: *Ishigaki 2010*

4 m

3 m

Tyrannosauripus pillmorei
CU-MWC 225.1
11.4 m / 5.8-6.9 t

Tyrannosaurus rex
UCMP 137538
12.3 m / 8.5 t

Unnamed footprint
Trackway J
4 m / 207 kg

Qianzhousaurus sinensis
IGM 100/1844
6.3 m / 750 kg

Characteristics and habits

Theropods with long or large heads, large and massive knife-like teeth, fused nostrils, short arms with two fingers on each hand and four on each foot, short necks, long or relatively short tails, and semirobust to robust bodies. The primitive tyrannosaurids descended from the primitive tyrannosauroids. Most were robust predators with strong jaws and very short arms. See more: *Senter 2007*
Diet: small to large game carnivores.

Time range

Tyrannosaurids are known to have existed from the early Upper Cretaceous to the late Upper Cretaceous (ca. 86.3–66 Ma), an approximate duration of 20.3 million years in the fossil record.

The smallest (Class: Large Grade I)

Qianzhousaurus sinensis ("lizard of China's Qian prefecture"): From the late Upper Cretaceous (lower Maastrichtian, ca. 72.1–69 Ma) of eastern Laurasia (present-day China). It could have been an advanced tyrannosauroid or a primitive tyrannosaurid. *Nanuqsaurus hoglundi* was shorter (6 m) but heavier (900 kg). Both were about 2 times shorter and between 11 and 9.5 times lighter than *Tyrannosaurus rex*. See more: *Holtz 2004; Brusatte et al. 2009; Loewen et al. 2013; Lu et al. 2014; Brusatte et al. 2015*

The smallest juvenile specimen

Aublysodon mirandus (ANSP 9535), from the Upper Cretaceous of western Laurasia (present-day Montana, USA), is a premaxillary tooth of a hatchling that might have measured 85 cm long and weighed 2 kg. See more: *Leidy 1868*

The oldest

The oldest tyrannosaurids are found at the beginning of the late Upper Cretaceous (Santonian, ca. 86.3–83.6 Ma). *"Tarbosaurus* aff. *Bataar"* IZK 33/MP-61 is an incomplete tooth from western Laurasia (present-day Kazakhstan). See more: *Khozatsky 1957; Nesov 1995; Averianov et al. 2012*

The most recent

Aublysodon sp., *"Albertosaurus" periculosus*, *Nanuqsaurus hoglundi*, cf. *Tarbosaurus* sp. and *Tyrannosaurus rex*, (present-day Canada, the United States, Mexico, Russia, and China, respectively): They were found in the upper layers of the late Upper Cretaceous (upper Maastrichtian, ca. 69–66 Ma). The *Tyrannosaurus rex* TMP 81.12.1 was located 10.5 m from the division that separates the Upper Cretaceous from the Paleogene. See more: *Osborn 1905; Russell & Sweet 1993; Eberth 1997; Itterbeeck et al. 2005; Bolotsky 2011; Fiorillo & Tykoski 2014*

The first published species

Deinodon horridus (1856) ("terrible and horrible tooth"): From late Upper Cretaceous of western Laurasia (present-day Montana, USA). It was the first to be named and perhaps the first tyrannosaurid to be found. Legend holds that in 1806, *Tyrannosaurus rex* ribs were found in the Montana expedition led by William Clark. Hadrosaurs abound there, however, and the aforementioned specimens were lost in time, so the legend can't be verified See more: *Leidy 1856; Saindon 2003*

The most recently published species

Nanuqsaurus hoglundi and *Qianzhousaurus sinensis* were described in 2014. The former was published in March and the latter in May. See more: *Lu et al. 2014; Fiorillo & Tykoski 2014*

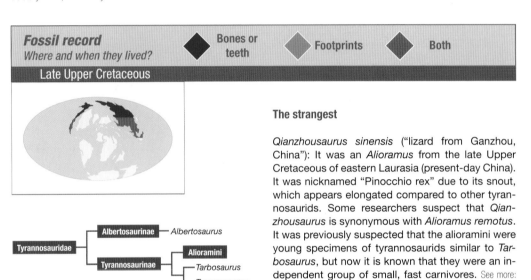

The strangest

Qianzhousaurus sinensis ("lizard from Ganzhou, China"): It was an *Alioramus* from the late Upper Cretaceous of eastern Laurasia (present-day China). It was nicknamed "Pinocchio rex" due to its snout, which appears elongated compared to other tyrannosaurids. Some researchers suspect that *Qianzhousaurus* is synonymous with *Alioramus remotus*. It was previously suspected that the alioramini were young specimens of tyrannosaurids similar to *Tarbosaurus*, but now it is known that they were an independent group of small, fast carnivores. See more: *Lu et al. 2014; Mortimer*

1:35

Qianzhousaurus sinensis
Lu et al. 2014

RECORD SMALLEST

Juvenile
Aublysodon mirandus
ANSP 9535

Specimen: IGM 100/1844
(Brusatte et al. 2009)

Length: 6.3 m
Hip height: 2 m
Body mass: 750 kg
Material: skull and partial skeleton
ESR: ●●●○

1 m   2 m   3 m   4 m   5 m   6 m   7 m

Deinocheirus mirificus
Osmólska & Roniewicz 1970

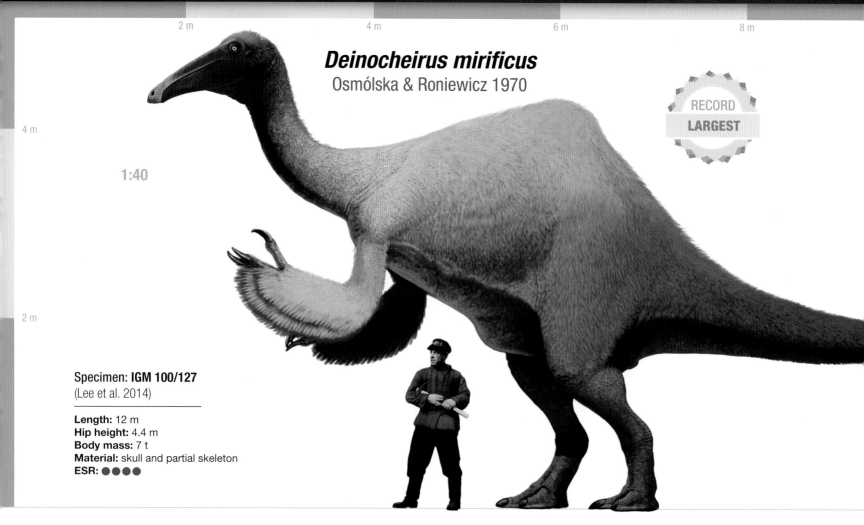

4 m

1:40

2 m

RECORD
LARGEST

Specimen: IGM 100/127

(Lee et al. 2014)

Length: 12 m
Hip height: 4.4 m
Body mass: 7 t
Material: skull and partial skeleton
ESR: ●●●●

The largest (Class: Very large Grade II)

Deinocheirus mirificus ("extraordinary terrible hand") lived during the Upper Cretaceous (lower Maastrichtian, ca. 72.1–69 Ma) in western Laurasia (present-day Mongolia). This incredible theropod possessed extraordinary anterior limbs more than 2.5 m in length. It may have used them to manipulate plants, which it fed on, or to defend itself. The great reach afforded by such arms and its weight (comparable to a large orca) would have made adults dangerous prey. For many years, only a couple of huge arms were known, but two new specimens were found recently. One of them is even bigger and significantly more complete. See more: *Osmólska & Roniewicz 1970; Lee et al. 2014*

Enigmatic and questionable giants

A fragmented tooth from South Korea has been called "*Deinocheirus* sp.," but the material does not belong to an ornithomimosaur and is also much older. A 1 m long ulna was reported in Mongolia. It is not formally described. If that piece was really an ulna and belonged to a *Deinocheirus*, the individual would be about 16 m long and weigh almost 16 t—a very doubtful assignation. It is likely another type of long theropod bone, or even a hadrosauromorph. See more: *Zhen et al. 1993; Suzuki et al. 2010*

Footprints

a)
37.8 cm   32.8 cm

b)
5.49 cm

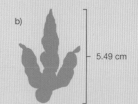

The largest

a) *Ornithomimipus* sp.: From the late Upper Cretaceous of western Laurasia (present-day Mexico). It has not been formally described yet. (It is in the process of being published). The longest footprint has the metatarsal marking, but the smaller footprint was of a larger specimen. On the other hand, no imprints have been found similar to those of the huge *Deinocheirus*. They would be similar to those of some ornithischian dinosaurs. See more: *Gudiño-Maussan**

Primitive ornithomimosaurs differ from basal coelurosaurs (*Regarosauropus*) on the basis of morphology and toe depth. The feet of the derivative species are very similar in appearance to those of *Ornithomimus* (*Ornithomimipus*), with a more robust center toe.

The smallest

b) Unnamed: A set of tiny to medium-size tracks from the late Lower Cretaceous of central Laurasia (present-day Spain). The largest is 25 cm long, so they are presumed to be hatchlings, as adults would be 3.4 m long and weigh 55 kg. See more: *Pascual-Arribas et al. 2011*

50 cm

4 m

Unnamed footprint	*Deinocheirus mirificus*	Unnamed footprint	*Nqwebasaurus thwazi*
AA-14	IGM 100/127	MNS-2005/111-2-3	AM 6040
6.7 m / 685 kg	12 m / 7 t	74 cm / 590 g	1.2 m / 2.5 kg

12 m   14 m   16 m   18 m

Characteristics and habits

Theropods with small heads, large eyes, tiny or absent teeth, long arms and legs, long necks and tails, graceful or semirobust bodies, and three fingers on each hand and three on each foot. *Ornithomimosaurs* descended from basal coelurosaurs similar to *Ornitholestes*. They fed on plants but may have been opportunists, taking advantage of available food sources. See more: *Kobayashi et al. 1999; Norell et al. 2001; Barrett 2005*

Diet: omnivores and/or herbivores

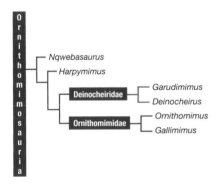

Time range

They were present throughout the Cretaceous (ca. 145–66 Ma), an approximate duration of 79 million years in the fossil record.

The smallest (Class: Very small Grade II)

Nqwebasaurus thwazi ("runner lizard from Nqweba"): From the early Lower Cretaceous (Berriasian, ca. 145–139.8 Ma) of south-central Gondwana (present-day South Africa). It is a very primitive subadult and the oldest ornithomimosaur. Equally long but heavier was *Hexing qingyi*, whose adult size was only 1.2 m long and weighed 4.5 kg. Both were 10 times shorter and 2,800 and 1,555 times, respectively, lighter than *Deinocheirus mirificus*. See more: *de Klerk et al. 1997; Lee et al. 2014*

O
r
n
i
t
h
o
m
i
m
o
s
a
u
r
i
a

— Nqwebasaurus
— Harpymimus
— **Deinocheiridae** — Garudimimus
                    — Deinocheirus
— **Ornithomimidae** — Ornithomimus
                     — Gallimimus

The smallest juvenile specimen

There are numerous juvenile ornithomimosaurs, but none seems to have been as small as the YPM PU 22416 from the late Upper Cretaceous of eastern Laurasia (present-day Delaware, USA). It could have been 70 cm long and weighed 1.1 kg. See more: *Baird 1986*

The oldest

Nqwebasaurus thwazi: From the early Lower Cretaceous (Berriasian, ca. 145–139.8 Ma) of south-central Gondwana (present-day South Africa). An even older one has been reported: *Lepidocheirosaurus natatilis*, from the Middle Jurassic of eastern Paleoasia (present-day western Russia), but it is suspected that they are actually ornithischian *Kulindadromeus zabaikalicus* bones. See more: *de Klerk et al. 2000; Choiniere et al. 2012; Alifanov & Saveliev 2015; Mortimer**

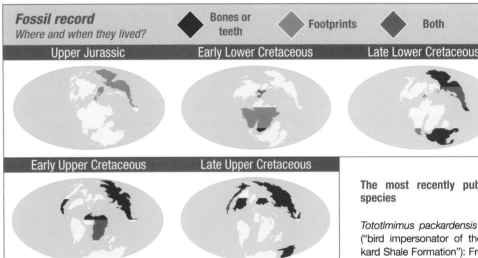

Fossil record
Where and when they lived?   ◆ **Bones or teeth**   ◆ **Footprints**   ◆ **Both**

Upper Jurassic | Early Lower Cretaceous | Late Lower Cretaceous

Early Upper Cretaceous | Late Upper Cretaceous

The first published species

Thecocoelurus daviesi (originally *Thecospondylus* in 1888) could have been a cervical vertebra of an ornithomimosaur or a basal therizinosaur. *"Megalosaurus" oweni* (currently *Valdoraptor*) are incomplete metatarsals that were described one year later. See more: *Seeley 1888; Lydekker 1889; Allain et al. 2014; Mortimer 2015*

The most recent

The species *Ornithomimus velox*, *Struthiomimus sedens*, and "*Orcomimus*" from Laurasia (present-day western United States) and GNS CD 579 from Oceania (present-day New Zealand) are found in more recent layers of the late Upper Cretaceous (upper Maastrichtian, ca. 69–66 Ma). See more: *Marsh 1890, 1892; Triebold 1997; Stilwell et al. 2006*

The strangest

Apart from its gigantic size, *Deinocheirus mirifcus* had a unique shape: a broad snout, proportionally short legs, a pygostile, and a hump—a unique feature among ornithomimosaurs. See more: *Lu et al. 2014*

The most recently published species

Tototlmimus packardensis (2016) ("bird impersonator of the Packard Shale Formation"): From the late Upper Cretaceous of western Laurasia (present-day Mexico). It was the first theropod to receive a name in Nahuatl. Some sources quote it from the year 2015, but it was published at the end of March 2016. This is because the accepted manuscript was published first online, a common practice nowadays. See more: *Serrano-Brañas et al. 2016; Olshevsky**

Nqwebasaurus thwazi
de Klerk et al. 1997

RECORD
SMALLEST

1:8

Specimen: AM 6040
(de Klerk et al. 1997)

Length: 1.2 m
Hip height: 35 cm
Body mass: 2.5 kg
Material: skull and partial skeleton
ESR: ●●●●

Juvenile
GNS CD 579

50 cm   100 cm   150 cm

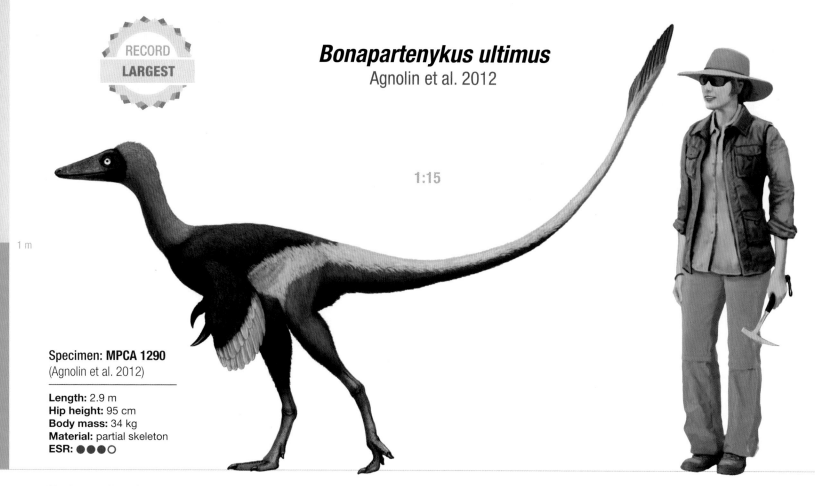

RECORD LARGEST

Bonapartenykus ultimus
Agnolin et al. 2012

1:15

1 m

Specimen: MPCA 1290
(Agnolin et al. 2012)

Length: 2.9 m
Hip height: 95 cm
Body mass: 34 kg
Material: partial skeleton
ESR: ●●●○

The largest (Class: Small Grade III)

Bonapartenykus ultimus ("last claw of the paleontologist José Fernando Bonaparte"): From the late Upper Cretaceous (Campanian, ca. 83.6–72.1 Ma) of western Gondwana (present-day Argentina). The only known specimen is a female that had two internal eggs called *Arraigadoolithus patagonicus*. Although it is not known precisely what their food source was, it has been suggested that some were either insectivores or omnivores. So it is not surprising that the largest alvarezsaurids were similar in size to today's largest mammal insectivore, the yurumí (*Myrmecophaga tridactyla*), or giant anteater. See more: *Jackson & Varricchio 2010; Agnolin et al. 2012*

A double or triple tie

Many alvarezsauroids are the smallest terrestrial theropods. Even the largest species do not stand out. *Patagonykus puertai* ("Patagonian claw, by Pablo Puerta") from the early Upper Cretaceous and *Achillesaurus manazzonei* ("Achilles lizard of Rafael Manazzone") from the late Upper Cretaceous, both from the western zone of Gondwana (present-day Argentina), were about 2.8 m long and weighed 30 kg. They were very close in size to *Bonapartenykus ultimus*, but it is unknown if they were adults. See more: *Novas 1993; Martinelli & Vera 2007*

Footprints

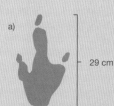

a)

─ 29 cm

b)

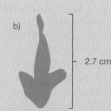

─ 2.7 cm

The largest

a) Unnamed: The largest identified alvarezsaurid footprints are from the early Upper Cretaceous of central Gondwana (present-day Morocco). See more: *Contessi & Fanti 2012*

Alvarezsaurid footprints are similar to those of other small theropods, except for the very developed central fingernail. The primitive forms (*Picunichnus*) had a thicker heel than those derived from Laurasia (*Neograllator*), because they did not have narrow metatarsals.

The smallest

b) *Neograllator emeiensis*: Despite its large nail, it is one of the smallest known theropod footprints. It is from the late Lower Cretaceous of eastern Laurasia (present-day China). Alvarezsaurid leg proportions vary greatly, so it's necessary to use a wide range when estimating the size of the animal that left the impression—that is, until other identifying characters are known. See more: *Zhen et al.1994; Lockley et al. 2008*

2 m

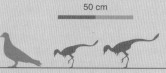

50 cm

Unnamed footprint
MGGC 21851
3.2 m / 42 kg

Bonapartenykus ultimus
MPCA 1290
2.9 m / 34 kg

Neograllator emeiensis
CFEC-C-1
37-46 cm / 75-133 g

Parvicursor remotus
PIN 4487/25
50 cm / 185 g

4 m   5 m   6 m   7 m

Characteristics and habits

Theropods with medium to large heads, large eyes, tiny teeth, short arms, short necks and tails, long legs, graceful bodies, and one to three fingers on each hand and three on each foot. Alvarezsauroids descended from primitive coelurosaurs similar to *Ornitholestes*. Perhaps they specialized in the consumption of insects, or they might have been herbivorous See more: *Longrich & Currie 2009*
Diet: insectivores and/or omnivores.

Time range

From the Upper Jurassic to the late Upper Cretaceous (ca. 163.5–66 Ma), an approximate duration of about 97.5 million years in the fossil record.

The smallest (Class: Dwarf Grade I)

Parvicursor remotus ("little ancient runner"): It dates from the late Upper Cretaceous (Campanian, ca. 83.6–72.1 Ma) and lived in eastern Laurasia (present-day Mongolia). Despite being smaller than a pigeon, it was an adult. Another alvarezsauroid from China, IVPP V20341, barely exceeded its size. *Parvicursor* was about 6 times shorter and about 185 times lighter than *Bonapartenykus ultimus*. See more: *Karhu & Rautian 1996; Pittman et al. 2015*

The smallest juvenile specimen

The smallest *Shuvuuia* deserti (MGI 100/1001) was a juvenile that was 70 cm long and weighed 550 g. Despite being the youngest juvenile found, it was larger than the subadults and adults of seven other species of alvarezsaurids. See more: *Xu et al. 2010; Pittman et al. 2014; Hendrickx et al. 2014*

The oldest

Haplocheirus sollers: From the Upper Jurassic (Oxfordian, ca. 163.5–157.3 Ma) of the eastern zone of Paleoasia (present-day China). It was so primitive that its appearance is quite different from that of other alvarezsaurids. The specimen V15849 is a contemporary tooth, which is similar but somewhat different from *Haplocheirus*. See more: *Choiniere et al. 2010; Han et al. 2011*

The most recent

Heptasteornis andrewsi from Romania, "Ornithomimus" minutus from Colorado, and the alvarezsaurid UCMP 154584 from Montana, along with other materials, date from the late Upper Cretaceous (upper Maastrichtian, ca. 69–66 Ma). See more: *Marsh 1892; Harrison & Walker 1975; Hutchinson & Chiappe 1998; Anduza et al. 2013*

The strangest

Mononykus olecranus ("single claw with prominent elbow") was believed to have only one finger on each hand when it was discovered. A later study showed that alvarezsaurids had two atrophied fingers, but they had not been preserved in that fossil specimen. *Linhenykus monodactylus*, which actually had only one finger on each hand, was eventually discovered. See more: *Xu et al. 2011*

The first published species

"*Ornithomimus*" minutus (1892): From the late Upper Cretaceous of western Laurasia (present-day Colorado, USA). It is based on a partial metatarsal and was considered an ornithomimosaur. Unfortunately, the piece was lost. See more: *Marsh 1892*

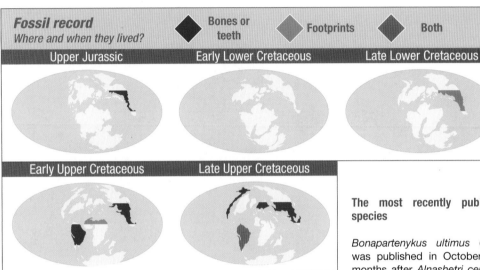

Fossil record
Where and when they lived?   ◆ Bones or teeth   ◆ Footprints   ◆ Both

Upper Jurassic | Early Lower Cretaceous | Late Lower Cretaceous

Early Upper Cretaceous | Late Upper Cretaceous

The most recently published species

Bonapartenykus ultimus (2012) was published in October, four months after *Alnashetri cerropoliciensis*. Both lived in western Gondwana (present-day Argentina). See more: *Agnolin et al. 2012; Mackovicky et al. 2012*

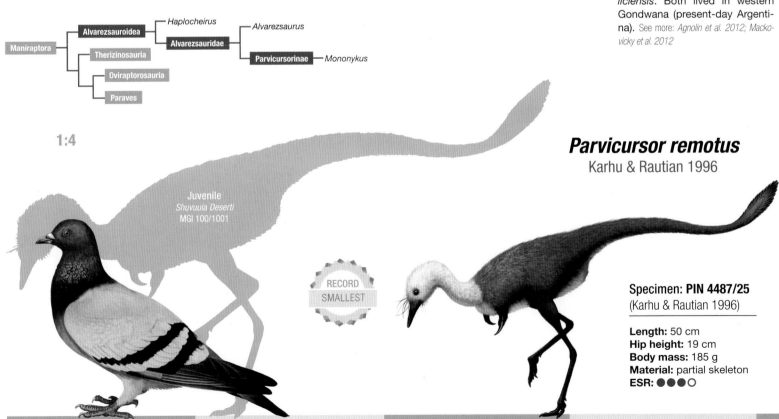

Maniraptora — Alvarezsauroidea — Haplocheirus
Alvarezsauroidea — Alvarezsauridae — Alvarezsaurus
Alvarezsauridae — Parvicursorinae — Mononykus
Therizinosauria
Oviraptorosauria
Paraves

1:4

Juvenile
Shuvuuia Deserti
MGI 100/1001

RECORD SMALLEST

Parvicursor remotus
Karhu & Rautian 1996

Specimen: PIN 4487/25
(Karhu & Rautian 1996)

Length: 50 cm
Hip height: 19 cm
Body mass: 185 g
Material: partial skeleton
ESR: ●●●○

20 cm   40 cm   60 cm   80 cm

RECORD
LARGEST

4 m

2 m

1:40

Therizinosaurus cheloniformis
Maleev 1954

Specimen: IGM 100/15
(Barsbold 1976)

Length: 9 m
Hip height: 3.1 m
Body mass: 4.5 t
Material: scapulocoracoid and arms
ESR: ●●○○

The largest (Class: Very large Grade I)

Therizinosaurus cheloniformis ("tortoise-shaped scythe lizard"): Lived during the late Upper Cretaceous (lower Maastrichtian, 72.1–69 Ma) in eastern Laurasia (present-day Mongolia). It was known only for its huge nails, which seemed to be part of the shell of a giant turtle over 3 m long. Later, a few more bones were found, including a practically complete anterior limb that was 2.5 m long. Along with *Deinocheirus*, they had the largest arms among the theropods. Their nails had the appearance of scythes, which must have been impressive when fully extended. When compared with other therizinosaurids, such as *Nothronychus*, it is concluded that this arm must have belonged to an animal bigger than an Asian elephant. See more: *Maleev 1954; Barsbold 1976; Kirkland & Wolfe 2001*

Footprints

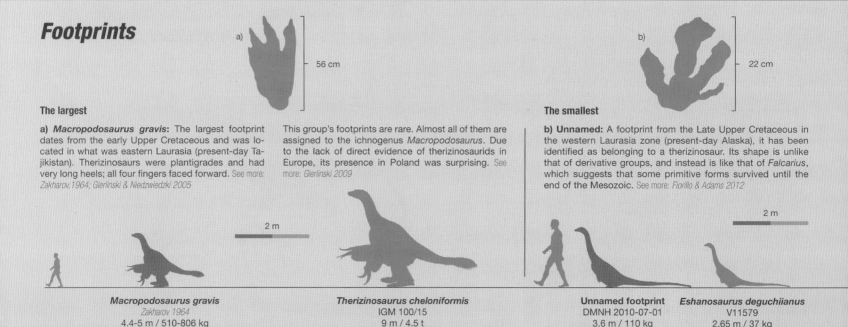

a)   56 cm

b)   22 cm

The largest

a) *Macropodosaurus gravis*: The largest footprint dates from the early Upper Cretaceous and was located in what was eastern Laurasia (present-day Tajikistan). Therizinosaurs were plantigrades and had very long heels; all four fingers faced forward. See more: *Zakharov,1964; Gierlinski & Niedzwiedzki 2005*

This group's footprints are rare. Almost all of them are assigned to the ichnogenus *Macropodosaurus*. Due to the lack of direct evidence of therizinosaurids in Europe, its presence in Poland was surprising. See more: *Gierlinski 2009*

The smallest

b) Unnamed: A footprint from the Late Upper Cretaceous in the western Laurasia zone (present-day Alaska), it has been identified as belonging to a therizinosaur. Its shape is unlike that of derivative groups, and instead is like that of *Falcarius*, which suggests that some primitive forms survived until the end of the Mesozoic. See more: *Fiorillo & Adams 2012*

2 m

2 m

2 m

Macropodosaurus gravis
Zakharov 1964
4.4-5 m / 510-806 kg

Therizinosaurus cheloniformis
IGM 100/15
9 m / 4.5 t

Unnamed footprint
DMNH 2010-07-01
3.6 m / 110 kg

Eshanosaurus deguchiianus
V11579
2.65 m / 37 kg

12 m   14 m   16 m   18 m

Characteristics and habits

Theropods with small heads, tiny serrated teeth, long arms and necks, heavy and massive bodies, large flat nails, short legs, long or short tails, and three fingers on each hand and four on each foot. Therizinosaurs descended from basal coelurosaurs similar to *Ornitholestes*. They slowly adapted to include plants into their diet. See more: *Barsbold & Perle 1980; Zanno 2010*

Diet: herbivores and/or omnivores.

Time range

From the Lower Jurassic or early Lower Cretaceous to the late Upper Cretaceous (ca. 201.3 or 145–66 Ma), a duration of either 135.3 million or 79 million years in the fossil record.

The smallest (Class: Small Grade III)

Eshanosaurus deguchiianus ("Eshan lizard of Hikaru Deguchi"): An enigmatic fossil, it resembles therizinosaurs but is too old and incomplete, arousing suspicion that it perhaps was a sauropodomorph. Alternatively, it might be less ancient than believed. Also not known is whether it was a juvenile or an adult. On the other hand, *Beipiaosaurus inexpectatus* is known to have been a subadult that was only 1.85 m long and weighed 43 kg. Both were 3.4–5 times shorter and 120–105 times lighter than *Therizinosaurus cheloniformis*. See more: *Xu et al. 1999; Barrett 2009; Mortimer**

The smallest juvenile specimen

Several therizinosaurid embryos are known in China, but one tooth and extremely tiny fragments from Kazakhstan belong to the smallest therizinosaur. Comparing these remains to the embryos, the small animal may have been just 9.2 cm long and weighed about 3.6 g. See more: *Manning et al 1997; Averianov 2007*

The oldest

Eshanosaurus deguchiianus (formerly known as "*Oshanosaurus youngi*"): From the Lower Jurassic (Hettangian, ca. 201.3–199.3 Ma) of northeastern Pangea (present-day China). It is exceptionally old for a therizinosaur. The remains are very incomplete, so their identification is problematic. Other very old material is a Moroccan tooth very similar to the therizinosaur embryos. It dates from the early Lower Cretaceous (Berriasian, ca. 145–139.8 Ma). See more: *Sigogneau-Russeil et al. 1998; Xu, Zhao & Clark 2001; Kirkland et al. 2005; Barrett 2009; Allain et al. 2014*

The most recent

Both the unnamed therizinosaurid RTMP 86.207.17 of western Laurasia (present-day west Canada) and WAM 90.10.2, an enigmatic fossil from Australia whose identity is in doubt, date from the late Upper Cretaceous (upper Maastrichtian, ca. 69–66 Ma). See more: *Long 1992; Ryan & Russell 2001*

The first published species

Thecospondylus daviesi (1888) could be the first to be described, although other researchers suggest that it was an ornithomimosaur. *Therizinosaurus cheloniformis* was described 66 years after *Thecospondylus*. See more: *Seeley 1888; Maleev 1954*

The strangest

Therizinosaurus cheloniformis is certainly the most extravagant. It looked like a giant sloth and had enormous nail phalanges (claws) almost 70 cm in length, the longest among all known animals. See more: *Maleev 1954*

The most recently published species

Jianchangosaurus yixianensis (2013) ("Jianchang's lizard from the Yixian Formation"): From the late Lower Cretaceous of eastern Laurasia (present-day China). It is a fairly complete juvenile that includes preserved filamentous feathers. It is the third therizinosaur to show feathers. The previous two are specimens of *Beipiaosaurus inexpectatus*. See more: *Xu et al. 1999, 2009; Pu et al. 2013*

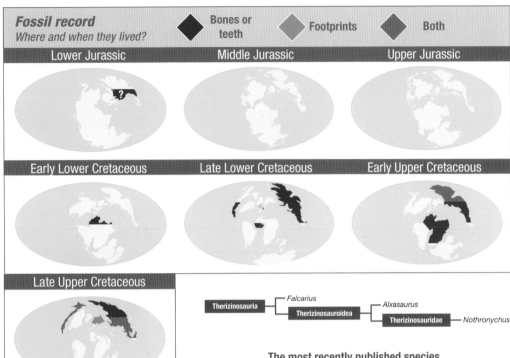

Fossil record — Where and when they lived?
Bones or teeth | Footprints | Both

Lower Jurassic | Middle Jurassic | Upper Jurassic
Early Lower Cretaceous | Late Lower Cretaceous | Early Upper Cretaceous
Late Upper Cretaceous

Therizinosauria — Therizinosauroidea — Falcarius — Alxasaurus — Therizinosauridae — Nothronychus

Eshanosaurus deguchiianus
Xu et al. 2001

RECORD SMALLEST

Neonate

1:15

Specimen: V11579
(Zhao & Xu 1998)

Length: 2.65 m
Hip height: 60 cm
Body mass: 37 kg
Material: incomplete jaw and teeth
ESR: ●○○○

1 m   2 m   3 m

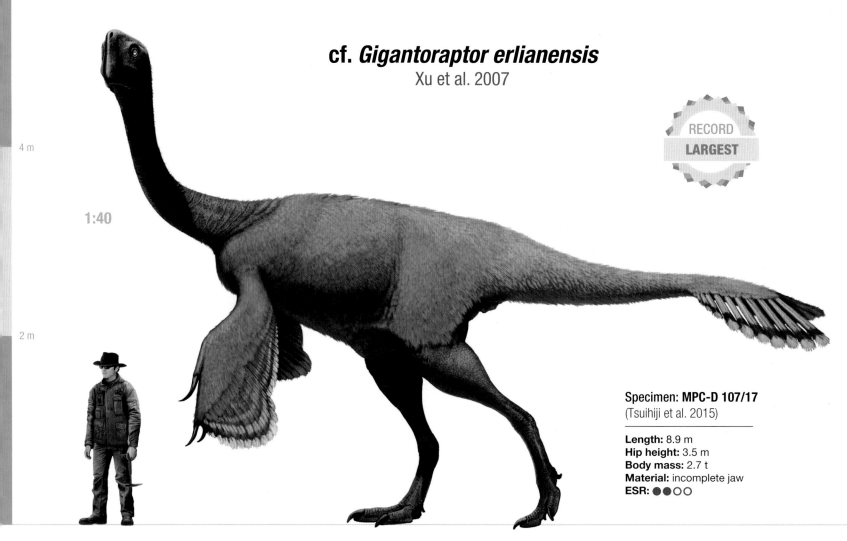

cf. *Gigantoraptor erlianensis*
Xu et al. 2007

2 m   4 m   6 m   8 m

4 m

2 m

1:40

RECORD
LARGEST

Specimen: MPC-D 107/17
(Tsuihiji et al. 2015)

Length: 8.9 m
Hip height: 3.5 m
Body mass: 2.7 t
Material: incomplete jaw
ESR: ●●○○

The largest (Class: Large Grade III)

MPC-D 107/17 consists of an incomplete mandible from a dinosaur similar to but somewhat larger than ***Gigantoraptor erlianensis*** ("giant thief from Erlian"). It is suspected that it could be a new individual of the same species. Both date from the late Upper Cretaceous (Campanian, ca. 83.6–72.1 Ma) and lived in the eastern part of Laurasia (present-day Mongolia and China). *Gigantoraptor* was either a herbivore or an omnivore and could have defended itself with its large claws or its considerable speed (its hind legs were much longer than those of its possible predators). It was the tallest theropod—its impressive height surpassed that of a giraffe. As far as body mass is concerned, it was somewhat heavier than a white rhinoceros. Large fossil eggs known as *Macroelongathoolithus* must have been laid by an oviraptorosaur as large as *Gigantoraptor*, which was more recent. See more: *Jin et al. 2007; Xu et al. 2007; Tsuihiji et al. 2015*

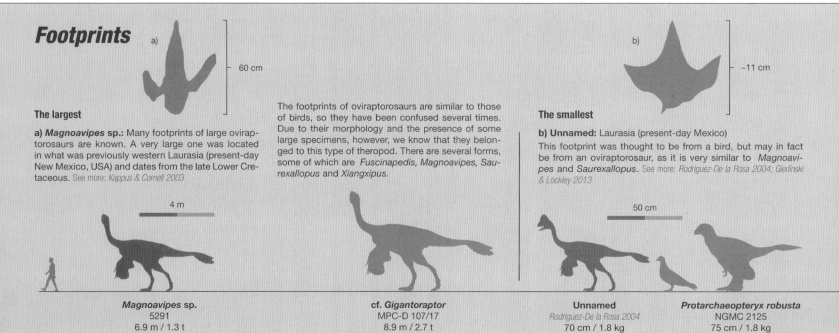

Footprints

a)

60 cm

The largest

a) *Magnoavipes* sp.: Many footprints of large oviraptorosaurs are known. A very large one was located in what was previously western Laurasia (present-day New Mexico, USA) and dates from the late Lower Cretaceous. See more: *Kappus & Cornell 2003*

The footprints of oviraptorosaurs are similar to those of birds, so they have been confused several times. Due to their morphology and the presence of some large specimens, however, we know that they belonged to this type of theropod. There are several forms, some of which are *Fuscinapedis*, *Magnoavipes*, *Saurexallopus* and *Xiangxipus*.

b)

~11 cm

The smallest

b) Unnamed: Laurasia (present-day Mexico)

This footprint was thought to be from a bird, but may in fact be from an oviraptorosaur, as it is very similar to *Magnoavipes* and *Saurexallopus*. See more: *Rodriguez-De la Rosa 2004; Gierlinski & Lockley 2013*

4 m

50 cm

Magnoavipes sp.
5291
6.9 m / 1.3 t

cf. *Gigantoraptor*
MPC-D 107/17
8.9 m / 2.7 t

Unnamed
Rodriguez-De la Rosa 2004
70 cm / 1.8 kg

Protarchaeopteryx robusta
NGMC 2125
75 cm / 1.8 kg

Characteristics and habits

Theropods with small round heads, with or without bony crests, large eyes, tiny or missing teeth, long arms and legs, long necks, short tails, graceful or light bodies, and three fingers on each hand and four on each foot. Oviraptorosaurs descended from basal coelurosaurs similar to *Ornitholestes*. We known that some were carnivores and others were herbivores, so some may have been omnivores. See more: *Osmólska et al. 2004*

Diet: carnivores of small game, omnivores, and/or phytophages.

Time range

From the early Lower Cretaceous to the late Upper Cretaceous (ca. 145 or 139.8–66 Ma), a duration of between 73.8 million and 79 million years in the fossil record.

The smallest (Class: Very small Grade II)

Protarchaeopteryx robusta ("prior to *Archeopteryx*, robust"): From the late Lower Cretaceous (Barremian, ca. 129.4–125 Ma) of eastern Laurasia (present-day China). It is an adult, with developed arm and tail feathers. It was only slightly bigger than a crow. BEXHM: 2008.14.1, from the late Lower Cretaceous of central Laurasia (present-day England), is even smaller. It was probably 45 cm long and weighed 420 g. It should be noted that this identification is in doubt. *Protarchaeopteryx* was 12 times shorter and 1,500 times lighter than MPC-D 107/17. See more: *Ji & Ji 1997; Naish & Sweetman 2011; Headden 2011*

The smallest juvenile specimen

There are more oviraptorosaur fossil embryos than any other type of theropod. The smallest, perhaps, is MPC-D100/1018 from Mongolia, assigned to cf. "*Ingenia*" sp. Due to its location, it was likely a *Rinchenia mongoliensis* or *Nemegtomaia barsboldi*. It may have measured about 21 cm long and weighed 52 g. See more: *Weishampel et al. 2008; Mortimer**

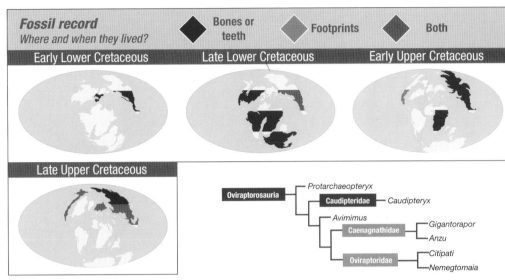

Fossil record
Where and when they lived? ◆ **Bones or teeth** ◆ **Footprints** ◆ **Both**

Early Lower Cretaceous | Late Lower Cretaceous | Early Upper Cretaceous

Late Upper Cretaceous

Oviraptorosauria
— *Protarchaeopteryx*
— Caudipteridae — *Caudipteryx*
— *Avimimus*
— Caenagnathidae — *Gigantorapor* / *Anzu*
— Oviraptoridae — *Citipati* / *Nemegtomaia*

The oldest

The oldest reported material is BEXHM: 2008.14.1, which dates from the early Lower Cretaceous (Berriasian, ca. 145–139.8 Ma) and was from central Laurasia (present-day England), but it is possible that it was not an oviraptorosaur. SBEI-167 (Valanginian-Hauterivian, 139.8–129.4 Ma) from Japan is more recent. It consists of one ungual phalanx from a hand. See more: *Manabe & Barrett 2000; Naish & Sweetman 2011; Headden**

The most recent

Anzu wyliei ("Anzû, feathered demon of Wylie J. Tuttle"): After *Gigantoraptor erliaensis*, it is the second largest oviraptorosaur. It dates from the late Upper Cretaceous (Maastrichtian, ca. 69–66 Ma). It has been nicknamed "the chicken from hell" because it comes from the Hell Creek Formation ("inferno gorge"). See more: *Lamanna et al. 2014*

The strangest

Avimimus portentosus ("amazing bird imitator"): If compared to other crested relatives, it looks simple, but its carpometacarpal, tibiotarsus, tarsometatarsus, and hip are fused in several places, something common in birds. Its metatarsus presents the most extreme arctometatarsal condition in any theropod. See more: *Kurzanov 1987*

The first published species

Oviraptor philoceratops (1924) ("ceratopsid egg thief"): From the late Upper Cretaceous of eastern Laurasia (present-day Mongolia). It would have been named "Fenestrosaurus," but this was reconsidered because of its relation to fossil eggs. See more: *Osborn 1924; Norell et al. 1994*

The most recently published species

Huanansaurus ganzhouensis ("Huanan County and Ganzhou City lizard"): From the late Upper Cretaceous of eastern Laurasia (present-day China). It was found in the Nanxiong Formation, in China. Six different genera of oviraptorosaurs were found in the same region. This suggests that each of the species would occupy different ecological niches. See more: *Lu et al. 2015*

Protarchaeopteryx robusta
Ji & Ji 1997

1:6

RECORD SMALLEST

Juvenile
MPC-D100/10

Specimen: NGMC 2125
(Ji & Ji 1997)

Length: 75 cm
Hip height: 40 cm
Body mass: 1.8 kg
Material: skull, partial skeleton, and feathers
ESR: ●●●●

RECORD
LARGEST

Austroraptor cabazai
Novas et al. 2008

1:30

2 m

Specimen: **MML-195**
(Novas et al. 2008)

Length: 6.2 m
Hip height: 1.5 m
Body mass: 340 kg
Material: skull and partial skeleton
ESR: ●●●●

The largest (Class: Medium Grade III)

Second place

Austroraptor cabazai ("Hector Cabaza's southern thief"): From the late Upper Cretaceous (lower Maastrichtian, ca. 72.1–69 Ma) of western Gondwana (present-day Argentina). It was longer than a saltwater crocodile and as heavy as two lions. It must have been very effective at catching fish, because its teeth were conical. The teeth of other unenlagians were flat and knife-shaped. It is interesting to note that paravians and their descendants did not reach gigantic sizes as their more primitive relatives did. This may be due to the fact that they weren't usually the dominant animals during the Mesozoic or Cenozoic. See more: *Novas et al. 2008; Cau**

Unenlagia paynemili ("half bird of Maximino Paynemila"): From the early Upper Cretaceous (Turonian, ca. 93.9–89.8 Ma) of western Gondwana (present-day Argentina). Measuring 3.5 m long and weighing 75 kg, it was the second largest, but it was considerably smaller than *Austroraptor cabazai*. Its developmental stage is not fully known. It has been suggested that UFMA 1.20.194-2, from the early Upper Cretaceous of Brazil, is a large tooth that belonged to a eudromaeosaur dromaeosaurid. Since that clade is not typically found there, it is likely that it was in fact from a different type of unenlagian dromaeosaurid or even an **abelisaurid**. See more: *Elias et al. 2007; Mortimer**

Footprints

a)

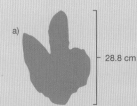

⌐ 28.8 cm

b)

⌐ 1.5 cm

The largest

a) *Dromaeopodus* isp.: Footprints from the early Upper Cretaceous of western Gondwana (present-day Bolivia). They undoubtedly belonged to an unenlagian. See more: *Apesteguia et al. 2011*

Due to the similarity of *Dromaeosauripus* footprints, they could represent primitive paravians similar to *Microraptor* and other small eudromaeosaurs.

The smallest

b) Unnamed: Possible didactyl footprints made in the Middle Jurassic in south-central Neopangea (present-day Morocco). They represent small juvenile paravians. See more: *Belvedere et al. 2011*

2 m

20 cm

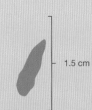

| *Dromaeopodus* isp.
Track 1
3.2 m / 75 kg | *Austroraptor cabazai*
MML-195
6.2 m / 340 kg | Unnamed footprint
Track H
33-36 cm / 70-77 g | Unnamed specimen
IVPP V22530
46 cm / 175 g |

8 m 10 m 12 m 14 m

Characteristics and habits

Theropods with medium heads, knife-like teeth, long arms and legs, relatively short tails, graceful bodies, three fingers on each hand and four on each foot, and very developed nails on toe II. Paravians descended from basal coelurosaurs and gave rise to eudromeaosaurs and troodontids. Some were opportunists, while others might have specialized in certain types of prey. See more: *Senter et al. 2004*
Diet: carnivores of small game, piscivores, and/or insectivores.

Time range

From the Middle Jurassic to the late Upper Cretaceous (ca. 168.3–66 Ma), an approximate duration of 102.3 million years in the fossil record.

The smallest (Class: Dwarf Grade I)

IVPP V22530 of the late Lower Cretaceous of eastern Laurasia (present-day China). It is a subadult or small adult dromaeosaurid, since it weighed barely half as much as a pigeon. *Epidexipteryx hui* ("colorful feathers of the paleontologist Hu Yaoming"), from the Middle Jurassic (Callovian, 166.1–163.5 Ma) of eastern Paleoasia (present-day China), was shorter (25 cm) but heavier (220 g). They were 13.5–25 times shorter and about 1,950–1,550 times lighter than *Austroraptor cabazai*. See more: *Harrison & Walker 1973; Naish 2011; Zhang et al. 2008; Pittman et al. 2015*

The smallest juvenile specimen

Epidendrosaurus ningchengensis and *Scansoriopteryx heilmanni* date from the Middle Jurassic and early Lower Cretaceous, respectively (present-day China). They were about 13 cm long and weighed 6.8 g. See more: *Czerkas & Yuan 2002; Zhang et al. 2002; Zhang et al. 2006; Kurochkin et al. 2013*

The oldest

Richardoestesia cf. *gilmorei* GLRCM G.50823 from the Middle Jurassic (Bathonian, ca. 168.3–166.1 Ma) of north-central Neopangea (present-day England). *Richardoestesia* sp. PIN 4767/5 of European Russia is from the same age but less ancient, as it is from the upper layers, while the former comes from the lower layers. See more: *Metcalf & Walker 1994; Novikov, Lebidev & Alifanov 1998; Alifanov & Sennikov 2001*

The most recent

There are numerous reports of paravians in the last layers of the late Upper Cretaceous (Maastrichtian, ca. 69–66 Ma), from which several teeth stand out: *Richardoestesia* cf. *gilmorei* (Wyoming and South Dakota), *Richardoestesia* sp. (Wyoming and New Mexico, western United States and western Canada), *Richardoestesia* cf. *isosceles* (Romania), and *Richardoestesia* sp. (eastern Russia). See more: *Currie, Rigby & Sloan 1990; Wroblewski 1995; Baszio 1997; Weil & Williamson 2000; Sankey 2001; Codrea et al. 2002; Stokosa 2005; Itterbeeck et al. 2005*

The strangest

Yi qi has the shortest name of all the dinosaurs. It is also the only one to present a styliform bone similar to that of flying squirrels and the greater gliders. See more: *Xu et al. 2015*

The first published species

"*Paleopteryx thomsoni*" (1989): From the Upper Jurassic of northwestern Neopangea (present-day Colorado, USA). It was an invalid name until its formal revision eight years later. Part of the assigned material turned out to be of the pterosaur *Mesadactylus ornithosphyos*. Another part belonged to a microraptor from the Upper Jurassic of Colorado, USA. See more: *Jensen 1981; Jensen & Padian 1989; Currie, Rigby & Sloan 1990*

The most recently published species

Yi qi (2015) ("strange wing"): From the late Lower Cretaceous of eastern Laurasia (present-day China). It is the first theropod to present clear evidence of a membrane similar to that of bats or pterosaurs. See more: *Xu et al. 2015*

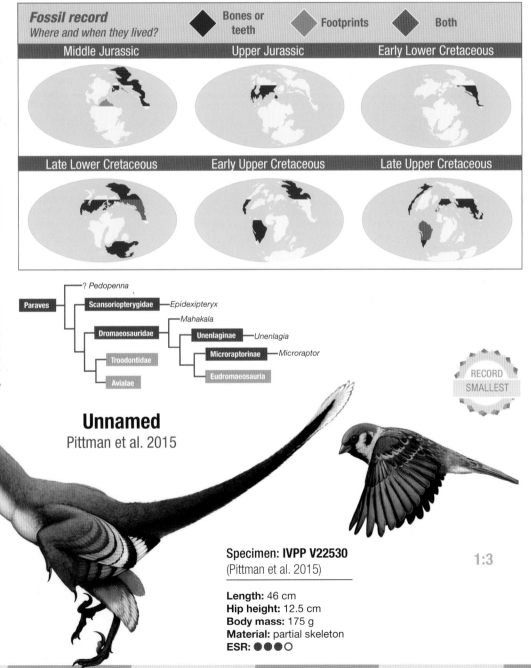

Fossil record
Where and when they lived?

◆ Bones or teeth ◆ Footprints ◆ Both

Middle Jurassic Upper Jurassic Early Lower Cretaceous

Late Lower Cretaceous Early Upper Cretaceous Late Upper Cretaceous

Paraves
? *Pedopenna*
Scansoriopterygidae — *Epidexipteryx*
Mahakala
Dromaeosauridae
Unenlaginae — *Unenlagia*
Microraptorinae — *Microraptor*
Troodontidae
Eudromaeosauria
Avialae

RECORD SMALLEST

Unnamed
Pittman et al. 2015

Breeding
Epidendrosaurus ningchengensis
Scansoriopteryx heilmanni

Specimen: IVPP V22530
(Pittman et al. 2015)

1:3

Length: 46 cm
Hip height: 12.5 cm
Body mass: 175 g
Material: partial skeleton
ESR: ●●●○

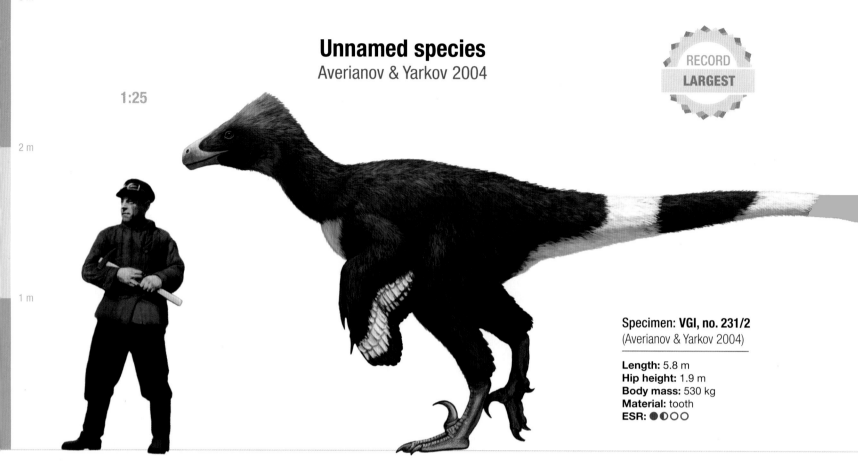

Unnamed species
Averianov & Yarkov 2004

1:25

RECORD
LARGEST

Specimen: VGI, no. 231/2
(Averianov & Yarkov 2004)

Length: 5.8 m
Hip height: 1.9 m
Body mass: 530 kg
Material: tooth
ESR: ●◐○○○

The largest (Class: Large Grade I)

The dromaeosaurid **VGI, no. 231/2**, from the late Upper Cretaceous (Maastrichtian, ca. 72.1–66 Ma) of central Laurasia (present-day European Russia). It is a large tooth that, compared to the other dromaeosaurids, must have been the largest. It was a raptor heavier than a brown bear, capable of killing prey even greater than itself. Recently it has been discovered that the body proportions of dromaeosaurids and velociraptorids were different. They had shorter faces, serrated teeth, longer legs, and shorter bodies and tails. All this substantially changes their former popular aspect. See more: *Averianov & Yarkov 2004; DePalma et al. 2015; Hartman**

The myth of the gigantic raptor

Utahraptor ostrommaysorum ("Utah thief of the paleontologist John Ostrom and DIC president Chris Mays"): From the late Lower Cretaceous (Berriasian, ca. 129.4–125 Ma) of western Laurasia (present-day western United States). The adult specimen BYU 15465 would have been about 4.9 m long and weighed 280 kg. An 11 m long specimen was reported. Apparently, it was a miscalculation. Nine individuals were mixed, so there could have been confusion when allocating and comparing pieces. See more: *Britt et al. 2001; Erickson et al. 2009*

Footprints

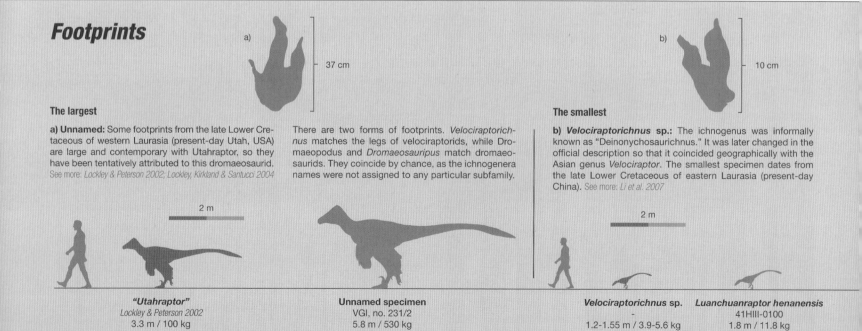

a)   ⌐ 37 cm

b)   ⌐ 10 cm

The largest

a) Unnamed: Some footprints from the late Lower Cretaceous of western Laurasia (present-day Utah, USA) are large and contemporary with Utahraptor, so they have been tentatively attributed to this dromaeosaurid. See more: *Lockley & Peterson 2002; Lockley, Kirkland & Santucci 2004*

There are two forms of footprints. *Velociraptorichnus* matches the legs of velociraptorids, while *Dromaeopodus* and *Dromaeosauripus* match dromaeosaurids. They coincide by chance, as the ichnogenera names were not assigned to any particular subfamily.

The smallest

b) *Velociraptorichnus* sp.: The ichnogenus was informally known as "Deinonychosaurichnus." It was later changed in the official description so that it coincided geographically with the Asian genus *Velociraptor*. The smallest specimen dates from the late Lower Cretaceous of eastern Laurasia (present-day China). See more: *Li et al. 2007*

2 m

2 m

"Utahraptor"
Lockley & Peterson 2002
3.3 m / 100 kg

Unnamed specimen
VGI, no. 231/2
5.8 m / 530 kg

Velociraptorichnus sp.
-
1.2-1.55 m / 3.9-5.6 kg

Luanchuanraptor henanensis
41HIII-0100
1.8 m / 11.8 kg

Characteristics and habits

Theropods with large heads, knife-like teeth, long arms and legs, rigid tails, graceful bodies, three fingers on each hand and four on each foot, advanced feathers, and very developed claws, primarily on toe II. Eudromaeosaurids descended from basal paravians. They were very specialized predators. Some may have hunted in packs to subdue larger prey. See more: *Senter et al. 2004*
Diet: carnivores of small or large game.

Time range

From the Middle Jurassic to the late Upper Cretaceous (ca. 168.3–66 Ma), a fossil duration of about 102.3 million years.

The smallest (Class: Small Grade I)

Luanchuanraptor henanensis: From the late Upper Cretaceous of eastern Laurasia (present-day China). It was a raptor similar to *Velociraptor mongoliensis* but smaller, although it is not known for certain if it was an adult. "*Dromaeosaurus*" *gracilis* from Maryland, "Ichabodcraniosaurus" from Mongolia, and *Ornithodesmus cluniculus* from England were smaller than *Luanchuanraptor*, but they are too questionable and incomplete. *Luanchuanraptor* was three times shorter and forty-five times lighter than the dromaeosaurid VGI, no. 231/2. See more: *Lull 1911; Novacek 1996; Allain & Taquet 2000; Lu et al. 2007; Company et al. 2009; Naish 2011; Mortimer**

The smallest juvenile specimen

Archaeornithoides deinosauriscus ("small dinosaur similar to a small bird"): From the late Upper Cretaceous of eastern Laurasia (present-day Mongolia). It is a dromaeosaurid hatchling about 50 cm long and weighing 90 g. See more: *Elzanowski & Wellnhofer 1992*

The oldest

Dromaeosaurid PIN 4767/6 dates from the Middle Jurassic (Bartonian, ca. 168.3–166.1 Ma) and lived in north-central Neopangea (present-day western Russia). See more: *Alifanov & Sennikov 2001*

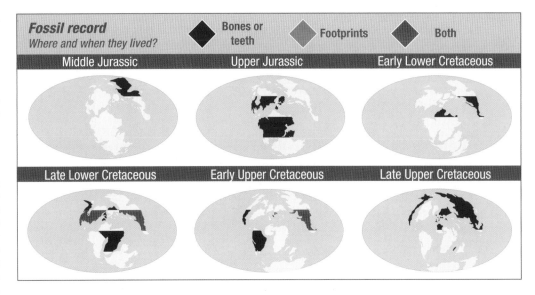

Fossil record *Where and when they lived?* — ◆ **Bones or teeth**   ◆ **Footprints**   ◆ **Both**

Middle Jurassic | Upper Jurassic | Early Lower Cretaceous
Late Lower Cretaceous | Early Upper Cretaceous | Late Upper Cretaceous

The most recent

Acheroraptor temertyorum (Montana, USA), cf. *Dromaeosaurus albertensis* FMNH PR2898 (South Dakota, USA), cf. *Dromaeosaurus albertensis* (Montana, USA), cf. *Dromaeosaurus albertensis* (Wyoming, USA), *Dromaeosaurus* sp. NMMNH P-32814 (New Mexico, USA), cf. *Zapsalis abradens* UA 132 (Wyoming, USA), cf. *Zapsalis abradens* UA 103 and cf. *Dromaeosaurus albertensis* UAKUA-1:106 (Canada), cf. *Dromaeosaurus* sp. and cf. *Saurornitholestes* sp. (eastern Russia), and MPZ98-68 (Spain) belong to the most recent layers of the late Upper Cretaceous (upper Maastrichtian, ca. 69–66 Ma). See more: *López-Martínez et al. 2001; Itterbeeck et al. 2005; Evans et al. 2013; Williamson & Brusatte 2014; Gates et al. 2015*

The strangest

Although it is not a basal genus, *Adasaurus mongoliensis* had sickle claws much smaller than those of other eudromosaurs See more: *Barsbold 1983*

The first published species

"*Laelaps*" *explanatus* and *L. laevifrons* (1876): From the Upper Cretaceous of western Laurasia (present-day Montana, USA). They are probably *Saurornitholestes langstoni* teeth (present-day Canada) but were described 102 years earlier. See more: *Cope 1876; Sues 1978*

The most recently published species

Acheroraptor temertyorum (2013) ("The thief of the underworld of James and Louise Temerty"): From the late Upper Cretaceous of western Laurasia (present-day Montana, USA). *Boreonykus certekorum* (2015), from western Canada, could be the last published species of this group, but some researchers believe that the material is not a dromaeosaurid. Only the claws are assigned to this species, so it is a paleontological chimera. See more: *Evans et al. 2013; Bell & Currie 2015; Cau**

Eudromaeosauria — Velociraptorinae — Saurornitholestes / Deinonychus / Velociraptor — Dromeosaurinae — Dromaeosaurus

Luanchuanraptor henanensis
Lu et al. 2007

RECORD SMALLEST

1:10

Specimen: 41HIII-0100
(Lu et al. 2007)

Length: 1.8 m
Hip height: 49 cm
Body mass: 11.8 kg
Material: skull and partial skeleton
ESR: ●●●●

Breeding
Archaeornithoides deinosauriscus

50 cm   100 cm   150 cm   200 cm

cf. *Troodon formosus*
Leidy 1856

RECORD
LARGEST

1:25

Specimen: AK498-V-001
(Erickson et al. 2009)
___

Length: 5.4 m
Hip height: 1.65 m
Body mass: 380 kg
Material: tooth
ESR: ●◑○○

The largest (Class: Medium Grade III)

AK498-V-001 assigned to cf. *Troodon formosus* ("hurting tooth") from the late Upper Cretaceous (Campanian, ca. 83.6–72.1 Ma) of western Laurasia (present-day Alaska). This material consists of a tooth that is very large for a troodontid. Perhaps it belonged to a large carnivore twice as heavy as a lion. Although it was assigned to *Troodon formosus*, it may be a much larger new species that lived in a very northern latitude where there would be less sunlight. Their teeth looked similar to some herbivorous lizards, but their dental crowns were very narrow and the apical denticle was hooked, as in carnivores. Perhaps they were opportunistic omnivores; such a diet would allow them to take advantage of different food sources. See more: *Holtz et al. 1998; Fiorillo 2008; Fiorillo et al. 2009; Headden*

Another likely contender

The teeth of *Troodon* cf. *formosus* ZIN PH 1/28, from the late Upper Cretaceous of eastern Laurasia (present-day eastern Russia), are also quite large, but their developmental stage is not known. The owner of these teeth could have been 4.4 m long and weighed 208 kg. See more: *Grigorescu et al. 1985; Nesov & Golovneva 1990*

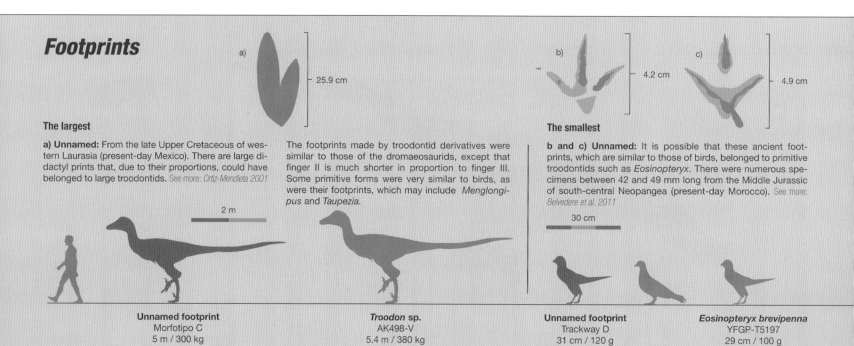

Footprints

a) — 25.9 cm

b) — 4.2 cm

c) — 4.9 cm

The largest

a) Unnamed: From the late Upper Cretaceous of western Laurasia (present-day Mexico). There are large didactyl prints that, due to their proportions, could have belonged to large troodontids. See more: *Ortiz-Mendieta 2001*

The footprints made by troodontid derivatives were similar to those of the dromaeosaurids, except that finger II is much shorter in proportion to finger III. Some primitive forms were very similar to birds, as were their footprints, which may include *Menglongipus* and *Taupezia*.

The smallest

b and c) Unnamed: It is possible that these ancient footprints, which are similar to those of birds, belonged to primitive troodontids such as *Eosinopteryx*. There were numerous specimens between 42 and 49 mm long from the Middle Jurassic of south-central Neopangea (present-day Morocco). See more: *Belvedere et al. 2011*

30 cm

2 m

Unnamed footprint
Morfotipo C
5 m / 300 kg

Troodon sp.
AK498-V
5.4 m / 380 kg

Unnamed footprint
Trackway D
31 cm / 120 g

Eosinopteryx brevipenna
YFGP-T5197
29 cm / 100 g

Characteristics and habits

Theropods with medium heads, large eyes, compact knife-like teeth, long arms and legs, relatively short tails, graceful bodies, and three fingers on each hand and four on each foot. Troodontids descended from primitive paravians and preceded birds. Their legs were well adapted for running, and they had large brains.

Diet: carnivores of small game, omnivores, and/or insectivores.

Time range

From the Middle Jurassic to the late Upper Cretaceous (ca. 168.3–66 Ma), an approximate duration of about 102.3 million years in the fossil record.

The smallest (Class: Tiny Grade III)

Eosinopteryx brevipenna ("ancient short feathers from China"): It could be a very primitive troodontid or a basal paravian. It dates from the Middle Jurassic of eastern Paleoasia (present-day China). It weighed little more than three sparrows. It has juvenile and mature characteristics, so it may be a subadult. Some authors consider it a primitive troodontid, and others a paravian. *Aurornis xui* ("dawn bird of the paleontologist Xu Xing") was somewhat larger; it was 40 cm long and weighed 260 grams. They are 18.5–3.5 times shorter and 3,800–1,460 times lighter than cf. *Troodon formosus* AK498-V-001. See more: *Godefroit et al. 2013a, 2013b; Lefevre et al. 2014*

The smallest juvenile specimen

Some *Troodon formosus* embryos from the late Upper Cretaceous (present-day Montana, USA) were originally identified as the ornithopod *Orodromeus makelai*. Specimens MOR 246-11 and MOR 246-1 were approximately 35 cm long and weighed slightly more than 80 g. See more: *Horner & Weishampel 1988*

The oldest

Anchiornis huxleyi, Aurornis xui, Eosinopteryx brevipenna and *Xiaotingia zhengi*: They lived during the Middle Jurassic (Bathonian, ca. 168.3–166.1 Ma) in eastern Paleoasia (present-day China). See more: *Xu et al. 2008, 2011; Godefroit et al. 2013a, 2013b*

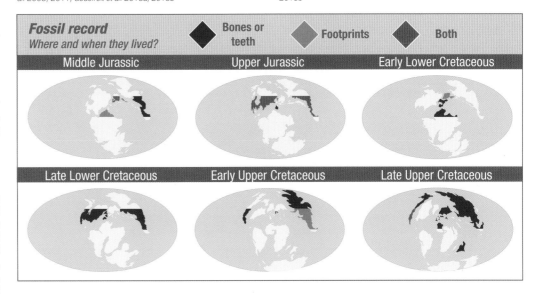

The most recent

cf. *Euronychodon* sp. (cf. *Paronychodon* sp.) from Spain, *Paronychodon* cf. *lacustris* UA MR-48 from Canada, *Paronychodon* sp. from South Dakota, *Paronychodon* sp. from Montana, *Paronychodon* sp. from Wyoming, *Paronychodon* sp. from Romania, *Troodon bakkeri* from Montana and Wyoming, *Troodon* sp. from Texas, *Troodon* sp. from Canada, *Troodon* sp. from Russia, the specimen in the OMNH collection from Utah, FGGUB R.1318, and the questionable *Elopteryx nopcsai* from Romania are among the most recent fossil remains from the late Upper Cretaceous (upper Maastrichtian, ca. 69–66 Ma). See more: *Andrews 1913; Estes et al. 1969; Bolotsky & Moiseenko 1988; Wroblewski 1995; Baszio 1997; Csiki & Grigorescu 1998; Carpenter 1982; Langston et al. 1989; Kirkland et al. 1998; López-Martínez et al. 2001*

The strangest

Eosinopteryx brevipenna: It had shorter and smaller feathers than its relatives. It also lacked rectrices, even in the lower limbs. It could be a basal paravian or a primitive troodontid. See more: *Godefroit et al. 2013a, 2013b*

The first published species

Troodon formosus (1856) ("well-formed hurting tooth"): From the late Upper Cretaceous of western Laurasia (present-day Montana, USA). It is based on a tooth that was considered a lizard until it was identified as a dinosaur about 45 years later. This species may lose its validity in the future, since the type specimen presents features that are very typical of troodontids, which makes it difficult to distinguish from other contemporary species. See more: *Leidy 1856; Nopcsa 1901; Gilmore 1924; Barsbold 1974*

The most recently published species

Gobivenator mongoliensis (2014) ("Gobi Desert hunter, Mongolia"): From the late Upper Cretaceous of eastern Laurasia (present-day Mongolia). It is one of the most complete troodontid derivatives known to date. It is possible that the type material of *Boreonykus certekorum* (2015) from western Canada belongs to a troodontid. See more: *Evans et al. 2013; Tsuihiji et al. 2014; Bell & Currie 2015; Cau*

Eosinopteryx brevipenna
Godefroit et al. 2013

```
? Anchiornis
Troodontidae ┤ Jinfengopteryx
             │ Sinovenator
             │ Byronosaurus
             └ Saurornithoides
```

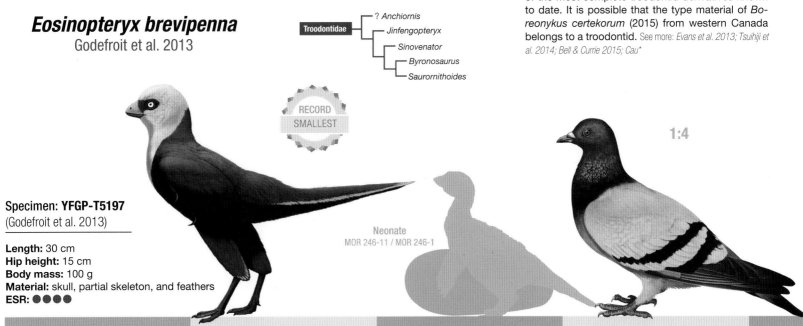

RECORD SMALLEST

1:4

Specimen: YFGP-T5197
(Godefroit et al. 2013)

Length: 30 cm
Hip height: 15 cm
Body mass: 100 g
Material: skull, partial skeleton, and feathers
ESR: ●●●●

Neonate
MOR 246-11 / MOR 246-1

Unnamed
Molnar 1999

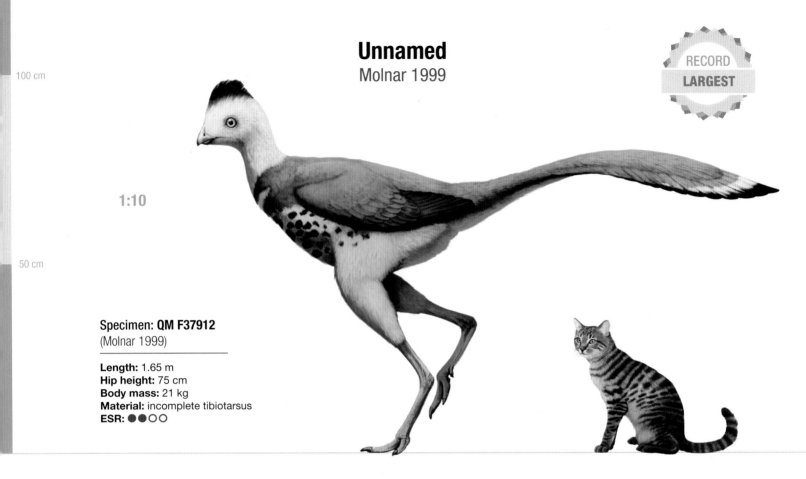

RECORD
LARGEST

1:10

Specimen: QM F37912
(Molnar 1999)

Length: 1.65 m
Hip height: 75 cm
Body mass: 21 kg
Material: incomplete tibiotarsus
ESR: ●●○○

The largest (Class: Small Grade II)

The avialan **QM F37912**, from the late Lower Cretaceous (Albian, ca. 113–100.5 Ma) of eastern Gondwana (present-day Australia), is based on a distal fragment of a tibiotarsus, which is similar to that of *Yandangornis longicaudus*. Their size and weight were comparable to those of a small rhea. It is possible that their way of life was similar, since perhaps it was a flightless terrestrial bird like *Yandangornis*. See more: *Molnar 1999*

The longest

Bauxitornis mindszentyae ("bird from bauxite mine of Andrea Mindszenty"): From the late Upper Cretaceous (Santonian, ca. 86.3–83.6 Ma) of central Laurasia (present-day Hungary). With its length of 1.9 m and weight of 18 kg, it was the longest avialan. It seems to be similar to *Balaur bondoc*, a strange avialan. Some authors consider *Bauxitornis* a large enantiornithe. See more: *Csiki et al. 2010; Dyke & Osi 2010; Cau**

Footprints

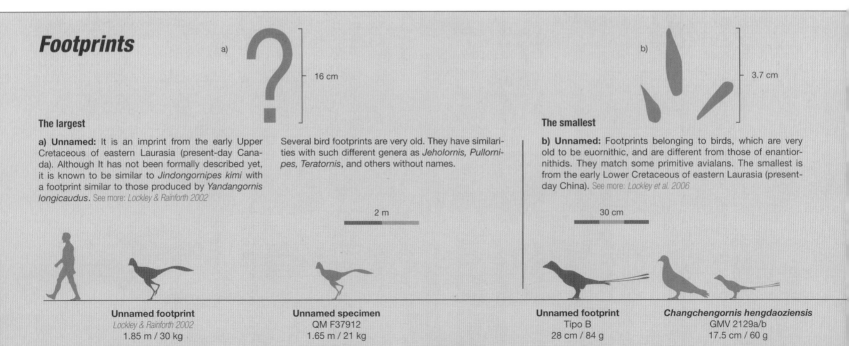

a)

? 16 cm

b)

3.7 cm

The largest

a) Unnamed: It is an imprint from the early Upper Cretaceous of eastern Laurasia (present-day Canada). Although It has not been formally described yet, it is known to be similar to *Jindongornipes kimi* with a footprint similar to those produced by *Yandangornis longicaudus*. See more: *Lockley & Rainforth 2002*

Several bird footprints are very old. They have similarities with such different genera as *Jeholornis*, *Pullornipes*, *Teratornis*, and others without names.

2 m

The smallest

b) Unnamed: Footprints belonging to birds, which are very old to be euornithic, and are different from those of enantiornithids. They match some primitive avialans. The smallest is from the early Lower Cretaceous of eastern Laurasia (present-day China). See more: *Lockley et al. 2006*

30 cm

Unnamed footprint
Lockley & Rainforth 2002
1.85 m / 30 kg

Unnamed specimen
QM F37912
1.65 m / 21 kg

Unnamed footprint
Tipo B
28 cm / 84 g

Changchengornis hengdaoziensis
GMV 2129a/b
17.5 cm / 60 g

Characteristics and habits

Birds with large or medium heads, toothless or with small or bulbous teeth, large eyes, very long arms, long legs, relatively short or very short tails, graceful bodies, and three fingers on each hand and four on each foot. Advanced feathers. Limited flight or terrestrial.

Avialans may descend from troodontids and predate enantiornithean birds and ornithuromorphs. They adapted to arboreal or terrestrial life. They weren't perfect fliers because the keel in the sternum was underdeveloped and they didn't have an alula, a structure that helps perform landing maneuvers. See more: *Chiappe et al 1999*

Diet: carnivores, piscivores, omnivores, and/or insectivores.

Time range

From the Upper Jurassic to the late Upper Cretaceous (ca.152.1–66 Ma), a duration of approximately 86.1 million years in the fossil record.

The smallest (Class: Tiny Grade III)

Changchengornis hengdaoziensis ("bird of the Great Wall of China (Changcheng) of Hengdaozi"): It was an adult confuciusornithid bird from the early Lower Cretaceous (Berriasian, ca. 145–139.8 Ma) of eastern Laurasia (present-day China). Because of its plumage, it would look similar to current turacos, albeit much smaller (as heavy as two sparrows). It was about 9 times shorter and 350 times lighter than the avialaen QM F37912. See more: *Ji, Chiappe & Ji 1999*

The smallest juvenile specimen

Zhongornis haoae ("intermediate bird of Ms. Hao"): It was a hatchling that was 8 cm long and weighed 6 g. Because of its age, it lacked a pygostyle. It had a short tail, consisting of 13 vertebrae. It was apparently a primitive confuciusornis bird, although some authors suspect that it could be a scansoriopterygid paravian. See more: *Gao et al. 2008; O'Connor & Sullivan 2014*

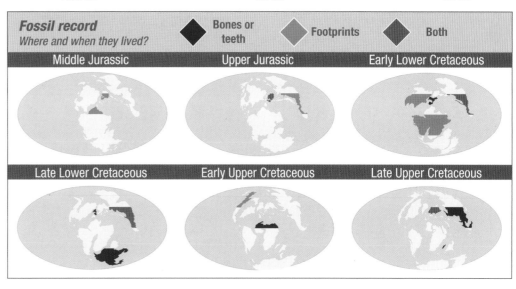

Fossil record
Where and when they lived? ◆ Bones or teeth ◆ Footprints ◆ Both

Middle Jurassic | Upper Jurassic | Early Lower Cretaceous
Late Lower Cretaceous | Early Upper Cretaceous | Late Upper Cretaceous

The strangest

Balaur bondoc ("strong Balaur dragon"): From the late Upper Cretaceous of central Laurasia (present-day Romania). It would look similar to dromaeosaurids, but they had two retractable nails and only two functional fingers on each hand. See more: *Godefroit et al. 2013a, 2013b*

The oldest

Archaeopteryx lithographica ("ancient wing imprinted on stone"): From the Upper Jurassic (Tithonian, ca. 152.1–145 Ma) of north-central Neopangea (present-day Germany). It has historically been considered the "oldest bird," although the term bird is currently very ambiguous. The anatomical characteristics with which birds are defined have gradually appeared in theropod dinosaurs. Unlike other theropods similar to birds, *Archeopteryx lithographica* and other avialaens have bodies covered with feathers. cf. *Archeopteryx* sp. is even older, from the Upper Jurassic (Kimmeridgian, ca. 157.3–152.1 Ma) (present-day Portugal), but some experts identify it as a probable basal troodontid. See more: *Weigert 1995; Xu et al. 2008, 2011; Godefroit et al. 2013a, 2013b; Mortimer**

The most recent

Rahonavis ostromi and *Balaur bondoc*, date from the late Upper Cretaceous (Maastrichtian, ca. 72.1–66 Ma) in what is now Madagascar and Romania, respectively. There is doubt if the former is an avialaen or a unenlagian paravian with long arms. See more: *Forster et al. 1998; Makovicky et al. 2005; Xu et al. 2008; Csiki et al. 2010; Cau et al. 2015*

The first published species

The first specimen of *Archaeopteryx lithographica* was a very incomplete fossil that was thought to be a pterosaur because of its aerodynamic shape. It was called *Pterodactylus crassipes* in 1857 and was identified as a bird only 113 years later. There are reports of even older fossil finds that may have belonged to Archeopteryx, but the specimens were not well described. See more: *Meyer 1857, 1861; Ostrom 1970; Moore 2014*

The most recently published species

Jeholornis curvipes (2014) ("bird with curved feet from the Jehol Province"): From the late Lower Cretaceous (present-day China). There is conflict over which of its two names has priority (*Jeholornis* or *Shenzhousaurus*), but most authors seem to use *Jeholornis*. See more: *Lefévre et al. 2014; Mortimer**

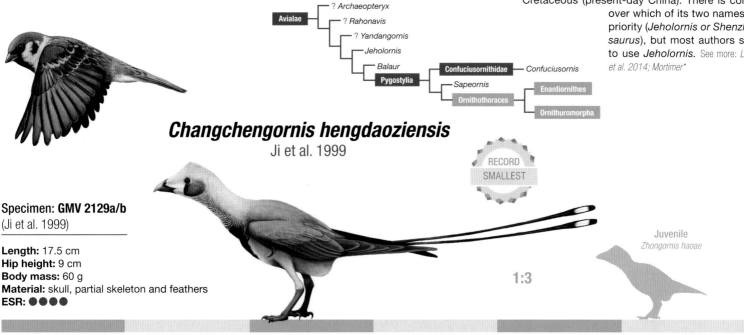

Avialae
— ? *Archaeopteryx*
— ? *Rahonavis*
— ? *Yandangornis*
— *Jeholornis*
— *Balaur*
Pygostylia
Confuciusornithidae — *Confuciusornis*
Sapeornis
Ornithothoraces
Enantiornithes
Ornithuromorpha

Changchengornis hengdaoziensis
Ji et al. 1999

RECORD SMALLEST

Specimen: GMV 2129a/b
(Ji et al. 1999)

Length: 17.5 cm
Hip height: 9 cm
Body mass: 60 g
Material: skull, partial skeleton and feathers
ESR: ●●●●

1:3

Juvenile
Zhongornis haoae

RECORD
LARGEST

cf. *Soroavisaurus australis*
Chiappe 1993

1:7

50 cm

Specimen: PVL-4033
(Walker 1981)

Length: 80 cm
Hip height: 35 cm
Body mass: 7.25 kg
Material: tibiotarsus
ESR: ●●○○

The largest (Class: Small Grade I)

PVL-4033, from the late Upper Cretaceous (lower Maastrichtian, ca. 72.1–69 Ma) of western Gondwana (present-day Argentina), is a tibiotarsus of a large bird that has been attributed to cf. *Martinavis* and, more recently, to *Soroavisaurus australis*. It might have weighed more than a cat and crow together. Enantiornitheans had a great diversity of habits, so it isn't known what its diet was. See more: *Walker 1981; Fuentes Chiappe 1993; Walker & Dyke 2009*

A problematic identification

Bauxitornis mindszentyae (discovered in modern Hungary) was originally interpreted as an enantiornithean bird. If so, it would have been even bigger than cf. *Soroavisaurus australis* PVL-4033, but other experts say that it is more like *Balaur bondoc*. The size of *Enantiornis leali* (present-day Argentina), 78 cm long and weighing 6.75 kg, was very close to cf. *Soroavisaurus*. See more: *Walker 1981; Dyke & Osi 2010; Cau**

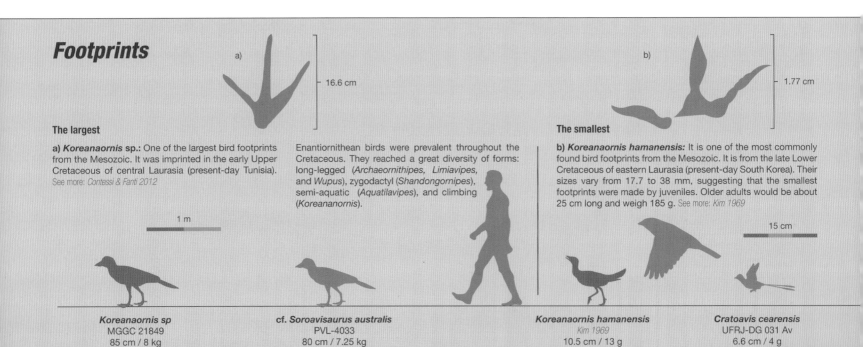

Footprints

a)
├ 16.6 cm

b)
┤ 1.77 cm

The largest

a) *Koreanaornis* sp.: One of the largest bird footprints from the Mesozoic. It was imprinted in the early Upper Cretaceous of central Laurasia (present-day Tunisia). See more: *Contessi & Fanti 2012*

Enantiornithean birds were prevalent throughout the Cretaceous. They reached a great diversity of forms: long-legged (*Archaeornithipes, Limiavipes,* and *Wupus*), zygodactyl (*Shandongornipes*), semi-aquatic (*Aquatilavipes*), and climbing (*Koreananornis*).

The smallest

b) *Koreanaornis hamanensis*: It is one of the most commonly found bird footprints from the Mesozoic. It is from the late Lower Cretaceous of eastern Laurasia (present-day South Korea). Their sizes vary from 17.7 to 38 mm, suggesting that the smallest footprints were made by juveniles. Older adults would be about 25 cm long and weigh 185 g. See more: *Kim 1969*

1 m

15 cm

***Koreanaornis* sp**
MGGC 21849
85 cm / 8 kg

cf. *Soroavisaurus australis*
PVL-4033
80 cm / 7.25 kg

Koreanaornis hamanensis
Kim 1969
10.5 cm / 13 g

Cratoavis cearensis
UFRJ-DG 031 Av
6.6 cm / 4 g

Characteristics and habits

Birds with large heads, absent or small bulbous teeth, large eyes, very long arms, very short tails, graceful bodies, and three fingers on each hand and four on each foot. Fliers and climbers. They were adapted to arboreal life, although some have been linked to semi-aquatic environments. These birds had a proportionally longer thigh bone (femur) compared to modern birds. It has been estimated that the size of *Alexornis antecedens* was similar to that of a sparrow, but because of the difference in proportions, it would actually be twice as light.

Diet: piscivores, omnivores, and/or insectivores.

Time range

From the early Lower Cretaceous to the late Upper Cretaceous (ca. 139.8–66 Ma), an approximate duration of about 73.8 million years in the fossil record.

The smallest (Class: Tiny Grade II)

Cratoavis cearensis ("bird of the Crato Formation, from the state of Ceará"): From the late Lower Cretaceous (Aptian, ca. 125–113 Ma) of western Gondwana (present-day Brazil). It was the smallest of all the Mesozoic dinosaurs. Only today's smallest bird, the bee hummingbird (*Mellisuga helenae*), is lighter. It was 12 times shorter and about 1,800 times lighter than cf. *Soroavisaurus australis* PVL-4033. See more: *Walker, Buffetaut & Dyke 2007; Carvalho et al. 2015*

The smallest juvenile specimen

Gobipipus reshetovi ("Doctor Valery Reshetov's Gobi Desert chick"): From the late Upper Cretaceous of eastern Laurasia (present-day Mongolia). It is known only from embryos. Only 4.7 cm long and weighing 2 g, the sample PIN 4492-4 was the smallest of all. See more: *Kurochkin et al. 2013*

The oldest

The enantiornithean SBEI 307 is a humerus with features similar to *Otogornis genghisi*. It dates from the Early Lower Cretaceous (Valanginian, approx. 139.8-132.9 Ma) and lived in the eastern area of Laurasia (present-day Japan). *Wyleyia valdensis* (present-day England) is the same age, although it is not known if it was a basal paravian or an enantiornithean. See more: *Harrison & Walker 1973; Unwin & Matsuoka 2000*

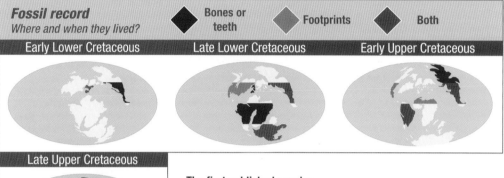

Fossil record
Where and when they lived?

◆ Bones or teeth   ◆ Footprints   ◆ Both

Early Lower Cretaceous   Late Lower Cretaceous   Early Upper Cretaceous

Late Upper Cretaceous

The most recent

Avisaurus archibaldi and *Avisaurus* sp. (present-day Montana, USA), EME V.314 and NVEN 1 (present-day Romania), and other possible remains from western Russia date from the late Upper Cretaceous (upper Maastrichtian, ca. 69–66 Ma). See more: *Brett-Surman & Paul, 1985; Nesov 1992; Hutchison 1993; Stidham 1999; Wang et al. 2011; Dyke et al. 2012*

The strangest

The zygodactyl fingers of *Dalingheornis liwei* were unusual in the Mesozoic. This bird is characterized by having zygodactyl feet (two fingers face forward, and two face backward), a characteristic that is common in some climbing birds. This feature is also present in the *Shandongornipes muxiai* footprints. See more: *Zhang et al. 2006; Li, Lockley & Liu 2005*

The first published species

Gobipteryx minuta (1974) is the first enantiornithean bird in which the absence of teeth was proven. A trend that was appearing independently in forms of birds, such as confuciusornithids, some euornithes, and all neornithes. See more: *Elzanowski 1974; Brodkorb 1976; Elzanowski 1976; Luochart & Viriot 2011*

The most recently published species

In 2015, *Dunhuangia cuii* (January), *Houornis caudatus* and *Yuanjiawaornis viriosus* (February), *Parapengornis eurycaudatus* (April), *Cratoavis cearensis* (May), *Holbotia ponomarenkoi* (June), *Pterygornis dapingfangensis* (August), and *Chiappeavis magnapremaxillo* (December) were published. See more: *Kurochkin 1991; Carvalho et al. 2015; Hu et al. 2015; Wang & Liu 2015; Wang et al. 2015*

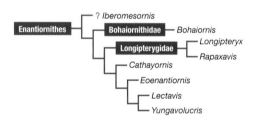

Enantiornithes
? *Iberomesornis*
Bohaiornithidae — *Bohaiornis*
Longipterygidae — *Longipteryx*
— *Rapaxavis*
Cathayornis
Eoenantiornis
Lectavis
Yungavolucris

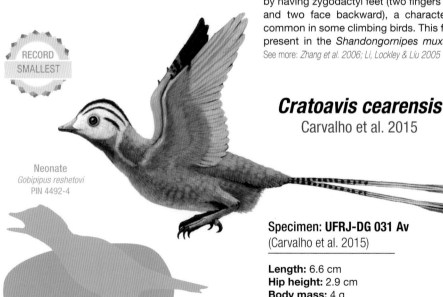

RECORD SMALLEST

Neonate
Gobipipus reshetovi
PIN 4492-4

Cratoavis cearensis
Carvalho et al. 2015

1:1

Specimen: UFRJ-DG 031 Av
(Carvalho et al. 2015)

Length: 6.6 cm
Hip height: 2.9 cm
Body mass: 4 g
Material: skull, partial skeleton, and feathers
ESR: ●●●●

Gargantuavis philoinos
Buffetaut & Le Loeuff 1998

RECORD
LARGEST

1 m

1:15

Specimen: MDE-A08
(Buffetaut & Le Loeuff 1998)

Length: 1.8 m
Hip height: 1.3 m
Body mass: 120 kg
Material: incomplete femur
ESR: ●●○○

The largest (Class: Medium Grade II)

Gargantuavis philoinos ("Gargantua wine-loving bird"), from the late Upper Cretaceous (Campanian, ca. 83.6–72.1 Ma) of central Laurasia (present-day France), was a large land bird the size of an ostrich but more robust, similar to *Patagopteryx* but with different bone proportions (a proportionally shorter and more massive femur). This feature makes it more similar to ostriches and cassowaries. Its discovery was surprising because its size far exceeds that of any other bird from the Mesozoic. It was suggested that perhaps they were fragmented bones of a large pterosaur, but studies have confirmed that it was indeed a

bird. See more: *Buffetaut & Le Loeuff 1998; Mayr 2009; Buffetaut & Le Loeuff 2010; Buffetaut & Angst 2013*

The group's largest flying bird

Palintropus sp. was the largest primitive ornithurine bird with the ability to fly. RTMP 86.36.126 from western Canada had an estimated length of 85 cm and a weight of about 6.5 kg. See more: *Hope 2002; O'Connor & Forster 2010*

Footprints

a)

~8.65 cm

b)

2 cm

The largest

a) **Unnamed:** The largest footprint that seems to have been left by euornithic bird is from the early Lower Cretaceous of central Laurasia (present-day Spain). See more: *Moratalla & Sanz 1992*

An important group of footprints coincides with some euornithes since they have lobed fingers like *Yanornis* (*Muguiornipes*), are robust like *Patagopteryx* (*Barrosopus* and *Paxavipes*), are symmetrically web-footed like *Changmaiornis* (*Goseongornipes, Gyeongsangornipes,* and *Ignotornis*), and are asymmetrically web-footed like *Gansus* (*Hwangsanipes*) and *Hongshanornis*. They are different from neornithes because they were from the Lower Cretaceous, while the latter did not diversify until the Upper Cretaceous

The smallest

b) **Unnamed:** A few tiny footprints from the late Lower Cretaceous of central Laurasia (present-day Spain) represent a bird whose length, when compared with more primitive birds, could be from 7.3 to 9.5 cm. See more: *Moratalla et al. 2003*

2 m

20 cm

Unnamed footprint
LCB-IA/3
35-46 cm / 770-1800 g

Gargantuavis philoinos
MDE-A08
1.8 m / 120 kg

Unnamed footprint
Los Cayos C (level C5).
7.3-9.5 cm / 13-30 g

Hongshanornis longicresta
IVPP V14533
12 cm / 58 g

4 m   5 m   6 m   7 m

Characteristics and habits

Birds with medium or large heads, small or absent teeth, short or very long arms, long necks, very short tails, graceful bodies, and three fingers on each hand and four on each foot. Flying, semi-aquatic, or terrestrial.
Diet: piscivores, omnivores, and/or insectivores.

Time range

From the late Lower Cretaceous to the late Upper Cretaceous (ca. 129.4–66 Ma). They lasted in the fossil record approximately 63.4 million years.

The smallest (Class: Tiny Grade III)

Hongshanornis longicresta ("long-crested bird of the ancient Hongshan Culture"): From the late Lower Cretaceous (Barremian, ca. 129.4–125 Ma) of eastern Laurasia (present-day China). It was a tiny adult barely twice a sparrow's weight. It was 15 times shorter and 2,070 times lighter than *Gargantuavis philoinos*. Another even smaller individual was "Ornithurine C" UALVP 47942 from the late Upper Cretaceous of Canada. Although its age is unknown, it was 10 cm long and weighed 30 g. See more: *Zhou & Zhang 2005; Longrich 2009*

The smallest juvenile specimen

MACN PV RN 1108, from the late Upper Cretaceous of the western zone of Gondwana (present-day Argentina), was a juvenile of *Alamitornis minutus* ("tiny bird of the Los Alamitos Formation"). It is based on an incomplete tibiotarsus whose owner was only 11 cm long and weighed 20 g. See more: *Agnolin & Martinelli 2009*

The oldest

Along with *Ambiortus dementjevi* from Mongolia and the K3-1 specimen from Thailand, *Archaeornithura meemannae*, *Hongshanornis longicresta*, and *Longicrusavis houi* from China date from the late Lower Cretaceous (Barremian, ca. 129.4–125 Ma). Their great diversity of forms suggests that they already existed in the early Lower Cretaceous. See more: *Kurochkin 1982; Buffetaut et al. 2005; Zhou & Zhang 2005; O'Connor et al. 2010; Wang et al. 2015*

The most recent

NHMM/RD 271 from Belgium and several more from the BHI and SDSM collections from Wyoming date from the last layers of the late Upper Cretaceous (upper Maastrichtian, ca. 69–66 Ma). See more: *Dyke et al. 2002; Longrich 2009*

The most recently published species

Juehuaornis zhangi and *Archaeornithura meemannae* (2015), both from the late Lower Cretaceous of eastern Laurasia (present-day China), were published in the same year (March and May, respectively). See more: *Wang, Wang & Hu 2015; Wang et al. 2015*

The strangest

Hollanda luceria ("dedicated to the Holland family and the punk band Lucero"): From the late Upper Cretaceous of eastern Laurasia (present-day Mongolia). It had quite long legs with an unusual toe arrangement and might have adapted to terrestrial life, being a very fast runner like the roadrunner (*Geoccocyx*). See more: *Bell et al. 2010*

The first published species

Ichthyornis anceps (formerly "Graculavus") ("dubious bird-fish"): From the late Upper Cretaceous of western Laurasia (present-day Kansas, USA). It was discovered in 1870 and described two years later. It is the first known bird with teeth, as the first *Archeopteryx* fossil didn't have a skull. See more: *Marsh 1872*

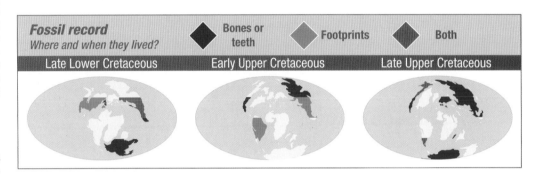

Fossil record
Where and when they lived? ◆ **Bones or teeth** ◆ **Footprints** ◆ **Both**

| Late Lower Cretaceous | Early Upper Cretaceous | Late Upper Cretaceous |

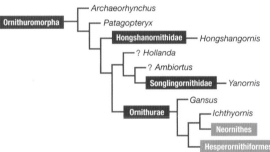

Ornithuromorpha
— Archaeorhynchus
— Patagopteryx
Hongshanornithidae — Hongshanornis
? Hollanda
? Ambiortus
Songlingornithidae — Yanornis
— Gansus
Ornithurae
— Ichthyornis
Neornithes
Hesperornithiformes

RECORD SMALLEST

Hongshanornis longicresta
Zhou & Zhang 2005

Specimen: IVPP V14533
(Zhou & Zhang 2005)

Length: 12 cm
Hip height: 4 cm
Body mass: 58 g
Material: skull, partial skeleton, and feathers
ESR: ●●●●

1:2

Juvenile
Alamitornis minutus
MACN PV RN 1110

10 cm   20 cm   30 cm   40 cm

RECORD LARGEST

Hesperornis rossicus
Nesov & Yarkov 1993

1:8

Specimen: ZIN PO 5463
(Nesov & Yarkov 1993)

Length: 1.6 m
Body mass: 30 kg
Material: tarsometatarsus and thoracic vertebra
ESR: ●●○○

The largest (Class: Small Grade III)

Hesperornis rossicus ("western bird of Russia"): From the late Upper Cretaceous (Campanian, ca. 83.6–72.1 Ma) of central and eastern Laurasia (present-day western Russia and Sweden). Its presence in both territories indicates that it may have migrated from both areas since they were closer back then. It was very similar in size to an emperor penguin, which makes it the largest aquatic bird of the Mesozoic. It was a fishing bird and a very skilled swimmer but quite clumsy on land. It could barely move on land, as is the case with modern *Podiceps*, which also have their legs at the rear of the body. See more: *Nesov & Yarkov 1993*

Of the same size

Hesperornis sp.: From the eastern part of Laurasia (present-day western Russia). It was almost the same size as *Hesperornis rossicus* but more recent. It dates from the late Upper Cretaceous (lower Maastrichtian, 72.1–69 Ma). See more: *Yarkov & Nesov 2000*

Footprints

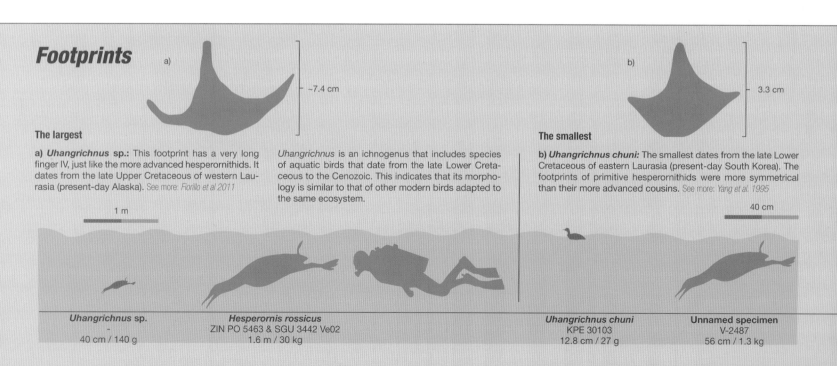

a) ~7.4 cm

b) 3.3 cm

The largest

a) *Uhangrichnus* sp.: This footprint has a very long finger IV, just like the more advanced hesperornithids. It dates from the late Upper Cretaceous of western Laurasia (present-day Alaska). See more: *Fiorillo et al 2011*

Uhangrichnus is an ichnogenus that includes species of aquatic birds that date from the late Lower Cretaceous to the Cenozoic. This indicates that its morphology is similar to that of other modern birds adapted to the same ecosystem.

The smallest

b) *Uhangrichnus chuni*: The smallest dates from the late Lower Cretaceous of eastern Laurasia (present-day South Korea). The footprints of primitive hesperornithids were more symmetrical than their more advanced cousins. See more: *Yang et al. 1995*

1 m

40 cm

Uhangrichnus sp.	*Hesperornis rossicus*	*Uhangrichnus chuni*	Unnamed specimen
-	ZIN PO 5463 & SGU 3442 Ve02	KPE 30103	V-2487
40 cm / 140 g	1.6 m / 30 kg	12.8 cm / 27 g	56 cm / 1.3 kg

250 cm   300 cm   350 cm

Characteristics and habits

Birds with medium or large heads, small and numerous teeth, atrophied arms, long necks, very short tails, hydrodynamic bodies, fingerless hands, and four fingers on each foot. Semi-aquatic or aquatic. **Diet:** piscivores.

Time range

From the late Lower Cretaceous to the late Upper Cretaceous (ca. 125–66 Ma), an approximate duration of 59 million years in the fossil record.

The smallest (Class: Very small Grade I)

V-2487, from the late Upper Cretaceous (Campanian, 83.6–72.1 Ma) of western Laurasia (present-day western Canada), is a small adult comparable in size to a ferret. It represents a new species yet to be named. It was almost three times shorter and twenty-three times lighter than *Hesperornis rossicus.* See more: *Tanaka et al. 2015*

The smallest juvenile specimens

There are several unexplained fragments of juvenile hesperornithids. Unfortunately, many are broken, and it is impossible to determine for now which is the smallest. The list of likely candidates is long:

• North America: *Pasquiaornis hardiei* SMNH P2077.125, *Canadaga arctica* NMC 41053 and *Hesperornis* cf. *regalis* from Canada YPM 5768 and KUVP 16112 of *Baptornis advenus* from Kansas, USA.
• Europe: Specimens BGS 87932, BGS 87936, and BMNH A483 of *Enaliornis barretti*; specimens BMNH A480, BMNH A480, SMC B55287, SMC B55289, SMC B55297, SMC B55301, YORYMG 581, and YORYMG 582 of *Enaliornis sedgwicki*; and specimens BMNH A481, BMNH A484, SMC B55290, SMC B55291, SMC B55292 and SMC B55293 of *Enaliornis seeleyi*. All are from the late Lower Cretaceous of England.
• Asia: A specimen from the ZIN PO collection of Kazakhstan and *"Hesperornis"* sp. Nov. from western Russia. The latter was 35 cm long and weighed **330 g.** See more: *Seeley 1876; Marsh 1877; Russell 1967; Walker 1967; Chauvire-Mourer 1991; Hou 1999, Kurochkin 2004*

The oldest

Enaliornis barretti, E. sedgwicki and *E. seeleyi*, from the late Lower Cretaceous (Albian, ca. 113–100.5 Ma) of England. An undescribed tooth of a probable hesperornithiform of Utah, USA, is older (Aptian, ca. 125–113 Ma). See more: *Seeley 1876; Kirkland et al. 1997; Cifelli et al. 1999*

The most recent

Brodavis americanus from western Canada, *Brodavis baileyi* from South Dakota, USA, *Potamornis skutchi* from Wyoming and Montana, USA, and cf. *Hesperornis* from Montana, USA, lived during the late stages of the late Upper Cretaceous (upper Maastrichtian, 69–66 Ma). See more: *Elzanowski et al. 2001; Hutchinson 2001; Martin et al. 2012*

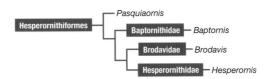

The strangest

A tooth with serrated edges from the late Lower Cretaceous of Utah, USA. Unusual in Mesozoic birds. See more: *Kirkland et al. 1997; Cifelli et al. 1999*

The first published species

Enaliornis barretti and *E. sedgwicki* were found in 1858 and reported in 1859. They have been known informally as "Paleocolyntus barretti" and "Pelagornis sedgwicki" since 1864, but they were not officially published until 1876. *Hesperornis* was published four years before them. See more: *Lyell 1859; Owen 1861; Seeley 1864, 1876; Marsh 1872; Mortimer**

The most recently published species

Fumicollis hoffmani (2015) ("Smoke Hill of Karen and Jim Hoffman"): From the late Upper Cretaceous of western Laurasia (present-day Kansas, USA). It was considered a specimen of *Baptornis advenus* for thirty-nine years. Then it was identified as a separate genus. See more: *Martin & Tate 1976; Bell & Chiappe 2015*

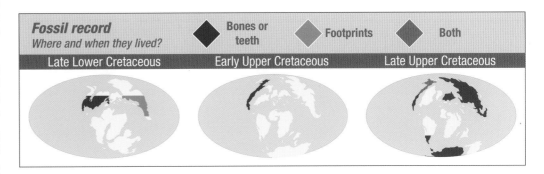

Fossil record — Where and when they lived?

Bones or teeth ◆   Footprints ◆   Both ◆

| Late Lower Cretaceous | Early Upper Cretaceous | Late Upper Cretaceous |

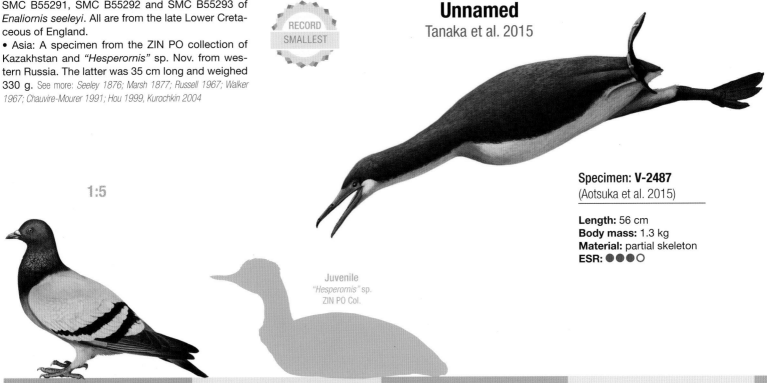

1:5

RECORD SMALLEST

Unnamed
Tanaka et al. 2015

Juvenile
"Hesperornis" sp.
ZIN PO Col.

Specimen: V-2487
(Aotsuka et al. 2015)

Length: 56 cm
Body mass: 1.3 kg
Material: partial skeleton
ESR: ●●●○

25 cm   50 cm   75 cm   100 cm

"Styginetta lofgreni"
Stidham 2001

RECORD
LARGEST

1:10

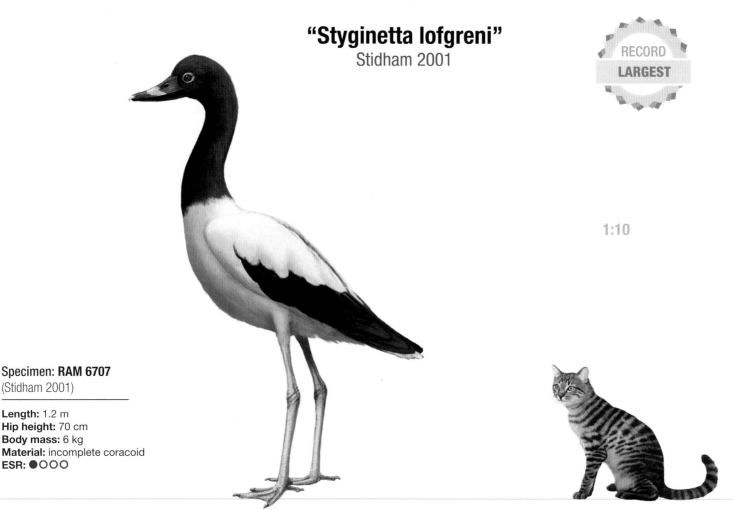

100 cm

50 cm

Specimen: RAM 6707
(Stidham 2001)
————————————
Length: 1.2 m
Hip height: 70 cm
Body mass: 6 kg
Material: incomplete coracoid
ESR: ●○○○

The largest (Class: Small Grade II)

RAM 6707 was a late Upper Cretaceous duck (Maastrichtian, 72.1–66 Ma) from western Laurasia (present-day Montana, USA). It belongs to the extinct presbyornithid family. It received the name of **"Stygenetta logfreni"** ("hell duck of Doctor Donald Lofgren") in a thesis. It has not yet been presented in any formal publication. It was taller than a stork and as heavy as a swan, making it the largest bird of the Mesozoic.
See more: *Stidham 2001, 2009*

Another likely contender

UCMP 53964 was another duck of similar size, measuring 1.15 m tall and weighing 5.2 kg. Initially, its remains were attributed to the rare *Cimolopteryx*, but it is now known that it was a late Upper Cretaceous duck from Wyoming, USA See more: *Brodkorb 1963*

Footprints

a)

8 cm

b)

2 cm

The largest

a) *Yacoraitichnus avi:* A footprint from the late Upper Cretaceous of western Gondwana (present-day Argentina). Its shape is similar to that of modern cariamas, only more robust. This type of bird was identified as being present at the end of the Mesozoic in Antarctica.
See more: *Alonso & Marquillas 1986; Case et al. 2006*

Modern birds or neornithes successfully diversified during the late Upper Cretaceous. The semi-aquatic footprints are known as *Dongyangornipes*, *Patagonichornis*, and *Sarjeantopodus*, while the long-legged birds are known as *Gruipeda*, "*Uhangrichnus*," and *Yacoraitichnus*.

The smallest

b) Unnamed: From the late Upper Cretaceous of western Laurasia (present-day Wyoming, USA). This is the smallest footprint of a neornithic bird we have found. See more: *Lockley & Rainforth 2002; Lockley, Nadon & Currie 2003*

20 cm

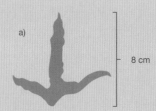

2 m

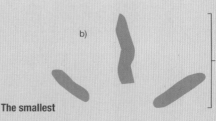

Yacoraitichnus avis	**"Styginetta lofgreni"**	**Unnamed footprint**	***Lamarqueavis australis***
Alonso & Marquillas 1986	RAM 6707 (V94078)	Morfotipo C	MML 207
56 cm / 2.9 kg	1.2 m / 6 kg	14 cm / 12 g	8 cm / 10 g

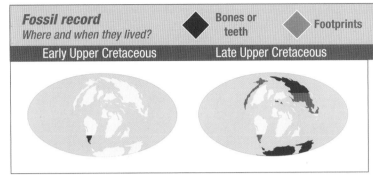

Characteristics and habits

Modern birds of the Mesozoic with large or medium heads, no teeth, large eyes, very long arms, long legs, very short tails, graceful bodies, and three fingers on each hand and four on each foot. Advanced feathers. Flying or terrestrial.
Diet: insectivores, filter feeders, granivores, and/or piscivores.

Time range

From the early Upper Cretaceous to the present, a duration of some 93.9 million years in the fossil record.

The smallest (Class: Tiny Grade I)

Lamarqueavis australis ("bird from southern Lamarque"): From the late Upper Cretaceous (Maastrichtian, 72.1–66 Ma) of western Gondwana (present-day Argentina). It may have been the smallest seabird of the Mesozoic; even a sparrow is twice as large. It was about 12.5 times shorter and 333 times lighter than "Stygenetta logfreni." Due to the fragmentary nature of almost all Mesozoic neornithic birds, it is difficult to know the ontogenetic stage of the specimens. See more: *Agnolin 2010*

The oldest

The specimen PVPH 237 (present-day Argentina) and another specimen of doubtful identification (present-day Uzbekistan) date from the early Upper Cretaceous (Turonian, ca. 93.9–89.8 Ma). Another fossil whose age is in doubt is *Gallornis straeleni* from France. It is estimated that it lived from the early Lower Cretaceous to the late Upper Cretaceous (Berriasian-Maastrichtian, ca. 145–66 Ma), the latter being more likely because neornithic birds do not seem to have been so ancient. Its location is in doubt, but it is possible that it belongs to the Hateg Basin Formation, which corresponds to the end of the Mesozoic. On the other hand, a genetic study suggests that neornithic birds appeared in South America. See more: *Lambrecht, 1931; Marshall 1960; Chiappe & Dyke 2002; Hope 2002; Agnolin et al. 2006; Claramunt & Cracraft 2015*

The most recent

Neornithic birds survived the great mass extinction that occurred at the end of the Mesozoic, extinguishing more than 10,000 living species. Their fossils are found between the limits of the line that separates the Cretaceous and the Paleocene. As such, they are frequently dated from both. Among the neornithic birds that are closest to this geological separation is *Volgavis marina* from European Russia, "Stygenetta logfreni" from Montana, and "Stygenetta sp." from Wyoming (all are from the Cretaceous), while *Anatalavis rex, Graculavus velox, Laornis edvarsianus, Novacaesareala hungerfordi, Palaeotringa littoralis, P. vagans, P. vetus, Telmatornis priscus, Tytthostonyx glauconiticus* and one cormorant ANSP 15713 from New Jersey are from the Paleocene. See more: *Olson & Parris 1987; Nesov & Jarkov 1989; Parris & Hope 2002; Stidham 2009*

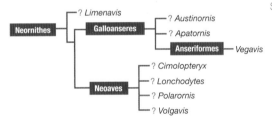

The strangest

These birds are very familiar to us, because they were similar to modern birds, but many were incomplete and many were interpreted based on anatomical similarities that could be wrong. A femur identified as a probable cariamid would demonstrate that cariamiformes (a group that contains the famous "birds of terror") were already present in the Mesozoic, but there is doubt due to its fragmentary nature. See more: *Case et al. 2006*

The first published species

Apatornis celer (1873) (before *Ichthyornis celer*) ("disappointing and rapid bird"): From the late Upper Cretaceous of western Laurasia (present-day Kansas, USA). It lived at the same time as *Ichthyornis* but was more derived. See more: *Marsh 1873*

The most recently published species

Lamarqueavis australis (2010) ("southern bird from Lamarque"): From the late Upper Cretaceous (present-day Argentina). In the same work in which it is described, the names *Cimolopteryx minima* and *C. petra* were changed to the genus *Lamarqueavis*. See more: *Agnolin 2010*

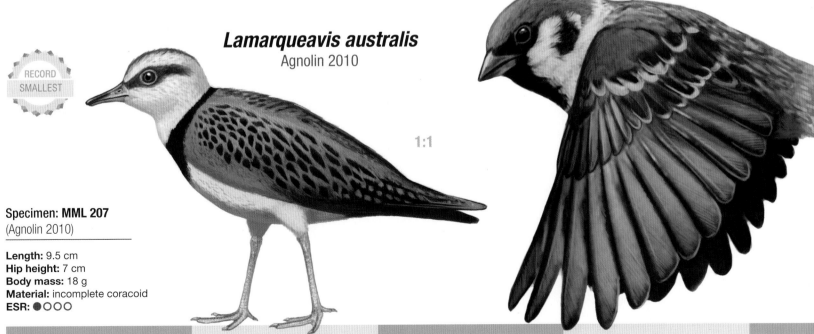

Lamarqueavis australis
Agnolin 2010

RECORD SMALLEST

1:1

Specimen: MML 207
(Agnolin 2010)

Length: 9.5 cm
Hip height: 7 cm
Body mass: 18 g
Material: incomplete coracoid
ESR: ●○○○

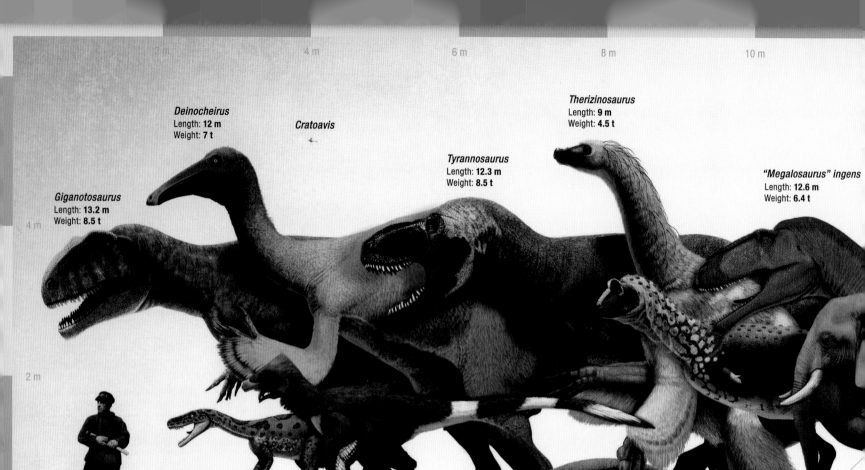

Deinocheirus
Length: **12 m**
Weight: **7 t**

Cratoavis

Therizinosaurus
Length: **9 m**
Weight: **4.5 t**

Tyrannosaurus
Length: **12.3 m**
Weight: **8.5 t**

"Megalosaurus" ingens
Length: **12.6 m**
Weight: **6.4 t**

Giganotosaurus
Length: **13.2 m**
Weight: **8.5 t**

4 m

2 m

1:50

cf. *Soroavisaurus*
Length: **80 cm**
Weight: **7.25 kg**

Frenguellisaurus
Length: **5.3 m**
Weight: **360 kg**

Giant dromaeosaurid
Length: **5.8 m**
Weight: **530 kg**

Giant silesaurid
Length: **3 m**
Weight: **85 kg**

Carnotaurus
Length: **7.7 m**
Weight: **1.85 t**

Spinosaurus
Length: **16 m**
Weight: **7.5 t**

Hesperornis sp.
Length: **56 cm**
Weight: **1.3 kg**

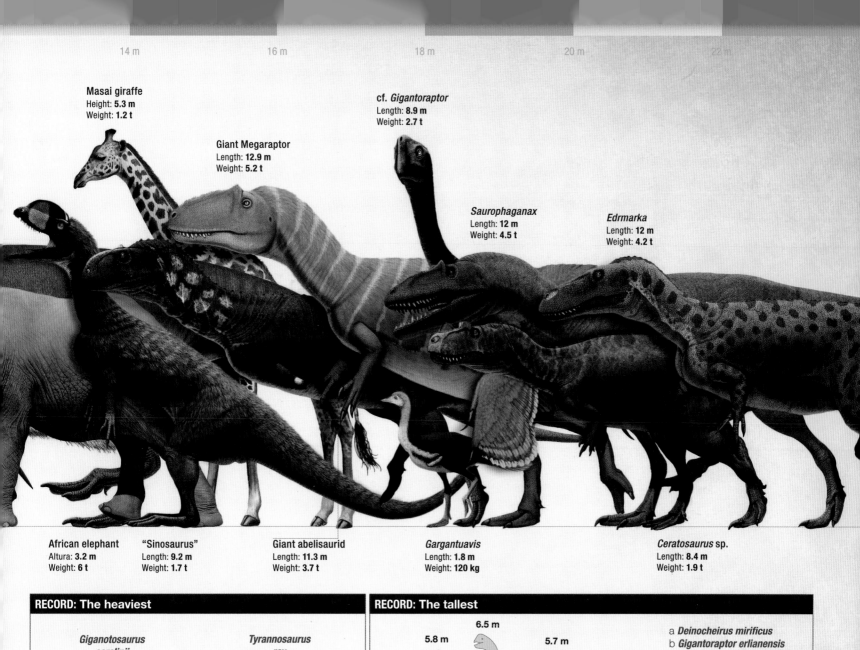

Masai giraffe
Height: **5.3 m**
Weight: **1.2 t**

Giant Megaraptor
Length: **12.9 m**
Weight: **5.2 t**

cf. *Gigantoraptor*
Length: **8.9 m**
Weight: **2.7 t**

Saurophaganax
Length: **12 m**
Weight: **4.5 t**

Edrmarka
Length: **12 m**
Weight: **4.2 t**

African elephant
Altura: **3.2 m**
Weight: **6 t**

"Sinosaurus"
Length: **9.2 m**
Weight: **1.7 t**

Giant abelisaurid
Length: **11.3 m**
Weight: **3.7 t**

Gargantuavis
Length: **1.8 m**
Weight: **120 kg**

Ceratosaurus sp.
Length: **8.4 m**
Weight: **1.9 t**

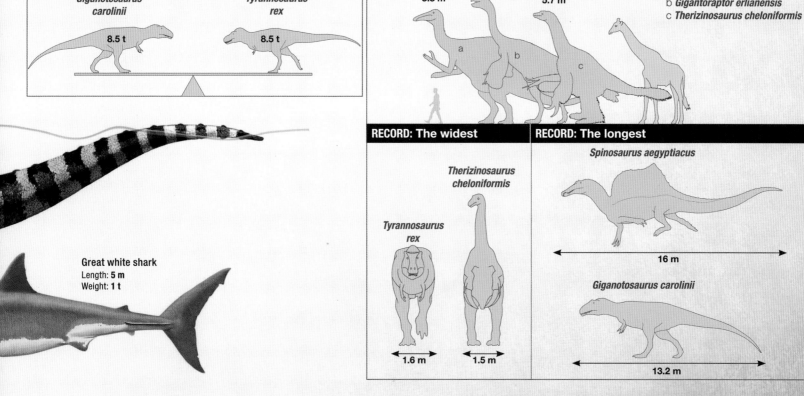

RECORD: The heaviest

Giganotosaurus carolinii

Tyrannosaurus rex

8.5 t   8.5 t

RECORD: The tallest

6.5 m

5.8 m

5.7 m

a *Deinocheirus mirificus*
b *Gigantoraptor erlianensis*
c *Therizinosaurus cheloniformis*

RECORD: The widest

Therizinosaurus cheloniformis

Tyrannosaurus rex

1.6 m   1.5 m

RECORD: The longest

Spinosaurus aegyptiacus

16 m

Giganotosaurus carolinii

13.2 m

Great white shark
Length: **5 m**
Weight: **1 t**

20 cm   30 cm   40 cm

Bee hummingbird
Length: **5 cm**
Weight: **2 g**

Cratoavis
Length: **6.6 cm**
Weight: **4 g**

Compsognathus (BSP AS I 536)
Length: **90 cm**
Weight: **550 g**
Historically considered the
smallest dinosaur

Aurornis
Length: **40 cm**
Weight: **260 g**

30 cm

Saltopus
(dinosauromorph)
Length: **50 cm**
Weight: **110 g**

Parvicursor
Length: **50 cm**
Weight: **185 g**

20 cm

10 cm

1:2

Scleromochlus
(avemetatarsalia)
Length: **18.5 cm**
Weight: **18 g**

Iberomesornis
Length: **8.7 cm**
Weight: **9.5 g**

Gobipipus (neonate)
Length: **4.7 cm**
Weight: **2 g**

Lamarqueavis australis
Length: **9.5 cm**
Weight: **18 g**

Changchengornis
Length: **17.5 cm**
Weight: **60 g**

Unnamed hesperorniform
Length: **56 cm**
Weight: **1.3 kg**

50 cm   60 cm   70 cm   80 cm   90 cm

Common sparrow
Length: **10 cm** (15 cm with feathers)
Weight: **30 g**

Epidexipteryx
Length: **25 cm**
Weight: **220 g**

Mystiornis
Length: **24.5 cm**
Weight: **165 g**

Shanweiniao
Length: **9 cm**
Weight: **10.5 g**

Hongshanornis
Length: **12 cm**
Weight: **58 g**

Common pigeon
Length: **25 cm** (35 cm with feathers)
Weight: **300 g**

Parvavis
Length: **7.5 cm**
Weight: **6 g**

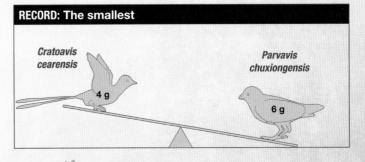

RECORD: The smallest

Cratoavis cearensis

Parvavis chuxiongensis

4 g

6 g

X 2,100,000
=

It would take a swarm of 2.1 million *Cratoavis* to equal the weight of *Tyrannosaurus*.

1686—England—The important bases of the current classification are created
John Ray publishes large volumes containing the description and classification of more than 18,000 species of plants, laying the foundation for the current classification. This work was later referenced in the classification of animals by Carl Linnaeus. See more: *Ray 1686*

1735—Netherlands—The work that gave rise to the current taxonomy
Carl Linnaeus writes his *Systema Naturae*, the work on which the classification of the species will be based. See more: *Linnaeus 1735*

1751—Sweden—The binomial nomenclature scheme is established
Carl Linnaeus establishes the modern foundations of binomial taxonomy. See more: *Linnaeus 1751*

1758—Sweden—The term *bird* is coined
Carl Linnaeus begins the official binomial nomenclature in the tenth edition of his book. He coined the word *bird* based on the Latin word *avis* (bird), along with the description of birds' external anatomy. See more: *Linnaeus 1758*

1768—Austria—The term *reptile* is coined
Josephus Nicolaus Laurenti creates the term *reptile* based on the Latin word *reptilium* (crawling). It is a clade that hosted the dinosaurs for many years. See more: *Laurenti 1768*

1800—France—The first dinosaur chimera
Two fossil collections were combined by mistake: the crocodile *Steneosaurus rostrominor* found by Abbot Bachelet (1770) and the theropod *Streptospondylus altdorfensis* found by the abbot Jean-François Dicquemare (1776). These two species were identified and reclassified in 1825 and 1832. See more: *Dicquemare 1776; Cuvier 1800; Geoffroy 1825; Meyer 1832*

1808—France—The first scientifically described theropod
Georges Cuvier describes and illustrates the vertebrae of *Streptospondylus*, which was thought to belong to a crocodile at that time. In 1825, they were divided into two species, and in 1832 one was identified as theropod remains. See more: *Meyer 1832; Geoffroy 1825*

1813—Switzerland—The term *taxonomy* is coined
Augustin Pyrame de Candolle comes up with this term based on the Greek words taxis (ordination) and *logos* (treaty or study) and the suffix *ia* (action). It is the discipline of biology that classifies living or fossil organisms. See more: *Candolle 1813*

1835—England—The first article on natural variation
This publication proved to be the forerunner of the theory of natural selection. See more: *Blyth 1835*

1837—USA—The theropod footprint with the most confusing species name
Anchisauripus minusculus ("tiny Anchisaurus footprint," formerly *Ornithichnites minusculus*). They are actually not *Anchisaurus* sauropodomorph footprints. On the other hand, it is the largest species of *Anchisauripus*, as the first specimens discovered were very small. See more: *Hitchcock 1837; Lull 1904*

1842—England—The term *Dinosauria* is coined
Richard Owen coined this term based on the Greek words deinos (terrible) and sauros (lizard or reptile). At that time, this group included only *Hylaeosaurus*, *Iguanodon*, and *Megalosaurus*. *Cardiodon*, *Cetiosaurus*, *Plateosaurus*, *Streptospondylus*, *Suchosaurus*, and *Thecodontosaurus* were already genera and were later recognized as part of the group. See more: *Owen 1842*

1843—Austria—The first taxonomic group for carnivorous dinosaurs
The first group created to contain *Megalosaurus* was *"Megalosauri."* As it was created in 1843 without considering the taxon Dinosauria, it is now an ancient synonym of Theropoda. See more: *Fitzinger 1843*

1853—Austria—The first theropod family
The term Megalosauridae (megalosaurids) appears in 1853, although it is usually assigned to Thomas Henry Huxley in 1869. See more: *Gervais 1853; Huxley 1869; Olshevsky 1991*

1859—England—The first theropod and ornithischian chimera and the one that lasted the longest without being correctly identified
Scelidosaurus harrisonii was an ornithischian dinosaur that was mixed with the leg bones of an unknown theropod that was recognized 109 years later. The carnivore was informally named "Merosaurus newmani." See more: *Owen 1859; Wells, Powell & Pickering 1994*

1871—England—The first family of birds from the Jurassic
Archaeopterygidae is a family that represents *Archeopteryx* and its close relatives. It is currently an inactive clade. See more: *Huxley 1871*

1872—USA—The first family of birds from the Cretaceous
The Hesperornithes family was created from *Hesperornis regalis*, an aquatic bird from the Upper Cretaceous of Kansas. See more: *Marsh 1872*

1875—England—The first sauropods to be confused with theropods
It was even suggested that the thoracic vertebrae of the sauropods *Bothriospondylus* and *Marmarospondylus* could be huge theropods. See more: *Owen 1875*

1881—USA—The term *Theropoda* is coined
The theropod group is still referred to today. Other previous attempts that were not successful were Megalosauri, Goniopoda, Harpagosauria, Therophagi, and Carnosauriformes. See more: *Fitzinger 1843; Cope 1866; Haeckel 1866; Marsh 1881; Jaekel 1914; Cooper 1985; Mortimer**

1883—Russia—The first theropod and sauropod chimera
Poekilopleuron schmidti, located in the Kursk Oblast, is a mixture of theropod bones with a metacarpal of a titanosauriform and other indeterminate fragments. See more: *Kiprijanow 1883; Mortimer**

1887—England—The term *Saurischia* is coined
Dinosaurs were divided into two large groups that are still used: ornithischia and saurischia. See more: *Seeley 1887*

1916—England—The term *Sauropsida* is coined
Sauropsida are a large group that currently includes tetrapod amniotes that aren't synapsids (mammals and their close relatives). It was a term that regained strength after the disappearance of the group "Reptilia," because the latter was a paraphyletic group. See more: *Goodrich 1916*

1923—India—The most recent theropod chimera
The dinosaur *Lametasaurus indicus* is based on sauropod osteoderms, crocodile teeth, and bones of the theropod *Rajasaurus narmadensis*. It dates from the late Upper Cretaceous (upper Maastrichtian, ca. 69–66 Ma). See more: *Matley 1923*

1974—USA—Dinosauria is separated from "reptiles"
The separation of dinosaurs from "reptiles" was proposed because the physiology of dinosaurs is more similar to that of mammals and birds, and dinosaurs have unique anatomical features. See more: *Bakker & Galton 1974*

1975—Argentina—The first family of dinosauromorphs
It was Lagosuchidae, although it is currently inactive. See more: *Paul 1975*

1985—England—The group "Dinosauromorpha" is created
It includes the dinosaurs and their closest relatives that have a reduction in the length of metatarsals I and V. See more: *Benton & Norman 1985*

1986—England—The group Ornithodira is created
They constitute a clade that groups both pterosaurs and dinosaurs. It is now almost replaced by Avemetatarsalia. See more: *Benton 1999; Gauthier 1986*

1988—USA—*Archeopteryx* is no longer considered a bird
It is proposed that the Archaeopterygidae family also include dromaeosaurs and troodontids. At the moment, controversy exists about if *Archeopteryx* is part of the line toward the origin of the birds or is just one of many branches. See more: *Paul 1988*

1991—Texas, USA—The most mixed theropod chimera
Protoavis texensis was reconstructed from several animal fossils that were intermixed. All were considered to be several specimens of the same species. The pieces had to be analyzed to identify each of them. There were juvenile cellophoids, a basal pterosauromorph, a megalancosaur (*Dolabrosaurus*), a lepidosauromorph (*Clevosaurus*), and other undetermined archosauriforms. See more: *Chatterjee 1987; Chatterjee 1991; Chiappe 1995; Wilford 1986; Ostrom 1987; Sereno 1997; Witmer 2001; Renesto 2000; Atanassov 2002; Harris & Downs 2002; Nesbitt et al. 2005*

1999—England—The Avemetatarsalia clade is created
The Avemetatarsalia group includes the pterosaurs and dinosaurs, in addition to their direct ancestors that have long and compressed metatarsals. See more: *Benton 1999*

2003—USA—Theropods with the shortest nicknames
Two specimens of Tyrannosaurus rex have short aliases: "Sue" for FMNH PR2081 and "007" for BHI 6219. See more: *Brochu 2003; Larson 2008*

2009—The last taxonomic group of dinosauromorphs
As of this edition, "Silesauridae" is the last clade created for the dinosauromorphs. See more: *Wang et al. 2014*

2014—The last taxonomic group of Mesozoic birds
Azu wyliei is known as "Chicken from hell." See more: *Lamanna et al. 2014*

2014—The last taxonomic group of Mesozoic birds
As of this edition, "Pengornithidae" is the last avian theropod clade. See more: *Wang et al. 2014*

2014—France—The last taxonomic theropod group
As of this edition, "Majungasaurinae" is the last clade created for the theropods. See more: *Tortosa et al. 2014*

2015—USA—The theropod with the most nicknames
Several specimens of *Tyrannosaurus rex* have received nicknames. Among them are "007"; "B-rex"; "Belle"; "Black Beauty"; "Bloody Mary"; "Bob"; "Bucky"; "C-rex"; "Celeste"; "County rex"; "Cupcake"; "Duffy"; "Fox"; "E.D. Cope"; "F-rex"; "Frank"; "G-rex"; "H-rex"; "Hager rex"; "J-rex"; "J-rex2"; "Jane"; "Jen-rex"; "King Kong"; "L-rex"; "Monty"; "Mr. Zed"; "Mud Butte T. rex"; "N-rex"; "Nathan"; "Peck's rex"; "Pete"; "Petey"; "Regina"; "Rex B"; "Rex C"; "Rigby specimen"; "Samson"; "Scotty"; "Stan"; "Steven"; "Sue"; "Thomas"; "Wankel rex"; "Wayne"; "Wyrex"; "Z-rex." See more: *Mortimer**

THE NOMENCLATURE OF THEROPOD DINOSAURS AND MESOZOIC BIRDS

1763—England—The first binomial name for a theropod
Richard Brookes creates the name *Scrotum humanum* from the specimen described by Robert Plott for descriptive but noninterpretive purposes. It isn't a valid name because it was before Linnaean classification. See more: *Brookes 1763*

1822—England—The first creator of a theropod name
William Daniel Conybeare was the one who named "*Megalosaurus*," based on the Greek words megas (big) and sauros (lizard or reptile). See more: *Parkinson 1822; Buckland 1824*

1822—England—The first published theropod genus
James Parkinson publishes the name *Megalosaurus*, estimating its length to be 12 m and its height 2.4 m. See more: *Parkinson 1822; Spalding & Sarjeant 2012*

1825—France—The first binomial name of a theropod
Back then, *Steneosaurus rostromajor* was considered a fossil crocodile. In 1932, it was made synonymous with the megalosauroid *Streptospondylus altdorfensis*. See more: *Meyer 1832*

1826—England—The first theropod species named in someone's honor
"*Megalosaurus conybearei*" was informally named after the geologist and paleontologist William Daniel Conybeare, so it was not valid. A year later, the combination *Megalosaurus bucklandii* was created to honor the naturalist, geologist, and paleontologist William Buckland See more: *Ritgen 1826; Mantell 1827*

1826—England—The first writing error that refers to a theropod
"*Melagosaurus*" is the first known typographical error. He meant *Megalosaurus*. See more: *Buckland 1824; Ritgen 1826*

1827—England—The first species name officially assigned to a theropod
Megalosaurus bucklandii honors the naturalist, geologist, and paleontologist William Buckland. See more: *Mantell 1827*

1832—England—The first attempt to correct a name of a theropod
Christian Erich Hermann von Meyer uses the name *M. bucklandi* instead of *M. bucklandii*, which was accepted by numerous authors. However, the use of double *ii*'s when it is dedicated to an author is valid according to the rules of nomenclature, so the attempt to change was not successful. *See more: Meyer 1832*

1832—France—The first theropod named after a place
The name *Streptospondylus altdorfensis* is dedicated to the village of Altdorf in Switzerland. See more: *Meyer 1832*

1834—England—The first pterosaur mistaken for a Mesozoic bird
Due to the great similarity between the broken bones of *Palaeornis clifti* and the pneumatic bones of modern birds, they were identified as being of a large bird. Since the name already existed for a fossil bird, it became a species of Ornithocheirus. Subsequently, this happened again with the remains of other pterosaurs: "*Osteornis (Lithosteornis) ardeaceus*," *O. diomedeus*, *O. scolopacinus*, *Cimoliornis diomedeus*, *Cretornis hlavaci*, *Laopteryx prisca (priscus)* and *Samrukia nessovi*. See more: *Vigors 1825; Mantell 1835; Gervais 1844; Owen 1846; Fritch 1880; Marsh 1881; Brodkorb 1971, 1978; Naish et al. 2012*

1835—England—The first fossil species confused with a Mesozoic bird named after someone
Gideon Algernon Mantell named the fragmentary remains of what he thought was a great bird *Palaeornis clifti* in honor of William Clift. It turned out to be the first large pterosaur that had been found. See more: *Mantell 1835; Martill 2010*

1837—France—The first name reused for a theropod
Poekilopleuron bucklandii is dedicated to William Buckland, as happened with *Megalosaurus*. At that time, it was thought to be a marine species able to rest on the beach. "Poecilopleuron bucklandi" was included in an unofficial publication one year earlier. See more: *Bronn 1837; Eudes-Deslongchamps 1838*

1838—France—The theropod name with the most misspellings
Poekilopleuron bucklandii has the most incorrect names. One year before its formal description, it was mentioned as "*Poecilopleuron bucklandi.*" In several publications, it is referred to as:"Peukilopleuron," "Pikilopleuron," "Poecilopleuron," "Poeiclopleuron," "Poecilopleurum," "Poekilipleuron," "Poekilopleuron," "Poelicopleurum," or "Poicllopleuron." See more: *Bronn 1837; Eudes-Deslongchamps 1838; Owen 1841; Fitzinger 1843; Agassiz 1846; Leidy 1870; Stromer 1934; Kuhn 1965; Olshevsky 1991*

1841—USA—The first species name from a footprint attributed to a theropod
Edward Hitchcock names *Ornithoidichnites deanii* (now *Platypterna deanianus*) in honor of James Dean, who was interested in his work and corresponded with him about it. Currently, it is not considered a theropod. See more: *Hitchcock 1841; Miller 2007*

1843—USA—The first species name in a theropod footprint
Edward Hitchcock dedicates the species *Ornithoidichnites sillimani* (now *Anchisauripus* or *Eubrontes sillimani*) to the science teacher Benjamin Silliman, for his contribution on the footprints found in the Connecticut River valley, Massachusetts. See more: *Hitchcock 1843; Miller 2007*

1847—Russia—The genus *"Dinosaurus"* is created
The genus *Dinosaurus* was not used for a dinosaur, but for a therapsid. Currently, it is not a valid name; it is synonymous with *Brithopus priscus*. See more: *Fischer 1847*

1857—USA—The first primitive archosaur mistaken for a bird
A sacrum of *Palaeonornis struthionoides* from North Carolina was identified as a large Triassic bird but belongs to a phytosaur synonymous with *Rutiodon*. See more: *Emmons 1857*

1862—Germany—The first Mesozoic bird with a second name
The specimen type of *Archeopteryx lithographica* ("ancient wing carved in stone") was renamed *Griphosaurus problematicus* ("problematic enigmatic lizard") and *Griphornis longicaudatus* ("long-tailed enigmatic bird"). This was because some authors did not accept that it was a bird. See more: *Meyer 1861; Wagner 1862; Woodward 1862*

1862—Germany—The first forgotten name of a Mesozoic bird
Archeopteryx macrura was a proposed name for *Archeopteryx lithographica*, but, ultimately, it did not stick. Currently, the name is protected. See more: *Owen 1862*

1863—England—The first corrected name of a Mesozoic bird
The change of the species *Archeopteryx macrurus* to *A. macrura* was proposed. This correction is no longer necessary, since it is a synonym of *A. lithographica*. See more: *Owen 1863a, 1863b*

1864—England—The first Mesozoic bird named in a thing's honor
"*Pelagornis sedgwicki*" (now *Enaliornis*) was named after the Sedgwick Museum collection. It was formalized twelve years later. See more: *Seeley 1864, 1876; Galton & Martin 2002*

1865—USA—The theropod footprint name most often used
Leptonyx changed to *Stenonyx* because the name was already occupied by a seal (currently called *Leptonychotes*) twenty-eight years before. On the other hand, this name has several homonyms since it was used for the birds *Coryphaspiza* and *Pteroptochos*, the mammal *Aonyx*, the beetles *Theone* and *Leptosonyx*, as well as the snail *Homalopoma*. See more: *Swainson 1832; Gray 1837; Swainson 1837; Lesson 1842; Carpenter 1864; Hitchcock 1865; Gill 1872; Weise 1886; Jacobson 1895; Lull 1904; Stephen & Groves 2015*

1865—USA—The first name of a theropod that was previously used
Coelosaurus is a genus created in 1865 for an American theropod. The name already existed eleven years earlier, but as it is currently forgotten, it is not necessary to replace the second name. See more: *Owen 1854; Leidy 1865; Spamer & Daeschler 1995*

1866—USA—The first name of a theropod that was previously used
Dryptosaurus aquilunguis was first known as *Laelaps aquilunguis*. The name was changed in 1877, because a mite had been named that way since 1836. See more: *Koch 1836; Cope 1866; Marsh 1877*

1866—USA—The first North American theropod named after a mythological being
Dryptosaurus was originally named *Laelaps*, which means "storm wind" in Greek. In the Greek mythology Laelaps or Lelaps was a dog that never failed to catch its prey and was turned into stone by Zeus. See more: *Cope 1866*

1869—USA—The first incorrect combination when quoting a theropod name
Compsognathus longipes has been mistakenly referred to as "*Compsognathus gracilis*." See more: *Cope 1969*

1876—USA, France—The first theropod name correction
The paleontologist Henri Émile Sauvage retired the dieresis in *Tröodon*, so now it is *Troodon*. See more: *Leidy 1856; Sauvage 1876*

1882—Tajikistan—The first name of a theropod footprint to refer to a locality
Eubrontes tianschanicum refers to the Tian (Tian Shan) mountains. Later, the name was changed to *Gabirutosaurus* and later to *Gabirutosaurichnus*. See more: *Romanovsky 1882; Gabunia & Kurbatov 1988*

1892—USA—The first forgotten name of a theropod
Manospondylus gigas was the first name given to the first incomplete remains of *Tyrannosaurus rex*. Because it was not used or recognized for so long, and because the name *Tyrannosaurus* is protected, *Manospondylus gigas* was forgotten. See more: *Cope 1962; Osborn 1905*

1902—USA—The first name correction of a Mesozoic bird
Cimolopteryx rarus was changed to *C. rara* by the paleontologist Oliver Perry Hay. See more: *Marsh 1892; Hay 1902*

1905—USA—The theropod with the most immediate synonym
Tyrannosaurus and *Dynamosaurus* were described in the same work. (*Tyrannosaurus* appeared only one page before *Dynamosaurus*.) Today, the first name is synonymous with the second. Both were considered different species because *Dynamosaurus* had minor differences and was assigned the osteoderms of ankylosaurs. See more: *Osborn 1905*

1906—England—First hypothetical theropod name
The name "*Proavis*" is created, a hypothetical form of the probable ancestor of birds. See more: *Pycraft 1906*

1907—England—First and only primitive avemetatarsalian
Scleromochlus taylori was a close relative of the ancestors that gave rise to pterosaurs and dinosaurs. See more: *Woodward 1907*

1907—Scotland—The first basal avemetatarsalian species named after a person
Scleromochlus taylori is named after William Taylor, who discovered it. See more: *Woodward 1907*

1910—Scotland—The first dinosauromorph species name to refer to a place
The name *Saltopus elginensis* honors the city of Elgin. See more: *Huene 1910*

1912—USA—The first wrong combination when citing a name of a Mesozoic bird
Cimolopteryx rara has one of the most complicated mistakes to identify: "Timolopteryx ogiluie". See more: *Marsh 1892; Grant 1912*

1912—Germany—The first name of a discarded theropod
"Hallopus celerrimus" was the first name proposed for *Procompsognathus triassicus*. See more: *Fraas 1912, 1913*

1913—Germany—The first theropod species name to refer to the Triassic
Procompsognathus triassicus was long considered one of the oldest dinosaurs. See more: *Fraas 1913*

1917—USA—The theropod with the most confusing name
Ornitholestes ("bird thief"). Its arms were more suitable for capturing small mammals than for catching birds. See more: *Osborn 1917; Bakker 1981*

1922—Germany—The first punctuation error when citing a name of Mesozoic bird
For language reasons, *Archaeopteryx* was written as "Archaöpteryx." See more: *Meyer 1861; Steiner 1922*

1924—Mongolia—The theropod with the most confusing genus and species name
The name *Oviraptor philoceratops* ("ceratopsid egg thief") was given because the specimen was found near a presumed *Protoceratops* nest. It was actually his; he may have been taking care of the eggs. Anyway, the possibility that he may have consumed the eggs of other dinosaurs has not been ruled out. See more: *Osborn 1924*

1924—Mongolia—The first typographical error in a theropod name in its first publication
Prodeinodon appears with the names *P. mongoliense* and *P. mongoliensis* in its first publication, so some sources choose one of the two alternatives as the official one. See more: *Osborn 1924*

1931—France—The first Mesozoic bird named after a people
Gallornis for the Gauls, the people who inhabited what is now Germany, Belgium, Spain, France, Holland, and Switzerland. See more: *Lambrecht 1931*

1932—Canada—The first theropod named after a country
The species name *Macrophalangia canadensis* refers to Canada. Perhaps it is a synonym of *Chirostenotes pergracilis*. It was preceded by one year by *Brazileosaurus pachecoi*, which refers to Brazil but is now known to be a crocodile. "Laevisuchus indicus," "Indosaurus matleyi," and "Indosuchus raptorius" were also described in 1931 but were not formalized until 1933. The name *Megalosaurus hungaricus* existed since 1902 but refers to the Hungarian area, not to Hungary, since it comes from Romania. See more: *Nopcsa 1902; Huene 1931, 1932; Sternberg 1932*

1933—China—The first theropod species name to refer to the Asian continent
Archaeornithomimus asiaticus (originally *Ornithomimus*) did not have a holotype specimen for 57 years. See more: *Gilmore 1933; Russell 1972; Smith & Galton 1990*

1940—Canada—The first non-avian theropod mistaken for a bird
A toothless and pointed lower jaw, similar to that of present-day

birds, was named *Caenagnathus collinsi*. Other theropods that were initially described as birds were *Bradycneme, C. sternbergi, Heptasteornis, Jinfengopteryx, Paleopteryx, Protarchaeopteryx, Protoavis* (in part) and *Sinosauropteryx*. See more: *Stenberg 1940*

1942—Brazil—The oldest fossil dinosaur chimera
Some paleontologists suggest that *Spondylosoma absconditum* is based on the bones of a basal saurischian mixed with that of an archosauromorph rauisuchid. It dates from the Middle Triassic (Ladinian, ca. 242–237 Ma). See more: *Huene 1942; Langer 2004*

1947—USA—The first theropod from a thesis
"Acrocanthus atokaensis" was published informally three years before the official description of *Acrocanthosaurus atokensis*. See more: *Langston 1947; Stovall & Langston 1950; Czaplewski, Cifelli & Langston 1994*

1948—China—The theropod with the most confusing species name
When it was discovered, *Sinosaurus triassicus* ("Chinese lizard of the Triassic") was considered to belong to the Triassic. It is now known that it dates from the Lower Jurassic. See more: *Young 1948*

1952—Germany—The dinosauromorphic footprint with the name with the simplest meaning
Dinosauripus means "dinosaur footprint." This name did not receive a species name. Thirty-one years later, it was changed to *Coelurosaurichnus arntzeniusi* by the same author. See more: *Rehnelt 1952, 1983*

1956—Algeria—The first theropod species name to refer to the African continent
"Megalosaurus africanus" arises from an error made by the paleontologist Friedrich von Huene in listing *Megalosaurus saharicus*. As a result, the official record belongs to *Afrovenator* ("African hunter"). See more: *Deperet & Savornin 1925; Huene 1956; Sereno et al. 1994*

1956—Georgia—The first error when referring to a theropod footprint
The author who named the species *Satapliasaurus* wrote "*Sathaplisaurus*" in a later reference. See more: *Gabunia 1951, 1956*

1963—Argentina—The first primitive saurischian named after a place
The name *Herrerasaurus ischigualastensis* refers to the provincial park of Ischigualasto. It is preceded by "*Zanclodon*" *silesiacus*, which refers to the region of Silesia, an area that currently occupies part of Germany, Poland, and the Czech Republic. It is not known if it was a very old dinosaur or an archosaur. See more: *Jaekel 1910; Reig 1963; Carrano et al. 2012*

1969—South Korea—The first Mesozoic bird named after a country
Koreanaornis refers to South Korea. See more: *Kim 1969*

1969—Kazakhstan—The first theropod name to refer to the Cretaceous
Cretaaviculus sarysuensis is an asymmetrical feather that was initially attributed to a bird, but it is now known that it could also belong to a paravian theropod. See more: *Bazhanov 1969; Nesov 1992*

1971—Arizona, USA—The first theropod footprint dedicated to a people
Kayentapus hopii is based on the ancient Hopis, as their clothes include these footprints as adornment. Another is *Hopiichnus shingi* (now *Anomoepus*), which was considered a theropod but now is known to be from an ornithischian dinosaur. See more: *Welles 1971*

1972—Argentina—The oldest fossil chimera dinosauromorph
Lewisuchus admixtus was originally described from mixed remains of a proterochampsid archosauriform and the bones of the specimen. Thanks to a review, *Lewisuchus* is currently considered a

primitive dinosauromorph close to *Marasuchus*. It dates from the Middle Triassic (Ladinian, ca. 242–237 Ma). See more: *Romer 1972; Nesbitt et al. 2010*

1973—USA—The first reference error in a family of Mesozoic birds
Baptornis advenus was mistakenly referred to as "Bathornis veredus." This fact caused another author to cite the Baptornitidae family as "Bathornithidae." See more: *Marsh 1877; Walker 1967; Adolphson 1973*

1973—Tanzania—The first misspelling of a dinosauromorph name
"Nyasaurus" is a mistaken writing of "Nyasasaurus." See more: *Charig 1967; White 1973; Nesbitt et al. 2013*

1974—Mongolia—The first Mesozoic bird named after a place
Gobipteryx alludes to the Gobi Desert, which lies between northern China and southern Mongolia. *Cretaaviculus sarysuensis* from Kazakhstan and *Wyleyia valdensis* from England preceded it by five years and one year, respectively, but the first is an asymmetrical feather that may have belonged to an avialan theropod, and the second is an incomplete humerus whose identification is uncertain. It may be an enantiornithean bird or a juvenile theropod. See more: *Bazhanov 1969; Harrison & Walker 1973; Elzanowski 1974*

1974—Argentina—The first pterosaur mistaken for a theropod
The incomplete remains of *Herbstosaurus pigmaeus* were interpreted as the remains of a small carnivorous dinosaur. See more: *Casamiquela 1974*

1975—Romania—The first European theropod named after a fictitious person
Bradycneme draculae is based on the name Dracula, from the homonymous novel by Bram Stoker. See more: *Stoker 1897; Harrison & Walker 1975*

1975—China—The first theropod fossil egg named after a place
Nanshiungoolithus refers to the city of Nanjing. A name that precedes it is *Oolithes nanshiungensis*, but some authors doubt that it belongs to a theropod. Another name that also preceded it was *Macroelongathoolithus carlylensis* (formerly known as *Boletuoolithus*), but as it referred to a person and not to a locality, it was corrected to *carlylei*. See more: *Young 1965; Jensen 1970; Zhao 1975; Hanson 2006*

1975—China—The theropod egg with the most similar genus and species name
An *Elongatoolithus elongatus* ("elongated egg petrified, elongated") was known as *Oolithes elongatus* twenty-one years earlier. See more: *Young 1954; Zhao 1975*

1976—USA—The first theropod name with a festive motif
The name *Marshosaurus bicentesimus* refers to the bicentennial of the independence of the United States, which took place on Sunday, July 4, 1976. See more: *Madsen 1976*

1977—Germany—The first protected name of a Mesozoic bird
Archaeopteryx lithographica had to be protected because it was not the first name created for the species; it was preceded by *Pterodactylus crassipes*. Due to its great impact on history and the incompleteness of the previous materials, however, the name is protected against any eventuality. See more: *Meyer 1857; Meyer 1861, ICZN 1977*

1978—France—The theropod name that has changed the most times for the same reason
The name *Saurornithoides* had to be changed to *Caudocoelus* because another theropod had been given that name four years earlier. But the name *Caudocoelus* was also previously occupied, so it was renamed as *Teinurosaurus*. See more: *Osborn 1924; Nopcsa 1928; Huene 1932; Olshevsky 1978*

1980—Argentina—The theropod footprint with the most confusing name

The footprint *Hadrosaurichnus* ("hadrosaur footprint") does not belong to hadrosaur dinosaurs; it's actually from a theropod. See more: *Alonso 1980; Huh et al. 2003*

1980—Australia—The first theropod name from Oceania based on a mythological being

Kakuru means "snake of light." The name comes from the rainbow serpent of Australian mythology. The bones were preserved in opal, a mineral that reflects light like a rainbow. See more: *Molnar & Pledge 1980*

1980—Argentina—The first acronym name for a theropod

Noasaurus comes from NOA, the abbreviation of Northwest Argentina. See more: *Bonaparte & Powell 1980*

1981—Mongolia—The first Asian theropod name based on a mythological being

Erlikosaurus – Erlik, Erlig, Erklik, or Erklikhan was the death god of Siberian, Turkish, and Mongolian mythology. See more: *Perle 1981*

1982—Uzbekistan—The first theropod footprint named after a country

Megalosauropus uzbekistanicus refers to Uzbekistan. Currently, it is known as *Megalosauripus*, because the name *Megalosauropus* belongs to another type of footprint previously found in Australia. See more: *Colbert & Merrilees 1967; Gabuniya & Kurbatov 1982; Lockley, Meyer & Santos 1998*

1983—Mongolia—The first Mesozoic bird named after the Cretaceous

Horezmavis eocretacea was preceded by 102 years by *Cretornis hlavaci*, but the latter turned out to be a pterosaur. Another name is *Cimolopteryx*, meaning "wing of white earth," referring to Cretaceous sediment, so it is an indirect reference that predates *Horezmavis* by more than ninety years. See more: *Frits 1881; Marsh 1889, 1892; Nesov & Borkin 1983*

1984—USA—The first dinosauromorph named after someone

Technosaurus smalli is dedicated to Bryan J. Small for his help, both in the field and in the laboratory. See more: *Chatterjee 1984*

1984—Germany—The theropod with the most similar genus and species name

Halticosaurus liliensterni changed to *Liliensternus liliensterni* fifty years later. A similar case may occur in the future if *"Tyrannosaurus" zhuchengensis* is shown to be synonymous with *Zhuchengtyrannus magnus*. This would produce the new combination *"Zhuchengtyrannus zhuchengensis."* See more: *Huene 1934; Welles 1984; Hu et al. 2001; Hone et al. 2011*

1984—USA—The first dinosauromorph named after an academic institution

Technosaurus refers to the Texas Tech University. See more: *Chatterjee 1984*

1985—China—The first theropod that was casually named

Gasosaurus constructus was discovered during the construction of a gas mine. See more: *Dong & Tang 1985*

1985—USA—The first bird mistaken for a non-avian theropod

The metatarsus of *Avisaurus archibaldi* appeared to be from a terrestrial theropod with characteristics similar to those of a bird. New fossils revealed that it actually belonged to an enantiornithean bird in Montana. See more: *Brett-Surman & Paul 1985*

1985—Spain—The most difficult mistake to detect in the name of a Mesozoic bird

Ilerdopteryx is often referred to as "Llerdopteryx" because of the similarity of both letters. See more: *Lacasa-Ruiz 1985; Palaeobiology database**

1985—Germany—The first name of a Mesozoic bird to refer to the Jurassic

Jurapteryx recurva is a juvenile specimen of *Archaeopteryx lithographica*. See more: *Howgate 1984; Howgate 1985.*

1985—Japan—The first chironomic names for theropods

"Mifunesaurus" and "Sanchusaurus" come from other informal names, such as "Mifune-ryu" or "Sanchu-ryu." These were changed to Latinize them, but both are invalid. See more: *Hisa 1985, Lambert 1990*

1986—India—The first previously used primitive saurischian name

Walkeria changed to *Alwalkeria*. The original name had been used for a bryozoan 163 years earlier. See more: *Fleming 1823; Chatterjee 1986; Chatterjee & Creisler 1994*

1987—Mongolia—The first theropod name based on a character from a poem

Borogovia, a thin bird with feathers bristling like a feather duster, was mentioned in the poem "Jabberwocky," by Lewis Carroll in *Through the Looking-Glass.* See more: *Osmólska 1987*

1988 (1882)—Tajikistan—Footprint that had the wrong genus for the longest period

Brontozoum tianschanicum was changed to *Gabirutosaurus tianschanicus* and later to *Gabirutosaurichnus tianschanicus* after 106 years. See more: *Romanovsky 1882; Gabuniya & Kurbanov 1988*

1989—China—The first Mesozoic bird footprint named after a place

Aquatilavipes anhuiensis was found in Anhui Province, China. "Aquatilavipes sinensis," which refers to the People's Republic of China, precedes it, but the name was not formalized until seven years later. See more: *Yin & Yin 1989; Zhen et al. 1987; Jin & Yan 1994; Zhen et al. 1995*

1991—Portugal—The first theropod named after Europe

Euronychodon portucalensis is based on one tiny or incomplete tooth similar to *Paronychodon*. An earlier name is *Eurolimnornis corneti*, a bird that turned out to be the remains of pterosaurs. See more: *Jurcsak & Kessler 1986; Antunes & Sigogneau-Russell 1991*

1991—Kazakhstan—The first theropod named after Asia

Asiahesperornis bazhanovi. Some researchers suspect that it belongs to the genus *Hesperornis*. See more: *Nesov & Prizemlin 1991; Mortimer**

1992—China—The first Mesozoic bird named after a country

Sinornis refers to the People's Republic of China. See more: *Sereno & Rao 1992*

1992—Spain—The first error in a Mesozoic bird's name with the same meaning

Iberomesornis has a wrong reference with the letters exchanged, "Mesoiberornis." See more: *Sanz & Bonaparte 1992; Lockley et al. 1992*

1993—Australia—The first theropod named after a child

Timimus was named by Thomas Rich and Patricia Vickers-Rich in honor of their son, Tim Rich. Coincidentally, in the novel *Jurassic Park*, the grandson of Hammond is called Timothy Murphy ("Tim"). See more: *Crichton 1990; Rich & Vickers-Rich 1993*

1993—China—The first Mesozoic bird footprint named after a girl

The species *Shandongornipes muxiai* refers to Muxía, the daughter of the paleontologist Rihui Li, who participated in the discovery of the tracks. See more: *Li et al. 2005*

1993—Mongolia—The fastest name change in a theropod

Mononychus changed to *Mononykus* after only twenty-nine days (from April 15 to May 13). This was because a beetle had received that name 169 years earlier. See more: *Schuppel 1824; Perle et al. 1993a, 1993b*

1994—Niger—The first theropod named after Africa

Afrovenator ("African hunter"). It comes before another name that usually appears in the lists of dinosaur species, "Megalosaurus africanus," but that name was a mistaken reference to *Megalosaurus saharicus*. See more: *Deperet & Savornin 1925; Huene 1956; Sereno et al. 1994*

1994—Mongolia—The strangest typographical error in a Mesozoic bird name

Holbotia is referred to as "Kholbotiaka." See more: *Kurochkin 1892, 1994; Zelenkov & Averianov 2015*

1994—Mongolia—The theropod egg with the most confusing name

Protoceratopsirovum ("petrified egg of *Protoceratops*") are theropod eggs that were mistaken for *Protoceratops* eggs. Changing the name to "Oviraptoroolithus" was proposed without success. See more: *Mikhailov 1994; Fang et al. 2009*

1995—Uzbekistan—The first theropod name to refer to the American continent

Asiamericana asiatica (currently *Richardoestesia*) refers to the continents Asia and North America. See more: *Nesov 1995; Sues & Averianov 2013*

1995—India—The first Mesozoic bird fossil egg named after a place

Subtiliolithus kachchhensis is dedicated to the city Kachchen. See more: *Khosla & Sahni 1995*

1995—China—The first Mesozoic bird named after a philosopher

Confuciusornis sanctus is dedicated to Confucius, a renowned Chinese thinker. See more: *Hou et al. 1995*

1995—China—The first theropod named after a conqueror

Jenghizkhan (a synonym of *Tarbosaurus*) was named after Genghis Khan, a Mongol warrior and conqueror. See more: *Olshevsky, Ford & Yamamoto 1995*

1995—Portugal—The first theropod named after a writer

Torvosaurus gurneyi is dedicated to writer and artist James Gurney, creator of the best seller *Dinotopia: A Land outside Time*, which describes a fantastic world where dinosaurs and humans cohabit. See more: *Hendrickx & Mateus 2014*

1996—Brazil—First South American theropod named after a fictitious person

The species *Irritator challengeri* was named by Professor George Edward Challenger after the novel *The Lost World*, by Sir Arthur Conan Doyle. The genus name ("irritant") refers to his aggressive and dominant personality. See more: *Martill et al. 1996*

1996—Canada—The theropod fossil egg named after a country

Continuoolithus canadensis is the only one whose species name refers to a nation: Canada. Later specimens of this species were found in Montana, USA. See more: *Zelenitsky et al. 1996; Schaff 2012; Zelenitsky & Therrien 2008*

1996—USA—The first theropod name dedicated to a movie monster

Gojirasaurus is based on Gojira, the Japanese name for Godzilla. See more: *Carpenter 1997*

1996—Mongolia—The first Asian theropod named after a fictitious person

The informal name of "Ichabodcraniosaurus" was assigned to a skeleton without a skull of a dromaeosaurid IGM 100/980. The name refers to Ichabod Crane, a character from Washington Irving's story "The Legend of Sleepy Hollow." See more: *Novacek 1996*

1997—China—The first Mesozoic bird name based on a poem

Cathayornis refers to the book *Cathay*, a composite of transcriptions of poems from ancient China. See more: *Mathews 1915; Hou 1997*

1998—France—First Mesozoic bird name dedicated to a fictitious person

Gargantuavis is based on Gargantua, a character in the French satirical novel *Pantagruel*, by François Rabelais. See more: *Rabelais 1532; Buffetaut & Le Loeuff 1998*

1998—Mongolia—The fastest name change of a Mesozoic bird

Rahona changed to *Rahonavis* after only thirty-one days (from March 20 to April 10) because the name was already occupied by the *Rhaona* butterfly eighty-three years earlier. See more: *Griveaud 1975; Forster et al. 1998a, 1998b*

1998—Italy—The first theropod name with two meanings

The name *Scipionyx* is based on the geologist Scipione Breislak and on Publius Cornelius Scipio, a politician of the Roman Republic. See more: *Dal Sasso & Signore 1998*

1998—USA—The first theropod named by a contest

Nedcolbertia justinhofmanni was named after Justin Hofmann, the winner of a Discover Card contest. Originally, the species was to be called "N. whittlei". See more: *Kirkland et al. 1995; Kirkland et al. 1998*

1999—China—Mesozoic bird described and named twice

Liaoxiornis delicatus and *Lingyuanornis parvus* were described one month apart, but they are the right and left halves of the same fossil. These species are doubtful because they represent a juvenile individual. See more: *Hou & Chen 1999; Ji & Ji 1999*

2000—South Africa—Only theropod name with a click in its pronunciation

Nqwebasaurus comes from the Nqewba dam, which is pronounced by clicking with an N before KWE-bah, a sound typical of South African languages. See more: *de Klerk et al. 2000*

2000—USA—The first North American theropod named after a fictional character

Bambiraptor is based on the character of the novel *Bambi, A Life in the Woods*, by Felix Salten. This is because it was a juvenile specimen discovered by children. See more: *Salten 1923; Burnham et al. 2000*

2001—Mongolia—The first Asian Mesozoic bird named after a mythological being

Apsaravis comes from apsara, an aquatic nymph of Hindu mythology that could change shape at will. See more: *Norell & Clarke 2001*

2001—Argentina—The first theropod named after the people of a region

Quilmesaurus after the Quilmes people, an Argentine population that came from an area that is currently in Chile. It precedes the informal name "Comanchesaurus," which refers to the fact they are Native Americans. See more: *Hunt 1994; Coria 2001; Nesbitt, Irmis & Parker 2007*

2001—Madagascar—The first theropod named after a singer

Masiakasaurus knopfleri means "Mark Knopfler's vicious lizard," in honor of the front man of the rock band Dire Straits. The discoverers were listening to the song "Sultans of Swing" when they found it. A similar case is *Cryolophosaurus*. Thanks to its crest, the magazine *Prehistoric Times* proposed the name "Elvisaurus" in 1993. See more: *Holmes 1933; Sampson et al. 2014*

2002 (1838)—The longest elapsed time between two species of the same theropod genus (not currently valid)

Poekilopleuron bucklandii was named 164 years before *P. valsdunensis*, although the latter has changed to a new genus, *Dubreuillosaurus*. There is another, even greater time difference: 177 years between *Megalosaurus bucklandii* and "M. cachuensis," but the latter is perhaps an error of "M. dapukaensis." Both are informal. See more: *Mantell 1827; Eudes-Deslongchamps 1838; Allian 2002, 2005; Weishampel et al. 2004*

2002—China—The first modern conflict of theropod names

The discoverers of *Epidendrosaurus ningchengensis* and *Scansoriopteryx heilmanni* fought over which of the two descriptions had priority, since both were considered the same species. *Epidendrosaurus* was described in an electronic format on August 21 and in print on September 30, 2002. Meanwhile, *Scansoriopteryx heilmanni* was described on August 1, although its publication was not distributed until September 2. They may not be synonymous, since they may have an age difference of several million years. See more: *Czerkas & Yuan 2002; Zhang et al. 2002; Harris 2004; Paleobiology Database**

2002—China—The first modern conflict of Mesozoic bird names

Shenzhouraptor was published on July 23, just two days before *Jeholornis*. Upon discovering that they were the same type of bird, some authors recognized the names as a minor synonym of the first, others of the second. Finally, the ICZN (International Commission on Zoological Nomenclature) declared the priority of *Shenzhouraptor*, although the decision was complicated because it was published in a monthly magazine with no specific date. A second species of *Jeholornis* (*J. palmapenis*) was described, a fact that supports the name *Jeholornis*. See more: *Ji et al. 2002; Zhou & Zhang 2002; Ji et al 2003; O'Connor et al. 2012; Mortimer**

2002—The most complicated dinosauromorph name

Agnosphytis ("of unknown lineage") is phonetically complicated. It was even written as "Agnostiphys" three times in its description. See more: *Fraser et al. 2002*

2002—China—The first Mesozoic bird name acronym

Sapeornis comes from the Society of Avian Paleontology and Evolution (SAPE). See more: *Zhou & Zhang 2002*

2002—Mongolia—The strangest typo in a theropod name

Byronosaurus was written as "Byranjaffa". See more: *Norell, Makovicky & Clark 2000; Novaceck 2002*

2004—China—The publication with the largest number of theropod typographical errors

Nemegtia barsboldi is referred to with multiple variants: "Nemrgtia barsholdi," "Nemrgtia harsboldi," "Nemegita harsboldi," "Nemegtiu barsboldi," "Nemrgtiu barsbolidi," and "Nemegtia hursholdi." See more: *Lu et al. 2004*

2004—Spain—The first error in a theropod footprint name with identical meaning

Gigandipus was called "Grandipus," a term that means the same thing. See more: *Hitchcock 1855; Bakker 2004*

2005—Antarctica—The first Mesozoic bird name acronym

The species *Vegavis iaai* is named after the Instituto Antártico Argentino (IAA). See more: *Clarke et al. 2005*

2005—Antarctica—The first Mesozoic bird with a palindrome name

The species name of *Vegavis iaai* is a palindrome—that is, it reads the same from left to right as from right to left. See more: *Clarke et al. 2005*

2005–2006—South Africa—The first African theropod named after a mythological being

Dracovenator was discovered in Drakensberg, or "dragon mountains." Although it appears in the journal *African Paleontology* dated December 2005, it was published in 2006. See more: *Yates 2006*

2006—Brazil—The first dinosauromorph based on a mythological being

Sacisaurus comes from Sací, a lame goblin who in Brazilian mythology smoked a pipe. He is a trickster that grants wishes to whoever manages to catch him. See more: *Ferigolo & Langer 2006*

2006—Japan—The first expired theropod name

"Futabasaurus" was created for a remnant of a possible tyrannosaurid, but the name was never formalized. Now it is used for a plesiosaurus. A similar case happened with "Koreanosaurus," the informal theropod name. It is now used by the ornithischian *Koreanosaurus boseongensis*. See more: *Lambert 1990; Kim 1979; Sato et al. 2006; Min et al. 2011*

2006—Argentina—The first theropod name accidentally revealed

"Bayosaurus" appears accidentally in a cladogram. It refers to the example MCF-PVPH-237, a very incomplete abelisaurid. See more: *Coria, Currie & Carbajal 2006; Gasparini, Salgado & Coria 2007*

2007—Mongolia—The most similar theropod and fossil bird names

Mahakala is a dromaeosaurid, and *Makahala* is an extinct bird from the Cenozoic. See more: *Turner et al. 2007; Mayr 2015*

2009—China—The first Mesozoic bird named after a spiritual leader

Zanabazar refers to Jebtsundamba Khutuktu, a Tibetan Buddhist spiritual leader in Mongolia. See more: *Norell et al. 2009*

2009—China—The first theropod name with a corporate motive

Shidaisaurus jinae refers to the company Jin-Shadai ("Golden Age"), a company that worked in China's Jurassic World Park. See more: *Wu et al. 2009*

2010—China—The first Mesozoic bird named with an academic motive

Shenshiornis is based on Shenyang Normal University. See more: *Hu et al. 2010*

2010—China—The first Mesozoic bird named after a celebration

The naming of *Shenqiornis* was part of the celebration of the successful launch of *Shenzhou 7*, the third Chinese manned space mission. See more: *Wang et al. 2010*

2010—USA—The theropod name that has most inspired other species names

Tyrannosaurus rex has inspired many names, among which are the theropod *Edmarka rex* ("hunchback king tyrant"), an extinct beetle; *Tyrannobdella rex* ("queen leech tiger"), a leech with large "teeth"; *Tyrannoberingius rex* ("Beringius tyrant king"), an extinct conch shell; and *Tyrannomyrmex rex* ("ant tyrant queen"), an extinct ant. In addition, there are *Tyrannosauripus* and *Tyrannosauropus* footprints. See more: *Osborn 1905; Haubold 1971; Maricovich 1981; Bakker et al. 1992; Lockley & Hunt 1994; Ratcliffe & Ocampo 2001; Fernández 2003; Phillips et al. 2010*

2010—Romania—First European theropod name based on a mythological being

Balaur is the name of a great dragon from Romanian folklore. When Balaur opened its mouth, one jaw touched the earth and the other the sky. It had fins, legs, and multiple serpent heads. See more: *Csiki et al. 2010*

2011—China—The first Mesozoic bird named after a singer

Quiliania graffini was named in honor of Greg Graffin, a paleontologist and also a member of the punk rock band Bad Religion. See more: *Ji et al. 2011*

2011 (1957)—England—The name of dinosaur with the most time between its informal and formal publications

"Nyasasaurus" was described in a doctoral thesis. Fifty-six years later, it appeared in a formal publication. The name of the species was changed from "N. cromptoni" to *Nyasasaurus parringtoni*. See more: *Charig 1957; Appleby et al. 1967; Nesbitt et al. 2013*

2012 (1920)—Tanzania—The theropod with the longest time between its discovery and its scientific nomination

Ostafrikasaurus crassiserratus was thought to be a specimen of *Labrosaurus stechowi*. Its identification took ninety-two years. See more: *Janensch 1920; Buffetaut 2012*

2012—Canada—The first Mesozoic bird named after America

Brodavis americanus is the type species of the genus that includes three other probable species: *B. baileyi, B. mongoliensis,* and *B. varneri*. See more: *Martin, Kurochkin & Tokaryk 2011*

2012—Morocco—The first African theropod named after a fictional character

Sauroniops means "eye of Sauron," a character from the epic novel *The Lord of the Rings*, by J. R. R. Tolkien. The name *Sauron* means "the creepy or the hated," according to the fictional Quenya language. See more: *Tolkien 1954; Cau et al. 2012*

2012—Morocco—The first name of a dinosauromorph with two meanings

Diodorus scytobrachion—The species name means "leathery leather" in Greek, and it also honors Dionisio Scytobrachion, a mythographer from North Africa. See more: *Kammerer, Nesbitt & Shubin 2012*

2012—Argentina—The first acronym theropod name

Eoabelisaurus mefi refers to the Museo Paleontologico Egidio Feruglio (MEF) of Argentina. See more: *Pol & Rauhut 2012*

2013—China—The first theropod name to refer to a paleocontinent

Panguraptor has two similar meanings. One refers to the supercontinent Pangea, the other to Pangu, the creator god of Chinese mythology. See more: *You et al. 2014*

2013—USA—The most terrifying theropod name

There are several names that refer to a predatory nature or an intimidating aspect. Perhaps the most extreme is *Lythronax*, which means "bloody king." See more: *Loewen et al. 2013*

2013—Zimbabwe—Same word used for a theropod name and for the species name of an ornithischian

The name *Syntarsus* ("ankle together") is currently disused because a beetle received the name 100 years earlier. *Albertadromeus syntarsus* is an ornithischian that repeats the term—and for the same reason: the fusion of bones in the legs. See more: *Fairmaire 1869; Raath 1969, Ivie et al. 2001, Brown et al. 2013*

2014 (1940)—China—The most-mentioned country in theropod names

The People's Republic of China is mentioned in twelve theropods *(Quianzhousaurus sinensis, Shenzhouraptor sinensis, Sinocalliopteryx, Sinocoelurus, Sinornis, Sinornithoides, Sinornithomimus, Sinornithosaurus, Sinosauropteryx, Sinosaurus, Sinotyrannus* and *Sinovenator),* and two tracks *(Aquatilavipes sinensis* and *Menglongipus sinensis),* as well as two currently invalid names *("Dilophosaurus" sinensis* and *Omnivoropteryx sinosaorum).*

2014 (1884)—Longest time elapsed between the description of two species of the same genus of theropod

There is a difference of 137 years between *Allosaurus fragilis* and *A. lucasi*, although some researchers suspect that the second species is synonymous with the first. *"Labrosaurus" meriani* (now *Ceratosaurus*) is 130 years apart from *C. dentisulcatus* and *C. magnicornis*. In case the species from Switzerland does not belong to the genus *Ceratosaurus*, there is a separation of 116 years between *C. nasicornis* and the last two species. See more: *Greppin 1870; Marsh 1884; Madsen & Welles 2000; Dalman 2014; Mortimer**

2014 (1924)—The country (Mongolia) most mentioned in the names of theropod species

The *mongoliensis* species name has been applied in nine genera *(Adasaurus, Brodavis, Enigmosaurus, Gallimimus, Gobivenator, Rinchenia, Prodeinodon, Saurornithoides* and *Velociraptor).*

2015 (1982)—Mongolia—The name of the Mesozoic bird with the most time before its scientific name validation

"Holbotia ponomarenkoi" was created thirty-three years before being published formally. See more: *Kurochkin 1982 in Kurochkin 1994; Zelenkov & Averianov 2015*

—Theropod names with the simplest meaning

Falcarius comes from the Latin word *falcatus*. A falcata is a kind of very sharp sword.
Hexing: similar to a crane in the Chinese language.
Khaan: from the Mongolian language, meaning "lord" or "ruler."
Kileskus means "lizard" in the Turkic language "Khakas" spoken by the Khakassans.
Shuvuuia is "bird" in the Mongolian language.
Yurgovuchia means "coyote" in the language of the Ute tribe.
See more: *Chiappe et al. 1998; Averianov et al. 2010; Kirkland et al. 2005; Averianov et al. 2010; Jin et al. 2012; Senter et al. 2012*

—The names of Mesozoic birds with the simplest meaning

Ambiortus: It means "ambiguous."
Canadaga: Refers to its origin in Canada.
Gansus: It is the name of a province in China.
Qiliania: It means "sky."
Vorona: It means "bird" in the Malgalache language.
See more: *Kurochkin 1982; Forster et al. 1996; Hou 1999; Hou & Liu 1984; Forster et al. 1996; Ji et al. 2011*

—The theropod species with the most synonyms

Tyrannosaurus rex (thirteen): Five formal synonyms *(Aublysodon molnari, Dynamosaurus imperiosus, Dynotyrannus megagracilis, Manospondylus gigas* and *Nannotyrannus lancensis),* two questionable synonyms *(Aublysodon amplus* and *A. cristatus),* and six informal synonyms ("Chicagotyrannus," "Clevelanotyrannus," "Nanotyrannes," Tyrannosaurus "imperator," T. "stanwinstonorum," T. "gigantus," and T. "vannus"). See more: *Mortimer**

—The theropod species with the most name combinations

Tyrannosaurus rex (twenty-five): *(Albertosaurus lancensis, Albertosaurus "megagracilis," Aublysodon amplus, Aublysodon cristatus, Aublysodon lancensis, Aublysodon molnari, Aublysodon molnaris, Gorgosaurus lancensis, Deinodon amplus, Deinodon cristatus, Deinodon lancensis, Dynamosaurus imperiosus, Dinotyrannus megagracilis, Manospondylus amplus, Manospondylus gigas, Nanotyrannus lancensis, Stygivenator amplus, Stygivenator cristatus, Stygivenator molnari, Tyrannosaurus amplus, Tyrannosaurus imperiosus, Tyrannosaurus "gigantus," Tyrannosaurus "imperator," Tyrannosaurus "stanwinstonorum,"* and *Tyrannosaurus "vannus").* See more: *Mortimer**

—The species of Mesozoic bird with the most synonyms

Archaeopteryx lithographica (ten): Eight formal synonyms *(Archaeopteryx bavarica, crassipes, macrura, A. oweni, A. siemensii, Jurapteryx recurva, Pterodactylus [Rhamphorhynchus] crassipes,* and *Wellnhoferia grandis)* and two informal ("Griphornis

longicaudatus" and "Griphosaurus problematicus").

—The Mesozoic bird species with the most name combinations

Archaeopteryx lithographica (fourteen): *(Archaeopteryx bavarica, crassipes, macrura, A. oweni, A. recurva, A. siemensii, Archaeornis siemensii, Griphornis longicaudatus, Griphosaurus longicaudatus, Griphosaurus problematicus, Jurapteryx recurva, Pterodactylus (Rhamphorhynchus) crassipes, Scaphognathus crassipes* and *Wellnhoferia grandis).* See more: *Mortimer**

—The theropod footprint with the most synonyms

Grallator (forty): *(Anchisauripus, Brontozoum, Changpeipus, Chonglongpus, Chongqingpus, Chuangchengpus, Coelurosaurichnus, Cybele, Deuterotrisauropus, Dilophosauripus, Eubrontes, Hitchcokia, Gabirutosaurichnus, Gabirutosaurus, Gigandipus, Gigantitherium, Ichnites, Jeholosauripus, Jinlijingpus, Kainotrisauropus, Kayentapus, Masitisisauropus, Neograllator, Neotrisauropus, Neotripodiscus, Ornithoidichnites, Otouphepus, Paracoelurosaurichnus, Paragrallator, Plastisauropus, Platysauropus, Plesiornis, Plesiotornipos, Prototrisauropodiscus, Prototrisauropus, Qemetrisauropus, Saurischnium, Schizograllator, Weiyuanpus,* and *Youngichnus).* Some authors recommend the synonymy of *Grallator* and *Anchisauripus* with *Eubrontes* because of differences that may be due to behavioral or ontogenetic factors instead of osteological factors. Other authors add other genera, which causes many combinations. It is possible that in the future, under more detailed criteria and analysis, some names will be revalidated. See more: *Rainforth 2007*

—The theropod with the most assigned species

Megalosaurus was used to define multiple theropods, since there were not many references to the differences between the materials. It is known as a "tailor-made" genus because it included multiple species that, over time, were identified as specimens of different taxa, and some were not even theropods. Actually, the genus *Megalosaurus* includes only the species *M. bucklandii* and perhaps the questionable *M.* "phillipsi." *Megalosaurus* (fifty-four): One valid *(M. bucklandii)* and one doubtful (M. "phillipsi"), plus one synonym ("M. conybearei"), and forty-nine not belonging *("M." andrewsi, "M." aquilunguis, "M." argentinus, "M." bradleyi, "M." bredai, "M." cambrensis, "M." campbelli, "M." chubutensis, "M." cloacinus, "M." crenatissimus, "M." cuvieri, "M." destructor, "M." dunkeri, "M." gracilis, "M." hesperis, "M." horridus, "M." hungaricus, "M." incognitus, "M." inexpectatus, "M." ingens, "M." insignis, "M." lydekkeri, "M." lonzeensis, "M." matleyi, "M." maximus, "M." mercensis, "M." meriani, "M." monasterii, "M." nasicornis, "M." nethercombensis, "M." nicaeensis, "M." obstusus, "M." oweni, "M." pannoniensis, "M." parkeri, "M." poekilopleuron, "M." pombali, "M." rawesi, "M." aharicus, "M." schmidti, "M." schnaitheimii, "M." silesiacus, "M." superbus, "M." tanneri, "M." terquemi, "M." trihedrodon, "M." valens,* and *"M." woodwardi).* Four invalids ("M. africanus," "M. cachuensis," "M. dabukaensis" or "M. dapukaensis," and "M. tibetensis").

—The theropod with the most valid species

Ceratosaurus (seven): Three valid *(C. dentisulcatus, C. magnicornis,* and *C. nasicornis),* three doubtful *(C. meriani, C. roechlingi,* and *C. sulcatus),* one not valid (C. "willisobrienorum"), and one different genus *("C." ingens).*

—The Mesozoic bird with the most valid species

Hesperornis (ten): Five valid *(H. chowi, H. crassipes, H. gracilis, H. regalis,* and *H. rossicus),* one possible *(H. bazhanovi),* four doubtful *(H. altus, H. bairdi, H. mengeli,* and *H. montanus)* and one different genus *("H." macdonaldi).*

—The species name most used for theropods

mongoliensis—Fourteen valid: *Adasaurus, Altispinax, Brodavis, Citipati, Deinodon, Enigmosaurus, Gobivenator, Nomingia, Oviraptor, Prodeinodon, Rinchenia, Saurornithoides, Troodon,* and *Velociraptor;* and two not valid ("Gallimimus" and "Tonouchisaurus").

CURIOSITIES CONCERNING BINOMIAL NOMENCLATURE USED FOR DINOSAUROMORPHS, THEROPODS, AND MESOZOIC BIRDS

SCIENTIFIC NAMES

In order to identify and talk about dinosaurs and all living beings (current or extinct) without confusion, each receives a scientific name. The rules of taxonomy are governed by the International Code of Nomenclature, which dictates that the scientific name for each taxon must be a unique binomial name, composed of a genus name and a species name (specific name).

LONGEST AND SHORTEST BINOMIAL AND TRINOMIAL NAMES

The dinosauromorphs with the longest and shortest names (number of letters): *Agnosphitys cromhallensis* (24), *Dromomeron romeri* (16)

The primitive saurischians with the longest and shortest names: *Frenguellisaurus ischigualastensis* (33, possible synonym of *Herrerasaurus ischigualastensis*), *Herrerasaurus ischigualastensis* (30). *Eoraptor lunensis* (16, possible basal sauropodomorph), *Ischisaurus cattoi* (17, possible synonym of *Herrerasaurus ischigualastensis*).

Theropods with the longest and shortest names: *Appalachiosaurus montgomeriensis* (31), *Lisbosaurus mitracostatus mitracostatus* (37), is not a dinosaur, but a crocodile. The combination (*Metriacanthosaurus shangyouensis* (31) is currently not valid. *Yi qi* (4).

The Mesozoic birds with the longest and shortest names: *Largirostrisornis sexdentornis, Parahongshanornis chaoyangensis, Changchengornis hengdaoziensis* and *Dapingfangornis sentisorhinus* (30), *Nanantius eos* (12).

Dinosauromorph footprints with the longest and shortest names: *Coelurosaurichnus schlauersbachense* (34). "Coelurosaurichnus schlehenbergensis" (35) is a misspelling of *Coelurosaurichnus schlehenbergense*. *Parachirotherium postchirotheroides* (34) is not a dinosauromorph, but a primitive archosaur. *Atreipus metzneri* and *Atreipus sulcatus* (16 each). *Sphingopus ferox* (15) possibly is not from a dinosauromorph.

Primitive saurischian footprints with the longest and shortest names: *Prototrisauropus angustidigitus* (30). *Qemetrisauropus princeps* (23). "Saurichnium anserinum" (20) is an invalid name. *Prototrisauropus rectilineus lentus* and *Prototrisauropus rectilineus gravis* (33).

The theropod footprints with the longest and shortest names: *Gabirutosaurichnus tianschanicus* (31). *Megalosauripus titanopelobatidus* (31) is perhaps not a theropod footprint. *Gigandipus hei* (13, formerly *Chonglongpus hei*). *Masitisisauropus angustus cursor* (30).

Mesozoic bird footprints with the longest and shortest names: *Sarjeantopodus semipalmatus* (26). *Wupus agilis* (11).

Theropod eggs with the longest and shortest names: *Macroelongatoolithus goseongensis* (32). Dinosauriovum Grumuliovum tuberculatum (36) is not a valid name. *Sankofa pirenaica* (17). Oolithes rugustus (16) currently is Macroolithus rugustus.

Mesozoic bird eggs with the longest and shortest names: *Tristraguloolithus cracioides* (28). *Neixiangoolithus yani* (20).

LONGEST AND SHORTEST GENUS NAMES

Dinosauromorph with the longest and shortest names (number of letters): *Pseudolagosuchus* (16). *Avipes* (6).

Primitive saurischians with the longest and shortest names: *Frenguellisaurus* (16 possible synonym of *Herrerasaurus*). *Herrerasaurus* (13). *Eoraptor* (8, probable primitive Sauropodomorph).

Theropods with the longest and shortest names: *Carcharodontosaurus* (19), "Kumamotomifunesaurus" (20) and "Ichabodcraniosaurus" (19) are invalid names. *Yi* (2).

Mesozoic birds with the longest and shortest names: *Parahongshanornis* (18). *Gansus* and *Vorona* (6).

Dinosauromorph footprints with the longest and shortest name: *Coelurosaurichnus* (17). *Atreipus* (8). *Calopus* (7) was not a dinosaur footprint.

Saurischian footprints with the longest and shortest names: *Prototrisauropodiscus* (21). *Qemetrisauropus* (15). *Saurichnium* (11) is a not valid name.

Theropod footprints with the longest and shortest names: *Paracoelurosaurichnus* and *Prototrisauropodiscus* (23). *Chapus* (6).

Footprints of Mesozoic birds with the longest and shortest names: *Archaeornithipus* (16). *Wupus* (5).

Theropod eggs with the longest and shortest names: *Spheruprismatoolithus* (21). *Sankofa* (7).

Mesozoic bird eggs with the longest and shortest names: *Dispersituberoolithus* (21). *Agerolithus* (12).

LONGEST AND SHORTEST SPECIES NAMES

Dinosauromorphs with the longest and shortest species names (number of letters): *dillstedtianus* (14, *Avipes*), *talampayensis* and *Diodorus scytobrachion* (13). *Pseudolagosuchus major* (5).

Primitive saurischians with the longest and shortest species names: *ischigualastensis* (17, *Frenguellisaurus and Herrerasaurus*).*pricei* (6, *Staurikosaurus*) *primus* ("*Thecodontosaurus*" is doubtful), *cattoi* (6, *Ischisaurus*, a synonym for *Herrerasaurus*).

Theropods with the longest and shortest species names: *aguadagrandensis* (16, *Ilokelesia*), *cerropoliciensis* (16, *Alnashetri*), *ornitholestoides* (16, *Rapator*). "lydekkerhuenerorum" and "lydekker–huenensis" (19) are invalid names. *qi* (2, Yi).

Mesozoic birds with the longest and shortest species names: *dapingfangensis* (15, *Pterygornis*). magnapremaxillo (15) is synonymous with Pengornis houi. *dui* (3 *Confuciusornis*), *eos* (3, *Nanantius*), *jii* (3, *Didactylornis* synonym of Sapeornis chaoyangensis), *lii* (3, *Piscivoravis*), *lii* (3, *Schizoura*), "*lii*" (3, "*Eoornithura*" is an invalid name). rex (3, *Anatalavis*) was not of the Mesozoic but dates to the early Cenozoic. *wui* (3, *Aberratiodontus*).

Dinosauromorph footprints with the longest and shortest species names: *schlauersbachensis* (*Dinosaurichnium*), *schlehenbergensis* (17, *Coelurosaurichnus*). *postcheirotheroides* (17). *Dinosaurichnium* is not a dinosaur footprint. *rati* (4, *Coelurosaurichnus*).

Primitive saurischian footprints with the longest and shortest species names: *angustidigitus* (14, *Prototrisauropus*). *minor* (5, *Qemetrisauropus*).

Theropod footprints with the longest and shortest species names: *nianpanshanensis* (16, *Jinlijingpus*). "titanopelobatidus"

(17, "*Eubrontes*") probably was not from a theropod. *hei* (3, *Gigandipus*). *xui* (4, *Siamopodus* possibly is from an ornithischian).

Mesozoic bird footprints with the longest and shortest species names: *semipalmatus* (12, *Sarjeantopodus*). *lithographicum* (14, *Ichnites*) were not bird tracks; they were of an Atlantic horseshoe crab. *avis* (4, *Yacoraitichnus*). *kimi* (4, *Jindongornipes*).

Theropod eggs with the longest and shortest species names: *fengguangcunensis* (18, *Dendroolithus*). *levis* (5, *Prismatoolithus*). "xixia" (5, *Macroelongatoolithus* typographical error for *xixiaensis*).

Theropod eggs with the longest and shortest species names: *fontllongensis* (14, *Ageroolithus*). *yani* (4, *Neixiangoolithus*).

THEROPOD ALPHABET

If all the scientific names were ordered alphabetically in categories, these would be the first and last names in the row.

BINOMIAL NAMES

Dinosauromorphs: *Agnosphitys cromhallensis, Technosaurus smalli*, or *Teyuwasu barberenai* (it isn't known if the latter was a dinosauromorph or a primitive dinosaur).

Primitive saurischians: *Alwalkeria maleriensis* (may be a primitive sauropodomorph), *Staurikosaurus pricei* or *"Zanclodon" silesiacus*. (The latter may be a primitive archosaur).

Theropods: *Abelisaurus comahuensis – Zupaysaurus rougieri*

Mesozoic birds: *Abavornis bonaparti – Zhyraornis logunovi*

Dinosaur footprints: *Agialopus wyomingensis – Prorotodactylus mirus*

Primitive saurischian footprints: *Coelurosaurichnus grancieri – Qemetrisauropus princeps*

Theropod footprints: *Abelichnus astigarrae – Zhengichnus jinningensis*

Bird footprints: *Aquatilavipes anhuiensis* (currently *Koreanaornis anhuiensis*), *Aquatilavipes curriei – Yacoraitichnus avis*

Theropod eggs: "*Apheloolithus shuinanensis*" (possibly not valid name), *Arraigadoolithus patagonicus – Trigonoolithus amoae*

Mesozoic bird egg: *Ageroolithus fontllongensis – Tubercuoolithus tetonensis*

GENUS NAMES

Dinosauromorphs: *Agnosphitys, Technosaurus*, or *Teyuwasu*? (it is not known if it was a dinosauromorph or a primitive dinosaur).

Primitive saurischians: *Alwalkeria* (sauropodomorph?), *Staurikosaurus* or *"Zanclodon"* (could be a primitive archosaur)

Theropods: *Abelisaurus – Zupaysaurus*

Mesozoic birds: *Abavornis – Zhyraornis*

Dinosauromorph footprints: *Agialopus – Prorotodactylus*

Primitive saurischian footprints: ***Coelurosaurichnus – Qemetrisauropus***

Theropod footprints: ***Abelichnus – Zhengichnus***

Bird footprints: (*Alaripeda* sp. from the Cenozoic), ***Aquatilavipes – Yacoraitichnus***

Theropod eggs: "**Apheloolithus**" (possibly not a valid name), ***Arraigadoolithus – Trigonoolithus***

Mesozoic bird egg: ***Ageroolithus – Tubercuoolithus***

SPECIES NAME (SPECIFIC NAME)

Dinosauromorphs: ***admixtus*** (*Lewisuchus*), ***sitwensis*** (*Asilisaurus*), *talampayensis* (*Lagosuchus*, name doubtful), or *taylori* (*Scleromochlus* is a primitive avemetatarsalian)

Primitive saurischians: ***absconditum*** (*Spondylosoma*), ***pricei*** (*Staurikosaurus*), ***primus*** ("*Thecodontosaurus*" is doubtful), or ***silesiacus*** ("*Zanclodon*" is doubtful)

Theropods: ***abakensis*** (*Afrovenator*) – ***zoui*** (*Caudipteryx*)

Mesozoic birds: ***aberransis*** ("*Cathayornis*") – ***zhengi*** (*Boluochia*) – ***zhengi*** (*Eoconfuciusornis*) or ***zheni*** (*Gansus* – synonymous with *Iteravis huchzermeyeri*?)

Dinosauromorph footprints ***acadianus*** (*Atreipus*) – ***ziegelangernensis*** (*Coelurosaurichnus*)

Primitive saurischian footprints: ***"anserinum"*** ("*Saurichnium*" invalid), ***minor*** (*Qemetrisauropus*) – ***princeps*** (*Qemetrisauropus*)

Theropod footprints: ***abnormis*** (*Typopus* or *Sauroidichnites* – does not belong to a dinosaur), ***acutus*** (*Irenesauripus*) – ***zvierzi*** (*Grallator*)

Mesozoic bird footprints: ***anhuiensis*** (*Koreanaornis*) – ***yangi*** (*Ignotornis*)

Theropod eggs: ***achloujensis*** (*Tipoolithus*) – ***zhaoyingensis*** (*Dendroolithus*)

SIMILAR NAMES

If you consider that *Sinosauropteryx* and *Sinocallioteryx*, and *Austroraptor* and *Australoraptor*, are very similar names, you should be familiar with these matches:

The most similar theropod names: *Richardoestesia* and *Ricardoestesia:* 15 letters, 1 difference (93.34%). (It arose because of a name correction).

The most similar theropod and sauropodomorph names: *Giganotosaurus* and *Gigantosaurus:* 14 letters, 1 difference (92.86%).

The most similar theropod and arcosaur names: *Teratosaurus* and *Ceratosaurus:* 12 letters, 1 difference (91.67%).

The most similar theropod and ornithischian names: *Ganzhousaurus* and *Lanzhousaurus:* 13 letters, 1 difference (92.37%).

The most similar species names of the same genus: *Coelurus agilis, C. fragilis* and *C. gracilis:* 16 letters, 1–2 differences (93.75%–87.5%).

The most similar theropod footprint names: *Megalosauripus* and *Megalosauropus:* 14 letters, 1 difference (92.86%).

The most similar theropod footprint binomial names: *Megalosauripus brionensis* and *Megalosauropus broomensis:* 23 letters, 3 differences (86.96%).

The most similar theropod and ornithischian names: *Saurichnium* and *Staurichnium:* 12 letters, 1 difference (91.67%).

The most similar theropod and sauropodomorph footprint names: *Neosauropus* and *Eosauropus:* 11 letters, 1 difference (90.9%).

The most similar Mesozoic bird names: *Eoconfuciusornis* and *Confuciusornis:* 16 letters, 2 differences (87.5%).

The most similar Mesozoic bird species names: *chengi* and *zhengi:* 6 letters, 1 difference (83.3%, from *Tianyuornis* vs. *Boluochia, Eoconfuciusornis* and *Xiaotinga*), ***zhengi*** and ***zhangi:*** 6 letters, 1 difference (83.3%, from *Juehuaornis* vs. *Boluochia, Eoconfuciusornis* and *Xiaotinga*), ***zhengi*** and ***zheni:*** 5–6 letters, 1 difference (83.3%, from *Boluochia, Eoconfuciusornis,* and *Xiaotinga* vs. *Gansus*).

The most similar theropod footprint species names: *robusta* and *robustus:* 7–8 letters, 3 differences (62.5%, from *Ignotornis* and *Paragrallator* vs. *Velociraptorichnus*).

The most similar theropod and ornithischian footprint species names: *wangi* and *yangi:* 5 letters, 1 difference (80%, from *Shenmuichnus* vs. *Ignotornis* and *Paragrallator*).

INTEGRATED SPECIES NAMES

Theropod names with the highest degree of integration:
Sinosaurus and *Inosaurus:* 10 letters, 9 integrated (90%), and *Eustreptospondylus* and *Streptospondylus:* 18 letters, 16 integraded (88.88%).

Theropod and a sauropod names with the highest degree of integration:
Calamosaurus and *Alamosaurus:* 12 letters, 11 integrated (91.67%).

Theropod and dinosauromorph names with the highest degree of integration:
Eucoelophysis and *Coelophysis:* 13 letters, 11 integrated (84.62%).

Theropod and ornithischian names with the highest degree of integration:
Stenonychosaurus and *Onychosaurus:* 16 letters and 12 integrated (75%).

Theropod and former theropod names with the highest degree of integration:
Rhadinosaurus and *Inosaurus:* 12 letters, 9 integrated (75%).

Names among former theropods with the highest degree of integration:
Avalonianus and *Avalonia:* 11 letters, 8 integrated (72.73%).

Mesozoic bird names with the highest degree of integration:
Zhouornis and *Houornis:* 9 letters, 8 integrated (88.88%).

Theropod egg names with the highest degree of integration:
Triprismatoolithus and *Prismatoolithus:* 18 letters, 15 integrated (83.33%).

Theropod footprint names with the highest degree of integration:
Neograllator and *Grallator:* 12 letters, 9 integrated (75%).

Dinosauromorph footprint names with the highest degree of integration:
Paracoelurosaurichnus and *Coelurosaurichnus:* 21 letters, 17 integrated (80.95%).

Former dinosaur footprint names with the highest degree of integration:
Parachirotherium and *Chirotherium:* 16 letters, 12 integrated (75%).

The theropod name most often integrated with other dinosaurs:
Inosaurus (16 occasions): It is repeated in four different theropods (*Sinosaurus, Spinosaurus, Therizinosaurus* and *Wakinosaurus*), in seven sauropodomorphs (*Argentinosaurus, Dinosaurus, Jainosaurus, Lirainosaurus, Morinosaurus, Orinosaurus* and *Yibinosaurus*), in two ornithischians (*Gigantspinosaurus* and *Pachyrhinosaurus*), and in three names that were once considered to be dinosaurs (*Pekinosaurus, Rhadinosaurus* and *Tapinosaurus*).

The name of a theropod footprint most often integrated into other theropods:
Grallator is integrated in *Neograllator, Paragrallator* and *Schizograllator*.

The name of a theropod egg integrated most often into other dinosaurs:
"*Oolithus,*" 78 times (name currently invalid). It is repeated in 36 genera of theropod fossil eggs, in 16 Mesozoic bird fossil eggs, in 13 sauropodomorph fossil eggs, and in 13 ornithischian fossil eggs. It is currently not a valid genus.

Prismatoolithus is integrated into *Preprismatoolithus, Triprismatoolithus* and *Spherurprismatoolithus*.

Full name sectioned between two theropods:
Huaxiaosaurus and *Huaxiasaurus:* 13 letters, 12 integrated (92.31%).

The most sectioned complete theropod name inside another dinosaur:
Spinosaurus is divided into *Spinophorosaurus:* 16 letters, 12 integrated (68.75%).

Chronology of theropods considered largest historically

Year	Species and specimen	Actual size (length – weight)	Interesting facts and country	Reference
1824	*Megalosaurus bucklandii* OUM J13505	8.2 m - 1.1 t	The first "great carnivorous lizard," it was an estimated 12 m in length. England.	Buckland 1824
1856	*Deinodon horridus* Holotipo	7.5 m - 1.8 t	Its teeth were larger than those of *Megalosaurus*. USA.	Leidy 1856
1869	*Ornithotarsus immanis* YPM 3221	-	Mistakenly thought to be a giant theropod but was actually a large hadrosaurid. USA.	Cope 1869
1878	*Epanterias amplexus* AMNH 5767	10 m - 2.9 t	Possibly an adult *Allosaurus fragilis*. It was estimated at 13-14 m in length. USA.	Cope 1878
1905	*Tyrannosaurus rex* CMN 9380 or AMNH 973	11.6 m - 6 t	Gained unprecedented popularity. It was estimated at around 14 m in length. USA.	Osborn 1905
1913	*Aggiosaurus nicaeensis*	-	It was thought to be a theropod similar to *Tyrannosaurus* but was actually a saltwater crocodile. France.	Ambayrac 1913
1915	*Spinosaurus aegyptiacus* IPHG 1912 VIII 19	13.4 m - 3.8 t	Elongated vertebrae gave this creature an estimated length of 15 m. Egypt.	Stromer 1915
1920	*Megalosaurus ingens* MB R 1050	11.5 m - 5.7 t	Based on a single tooth. It is now known to be a carcharodontosaurid. Tanzania.	Janensch 1920
1955	*Tarbosaurus bataar* PIN 551-1	10 m - 4.5 t	Because of its long face, it was thought to be as big as *Tyrannosaurus*. Mongolia.	Maleev 1955
1960	*Carcharodontosaurus* sp. MNNHN col	12.7 m - 7.8 t	The largest *Carcharodontosaurus* tooth at that time. Niger.	Lapparent 1960
1970	*Deinocheirus mirificus* ZPAL MgD-I/6	11 m - 5.4 t	Its wieght was estimated at12 t. Mongolia.	Osmóska & Roniewicz 1970
1988	*Tyrannosaurus rex* UCMP 118742 m	10.4 m - 5.7 t	Its length was calculated at 13.6 m and its weight at 12 t. USA.	Molnar 1991; Paul 1988
1989	*Bruhathkayosaurus matleyi*	-	Originally considered a theropod 20 m in length and weighing 15 t, the remains are now believed to be of a sauropod or petrified tree trunks. India.	Yadagiri & Ayyasami 1989
1990	*Tyrannosaurus rex* FMNH PR2081	12 m - 8.3 t	Referred to in a magazine as "Tyrannosaurus gigantus." USA	Harlan 1990
1993	"Kelmayisaurus gigantus"	-	Mistakenly classified in an unofficial book. The remains may be bones of a 22 m-long sauropod.	Grady 1993
1995	*Giganotosaurus carolinii* MUCPv-Ch1	12.2 m - 7 t	Became famous for being "larger than *Tyrannosaurus rex*." Argentina.	Coria & Salgado 1995
1996	*Carcharodontosaurus saharicus* SGM-Din 1	12.7 m - 7.8 t	Its enormous size rivals that of *Tyrannosaurus*. Morocco	Sereno et al. 1996
1996	*Spinosaurus marocannus* CMN 41852	14.4 m - 6.5 t	At the time it was considered the largest theropod ever discovered. Morocco.	Russell 1996
2000	*Giganotosaurus carolinii* MUCPv-95	13.2 m - 8.5 t	A second specimen confirmed it as the largest therapod discovered up to that time. Argentina.	Calvo & Coria 2000
2000	*Tyrannosaurus rex*	-	Known as "Celeste." Its skeleton has not yet been formally described, but it is believed to be larger than FMNH PR2081. USA.	Horner 2000
2004	*Deinocheirus* sp.	15.7 m - 16 t?	A 1-meter long ulna was reported but there are doubts about its identification. Mongolia.	Suzuki et al. 2004
2005	*Spinosaurus aegyptiacus* MSNM V4047	16 m - 7.5 t	A length of almost 17 m has been suggested. Morocco.	Dal Sasso et al. 2005
2010	*Tyrannosaurus rex* UCMP 137538	12.3 m - 8.5 t	Its minimum estimated size makes this a record-setting specimen for this species. USA.	Longrich et al. 2010

Chronology of Theropods considered smallest historically.

Year	Species and Specimen	Actual size (length – weight)	Interesting facts and country	Reference
1824	*Streptospondylus rostromajor* MNHN 8900	5 m - 400 kg	Discovered in the 1770s, it was originally believed to be a crocodile. France.	Cuvier 1824
1854	*Nuthetes destructor* DORCM G 913	3.6 m - 110 kg	At the time it was thought to be a large monitor lizard. England.	Owen 1854
1856	*Troodon formosus* ANSP 9259	2.5 m - 39 kg	The smallest theropod discovered up to that time. USA.	Leidy 1856
1859	*Compsognathus longipes* BSP AS I 536	86 cm - 540 g	This juvenile specimen held the title of "smallest dinosaur ever" for many years. Germany.	Wagner 1859
1907	*Scleromochlus taylori* BMNH R3556	18.5 cm - 18 g	It was suggested that it was a dinosaur, but it is not. Scotland.	Woodward 1907
1910	*Saltopus elginensis* NHMUK R3915	50 cm - 120 g	It appeared to be a primitive theropod but is now known to be a dinosauromorph. Scotland.	Huene 1910
1971	*Lagosuchus talampayensis* UPLR 09	40 cm - 70 g	A dinosauromorph that was later considered a very primitive dinosaur. Argentina.	Romer 1971
1972	*Compsognathus longipes* MNHN CNJ 79	1.5 m - 2.3 kg	Described as *Compsognathus corallestris*, it was actually an adult *C. longipes*. France.	Bidar et al. 1972
1994	*Sinornithoides youngi* IVPP V9612	1.1 m - 2.5 kg	Shorter than *Compsognathus*. China.	Russell & Dong 1994
1996	*Parvicursor remotus* PIN 4487/25	50 cm - 187 g	An adult specimen despite its size. It remained the smallest up to 2013. Mongolia.	Karhu & Rautian 1996
2000	*Microraptor zhaoianus* IVPP V 12330	45 cm - 155 g	The specimen was a juvenile. Adults reached 95 cm in length and 1.45 kg in weight. China.	Xu et al. 2000; Czerkas et al. 2002
2008	*Anchiornis huxleyi* LPM-B00169	30 cm - 72 g	Considered the smallest dinosaur. Some individual specimens are 40 cm long and weigh 260 g. China.	Xu et al. 2008; Hu et al. 2009; Zheng et al. 2014
2008	*Epidexipteryx hui* IVPP V15471	25 cm - 220 g	Its body was much shorter than other non-avian theropods. China.	Zhang et al. 2008
2013	*Eosinopteryx brevipenna* YFGP-T5197	30 cm - 100 g	Thought to be a juvenile but was actually a subadult theropod. China.	Godefroit et al. 2013
2013	*Aurornis xui* YFGP-T5198	40 cm - 260 g	Same size as *Anchiornis*. China.	Godefroit et al. 2013
2015	Unnamed IVPP V22530	46 cm - 175 g	A subadult individual similar to *Sinornithosaurus*. China.	Pittman et al. 2015

Chronology of Mesozoic birds considered largest historically.

Year	Species and specimen	Actual size (length – weight)	Interesting facts and country	Reference
1864	*Pelagornis sedgwicki*	-	It was actually a pterosaur. England.	Seeley 1864
1866	*Pelagornis barretti*	-	A larger pterosaur than the above. England.	Seeley 1866
1870	*Laornis edvardsianus*	1 m - 5.5 kg	This bird was thought to be from the Cretaceous but turned out to be from the Early Paleocene. USA.	Marsh 1870
1871	*Struthiosaurus austriacus*	-	A nodosaurid cranium that appeared to be that of an ostrich-like bird. Austria.	Bunzel 1871
1872	*Hesperornis regalis* YPM 1476	1.4 m - 20 kg	The first large-sized Mesozoic bird ever discovered. USA.	Marsh 1872
1940	*Caenagnathus collinsi* CMN 8776	-	Appeared to be a bird because of its toothless beak, but was actually an oviraptorosaurid theropod. Canada.	Sternberg 1940
1981	cf. *Soroavisaurus australis* PVL-4033	80 cm - 7.25 kg	Possibly the largest flying bird of the Mesozoic. It was designated *Enantiornis leali*. Argentina.	Walker 1981
1992	*Patagopteryx deferrariisi* MACN-N-03	40 cm - 2 kg	The largest land bird discovered up to 1998. Argentina.	Alvarenga & Bonaparte 1992
1998	*Gargantuavis philoinos* MDE-A08	1.8 m - 120 kg	The largest bird of the Mesozoic. France.	Buffetaut & Le Loeuff 1998
2002	Unnamed CMN 50852	1.7 m - 8.8 kg	The largest soaring bird. Morocco.	Riff, Kellner, Mader & Russell 2002, 2004
2002	*Hesperornis chowi* PU 17208	1.4 m - 21 kg	The largest aquatic bird. USA.	Martin & Lim 2002
2004	*Hesperornis rossicus* ZIN PO 5463	1.6 m - 30 kg	It unseated *H. chowi* as the largest aquatic bird of the Mesozoic. Russia.	Panteleev, Popov & Averianov 2004
2005	*Hesperornis rossicus* SGU 3442 Ve02	1.6 m - 30 kg	New material turned out to be an individual identical in size to ZIN PO 5463. Sweden.	Rees & Lindgren 2005
2010	*Bauxitornis mindszentyae* MTM V 2009.38.1	1.9 m - 16 kg	Described as a flying enantiornithean, it may have been a land bird similar to *Balaur*. Hungary.	Dyke & Osi 2010
2012	*Samrukia nessovi* WDC Kz-001	-	A pterosaur that initially appeared to be a large flying bird. Kazakhstan.	Naish et al. 2012

Chronology of Mesozoic birds considered smallest historically.

Year	Species and specimen	Actual size (length – weight)	Interesting facts and country	Reference
1857	*Pterodactylus crassipes* MT 6928/29	42 cm - 220 g	A juvenile *Archaeopteryx*. Germany.	Meyer 1857
1861	*Archaeopteryx lithographica* BMNH 37001	45 cm - 310 g	An adult aged 1+ years. Germany.	Meyer 1861
1873	*Apatornis celer* YPM 1451	13 cm - 100 g	At that time, birds were not considered dinosaurs. USA	Marsh 1873
1880	*Ichthyornis tener* (act. *Guildavis*) YPM 1760	14 cm - 100 g	The smallest Mesozoic bird species described up to that time. USA.	Marsh 1880
1976	*Alexornis antecedens* LACM 33213	11.4 cm - 22 g	Its bones are similar to those of present-day motmots. Mexico.	Brodkorb 1976
1981	*Gobipipus reshetovi* PIN 4492-4	4.7 cm - 2 g	The smallest of several embryos. Mongolia.	Elzanowski 1981; Kurochkin et al. 2013
1992	*Iberomesornis romerali* LH-22	8.7 cm - 9.5 g	For a long time this specimen held the title of "smallest Mesozoic bird." Spain.	Sanz & Bonaparte 1992
1994	cf. *Iberomesornis* sp. LH-8200	10.1 cm - 15 g	Its classification as *Iberomesornis* is uncertain. Spain.	Sanz & Buscalioni 1994
1999	*Liaoxiornis delicatus* NIGP 130723	8.3 cm - 8 g	The specimen was thought to be an adult but is actually a juvenile. China.	Hou & Chen 1999
1999	*Lingyuanornis parvus* GMV-2156	8.3 cm - 8 g	Counterpart of *Liaoxiornis*. China.	Ji & Ji 1999
2006	*Dalingheornis liweii* CNU VB2005001	5.7 cm - 2.7 g	A juvenile specimen. China.	Zhang et al. 2006
2009	*Shanweiniao cooperorum* DMNH D1878	9 cm - 10.5 g	Smaller than the second specimen attributed to *Iberomesornis*. China.	O'Connor et al. 2009
2014	*Parvavis chuxiongensis* IVPP V18586	7.5 cm - 6 g	A subadult specimen. China.	Wang et al. 2014
2015	*Cratoavis cearensis* UFRJ-DG 031	6.6 cm - 4 g	Possibly a subadult. Brazil.	Carvalho et al. 2015

The longest estimated theropod specimens

The greatest length reported in a non-scientific book:
"Kelmayisaurus gigantus" 22 m - *Grady 1993*
(Possibly an unidentified sauropod)
The greatest length reported in an official scientific publication:
Spinosaurus aegyptiacus 17 m - *Dal Sasso et al. 2005*
(It may have measured 16 m or less in length)

The heaviest estimated theropod specimens

Highest estimated body mass reported in a historic cultural publication:
Tyrannosaurus rex 30 t
(It actually weighed less than one-third of that weight).
Highest estimated body mass reported in an official scientific publication:
Spinosaurus aegyptiacus 12-20.9 t - *Therrien & Henderson 2007*
(It actually weighed just over a third of that amount).

Mesozoic calendar
The chronology of dinosaurs

Records: The oldest and most recent.
The largest and smallest by period.

Dinosaurs appeared in the Mesozoic (middle life), a geological era that is between the Paleozoic (ancient life) and the Cenozoic (new life). The Mesozoic began approximately 252.17 million years ago, after the largest known mass extinction occurred. It ended 66 million years ago, with the second most drastic recorded extinction.

The Mesozoic is divided into three periods: Triassic, Jurassic, and Cretaceous. During this era, Earth went through massive changes, including the displacement of the continents, the formation of new mountain ranges and mountains, the appearance of numerous forms of animals and plants, and mass extinctions.

Current map. Earth during the Mesozoic was very different from what it is today. While the dinosaurs lived, the world's geography was transformed into the aspect we know today. In the following pages, you will appreciate this gradual evolution.

The Mesozoic began after the most devastating mass extinction in the history of Earth. At the end of the Permian, about 95%–96% of marine species and 75% of terrestrial species disappeared. It is believed that the extinction was caused by enormous volcanic activity in the northeast of the gigantic super-continent Pangea (present-day eastern Russia). Its magnitude caused a huge catastrophe. It is known that during that time, several meteorites hit the planet, increasing the effect of both cataclysms. See more: *Thomas 1977; Benton 2005; Knoll et al. 2007; Sahney & Benton 2008; Shen et al. 2011; Tohver et al. 2012; Hamilton 2014*

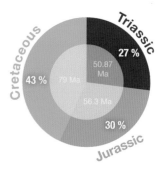

The second most devastating mass extinction in the history of Earth occurred between the Cretaceous and Paleocene. This marks the end of the Mesozoic and the beginning of the Cenozoic. Nearly 75% of terrestrial species disappeared, including all non-avian dinosaurs and most birds. Only a few modern birds survived, giving rise to today's birds. It has been suggested that three factors wholly or partially caused this disaster—several meteorite strikes, extreme volcanic activity, and a decrease in ocean levels—and a recent theory proposes that their effects could have combined. To all this were added the release of toxic substances, atmospheric changes, large tsunamis, the blocking of sunlight, the passage of ultraviolet light, the absence of oxygen, the ecological crisis, and many other side effects. See more: *Alvarez et al. 1980; McLean 1981, Duncan 1988; Chatterjee 1997, Sloan et al. 2002; Stewart & Allen 2002; Keller et al. 2008; Archibald & Fastovsky 2004; Jolley et al. 2010; Richard et al. 2015*

The Mesozoic (middle life) lasted 186.17 million years, or 34% of the Phanerozoic eon (which also includes the Paleozoic and Cenozoic). It is only 4% of Earth's total age, which is approximately 4.6 billion years.

Continental drift

To understand the biogeography of dinosaurs (their spatial distribution), it is necessary to know their migrations, the paleoclimate, the fauna, and the flora of each period. Each group of dinosaurs came from a point of origin, and they eventually spread to different parts of the world. This distribution was limited by mountains, oceans, climate, and other factors that functioned as natural barriers. See more: *González 2006*

First discoveries of each era

The first named sauropodomorphic dinosaur dating from the Triassic:
Thecodontosaurus antiquus

The first named theropod dinosaur from the Jurassic:
Megalosaurus bucklandii

The first named ornithischian dinosaur dating from the Cretaceous:
Iguanodon anglicus

The Mesozoic was also known as the "secondary era," "era of reptiles," and "era of cycads."

TRIASSIC

The Triassic lasted 50.87 million years, which is 27% of the Mesozoic.

The coming of avemetatarsalians, dinosaurs, and birds

Avemetatarsalians lived from the Lower Triassic to the present, a duration of about 251 million years, while dinosaurs may have been present since 247 Ma. The oldest remains, which belong to birds (avian theropods), are from the Middle and Upper Jurassic, a duration of about 157 million–170 million years.

Triassic days

Days in the Triassic lasted 23 hours, and each year had more than 380 days! This is because the speed of Earth's rotation is decreasing due to the distance between Earth and the Moon. This distance increases by 17 microseconds every year. *See more: Freistetter 2015*

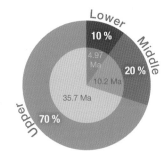
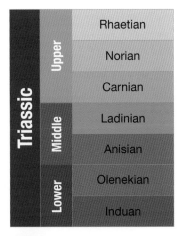

JURASSIC

The Jurassic lasted 56.3 million years, which is 30% of the Mesozoic.

Dinosaur dominance

The Jurassic is characterized by the almost complete dominance of dinosaurs in terrestrial ecosystems that herbivores had ruled at the end of the Triassic. Because several groups of archosaurs disappeared, carnivorous dinosaurs ruled this period.

Fragmented continents

At the beginning of this period, all the continents were still gathered in the supercontinent Pangea. With the arrival of the Mesozoic, it divided into two parts: Paleoasia (present-day central Asia, eastern and southeastern Asia) and Neopangea (North America, Europe, western Asia, Hindustan, South America, Africa, Antarctica, and Oceania). At the end of the Jurassic, North America and Europe separated from the rest of Neopangea, creating the paleocontinent Gondwana in the south.

CRETACEOUS

The Cretaceous spanned about seventy-nine million years, which is 43% of the Mesozoic, a longer duration than that of the Cenozoic. It is divided into two periods, Lower and Upper, and both are divided into two subdivisions.

Dinosaur revolution

Although the Cretaceous is associated with the extinction of the dinosaurs, numerous novelties developed that modified their lives, such as the predominance of flowering plants and the increase of epicontinental seas. Some theropods developed asymmetrical feathers with which they conquered the skies, while other groups developed diverse herbivorous forms.

Cretaceous jigsaw puzzle

In this period, the continents ended up fragmenting or contacting each other at different times. The northern area is known as Laurasia, the southern area as Gondwana. The central zone of Laurasia (present-day Europe) became a temporary bridge between North America, Paleoasia, and Gondwana. On the other hand, other areas were isolated several times. At the end of the Cretaceous, Madagascar and Hindustan became large islands.

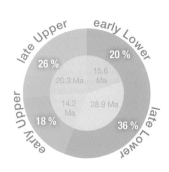

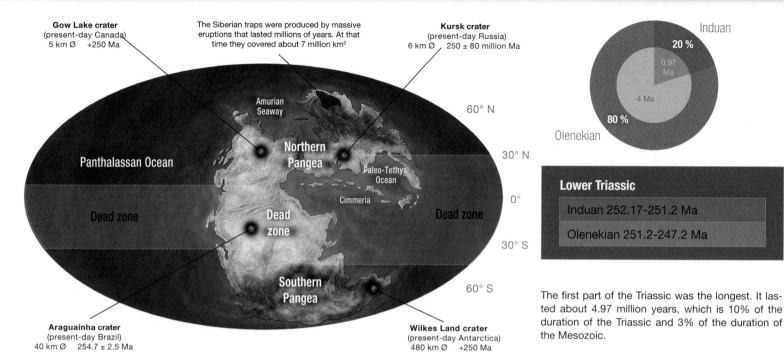

Gow Lake crater
(present-day Canada)
5 km Ø   +250 Ma

The Siberian traps were produced by massive eruptions that lasted millions of years. At that time they covered about 7 million km²

Kursk crater
(present-day Russia)
6 km Ø   250 ± 80 million Ma

Amurian Seaway

Northern Pangea

Panthalassan Ocean

Paleo-Tethys Ocean

Cimmeria

Dead zone

Dead zone

Dead zone

Southern Pangea

60° N
30° N
0°
30° S
60° S

Araguainha crater
(present-day Brazil)
40 km Ø   254.7 ± 2.5 Ma

Wilkes Land crater
(present-day Antarctica)
480 km Ø   +250 Ma

Induan
20 %
0.97 Ma
4 Ma
80 %
Olenekian

Lower Triassic

Induan 252.17-251.2 Ma	
Olenekian 251.2-247.2 Ma	

The first part of the Triassic was the longest. It lasted about 4.97 million years, which is 10% of the duration of the Triassic and 3% of the duration of the Mesozoic.

The Induan

The climate was extremely warm, as high as 50°C–60°C in the tropics and 40°C in the sea, which proved unbearable for abundant life. This created an extensive area, known as the Dead Zone, that practically lacked life. The environment was dominated by fungi due to the high increase of organic matter and an excessively dry environment. This heat was partly caused by the absence of vegetation on land. At sea, on the other hand, the oxygen level was low. Life recovered slowly on the banks of the polar regions (30° N and 40° S). The Induan was the period of the Triassic that lasted the least, barely 970,000 years. See more: *Parrish et al. 1986; Kutzbach & Gallimore 1989; Golonka et al. 1994; Visscher et al. 1996; Benton 2005; Sahney & Benton 2008; Ruhl et al. 2011; Shen et al. 2011; Sun et al. 2012*

The Olenekian

Life took almost a million years to win back the lost areas after the Permian extinction. Little by little, the appropriate conditions for life on all continents were restored. Because the oxygen level was still very low, however, there were no large animals during this time. The first avemetatarsalians and the dinosauro-morphs (identified only by tracks) appeared in northwestern and north-central Pangea. The animals were known as *Prorotodactylus* and *Rotodactylus*. See more: *Huene 1960; Berner et al. 2003; Berner 2006; Dyke et al. 2010*

The smallest of the Lower Triassic (Class: Tiny Grade III)

At present, there is no direct evidence of dinosauromorphs during the first part of the Triassic, but indirect remains have been preserved, such as tiny quadruped footprints with ectaxonic symmetry. A **Rotadactylus** isp. (b) ("rotated fingers") (present-day Poland) can be tentatively reconstructed by comparing it with Lagerpeton (a) from the Upper Triassic, although due to the characteristics that appear in these footprints, they would have had different proportions. The most notable aspect of the two is that the *Rotadactylus* metatarsals would be somewhat shorter, with more developed toes on the foot, longer arms, and a quadrupedal gait. It was really small, only about twice as heavy as a sparrow. It was two times shorter and eight times lighter than the largest specimens of *Prorotodactylus mirus*. See more: *Fichter & Kunz 2013*

40 cm

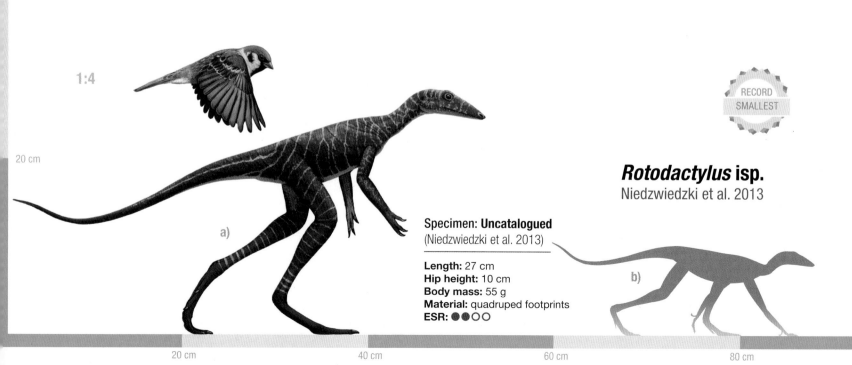

1:4

20 cm

a)

RECORD SMALLEST

Rotodactylus isp.
Niedzwiedzki et al. 2013

Specimen: Uncatalogued
(Niedzwiedzki et al. 2013)

Length: 27 cm
Hip height: 10 cm
Body mass: 55 g
Material: quadruped footprints
ESR: ●●○○

b)

20 cm   40 cm   60 cm   80 cm

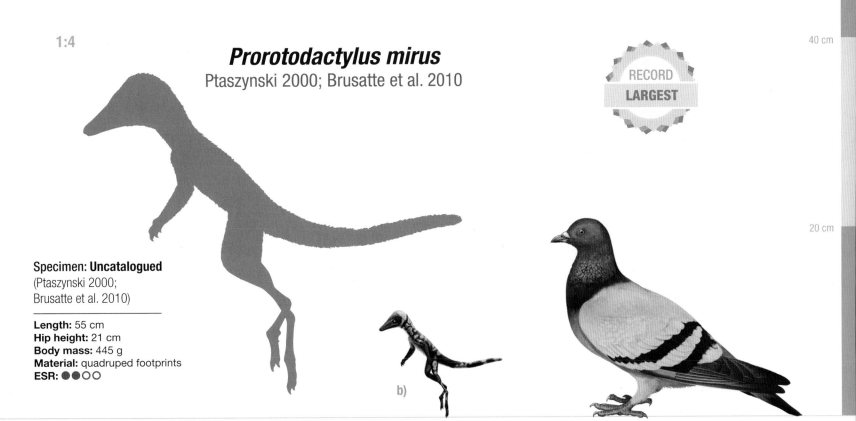

20 cm   40 cm   60 cm   80 cm

1:4

Prorotodactylus mirus
Ptaszynski 2000; Brusatte et al. 2010

40 cm

RECORD
LARGEST

20 cm

Specimen: Uncatalogued
(Ptaszynski 2000;
Brusatte et al. 2010)

Length: 55 cm
Hip height: 21 cm
Body mass: 445 g
Material: quadruped footprints
ESR: ●●○○

b)

The largest of the Lower Triassic (Class: Dwarf Grade III)

Prorotodactylus mirus ("previous to *Rotodactylus* and marvelous") (present-day Poland): It is an ichnogenus based on footprints that were located in north-central Pangea (present-day Poland). They may have belonged to an animal bigger than a pigeon. Here we compare the probable appearance of *Prorotodactylus mirus* with the diminutive avemetatarsalian *Scleromochlus* (b) of the Upper Triassic. This genus has a close relationship with pterosaurs and dinosaurs. It specialized in jumping rather than running. See more: *Brusatte et al. 2010; Niedzwiedzki et al. 2013*

The first clues to the origin of dinosaurs

Because of the lack of fossil bones, the earliest phase of dinosaur evolution is quite unknown. Thanks to indirect evidence (fossil footprints), however, we know that dinosaurs arose a few million years after the beginning of the Mesozoic.

At this time, almost all terrestrial vertebrates had quadrupedal locomotion, so it is believed that bipedalism in dinosauromorphs did not arise until the Middle Triassic. See more: *Brusatte et al. 2010*

The first published species of the Lower Triassic

Rotodactylus cursorius (1948) ("runner with rotated fingers"), from northwestern Pangea (present-day Arizona, USA): It is the oldest dinosauromorphic footprint in North America. See more: *Peabody 1948*

The last published species of the Lower Triassic

Prorotodactylus lutevensis (2000) ("previous to *Rotodactylus* from Lutev"), from north-central Pangea (present-day France): Initially, it was considered a new species of the ichnogenus *Rhynchosauroides*, which are footprints of some type of rhynchocephalic lizard. See more: *Demathieu 1984; Ptanzynski 2000*

Footprints

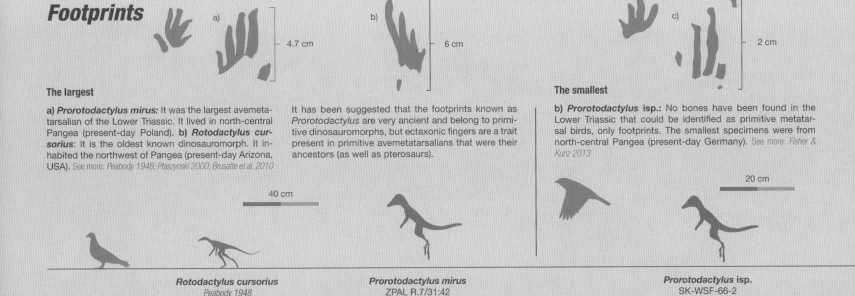

a) — 4.7 cm

b) — 6 cm

c) — 2 cm

The largest

a) *Prorotodactylus mirus*: It was the largest avemetatarsalian of the Lower Triassic. It lived in north-central Pangea (present-day Poland). **b) *Rotodactylus cursorius*:** It is the oldest known dinosauromorph. It inhabited the northwest of Pangea (present-day Arizona, USA). See more: *Peabody 1948; Ptaszynski 2000; Brusatte et al. 2010*

It has been suggested that the footprints known as *Prorotodactylus* are very ancient and belong to primitive dinosauromorphs, but ectaxonic fingers are a trait present in primitive avemetatarsalians that were their ancestors (as well as pterosaurs).

The smallest

b) *Prorotodactylus* isp.: No bones have been found in the Lower Triassic that could be identified as primitive metatarsal birds, only footprints. The smallest specimens were from north-central Pangea (present-day Germany). See more: *Fisher & Kunz 2013*

40 cm

20 cm

Rotodactylus cursorius
Peabody 1948
42 cm / 156 g

Prorotodactylus mirus
ZPAL R.7/31:42
55 cm / 445 g

Prorotodactylus isp.
SK-WSF-66-2
27 cm / 55 g

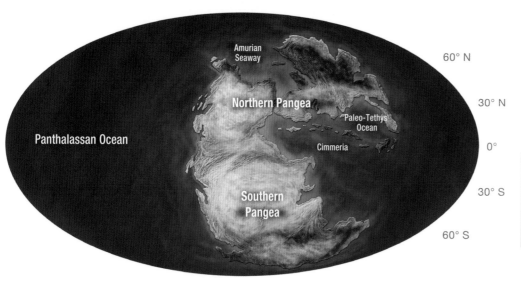

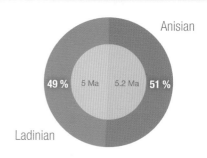

Middle Triassic

| Anisian 247.2-242 Ma |
| Ladinian 242-237 Ma |

The second part of the Triassic lasted about 10.2 million years, which is equivalent to 20% of the Triassic and 5% of the Mesozoic.

Anisian

Despite the extreme weather conditions, the Anisian was a great improvement over the Lower Triassic. The atmosphere was still hot and dry, especially in the interior. The Dead Zone disappeared and was replaced by a humid tropical climate with strong monsoons near the equator. Polar regions were humid and temperate, with very cold winters. Due to the high temperatures, which ranged between 20°C and 30°C, the sea level was low. In this period, dinosauromorphs were already present in the southern continents, and some were even specialized herbivores. Perhaps the first dinosaurs also existed, but they were not yet dominant, as the major predators were erythrosuchid archosaurs. See more: *Darwin 1834; Broom 1905; Szulc 1999; Veizer et al. 2000; Lucas 2005; Preto et al. 2010; Haas et al. 2012; Holtz 2015*

Ladinian

Wet episodes continued to increase in different parts of the world with heavy rains, but with seasonally dry forests. The largest carnivores were now the Rauisuchids, which looked like terrestrial crocodiles. The dinosauromorphs spread across the world. See more: *Mutti & Weissert 1995; Hounslow & Ruffell 2006; Marsicano 2008*

The smallest of the Middle Triassic (Class: Dwarf Grade III)

Avipes dillstedtianus ("feet of bird of the village Dillstedt"), from north-central Pangea (present-day Germany), is an incomplete metatarsal. Its shape is elongated, it is laterally compressed, and its symmetry suggests that it could have belonged to a derived (advanced) dinosauromorph or primitive dinosaur. Among the most likely first smaller dinosaurs is *"Thecodontosaurus" alophos*, which was about 2.45 m long and weighed 19 kg. It may be synonymous with *Nyasasaurus parringtoni*, which was slightly larger, with a length of 2.6 m and a weight of 23 kg. See more: *Haughton 1932; Huene 1932; Rauhut & Hungerbühler 2000; Nesbitt et al. 2013*

RECORD SMALLEST

Avipes dillstedtianus
Huene 1932

1:4

Specimen: Uncatalogued
(Huene 1932)

Length: 90 cm
Hip height: ~27 cm
Body mass: 630 g
Material: incomplete metatarsus
ESR: ●○○○

1 m   2 m   3 m

Unnamed
Peecook et al. 2013

RECORD
LARGEST

1:15

1 m

Specimen: Uncatalogued
(Peecook et al. 2013)

Length: 3 m
Hip height: 92 cm
Body mass: 83 kg
Material: femur
ESR: ●●○○

The largest of the Middle Triassic (Class: Medium Grade I)

Among the oldest dinosauromorphs, the largest of all is an undescribed specimen from south-central Pangea (present-day Zambia). It had a weight comparable to that of an adult person. It must have had quite derived features, suggesting that dinosauromorphs are older than what is known from direct remains. In addition, diverse footprints scattered all over the world (currently, Algeria, Arizona in the United States, Spain, France, Holland, Italy, Morocco, and Switzerland) show the presence of dinosauromorphs in the Middle Triassic. See more: *Peabody 1948; Demathieu & Gand 1975; Demathieu & Weidmann 1982; Demathieu 1984; Demathieu & Oosterink 1988; Demathieu et al. 1999; Kotanski et al. 2004; Gand et al. 2011; Klein et al. 2011; Peecook, Huttenlocker & Sidor 2013*

The first published species of the Middle Triassic

"Thecodontosaurus" primus (1905) ("alveoli-toothed ancient lizard"), from north-central Pangea (present-day Poland), was a primitive archosaur or a saurischian dinosaur. *"Zanclodon" silesiacus* (1910) is a tooth that could have been a rauisuchid or a very ancient theropod. *Avipes dillstedtianus* (1932) could be a dinosauromorph or a primitive dinosaur. See more: *Huene 1905; Huene 1932; Jaekel 1910; Carrano et al. 2012*

The last published species of the Middle Triassic

Lutungutali sitwensis (2013) ("high hip of the Sitwa village"), from south-central Pangea (present-day Zambia), is a silesaurid that was known as "N'tawere form." See more: *Peecook et al. 2013*

Footprints

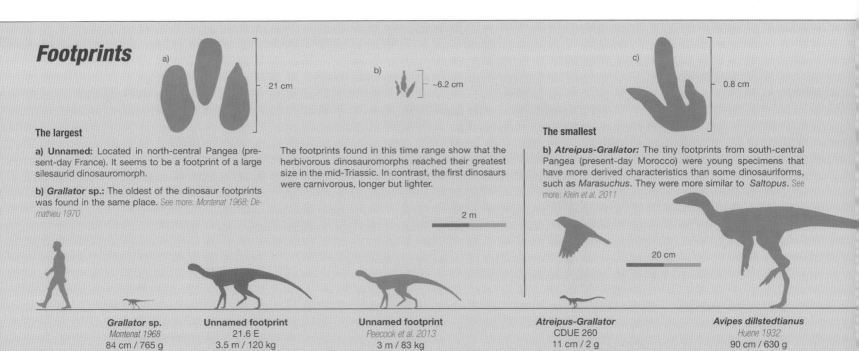

a)   21 cm

b)   ~6.2 cm

c)   0.8 cm

The largest

a) Unnamed: Located in north-central Pangea (present-day France). It seems to be a footprint of a large silesaurid dinosauromorph.

b) Grallator sp.: The oldest of the dinosaur footprints was found in the same place. See more: *Montenat 1968; Demathieu 1970*

The footprints found in this time range show that the herbivorous dinosauromorphs reached their greatest size in the mid-Triassic. In contrast, the first dinosaurs were carnivorous, longer but lighter.

The smallest

b) Atreipus-Grallator: The tiny footprints from south-central Pangea (present-day Morocco) were young specimens that have more derived characteristics than some dinosauriforms, such as *Marasuchus*. They were more similar to *Saltopus*. See more: *Klein et al. 2011*

2 m

20 cm

Grallator sp.	Unnamed footprint	Unnamed footprint	*Atreipus-Grallator*	*Avipes dillstedtianus*
Montenat 1968	21.6 E	*Peecook et al. 2013*	CDUE 260	*Huene 1932*
84 cm / 765 g	3.5 m / 120 kg	3 m / 83 kg	11 cm / 2 g	90 cm / 630 g

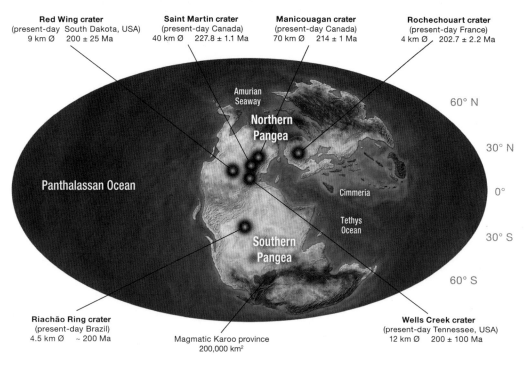

Red Wing crater
(present-day South Dakota, USA)
9 km Ø   200 ± 25 Ma

Saint Martin crater
(present-day Canada)
40 km Ø   227.8 ± 1.1 Ma

Manicouagan crater
(present-day Canada)
70 km Ø   214 ± 1 Ma

Rochechouart crater
(present-day France)
4 km Ø   202.7 ± 2.2 Ma

Amurian Seaway

Northern Pangea

Panthalassan Ocean

Cimmeria

Tethys Ocean

Southern Pangea

60° N

30° N

0°

30° S

60° S

Riachão Ring crater
(present-day Brazil)
4.5 km Ø   ~ 200 Ma

Magmatic Karoo province
200,000 km²

Wells Creek crater
(present-day Tennessee, USA)
12 km Ø   200 ± 100 Ma

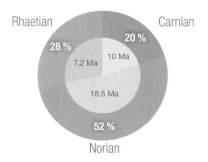

Rhaetian   Carnian

28 %   20 %

7.2 Ma   10 Ma

18.5 Ma

52 %

Norian

Upper Triassic

Carnian 237-227 Ma	
Norian 227-208.5 Ma	
Rhaetian 208.5-201.3 Ma	

The last part of the Triassic covered the most time, about 35.7 million years. This is equivalent to 70% of the duration of the Triassic and 20% of the duration of the Mesozoic.

Carnian

Due to the increase in Earth's flora, the amount of CO_2 increased in the atmosphere, creating conditions similar to the greenhouse effect. On the other hand, as a result of the separation of Pangea, rainfall increased significantly, especially near the equator. Meanwhile, the interior areas were arid. Finally, there are indications of a meteorite strike in northwestern Pangea (present-day western Canada). See more: *Simms & Ruffel 1989; Kent & Olsen 2000; Berner & Kothavala 2001; Sellwood & Valdés 2006; Roghi et al. 2010; Retallack 2013; Schmieder et al. 2014; Holtz 2015*

20 cm

Norian

Multiple meteorites fell in eastern Canada, creating a circle that currently forms Lake Manicouagan, or the "eye of Quebec." The largest terrestrial carnivores were the rauisuchids, the prestosuchids, and the phytosaurs. Dinosaurs still had a secondary role. It was the longest period of the Triassic, 18.5 million years in duration. See more: *Bonaparte 1981; Spray et al. 1998; Kent 1998; Ramezani et al. 2005*

Rhaetian

The highest concentration of Triassic CO_2 was reached during this age. At the end of the Triassic, a meteorite hit the north-central area of Pangea (present-day France), causing the fifth largest mass extinction, killing 76% of existing species. It must be said that the causes are not clear yet. See more: *Schmieder et al. 2010; Royer 2014*

The smallest of the Upper Triassic (Class: Tiny Grade I)

Scleromochlus taylori, from north-central Pangea (present-day Scotland), is the tiniest avemetatarsalian. It was 18.5 cm long and weighed 18 g. *Agnosphitys cromhallensis* (present-day England) and *Saltopus elginensis* (present-day Scotland) were the smallest dinosauromorphs. They were about 35 cm and 50 cm long and weighed 45 g and 70 g, respectively. Among the theropods, the smallest was *Procompsognathus triassicus* (present-day Germany), which was 1 m long and weighed 1.3 kg (comparable to a raven). See more: *Woodward 1907; Huene 1910; Fraas 1913; Fraser et al. 2002*

Smallest juveniles

The smallest specimens of *Scleromochlus taylori*; BMNH R3146B and BMNH 3146B, were barely 85% of the size of the type specimen **BMNH R3556**. They were about 15 cm long and weighed about 11 g. See more: *Benton 1999*

Scleromochlus taylori
Woodward 1907

10 cm

1:1,5

RECORD
SMALLEST

Specimen: BMNH R3556
(Woodward 1907)

Length: 18.5 cm
Hip height: 7 cm
Body mass: 18 g
Material: skull and partial skeleton
ESR: ●●●●

10 cm   20 cm   30 cm

Frenguellisaurus ischigualastensis
Reig 1963

1:25

Specimen: PVSJ 53
(Novas 1986)

Length: 5.3 m
Hip height: 1.55 m
Body mass: 360 kg
Material: skull and caudal vertebrae
ESR: ●●●○

The largest of the Upper Triassic (Class: Medium Grade III)

The saurischian dinosaur **PVSJ 53** was identified as the species *Frenguellisaurus ischigualastensis*, although it was twice as heavy as a lion and could be a large *Herrerasaurus ischigualastensis*. It lived in southwestern Pangea (present-day Argentina). Even longer but lighter, it could have been the possible coellophysoid SMNS 51958 (present-day Germany). Perhaps it reached a length of 6.8 m and a weight of 315 kg See more: *Novas 1986; Galton 1985*

The first published species of the Upper Triassic

Coelurus bauri, C. logicollis and *Tanystrophaeus willistoni* (now *Coelophysis*) (1887) from northwestern Pangea (present-day New Mexico, USA). Previous publications were *"Megalosaurus" cloacinus* (1858) and *"Megalosaurus" obtusus* (1876), from north-central Pangea (present-day Germany and France, respec-

tively), but there are doubts about their identification. The first dinosauromorph to receive a name was *Saltopus elginensis* ("leaping feet of Elgin") (1932) from north-central Pangea (present-day Scotland). See more: *Quenstedt 1858; Henry 1876; Cope 1887, 1889; Huene 1932; Reig 1963; Benton & Walker 2000*

The last published species of the Upper Triassic

Ignotosaurus fragilis (2013) ("unknown fragile lizard") from southwestern Pangea (present-day Argentina). The coelophysoid *Lepidus praecisio* (2015) from northwestern Pangea (present-day Texas, USA) was the last theropod to be named. On the other hand, the last country where a species of dinosauromorph was reported is Chile. The specimen is SGO.PV.22250. See more: *Martinez et al. 2013; Rubilar-Rugers et al. 2013; Sterling et al. 2015*

Footprints

a)

— 45 cm

b)

— 2cm

The largest

a) cf. *Eubrontes* **isp.:** From north-central Pangea (present-day Slovakia). They were large theropods similar to *Liliensternus* that were the largest carnivorous dinosaurs of the Triassic. See more: *Niedzwiedzki 2011*

The very asymmetrical footprints of the Triassic match better with those of *Liliensternus* than with any other coelophysoid, which had more even finger proportions.

The smallest

b) *Rotodactylus tumidus*: From northwestern Pangea (present-day Utah, USA). Initially considered to be of the ichnogenus *Rhynchosauroides*, it is now known that they were dinosauromorph footprints similar (but more primitive) to *Lagerpeton*. See more: *Morton 1897; Haubold 1971*

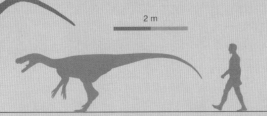

2 m

20 cm

cf. *Eubrontes* **isp.**
Niedzwiedzki 2011
8.4 m / 600 kg

Herrerasaurus ischigualastensis
PVSJ 53
5.3 m / 360 kg

Rotodactylus tumidus
Morton 1897
14 cm / 6 g

Saltopus elginensis
NHMUK R3915
50 cm / 110 g

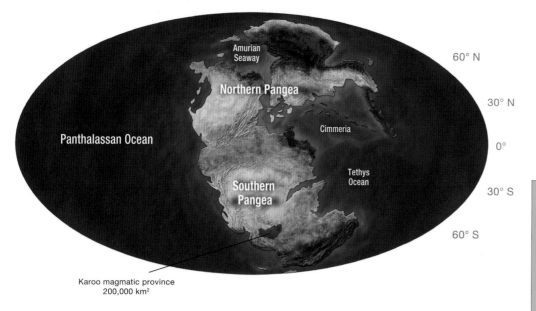

Hettangian

Toarcian

7 %

2 Ma

8.6 Ma   8.5 Ma   31 %   Sinemurian

32 %

8.1 Ma

30 %

Pliensbachian

Lower Jurassic
Hettangian 201.3–199.3 Ma
Sinemurian 199.3–190.8 Ma
Pliensbachian 190.8–182.7 Ma
Toarcian 182.7–174.1 Ma

The first part of the Jurassic lasted about 27.2 million years, which is 48% of the total duration of the Jurassic and 15% of the duration of the Mesozoic.

Hettangian

Once again, life began to recover from a huge mass extinction, resulting in the dominance of dinosaurs in terrestrial ecosystems. There were apparently no primitive saurischians or dinosauromorphs. The climate and atmospheric conditions were similar to those of the Upper Triassic; temperatures were between 5°C and 10° C higher than today. There were semi-arid deserts with dunes and oases near the equator, while in the south, there were cold and dark winters. The Hettangian and the Bajocian were the Jurassic periods that lasted the shortest time, just two million years. See more: *Chandler et al. 1992; Lucas 2006; Ruhl et al. 2011; Holtz 2015*

Sinemurian

The first large theropods, weighing more than 1 t, began to appear in the north and south of Pangea. See more: *Yadagiri 1982; Gierlinski 1991; Morales & Bulkey 1996: Niedzwedzki 2006*

Pliensbachian

Global warming that began again during this age was most likely the result of Karoo magmatism, a phenomenon that contributed to the separation of South America and Africa. There were also some cold periods. See more: *Jourdan et al. 2008; Krencker et al. 2014*

Toarcian

The sea lost oxygen due to high volcanic intensity in Karoo (present-day South Africa). It was one of the warmest periods of the Jurassic. There were waves of atmospheric cold produced by elements that disturbed solar radiation. The Toarcian was the Jurassic period that lasted the longest, 8.6 million years. See more: *Rampino & Self 1992; Dera et al. 2011; Kidder & Worsley 2012*

The smallest of the Lower Jurassic (Class: Small Grade I)

The coelophysoid **cf. *Megapnosaurus*** (informally called "Syntarsus mexicanum"), from northwestern Pangea (present-day Mexico), was a tiny adult specimen. It was barely 1 m long and weighed 1.1 kg. See more: *Munter 1999; Gudiño & Guzmán 2014*

The smallest juvenile

The specimen type of *Podokesaurus holyokensis* ("fast-footed lizard of the city Holyoke"), from northwestern Pangea (present-day Massachusetts, USA), was a juvenile coelophysoid that was 90 cm long and weighed 900 g. See more: *Talbot 1911*

---

40 cm

RECORD SMALLEST

cf. *Megapnosaurus* ("Syntarsus mexicanum")
Hernández-Rivera 2002

1:5

20 cm

Specimen: IGM 6624
(Munter 1999)

Length: 91 cm
Hip height: 26 cm
Body mass: 1.1 kg
Material: partial skeleton
ESR: ●●●○

20 cm   40 cm   60 cm   80 cm   100 cm

Dandakosaurus indicus
Yadagiri 1982

1:40

RECORD LARGEST

4 m

2 m

Specimen: GSI Collection
(Yadagiri 1982)

Length: 10 m
Hip height: 2.8 m
Body mass: 2.3 t
Material: tooth, vertebrae, and incomplete ischium
ESR: ●●○○

The largest of the Lower Jurassic (Class: Large Grade III)

Dandakosaurus indicus ("lizard of the Dandakaranya region and of India"), from south-central Pangea (present-day India), was the first predator dinosaur that weighed over 2 t. With this species, the new era of gigantic terrestrial carnivores began. See more: *Yadagiri 1982*

Worldwide distribution

At this time, theropod fossils were distributed all over the world. Direct evidence (bones) and indirect evidence (footprints) are found on all continents. See more: *Hitchcock 1836; Ellenberger 1970; Bittencourt & Langer 2011; Whiteside et al. 2015*

The first published species of the Lower Jurassic

Podokesaurus holyokensis (1911), from northwestern Pangea (present-day Massachusetts, USA), was the first to be identified. Material known as "Merosaurus newmani," from north-central Pangea (present-day England), was found 52 years earlier, but it was scrambled with the remains of the armed ornithischian *Scelidosaurus harrisonii*. "Merosaurus" was identified as the remains of a theropod 109 years later. See more: *Owen 1859; Talbot 1911; Newman 1968*

The most recently published species of the Lower Jurassic

Tachiraptor admirabilis (2014) ("admirable thief of the state Táchira"), from southwestern Pangea (present-day Venezuela), is the first theropod found in that country. It seems to be related to the trunk from which basal ceratosaurs derived. See more: *Larger et al. 2014*

Footprints

a)

65 cm

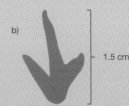

b)

1.5 cm

The largest

a) Unnamed: Due to its symmetrical shape, this footprint was not from a megalosaurid. It may have belonged to an animal similar to *Sinosaurus*, from north-central Pangea (present-day Poland). See more: *Gierlinski 1991; Niedzwedzki 2006*

The first theropods whose weight exceeded 1 t appeared in the Lower Jurassic, occupying the empty niches abandoned by the major predators of the Upper Triassic: the terrestrial archosaurs.

4 m

The smallest

b) *Grallator* sp.: A footprint belonging to a small juvenile coelophysoid from northwestern Pangea (present-day New Jersey, USA) is so tiny that there is no doubt that it belonged to a young specimen. See more: *Olsen 1995*

20 cm

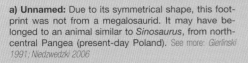

Unnamed footprint Gierlinski 1991; Niedzwedzki 2006 8.6 m / 1.35 t	***Dandakosaurus indicus*** GSI col. 10 m / 2.3 t	***Grallator* sp.** *Olsen 1995* 22 cm / 11 g	**cf. *Megapnosaurus*** IGM 6624 91 cm / 1.1 kg

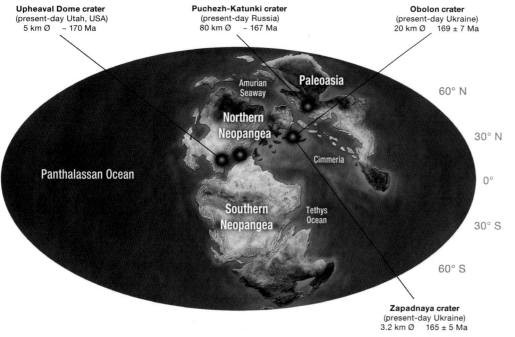

Upheaval Dome crater
(present-day Utah, USA)
5 km Ø   ~ 170 Ma

Puchezh-Katunki crater
(present-day Russia)
80 km Ø   ~ 167 Ma

Obolon crater
(present-day Ukraine)
20 km Ø   169 ± 7 Ma

Paleoasia

Amurian
Seaway

Northern
Neopangea

Cimmeria

Panthalassan Ocean

Southern
Neopangea

Tethys
Ocean

60° N

30° N

0°

30° S

60° S

Zapadnaya crater
(present-day Ukraine)
3.2 km Ø   165 ± 5 Ma

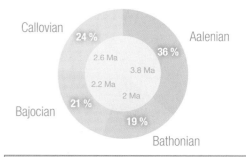

Callovian   24%   2.6 Ma   Aalenian   36%   3.8 Ma   2.2 Ma   2 Ma   Bajocian   21%   19%   Bathonian

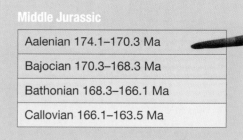

Middle Jurassic	
Aalenian 174.1–170.3 Ma	
Bajocian 170.3–168.3 Ma	
Bathonian 168.3–166.1 Ma	
Callovian 166.1–163.5 Ma	

The Middle Jurassic lasted about 10.6 million years, which is 19% of the duration of the Jurassic and 6% of the duration of the Mesozoic.

Aalenian

The forests were humid, and there were cold and dark winters. The deserts, on the other hand, began to retreat, and mangroves appeared on the coasts. See more: *Lucas 2006*

Bajocian

It was seasonally arid, with humid cycles. The sea maintained a surface temperature between 26°C and 30°C. This lasted until the late Lower Cretaceous (Aptian). Together with the Hettangian, the Bajocian was the Jurassic period that lasted the shortest time, just two million years. See more: *Hesselbo et al. 2003; Jenkyns et al. 2012*

Bathonian

During the Bajocian and Bathonian, meteorites fell in Ukraine, Russia, and Utah in the United States. Sea level began rising around the world, probably caused by a seabed expansion that flooded coastal areas. See more: *Dromart et al. 2003*

Callovian

The seas cooled drastically, possibly because of the presence of continental ice, which lasted about 2.6 million years. It was the coldest climate of the Mesozoic. See more: *Dromart et al. 2003, 2004*

The smallest of the Middle Jurassic (Class: Dwarf Grade III)

Epidexipteryx hui ("arboreal wing of the paleontologist Hu Yao Ming"), from eastern Paleoasia (present-day China) was a tiny and strange paravian with a very short tail and long rectrices. Because of its large forward-facing teeth, it is thought to be an insectivore. It was smaller than a pigeon. See more: *Zhang et al. 2008*

The smallest juvenile

Epidendrosaurus ningchengensis ("arboreal lizard of Ningcheng County"), from eastern Paleoasia (present-day China), was similar to *Epidexipteryx hui* but had a very long tail. It was barely 13 cm long and weighed 6.7 g. See more: *Zhang et al. 2002*

40 cm

20 cm

1:3

Epidexipteryx hui
Zhang et al. 2008

Specimen: **IVPP V15471**
(Zhang et al. 2008)

Length: 25 cm
Hip height: 14 cm
Body mass: 220 g
Material: skull, partial skeleton, and feathers
ESR: ●●●●

RECORD
SMALLEST

20 cm   40 cm   60 cm

cf. *Megalosaurus bucklandii*
Mantell 1827

RECORD LARGEST

Specimen: PVSJ 53
(Phillips 1871)

Length: 10.7 m
Hip height: 2.75 m
Body mass: 3 t
Material: incomplete femur
ESR: ●●○○

1:40

The largest of the Middle Jurassic (Class: Large Grade III)

Uncatalogued material from north-central Neopangea (present-day England) is referred to as ***Megalosaurus bucklandii*** but is more recent. Its weight was equivalent to that of two hippos together. See more: *Phillips 1871; Lydekker 1888*

The first published species of the Middle Jurassic

Megalosaurus (1824) ("big lizard") from north-central Neopangea (present-day England). It did not receive a species name until three years later (1827), although in a previous publication (1823), the name is mentioned. The skeleton

is based on materials found in 1815. See more: *Parkinson 1823; Buckland 1824; Mantell 1827; Buffetaut 1991; Allain 2001*

The most recently published species of the Middle Jurassic

Yi qi (2015) ("strange wing"), from eastern Paleoasia (present-day China), received the shortest known name for a theropod, even among modern birds. See more: *Xu et al. 2015*

Footprints

a)
— 72 cm

b)
— 1.78 cm

The largest

a) Unnamed: A 180 m long track of huge metriacantosaurid footprints from north-central Neopangea (present-day England). These theropods may have coexisted with megalosaurids of similar size, sharing or competing for the same prey. See more: *Day et al. 2004*

The first theropods to weigh over 3 t, including megalosauroids and the allosauroids, appeared in the Middle Jurassic.

The smallest

b) Unnamed: The smallest of several footprints located in north-central Neopangea (present-day Scotland) is recognized as the smallest dinosaur footprint in the world, although this record has been superseded. See more: *Clark et al. 2005*

4 m

20 cm

Unnamed footprint
T80
11.9 m / 3.2 t

cf. *Megalosaurus bucklandii*
Phillips 1871
10.7 m / 3 t

Unnamed footprint
Track H
28 cm / 18 g

Epidexipteryx hui
IVPP V15471
25 cm / 220 g

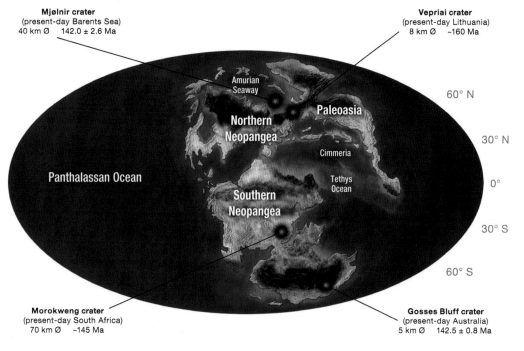

Mjølnir crater
(present-day Barents Sea)
40 km Ø   142.0 ± 2.6 Ma

Vepriai crater
(present-day Lithuania)
8 km Ø   ~160 Ma

Amurian
Seaway

Paleoasia

60° N

**Northern
Neopangea**

30° N

Cimmeria

Panthalassan Ocean

0°

Tethys
Ocean

**Southern
Neopangea**

30° S

60° S

Morokweng crater
(present-day South Africa)
70 km Ø   ~145 Ma

Gosses Bluff crater
(present-day Australia)
5 km Ø   142.5 ± 0.8 Ma

Tithonian   38%   7.1 Ma   6.2 Ma   34%   Oxfordian

5.2 Ma

28%

Kimmeridgian

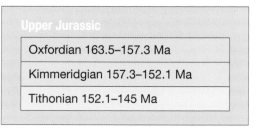

Upper Jurassic	
Oxfordian 163.5–157.3 Ma	
Kimmeridgian 157.3–152.1 Ma	
Tithonian 152.1–145 Ma	

The last part of the Jurassic lasted about 18.5 million years, which is 33% of the duration of the Jurassic and 8% of the duration of the Mesozoic.

Oxfordian

The climate was similar to that of the Callovian, although it was somewhat warmer, with open forests and denser vegetation in the interior. See more: *Dromart et al. 2003; Lucas 2006*

Kimmeridgian

There were arid conditions in northwestern and north-central Neopangea and cold climates in the high latitudes of Paleoasia (present-day Russia) and Gondwana (present-day Oceania). This might have caused some cool conditions in the global environment. See more: *Valdés & Sellwood 1992*

Tithonian

Rising sea levels isolated areas between continents and flooded large areas of Europe and North America. The global climate changed, causing periods of intense rain and an increase in temperature. Evidence of meteorite impacts is found near Norway, in southern Africa, and in central Australia. They may be

related to the extinction that occurred in the Cretaceous. See more: *Milton & Sutter 1987; Moore et al. 1992; Milton et al. 1996; Koeberl et al. 1997; Henkel et al. 2002; Maier et al. 2006; Keller 2008; Dypvik et al. 2010; Tsikalas 2005; Wierzbowski et al. 2013*

The smallest of the Upper Jurassic (Class: Dwarf Grade III)

The largest of the specimens of **Archeopteryx lithographica** ("ancient feather engraved in stone"), from the north-central area of Neopangea (present-day Germany), was three times lighter than a crow. The theropods *Compsognathus longipes* and *Koparion douglassi* weighed approximately 2.3 kg and were 1.5 m and 1 m long, respectively. See more: *Bidar et al. 1972; Chure 1994*

The smallest juveniles

An embryo of *Torvosaurus* sp. from north-central Neopangea (present-day Portugal) was hardly 22 cm long and weighed only 52 g. The youngest bird is the young *Archeopteryx lithographica* MJ SoS 2257 (present-day Germany), which was 28 cm long and weighed 60 g. See more: *Wellnhofer 1974; Araújo et al. 2013*

40 cm

20 cm

Archaeopteryx lithographica
Meyer 1861

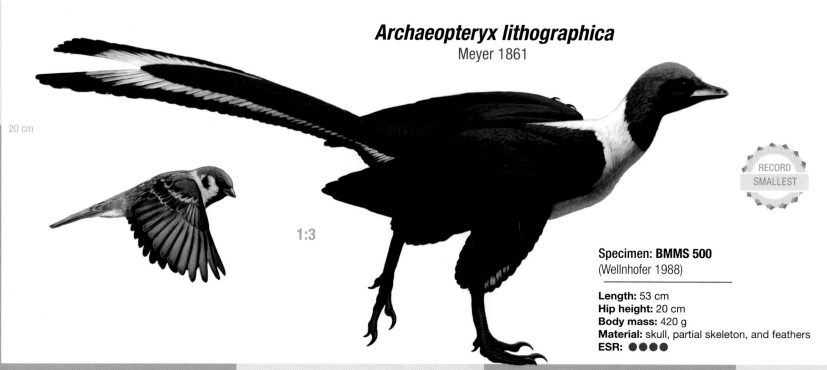

1:3

RECORD
SMALLEST

Specimen: BMMS 500
(Wellnhofer 1988)

Length: 53 cm
Hip height: 20 cm
Body mass: 420 g
Material: skull, partial skeleton, and feathers
ESR: ●●●●

20 cm   40 cm   60 cm

"Megalosaurus" ingens
Janensch 1920

1:50

4 m

2 m

Specimen: MNHUK R6758
(Charig 1979)

Length: 12.6 m
Hip height: 3.6 m
Body mass: 6.4 t
Material: Tooth
ESR: ●◑○○

The largest of the Upper Jurassic (Class: Very large Grade II)

"Megalosaurus" ingens, from south-central Gondwana (present-day Tanzania), is based on enormous teeth that have interdental grooves and other characteristics similar to those of derived carcharodontosaurids. Due to its age, however, this may be a convergent characteristic. MNHUK R6758 is the largest of the known pieces. It weighed more than an adult African elephant and was larger than the huge megalosaurs and allosaurs of its time. See more: *Charig 1979; Rauhut 1995*

The first published species of the Upper Jurassic

The theropod *Saurocephalus monasterii* (1846) ("head of lizard of Monaster"), from north-central Neopangea (present-day Germany), was mistaken for fossil fish teeth. Its true nature was discovered 151 years later. See more: *Münster 1846; Winfolf 1997*

The last published species of the Upper Jurassic

Chilesaurus diegosuarezi (2015) ("lizard from Chile of Diego Suárez"), from southwestern Gondwana (present-day Chile), is one of the strangest dinosaurs known. It had short arms and possibly didactyl hands, a long neck, a short head, and wide feet, and it was herbivorous. The specimen was almost complete, but due to its great specialization, and because close relatives have yet to be found, it is difficult to establish kinship with other dinosaurs. See more: *Novas et al. 2015*

Footprints

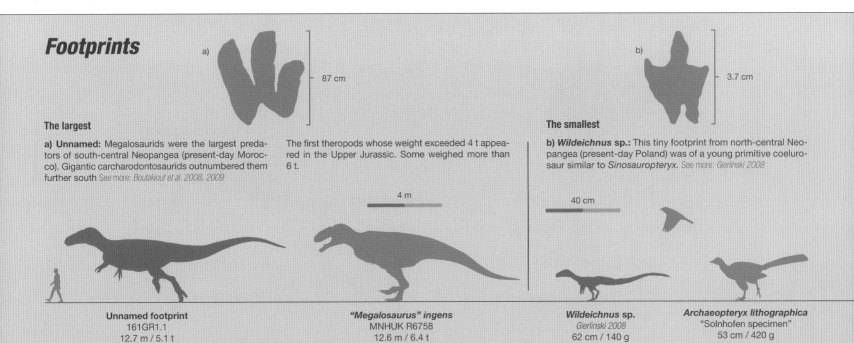

a) 87 cm

b) 3.7 cm

The largest

a) Unnamed: Megalosaurids were the largest predators of south-central Neopangea (present-day Morocco). Gigantic carcharodontosaurids outnumbered them further south See more: *Boutakiout et al. 2008, 2009*

The first theropods whose weight exceeded 4 t appeared in the Upper Jurassic. Some weighed more than 6 t.

The smallest

b) Wildeichnus sp.: This tiny footprint from north-central Neopangea (present-day Poland) was of a young primitive coelurosaur similar to *Sinosauropteryx*. See more: *Gierlinski 2008*

4 m

40 cm

Unnamed footprint
161GR1.1
12.7 m / 5.1 t

"Megalosaurus" ingens
MNHUK R6758
12.6 m / 6.4 t

Wildeichnus sp.
Gierlinski 2008
62 cm / 140 g

Archaeopteryx lithographica
"Solnhofen specimen"
53 cm / 420 g

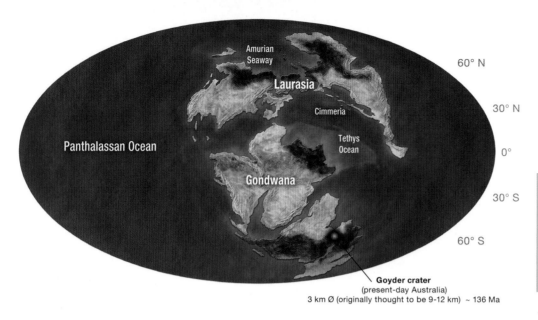

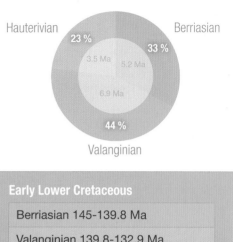

Hauterivian	Berriasian
23 %	33 %
3.5 Ma	5.2 Ma
6.9 Ma	
44 %	
Valanginian	

Goyder crater
(present-day Australia)
3 km Ø (originally thought to be 9-12 km) ~ 136 Ma

The early Lower Cretaceous lasted about 15.6 million years, which is equivalent to 20% of the duration of the Cretaceous and 8% of the duration of the Mesozoic.

Berriasian

The climate was relatively warm and arid near the equator and with open forests. In the interior, the vegetation was heavier and had a tropical monsoon climateSee more: *Lucas 2006; Hay & Floegel 2012*

Valanginian

Temperatures cooled, possibly due to temporary glaciation, but increased later. See more: *Wang et al. 2012*

Hauterivian

Temperatures rose and stabilized. It is suspected that the Goyder crater (present-day Australia) belongs to this age. Unfortunately, it is very eroded, and the terrain is difficult to access, so it is not easy to specify. See more: *Haines 1996, 2005; Wang et al. 2012*

The smallest of the early Lower Cretaceous (Class: Tiny Grade II)

The tiniest theropods of this period were avian. **Eopengornis martini** ("old *Pengornis* of Larry Martin"), from eastern Paleoasia (present-day China), was an enantiornithean subadult bird the size of a sparrow. Pengornitid enantiornithean birds had small teeth with low crowns, which indicates that they ate soft foods. SBEI 307 (present-day Japan) is the same size, but its developmental stage is unknown. Among theropods, the paravian ISMD-VP09 stands out, since it was 74 cm long and weighed 720 g, although there is no certainty of its developmental stage. See more: *Unwin & Matsuoka 2000; Zhang et al. 2008; O'Connor & Chiappe 2011; Prieto-Marquez et al. 2011; Wang et al. 2014*

20 cm

10 cm

1:2

Eopengornis martini
Wang et al. 2014

RECORD
SMALLEST

Specimen: STM24-1
(Wang et al. 2014)

Length: 14 cm
Hip height: 7 cm
Body mass: 39 g
Material: skull and partial skeleton
ESR: ●●●●

10 cm   20 cm   30 cm   40 cm

Unnamed
Galton & Molnar 2011

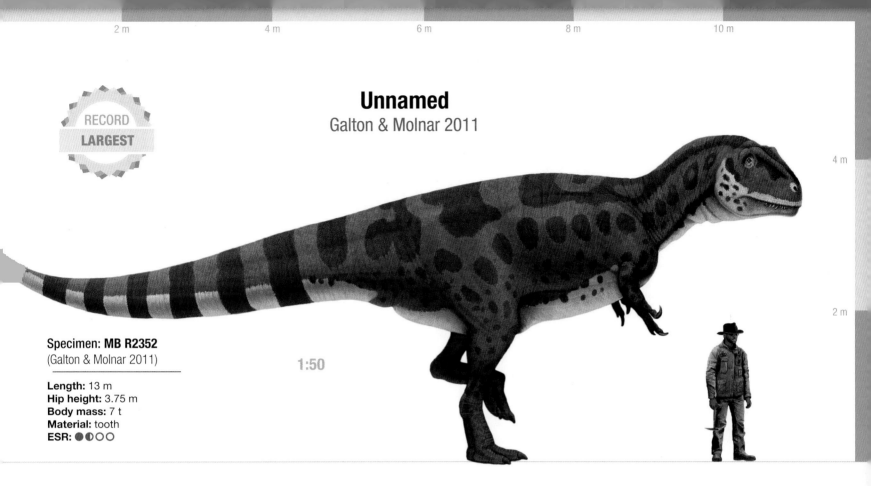

RECORD LARGEST

4 m

2 m

Specimen: **MB R2352**
(Galton & Molnar 2011)

1:50

Length: 13 m
Hip height: 3.75 m
Body mass: 7 t
Material: tooth
ESR: ●◐○○○

The largest of the early Lower Cretaceous (Class: Very large Grade II)

The carcharodontosaurid MB R2352, from south-central Gondwana (present-day South Africa), is known only by an incomplete tooth. Its anteroposterior part is one of the widest registered in theropods. It must have belonged to a predator as heavy as a killer whale. Among birds, *Eoconfuciusornis zhengi* was the largest at that time. It was only 18.5 cm long and weighed 70 g. See more: *Galton & Molnar 2011*

The first published species of the early Lower Cretaceous

The spinosaurid *Suchosaurus cultridens* (1841) ("crocodile lizard with sharp teeth") was the first of this age to receive a name. Due to its cylindrical teeth and

long face, it was initially considered a crocodile. See more: *Owen 1841; Lydekker 1889*

The most recently published species of the early Lower Cretaceous

Eopengornis martini (2014) is one of the few birds known from the early Lower Cretaceous. *Archaeornithura meemannae* (2015) ("old ornithuromorph of Doctor Meemann Chang"), from eastern Laurasia (present-day China), was found in the Huajiying Formation, which is thought to be from this time, but it may actually belong to the early Upper Cretaceous. See more: *Wang et al. 2014; Wang et al. 2015*

Footprints

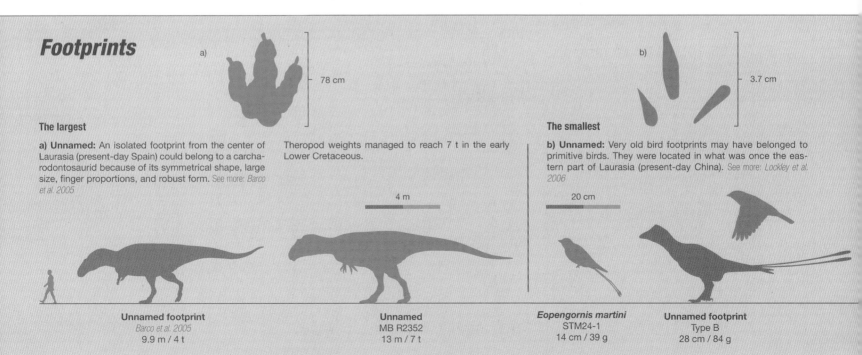

a)   — 78 cm

b)   — 3.7 cm

The largest

a) Unnamed: An isolated footprint from the center of Laurasia (present-day Spain) could belong to a carcharodontosaurid because of its symmetrical shape, large size, finger proportions, and robust form. See more: *Barco et al. 2005*

Theropod weights managed to reach 7 t in the early Lower Cretaceous.

The smallest

b) Unnamed: Very old bird footprints may have belonged to primitive birds. They were located in what was once the eastern part of Laurasia (present-day China). See more: *Lockley et al. 2006*

4 m

20 cm

Unnamed footprint
Barco et al. 2005
9.9 m / 4 t

Unnamed
MB R2352
13 m / 7 t

Eopengornis martini
STM24-1
14 cm / 39 g

Unnamed footprint
Type B
28 cm / 84 g

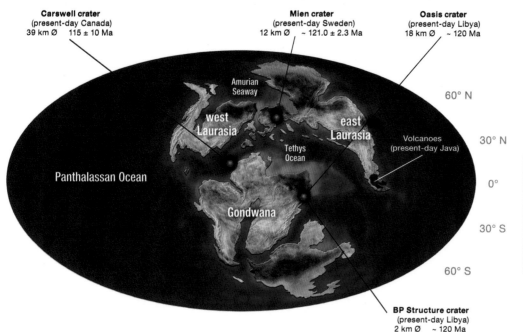

Carswell crater
(present-day Canada)
39 km Ø   115 ± 10 Ma

Mien crater
(present-day Sweden)
12 km Ø   ~ 121.0 ± 2.3 Ma

Oasis crater
(present-day Libya)
18 km Ø   ~ 120 Ma

Amurian Seaway

west Laurasia

east Laurasia

Volcanoes
(present-day Java)

Tethys Ocean

Panthalassan Ocean

Gondwana

60° N

30° N

0°

30° S

60° S

BP Structure crater
(present-day Libya)
2 km Ø   ~ 120 Ma

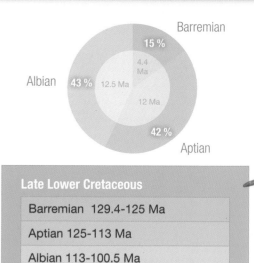

Barremian 15 %

Albian 43 %   12.5 Ma   4.4 Ma   12 Ma

Aptian 42 %

Late Lower Cretaceous

Barremian 129.4-125 Ma
Aptian 125-113 Ma
Albian 113-100.5 Ma

The Late Lower Cretaceous period lasted about 28.9 million years, which is equivalent to 36% of the duration of the Jurassic and to 15% of Mesozoic era.

Barremian

The climate was warm near the equator, with open forests and monsoon rains in the interior. Angiosperms proliferated around the world, a phenomenon known as an "abominable mystery." Near the start of the period, a meteorite probably hit Australia. See more: *Gorter et al. 1989; Gostin & Therriault 1997; Friedman 2009; Hay & Floegel 2012*

Aptian

The climate was warm until a temporary glaciation occurred and lasted 2.5 million years. The volcanic activity was intense in the paleocontinent Cimmeria (present-day Java). See more: *Kidder & Worsley 2012; Wang et al. 2012; McAnena et al. 2013*

Albian

Temperatures increased, especially near the equator, and CO_2 levels rose. Equatorial waters were too hot for marine life. The circulation of oxygen in the depths of the sea was interrupted, while large deserts spread on land. Midway through this age, a series of meteorites hit in Canada, Libya, and Sweden. It was the longest period of the Cretaceous, lasting 12.5 million years. See more: *Astrom 1998; Hay & Floegel 2012*

The smallest of the late Lower Cretaceous (Class: Tiny Grade II)

Cratoavis cearensis ("bird of the Crato Formation from the state of Ceará") from western Gondwana (present-day Brazil). The skeleton has characteristics of adults and juveniles, so it might be a subadult. Its size is smaller than that of *Regulus bulgaricus*, the smallest fossil bird of the Cenozoic, and perhaps smaller than Brace's emerald hummingbird, *Chlorostilbon bracei*, of the Bahamas, the smallest bird to go extinct in recent times. The smallest non-avian theropod of the time was perhaps IVPP V22530, an adult that was about 46 cm long and weighed 174 g. See more: *Lawrence 1877; Boev 1999; Carvalho et al. 2015; Pittman et al. 2015*

The smallest juvenile

Dalingheornis liweii ("bird from the Dalinghe River of the astronaut Yang Liwei") was a baby with zygodactyl feet, a specialization that allows birds to grasp branches more easily. It was possibly about 5.7 cm long and weighed just 2 g, the same as the bee hummingbird, the smallest known bird. See more: *Zhang et al. 2006*

10 cm

Cratoavis cearensis
Carvalho et al. 2015

1:1

5 cm

RECORD SMALLEST

Specimen: UFRJ-DG 031 Av
(Carvalho et al. 2015)

Length: 6.6 cm
Hip height: 2.9 cm
Body mass: 4 g
Material: skull, partial skeleton, and feathers
ESR: ●●●●

5 cm   10 cm   15 cm   20 cm

Carcharodontosaurus cf. *saharicus*
Lapparent 1960

RECORD
LARGEST

4 m

1:50

2 m

Specimen: MNNHN Collection
(Lapparent 1960)

Length: 12.7 m
Hip height: 3.75 m
Body mass: 7.8 t
Material: tooth
ESR: ●●○○○

The largest of the late Upper Cretaceous (Class: Huge Grade II)

Carcharodontosaurids were the dominant carnivores from the Late Jurassic to the late Upper Cretaceous, a reign that lasted approximately seventy-four million years. ***Carcharodontosaurus* cf. *saharicus*** ("lizard with shark teeth from the Sahara Desert"), of the MNNHN collection, is a colossal tooth, suggesting that its owner, which lived in central Gondwana (present-day Niger), must have been one of the largest theropods in history. This specimen was older and from farther south than *Carcharodontosaurus saharicus*. The largest bird of the time was QM F37912, which was 1.65 m long and weighed 21 kg (present-day Australia). See more: *Lapparent 1960; Mortimer**

The first published species of the late Lower Cretaceous

Calamospondylus oweni ("paleontologist Richard Owen's keel vertebra"), from central Laurasia (present-day England), are fragments of a possible tyrannosaurid. See more: *Fox 1866*

The most recently published species of the late Lower Cretaceous

Archaeornithura meemannae, Cratoavis cearensis, Dunhuangia cuii, Juehuaornis zhangi, Parapengornis eurycaudatus and *Yuanjiawaornis viriosus* were described in 2015. *Houornis caudatus* (2015) was a name change for *Cathayornis caudatus*. The last to be published, in 2015, was *Holbotia ponomarenkoi*. The name already existed 33 years prior but was not valid. See more: *Kurochkin 1982; Hou 1997; Carvalho et al. 2015; Hu et al. 2015a; Hu, et al. 2015b; Wang et al. 2015; Wang et al. 2015; Wang et al. 2015; Wang et al. 2015; Zelenov & Averianov 2015*

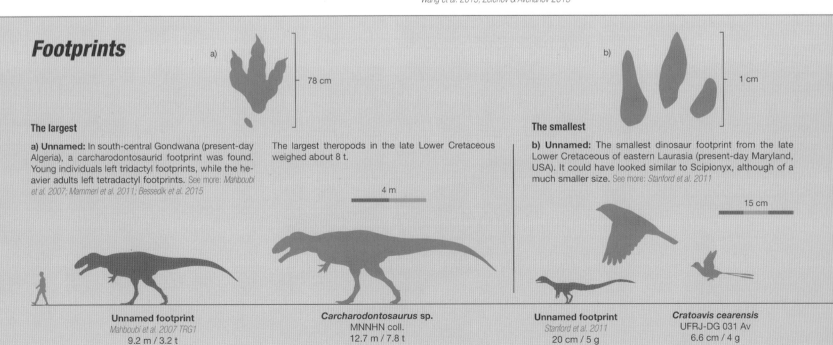

Footprints

a)

78 cm

The largest

a) Unnamed: In south-central Gondwana (present-day Algeria), a carcharodontosaurid footprint was found. Young individuals left tridactyl footprints, while the heavier adults left tetradactyl footprints. See more: *Mahboubi et al. 2007; Mammeri et al. 2011; Bessedik et al. 2015*

The largest theropods in the late Lower Cretaceous weighed about 8 t.

b)

1 cm

The smallest

b) Unnamed: The smallest dinosaur footprint from the late Lower Cretaceous of eastern Laurasia (present-day Maryland, USA). It could have looked similar to Scipionyx, although of a much smaller size. See more: *Stanford et al. 2011*

15 cm

4 m

Unnamed footprint
Mahboubi et al. 2007 TRG1
9.2 m / 3.2 t

***Carcharodontosaurus* sp.**
MNNHN coll.
12.7 m / 7.8 t

Unnamed footprint
Stanford et al. 2011
20 cm / 5 g

Cratoavis cearensis
UFRJ-DG 031 Av
6.6 cm / 4 g

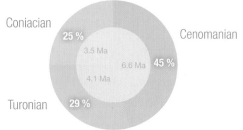

Deep Bay crater
(present-day Canada)
13 km Ø   99 ± 4 Ma

Steen River crater
(present-day Canada)
25 km Ø   91 ± 7 Ma.

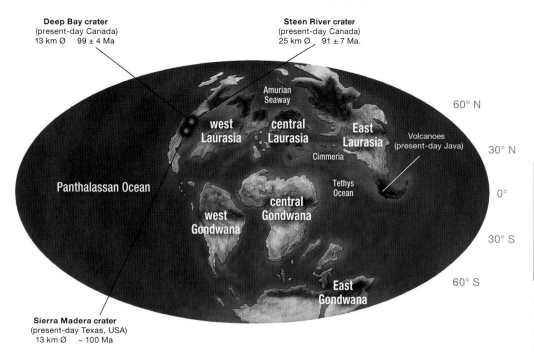

Amurian
Seaway

west
Laurasia

central
Laurasia

East
Laurasia

Volcanoes
(present-day Java)

60° N

30° N

Cimmeria

Panthalassan Ocean

Tethys
Ocean

0°

west
Gondwana

central
Gondwana

30° S

East
Gondwana

60° S

Sierra Madera crater
(present-day Texas, USA)
13 km Ø   ~ 100 Ma

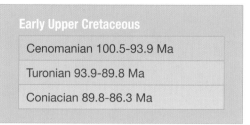

Coniacian

25 %
3.5 Ma

Cenomanian

6.6 Ma   45 %

4.1 Ma

Turonian   29 %

Early Upper Cretaceous

| Cenomanian 100.5-93.9 Ma |
| Turonian 93.9-89.8 Ma |
| Coniacian 89.8-86.3 Ma |

The early Upper Cretaceous lasted about 14.2 million years, which is equivalent to 18% of the duration of the Cretaceous and 8% of the duration of the Mesozoic.

Cenomanian

A gradual cooling began and continued until the end of the Mesozoic, although it is worth mentioning that there was still intense heat. Volcanic activity was renewed in the paleocontinent Cimmeria (present-day Java). At the poles, winters were dark and summers were bright. There were coastal mangroves in western Gondwana (present-day Brazil). Craters in Texas in the United States and in Canada show that meteorites struck during this time. See more: *Kidder & Worsley 2012; Wang et al. 2012; Zhou et al. 2012; McAnena et al. 2013; Wang et al. 2013*

Turonian

In east-central Laurasia (present-day Russia and Mongolia), temperatures could reach 45°C. In equatorial South America and Africa, they varied between 30°C and 42°C. Temperatures in the southern hemisphere were so low during some seasons that polar ice caps formed. A meteorite hit what is now Canada. See more: *Niccoli et al. 2005; Bice et al. 2006; Sellwood & Valdés 2006*

Coniacian

Temperatures near the equator rose. Forests expanded from the middle to the high latitudes. At the end of this age, volcanic activity in Cimmeria (present-day Java) finished. See more: *Hay & Floegel 2012; Zhou et al. 2012*

The smallest of the early Upper Cretaceous Cretaceous (Class: Tiny Grade II)

Parvavis chuxiongensis ("small bird of the city Chuxiong"), from eastern Laurasia (present-day China), was the smallest Mesozoic bird from Asia. It was really tiny; its weight was only three times that of the bee hummingbird (*Mellisuga helenae*). Among non-avian theropods, the alvarezsaurid *Xixianykus zhangi* (present-day China) was the smallest. It was 65 cm long and weighed 440 g. See more: *Xu et al. 2010; Wang, Zhou & Xu 2014*

10 cm

RECORD
SMALLEST

Parvavis chuxiongensis
Wang et al. 2014

1:1

5 cm

Specimen: **IVPP V18586**
(Wang et al. 2014)

Length: 7 cm
Hip height: 3.3 cm
Body mass: 6 g
Material: partial skeleton and feathers
ESR: ●●●○

5 cm   10 cm   15 cm   20 cm

2 m   4 m   6 m   8 m   10 m

Giganotosaurus carolinii
Coria & Salgado 1995

RECORD LARGEST

4 m

2 m

1:50

Specimen: MUCPv-95
(Calvo & Coria 2000)

Length: 13.2 m
Hip height: 3.85 m
Body mass: 8.5 t
Material: incomplete dentary with teeth
ESR: ●●○○

The largest of the early Upper Cretaceous (Class: Very large Grade II)

Giganotosaurus carolinii ("giant southern lizard of Rubén Carolini"), from western Gondwana (present-day Argentina), was the largest carcharodontosaurid and probably coexisted with one of the largest known dinosaurs, *Argentinosaurus huinculensis*. A healthy adult of these sauropods would have been invulnerable to their attacks, but they might have hunted their young or sick. The largest bird of that time was hesperornithiform FHSM VP-6318 (present-day Kansas, USA). It was 1.15 m long and weighed 12 kg. See more: *Calvo & Coria 2000*

The first published species of the early Upper Cretaceous

Poekilopleuron schmidti (1883) ("variable ribs of Friedrich Schmidt"), from eastern Laurasia (present-day western Russia), is a paleontological chimera. It is known now that it was formed from sauropod remains mixed with theropod material. See more: *Kiprijanov 1883*

The most recently published species of the early Upper Cretaceous

Parvavis chuxiongensis (2014) (present-day China) was the smallest Mesozoic theropod until the discovery of *Cratoavis cearensis*. The names "Marocannoraptor elbegiensis" and "Osteoporosia gigantea" (2015) are informal and belong to a private collection. The first is apparently a crocodile bone, and the second is possibly a carcharodontosaurid whose estimated size was originally exaggerated. See more: *Wang et al. 2014; Carvalho et al. 2015; Cau*; Mortimer*; Singer**

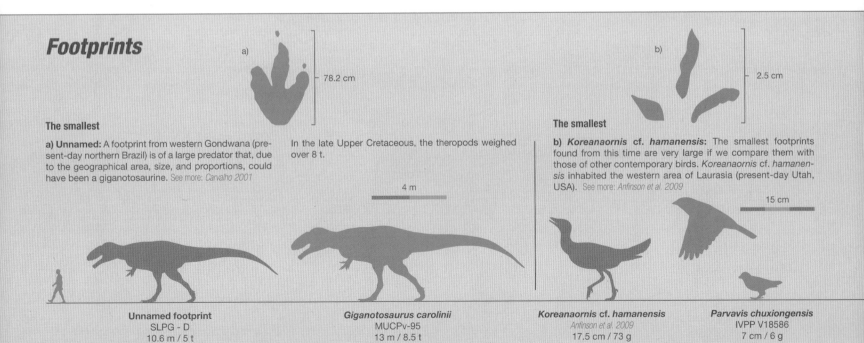

Footprints

a)

78.2 cm

The smallest

a) Unnamed: A footprint from western Gondwana (present-day northern Brazil) is of a large predator that, due to the geographical area, size, and proportions, could have been a giganotosaurine. See more: *Carvalho 2001*

In the late Upper Cretaceous, the theropods weighed over 8 t.

4 m

b)

2.5 cm

The smallest

b) *Koreanaornis* cf. *hamanensis*: The smallest footprints found from this time are very large if we compare them with those of other contemporary birds. *Koreanaornis* cf. *hamanensis* inhabited the western area of Laurasia (present-day Utah, USA). See more: *Anfinson et al. 2009*

15 cm

Unnamed footprint	*Giganotosaurus carolinii*	*Koreanaornis* cf. *hamanensis*	*Parvavis chuxiongensis*
SLPG - D	MUCPv-95	*Anfinson et al. 2009*	IVPP V18586
10.6 m / 5 t	13 m / 8.5 t	17.5 cm / 73 g	7 cm / 6 g

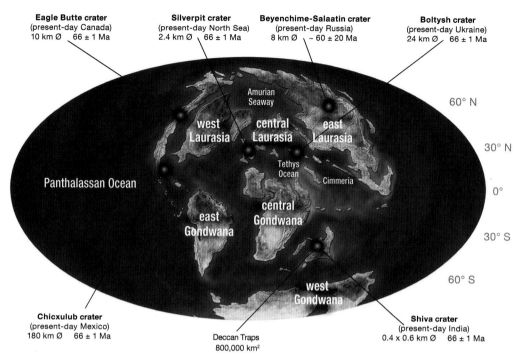

Eagle Butte crater
(present-day Canada)
10 km Ø   66 ± 1 Ma

Silverpit crater
(present-day North Sea)
2.4 km Ø   66 ± 1 Ma

Beyenchime-Salaatin crater
(present-day Russia)
8 km Ø   ~ 60 ± 20 Ma

Boltysh crater
(present-day Ukraine)
24 km Ø   66 ± 1 Ma

Amurian
Seaway

west
Laurasia

central
Laurasia

east
Laurasia

60° N

30° N

Tethys
Ocean

Cimmeria

0°

Panthalassan Ocean

east
Gondwana

central
Gondwana

30° S

60° S

west
Gondwana

Chicxulub crater
(present-day Mexico)
180 km Ø   66 ± 1 Ma

Deccan Traps
800,000 km²

Shiva crater
(present-day India)
0.4 x 0.6 km Ø   66 ± 1 Ma

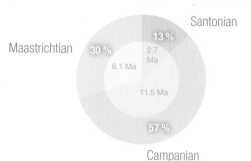

Santonian

Maastrichtian

13 %

2.7
Ma

30 %

6.1 Ma

11.5 Ma

57 %

Campanian

Late Upper Cretaceous
Santonian 86.3-83.6 Ma
Campanian 83.6-72.1 Ma
Maastrichtian 72.1-66 Ma

The late Upper Cretaceous lasted 20.3 million years, which is 26% of the duration of the Jurassic and 11% of the duration of the Mesozoic.

Santonian

There were cold seasons with a transitory glaciation. This period of the Cretaceous lasted the shortest time, just 2.7 million years. See more: *Wang et al. 2012*

Campanian

This era lasted almost two-thirds of the total duration of the late Upper Cretaceous. It had a great diversity of dinosaurs. Two meteorites struck during it, in what is now Finland and Iowa, in the United States. The Campanian ended with a series of meteorites that crashed in Russia and Algeria, perhaps triggering the massive eruption that arose in the Deccan Traps (present-day India). See more: *Izett et al. 1993; Badjukov et al. 2002; Lokrantz 2003; Wang & Dodson 2006; Schmieder & Jourdan 2013*

Maastrichtian

It began with a cold period, the last of four that occurred during the Cretaceous. There were marshes and coastal marshes in Madagascar. Falling sea levels caused an environmental crisis. Volcanic activity was intense in Hindustan (present-day India), and two meteorites struck in Russia and Algeria. It has been suggested that the combination of these three effects reduced the number of dinosaurs. They were once again decimated by meteorite strikes in Laramidia (present-day Mexico and Canada), eastern Europe (present-day Norway and Ukraine), and possibly in what is now India and Russia. See more: *Izett et al. 1993, Chatterjee 1997, Gurov 2002, Stewart & Allen 2002, Jolley et al. 2010, Wang et al. 2012, Lerbekmo 2014*

The smallest of the late Upper Cretaceous (Class: Tiny Grade I)

Alexornis antecedens ("ancestral bird of the ornithologist Alexander Wetmore"), from western Laurasia (present-day Mexico), was the smallest Mesozoic bird in North America; even a sparrow is larger. The alvarezsaurid *Parvicursor remotus* is the smallest non-avian theropod of that time; it was 50 cm long and weighed 185 g. See more: *Brodkorb 1976; Karhu & Rautian 1996*

The smallest juvenile

Gobipipus reshetovi PIN 4492-4 ("Gobi Desert chick of Doctor Valery Reshetov"), from eastern Laurasia (present-day Mongolia), was an enantiornithean bird embryo that was 4.7 cm long and weighed 2 g. See more: *Kurochkin et al. 2013*

10 cm

Alexornis antecedens
Brodkorb 1976

RECORD
SMALLEST

5 cm

1:1

Specimen: LACM 32213
(Brodkorb 1976)

Length: 11.4 cm
Hip height: 4 cm
Body mass: 22 g
Material: partial skeleton
ESR: ●●●○

5 cm   10 cm   15 cm   20 cm

Tyrannosaurus rex
Osborn 1905

RECORD
LARGEST

4 m

2 m

1:50

Specimen: UCMP 137538
(Longrich et al. 2010)

Length: 12.3 m
Hip height: 3.75 m
Body mass: 8.5 t
Material: pedal phalanx
ESR: ●○○○

The largest of the late Upper Cretaceous (Class: Very large Grade II)

Tyrannosaurus rex ("king of tyrant lizards"), from western Laurasia (present-day Canada, United States, and Mexico), was a huge predator that had significant sexual dimorphism. It is possible that the female was more robust. Although some studies have questioned this claim, a new analysis finally showed that the most massive forms belong to females. The largest bird of that time was *Gargantuavis philoinos*, which was 1.8 m tall and weighed 120 kg. See more: *Buffetaut & Le Loeuff 1998; Larson 2002; Erickson, Lappin & Larson 2005; Schweitzer & Wittmeyer 2006; Longrich et al. 2010*

The first published species of the late Upper Cretaceous

Deinodon horridus and *Troodon formosus* (1856), from western Laurasia (present-day Montana, USA), are the first two theropods from the end of the Cretaceous to be published. Both were based only on teeth, so they are not suffi-

ciently diagnosable to be kept as valid species. See more: *Leidy 1856*

The most recently published species of the late Upper Cretaceous

Dakotaraptor steini, Fumicollis hoffmani, Huanansaurus ganzhouensis, Saurornitholestes sullivani, Tototlmimus packardensis, Yi qi and *Zhenyuanlong suni* were published in 2015, as was *Boreonykus certekorum*, which was the last one, since it dates from December of that year. The species *Lepidocheirosaurus natatilis*, published in 2015, has been questioned, as some experts consider it material of the ornithischian *Kulindadromeus zabakalicus*. See more: *Alifanov & Salieliev 2015; Bell & Chiappe 2015; Bell & Currie 2015; Cau 2015; DePalna et al. 2015; Jasinski 2015; Lu et al. 2015; Lu & Brussate 2015; Novas et al. 2015, Serrano-Brañas et al. 2015; Mortimer**

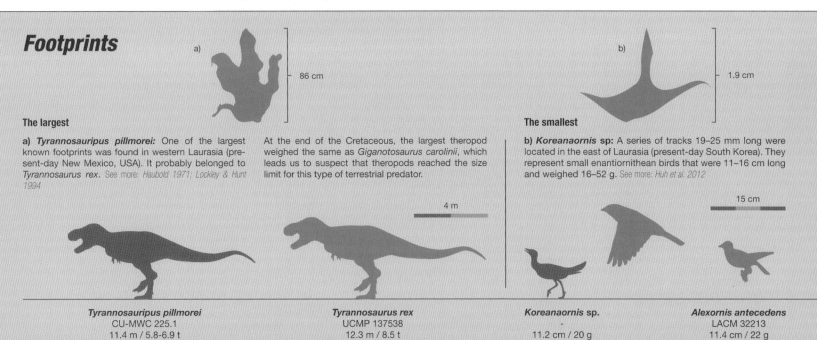

Footprints

a)

— 86 cm

b)

— 1.9 cm

The largest

a) Tyrannosauripus pillmorei: One of the largest known footprints was found in western Laurasia (present-day New Mexico, USA). It probably belonged to *Tyrannosaurus rex*. See more: *Haubold 1971; Lockley & Hunt 1994*

At the end of the Cretaceous, the largest theropod weighed the same as *Giganotosaurus carolinii*, which leads us to suspect that theropods reached the size limit for this type of terrestrial predator.

The smallest

b) Koreanaornis sp: A series of tracks 19–25 mm long were located in the east of Laurasia (present-day South Korea). They represent small enantiornithean birds that were 11–16 cm long and weighed 16–52 g. See more: *Huh et al. 2012*

4 m

15 cm

Tyrannosauripus pillmorei
CU-MWC 225.1
11.4 m / 5.8-6.9 t

Tyrannosaurus rex
UCMP 137538
12.3 m / 8.5 t

Koreanaornis sp.
-
11.2 cm / 20 g

Alexornis antecedens
LACM 32213
11.4 cm / 22 g

JANUARY (252.17–236.36 Ma)

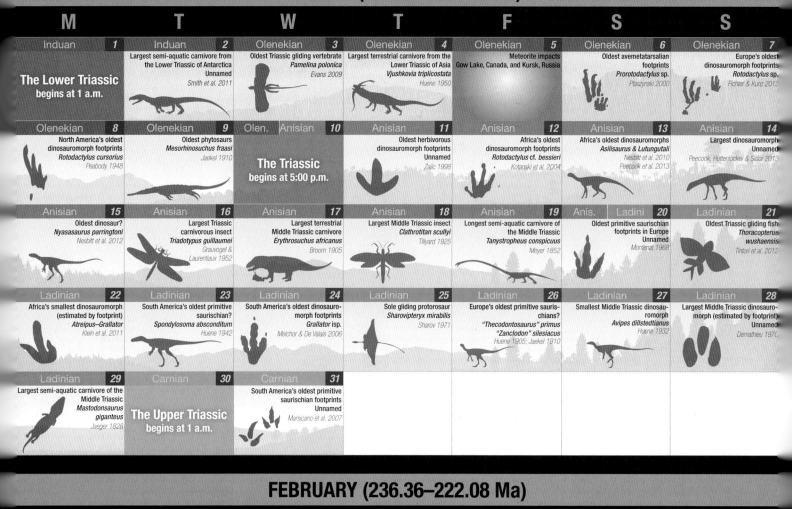

M	T	W	T	F	S	S
Induan **1** **The Lower Triassic** begins at 1 a.m.	Induan **2** Largest semi-aquatic carnivore from the Lower Triassic of Antarctica Unnamed *Smith et al. 2011*	Olenekian **3** Oldest Triassic gliding vertebrate *Pamelina polonica* *Evans 2009*	Olenekian **4** Largest terrestrial carnivore from the Lower Triassic of Asia *Vjushkovia triplicostata* *Huene 1950*	Olenekian **5** Meteorite impacts Gow Lake, Canada, and Kursk, Russia	Olenekian **6** Oldest avemetatarsalian footprints *Prorotodactylus* sp. *Ptaszynski 2000*	Olenekian **7** Europe's oldest dinosauromorph footprints *Rotodactylus* sp. *Fichter & Kunz 2013*
Olenekian **8** North America's oldest dinosauromorph footprints *Rotodactylus cursorius* *Peabody 1948*	Olenekian **9** Oldest phytosaurs *Mesorhinosuchus fraasi* *Jaekel 1910*	Olen. | Anisian **10** **The Triassic** begins at 5:00 p.m.	Anisian **11** Oldest herbivorous dinosauromorph footprints Unnamed *Zajic 1998*	Anisian **12** Africa's oldest dinosauromorph footprints *Rotodactylus* cf. *bessieri* *Kotanski et al. 2004*	Anisian **13** Africa's oldest dinosauromorphs *Asilisaurus & Lutungutali* *Nesbitt et al. 2010; Peecook et al. 2013*	Anisian **14** Largest dinosauromorph *Peecook, Huttenlocker & Sidor 2013*
Anisian **15** Oldest dinosaur? *Nyasasaurus parringtoni* *Nesbitt et al. 2012*	Anisian **16** Largest Triassic carnivorous insect *Triadotypus guillaumei* *Grauvogel & Laurentiaux 1952*	Anisian **17** Largest terrestrial Middle Triassic carnivore *Erythrosuchus africanus* *Broom 1905*	Anisian **18** Largest Middle Triassic insect *Clathrotitan scullyi* *Tillyard 1925*	Anisian **19** Longest semi-aquatic carnivore of the Middle Triassic *Tanystropheus conspicuus* *Meyer 1852*	Anis. | Ladini **20** Oldest primitive saurischian footprints in Europe Unnamed *Montenat 1968*	Ladinian **21** Oldest Triassic gliding fish *Thoracopterus wushaensis* *Tintori et al. 2012*
Ladinian **22** Africa's smallest dinosauromorph (estimated by footprint) *Atreipus–Grallator* *Klein et al. 2011*	Ladinian **23** South America's oldest primitive saurischian? *Spondylosoma absconditum* *Huene 1942*	Ladinian **24** South America's oldest dinosauromorph footprints *Grallator* isp. *Melchor & De Valais 2006*	Ladinian **25** Sole gliding protorosaur *Sharovipteryx mirabilis* *Sharov 1971*	Ladinian **26** Europe's oldest primitive saurischians? "*Thecodontosaurus*" *primus* "*Zanclodon*" *silesiacus* *Huene 1905; Jaekel 1910*	Ladinian **27** Smallest Middle Triassic dinosauromorph *Avipes dillstedtianus* *Huene 1932*	Ladinian **28** Largest Middle Triassic dinosauromorph (estimated by footprint) Unnamed *Demathieu 1970*
Ladinian **29** Largest semi-aquatic carnivore of the Middle Triassic *Mastodonsaurus giganteus* *Jaeger 1828*	Carnian **30** **The Upper Triassic** begins at 1 a.m.	Carnian **31** South America's oldest primitive saurischian footprints Unnamed *Marsicano et al. 2007*				

FEBRUARY (236.36–222.08 Ma)

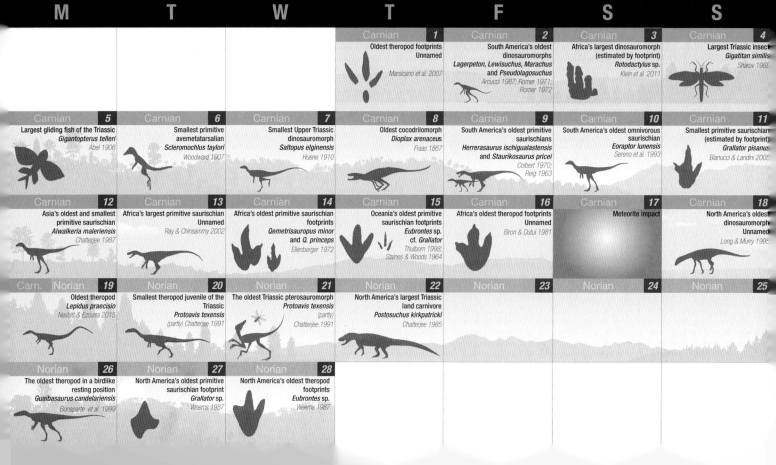

M	T	W	T	F	S	S
			Carnian **1** Oldest theropod footprints Unnamed *Marsicano et al. 2007*	Carnian **2** South America's oldest dinosauromorphs *Lagerpeton, Lewisuchus, Marachus* and *Pseudolagosuchus* *Arcucci 1987; Romer 1971; Romer 1972*	Carnian **3** Africa's largest dinosauromorph (estimated by footprint) *Rotodactylus* sp. *Klein et al. 2011*	Carnian **4** Largest Triassic insect *Gigatitan similis* *Sharov 1968*
Carnian **5** Largest gliding fish of the Triassic *Gigantopterus telleri* *Abel 1906*	Carnian **6** Smallest primitive avemetatarsalian *Scleromochlus taylori* *Woodward 1907*	Carnian **7** Smallest Upper Triassic dinosauromorph *Saltopus elginensis* *Huene 1910*	Carnian **8** Oldest cocodrilomorph *Dioplax arenaceus* *Fraas 1867*	Carnian **9** South America's oldest primitive saurischians *Herrerasaurus ischigualastensis* and *Staurikosaurus pricei* *Colbert 1970; Reig 1963*	Carnian **10** South America's oldest omnivorous saurischian *Eoraptor lunensis* *Sereno et al. 1993*	Carnian **11** Smallest primitive saurischian (estimated by footprint) *Grallator pisanus* *Bianucci & Landini 2005*
Carnian **12** Asia's oldest and smallest primitive saurischian *Alwalkeria maleriensis* *Chatterjee 1987*	Carnian **13** Africa's largest primitive saurischian Unnamed *Ray & Chinsammy 2002*	Carnian **14** Africa's oldest primitive saurischian footprints *Qemetrisauropus minor* and *Q. princeps* *Ellenberger 1972*	Carnian **15** Oceania's oldest primitive saurischian footprints *Eubrontes* sp. cf. *Grallator* *Thulborn 1998; Staines & Woods 1964*	Carnian **16** Africa's oldest theropod footprints Unnamed *Biron & Dutuit 1981*	Carnian **17** Meteorite impact	Carnian **18** North America's oldest dinosauromorph Unnamed *Long & Murry 1995*
Carn. | Norian **19** Oldest theropod *Lepidus praecisio* *Nesbitt & Ezcurra 2015*	Norian **20** Smallest theropod juvenile of the Triassic *Protoavis texensis* *(partly) Chatterjee 1991*	Norian **21** The oldest Triassic pterosauromorph *Protoavis texensis* *(partly) Chatterjee 1991*	Norian **22** North America's largest Triassic land carnivore *Postosuchus kirkpatricki* *Chatterjee 1985*	Norian **23**	Norian **24**	Norian **25**
Norian **26** The oldest theropod in a birdlike resting position *Guaibasaurus candelariensis* *Bonaparte et al. 1999*	Norian **27** North America's oldest primitive saurischian footprint *Grallator* sp. *Weems 1987*	Norian **28** North America's oldest theropod footprints *Eubrontes* sp. *Weems 1987*				

On this scale, there are 510,000 years per day, 21,000 years per hour, 350 years per minute, and 5.8 years per second.

The model of the cosmic calendar offers a way to visualize a time range by depicting it on an annual calendar. It helps us compare times of events that happened millions of years ago. Extrapolated into 365 days of the year are events and records related to dinosauromorphs, primitive saurischians, theropods, Mesozoic birds, other terrestrial predators, and gliding or flying animals that shared the world over the course of the Mesozoic. The proportions of each event's time can be compared, as well as the appearance and disappearance of new forms, massive catastrophes, and the end of those forms' dominion over Earth. See more: *Sagan 1978; de Grasse-Tyson 2014*

Carnivorous dinosaurs were not the dominant predators in terrestrial environments until the beginning of the Jurassic. This niche was occupied by other large archosaurs, close relatives similar to crocodiles, such as ornithosuchids, prestosuchids, rauisuchids, etc.

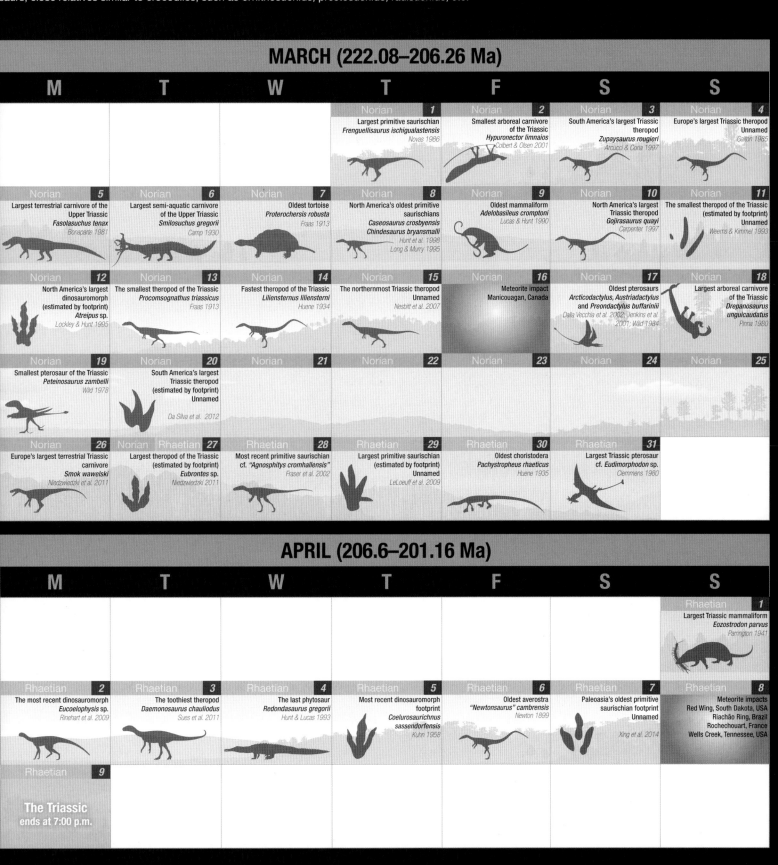

MARCH (222.08–206.26 Ma)

M	T	W	T	F	S	S
			Norian **1** Largest primitive saurischian *Frenguellisaurus ischigualastensis* Novas 1986	Norian **2** Smallest arboreal carnivore of the Triassic *Hypuronector limnaios* Colbert & Olsen 2001	Norian **3** South America's largest Triassic theropod *Zupaysaurus rougieri* Arcucci & Coria 1997	Norian **4** Europe's largest Triassic theropod Unnamed Galton 1985
Norian **5** Largest terrestrial carnivore of the Upper Triassic *Fasolasuchus tenax* Bonaparte 1981	Norian **6** Largest semi-aquatic carnivore of the Upper Triassic *Smilosuchus gregorii* Camp 1930	Norian **7** Oldest tortoise *Proterochersis robusta* Fraas 1913	Norian **8** North America's oldest primitive saurischians *Caseosaurus crosbyensis* Hunt et al. 1998 *Chindesaurus bryansmalli* Long & Murry 1995	Norian **9** Oldest mammaliform *Adelobasileus cromptoni* Lucas & Hunt 1990	Norian **10** North America's largest Triassic theropod *Gojirasaurus quayi* Carpenter 1997	Norian **11** The smallest theropod of the Triassic (estimated by footprint) Unnamed Weems & Kimmel 1993
Norian **12** North America's largest dinosauromorph (estimated by footprint) *Atreipus* sp. Lockley & Hunt 1995	Norian **13** The smallest theropod of the Triassic *Procompsognathus triassicus* Fraas 1913	Norian **14** Fastest theropod of the Triassic *Liliensternus liliensterni* Huene 1934	Norian **15** The northernmost Triassic theropod Unnamed Nesbitt et al. 2007	Norian **16** Meteorite impact Manicouagan, Canada	Norian **17** Oldest pterosaurs *Arcticodactylus, Austriadactylus* and *Preondactylus buffarinii* Dalla Vecchia et al. 2002; Jenkins et al. 2001; Wild 1984	Norian **18** Largest arboreal carnivore of the Triassic *Drepanosaurus* *unguicaudatus* Pinna 1980
Norian **19** Smallest pterosaur of the Triassic *Peteinosaurus zambelli* Wild 1978	Norian **20** South America's largest Triassic theropod (estimated by footprint) Unnamed Da Silva et al. 2012	Norian **21**	Norian **22**	Norian **23**	Norian **24**	Norian **25**
Norian **26** Europe's largest terrestrial carnivore *Smok wawelski* Niedzwiedzki et al. 2011	Norian **27** Largest theropod of the Triassic (estimated by footprint) *Eubrontes* sp. Niedzwiedzki 2011	Rhaetian **28** Most recent primitive saurischian cf. *"Agnosphitys cromhallensis"* Fraser et al. 2002	Rhaetian **29** Largest primitive saurischian (estimated by footprint) Unnamed LeLoeuff et al. 2009	Rhaetian **30** Oldest choristodera *Pachystropheus rhaeticus* Huene 1935	Rhaetian **31** Largest Triassic pterosaur cf. *Eudimorphodon* sp. Clemmens 1980	

APRIL (206.6–201.16 Ma)

M	T	W	T	F	S	S
						Rhaetian **1** Largest Triassic mammaliform *Eozostrodon parvus* Parrington 1941
Rhaetian **2** The most recent dinosauromorph *Eucoelophysis* sp. Rinehart et al. 2009	Rhaetian **3** The toothiest theropod *Daemonosaurus chauliodus* Sues et al. 2011	Rhaetian **4** The last phytosaur *Redondasaurus gregorii* Hunt & Lucas 1993	Rhaetian **5** Most recent dinosauromorph footprint *Coelurosaurichnus* *sassendorfensis* Kuhn 1958	Rhaetian **6** Oldest averostra *"Newtonsaurus" cambrensis* Newton 1899	Rhaetian **7** Paleoasia's oldest primitive saurischian footprint Unnamed Xing et al. 2014	Rhaetian **8** Meteorite impacts Red Wing, South Dakota, USA Riachão Ring, Brazil Rochechouart, France Wells Creek, Tennessee, USA
Rhaetian **9** **The Triassic** ends at 7:00 p.m.						

Theropods dominated the terrestrial ecosystems and had practically no direct competition during the Jurassic. Some very large crocodiles were the only exception. A mass extinction ended the Triassic and began the Jurassic. This event made numerous species disappear, including most archosaurs. Only some dinosaurs, pterosaurs, and crocodiles were left. This allowed theropods to assume the dominant role and to diversify into ecological niches.

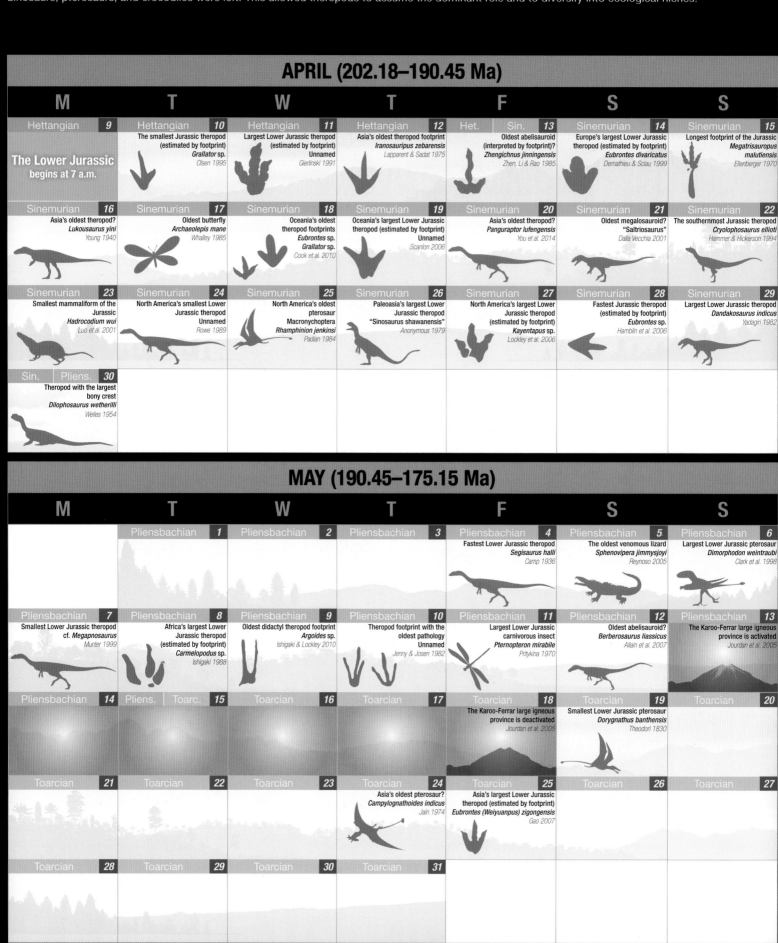

APRIL (202.18–190.45 Ma)

M	T	W	T	F	S	S
Hettangian **9** **The Lower Jurassic begins at 7 a.m.**	Hettangian **10** The smallest Jurassic theropod (estimated by footprint) *Grallator* sp. *Olsen 1995*	Hettangian **11** Largest Lower Jurassic theropod (estimated by footprint) Unnamed *Gierlinski 1991*	Hettangian **12** Asia's oldest theropod footprint *Iranosauripus zebarensis* *Lapparent & Sadat 1975*	Het. | Sin. **13** Oldest abelisauroid (interpreted by footprint)? *Zhengichnus jinningensis* *Zhen, Li & Rao 1985*	Sinemurian **14** Europe's largest Lower Jurassic theropod (estimated by footprint) *Eubrontes divaricatus* *Demathieu & Sciau 1999*	Sinemurian **15** Longest footprint of the Jurassic *Megatrisauropus malutiensis* *Ellenberger 1970*
Sinemurian **16** Asia's oldest theropod? *Lukousaurus yini* *Young 1940*	Sinemurian **17** Oldest butterfly *Archaeolepis mane* *Whalley 1985*	Sinemurian **18** Oceania's oldest theropod footprints *Eubrontes* sp. *Grallator* sp. *Cook et al. 2010*	Sinemurian **19** Oceania's largest Lower Jurassic theropod (estimated by footprint) Unnamed *Scanlon 2006*	Sinemurian **20** Asia's oldest theropod? *Panguraptor lufengensis* *You et al. 2014*	Sinemurian **21** Oldest megalosauroid? "*Saltriosaurus*" *Dalla Vecchia 2001*	Sinemurian **22** The southernmost Jurassic theropod *Cryolophosaurus elliotti* *Hammer & Hickerson 1994*
Sinemurian **23** Smallest mammaliform of the Jurassic *Hadrocodium wui* *Luo et al. 2001*	Sinemurian **24** North America's smallest Lower Jurassic theropod Unnamed *Rowe 1989*	Sinemurian **25** North America's oldest Macronychoptera *Rhamphinion jenkinsi* *Padian 1984*	Sinemurian **26** Paleoasia's largest Lower Jurassic theropod "*Sinosaurus shawanensis*" *Anonymous 1979*	Sinemurian **27** North America's largest Lower Jurassic theropod (estimated by footprint) *Kayentapus* sp. *Lockley et al. 2006*	Sinemurian **28** Fastest Jurassic theropod (estimated by footprint) *Eubrontes* sp. *Hamblin et al. 2006*	Sinemurian **29** Largest Lower Jurassic theropod *Dandakosaurus indicus* *Yadagiri 1982*
Sin. | Pliens. **30** Theropod with the largest bony crest *Dilophosaurus wetherilli* *Welles 1954*						

MAY (190.45–175.15 Ma)

M	T	W	T	F	S	S
	Pliensbachian **1**	Pliensbachian **2**	Pliensbachian **3**	Pliensbachian **4** Fastest Lower Jurassic theropod *Segisaurus halli* *Camp 1936*	Pliensbachian **5** The oldest venomous lizard *Sphenovipera jimmysjoyi* *Reynoso 2005*	Pliensbachian **6** Largest Lower Jurassic pterosaur *Dimorphodon weintraubi* *Clark et al. 1998*
Pliensbachian **7** Smallest Lower Jurassic theropod cf. *Megapnosaurus* *Munter 1999*	Pliensbachian **8** Africa's largest Lower Jurassic theropod (estimated by footprint) *Carmelopodus* sp. *Ishigaki 1988*	Pliensbachian **9** Oldest didactyl theropod footprint *Argoides* sp. *Ishigaki & Lockley 2010*	Pliensbachian **10** Theropod footprint with the oldest pathology Unnamed *Jenny & Josen 1982*	Pliensbachian **11** Largest Lower Jurassic carnivorous insect *Pternopteron mirabile* *Pritykina 1970*	Pliensbachian **12** Oldest abelisauroid? *Berberosaurus liassicus* *Allain et al. 2007*	Pliensbachian **13** The Karoo-Ferrar large igneous province is activated *Jourdan et al. 2005*
Pliensbachian **14**	Pliens. | Toarc. **15**	Toarcian **16**	Toarcian **17**	Toarcian **18** The Karoo-Ferrar large igneous province is deactivated *Jourdan et al. 2005*	Toarcian **19** Smallest Lower Jurassic pterosaur *Dorygnathus banthensis* *Theodori 1830*	Toarcian **20**
Toarcian **21**	Toarcian **22**	Toarcian **23**	Toarcian **24** Asia's oldest pterosaur? *Campylognathoides indicus* *Jain 1974*	Toarcian **25** Asia's largest Lower Jurassic theropod (estimated by footprint) *Eubrontes (Weiyuanpus) zigongensis* *Gao 2007*	Toarcian **26**	Toarcian **27**
Toarcian **28**	Toarcian **29**	Toarcian **30**	Toarcian **31**			

JUNE (175.15–159.85 Ma)

M	T	W	T	F	S	S
				Toarcian `1`	**Toarcian** `2`	Toar. | Aaleni. `3`
					Hindustan's oldest theropod footprints *Eubrontes* cf. *giganteus* *Grallator tenuis* *Pienskowski et al. 2015*	**The Middle Jurassic** begins at 1 am.
Aalenian `4`	Aalenian `5`	Aalenian `6`	Aalenian `7`	Aalenian `8`	Aalenian `9`	Aalen. | Baj. `10`
	Smallest Middle Jurassic pterosaur Unnamed *Bakhurina & Unwin 1995*	Oldest abelisaurid? *Eoabelisaurus mefi* *Pol & Rauhut 2012*	Oldest alosauroid *Shidaisaurus jinae* *Wu et al. 2009*	Oceania's oldest theropod *Ozraptor subotaii* *Long & Molnar 1998*	Oldest theropod footprint with widespread fingers Unnamed *Abbassi & Madanipour 2014*	Oceania's largest Jurassic theropod (estimated by footprint) Unnamed *Long 1998*
Bajocian `11`	Bajocian `12`	Bajocian `13`	Baj. | Bathon. `14`	Bathonian `15`	Bathonian `16`	Bathonian `17`
Meteorite impacts Obolon, Ukraine Upheaval Dome, Utah, USA	Asia's oldest pterosaur? *Angustinaripterus longicephalus* *He et al. 1983*	Oldest coelurosaur *Gasosaurus constructus* *Dong & Tang 1985*	The largest semi-aquatic carnivore of the Middle Jurassic *Razanandrongobe sakalavae* *Maganuco et al. 2006*	Largest Middle Jurassic theropod cf. *Megalosaurus bucklandii* *Phillips 1871*	Meteorite impacts Puchezh-Katunki, Russia Zapadnaya, Ukraine	Largest carnivorous insect of the Middle Jurassic *Hemerobioides giganteus* *Westwood 1845*
Bathon. | Call. `18`	Callovian `19`	Callovian `20`	Callovian `21`	Callovian `22`	Call. | Oxfor. `23`	Oxfordian `24`
Largest Middle Jurassic pterosaur *Rhamphocephalus* sp. *Unwin 1996*	Oldest gliding mammal *Volaticotherium antiquus* *Meng et al. 2006*	The smallest Middle Jurassic theropod *Epidexipteryx hui* *Zhang et al. 2008*	Largest Middle Jurassic theropod (estimated by footprint) *Turkmenosaurus kugitangensis* *Amanniazov 1985*	Fastest Middle Jurassic theropod *Eustreptospondylus oxoniensis* *Walker 1864*	**The Upper Jurassic** begins 9 p.m.	Largest Jurassic arachnid *Mongolarachne jurassica* *Selden, Shih & Ren 2011*
Oxfordian `25`	Oxfordian `26`	Oxfordian `27`	Oxfordian `28`	Oxfordian `29`	Oxfordian `30`	
Largest dinosaur parasite of the Jurassic? *Pseudopulex jurassicus* *Gao et al 2012*	Oldest alvarezsauroid *Haplocheirus sollers* *Choiniere et al. 2010*	Asia's oldest herbivorous theropod *Limusaurus inextricabilis* *Xu et al. 2009*	Asia's largest Jurassic theropod (tie) Unnamed *Xing & Clark 2008*	Asia's largest Jurassic theropod (tie) *Sinraptor dongi* *Currie & Zhao 1994*	Meteorite impact Vepriai, Lithuania	

JULY (159.85–145.06 Ma)

M	T	W	T	F	S	S
						Oxfordian `1`
						Central America's oldest pterosaur *Cacibupteryx caribensis* *Gasparini et al. 2004*
Oxfordian `2`	Oxfordian `3`	Oxfordian `4`	Oxfordian `5`	Kimmeridgian `6`	Kimmeridgian `7`	Kimmeridgian `8`
Smallest Upper Jurassic theropod (estimated by footprint) Unnamed *Gierlinski et al. 2009*	West Asia's oldest pterosaurs *Batrachognathus volans* *Riabinin 1948* *Sordes pilosus* *Sharov 1971*			Largest Upper Jurassic theropod (estimated by footprint) Unnamed *Boutakiout et al. 2009*		Oldest theropod egg *Preprismatoolithus coloradensis* *Hirsch 1994*
Kimmeridgian `9`	Kimmeridgian `10`	Kimmeridgian `11`	Kimmeridgian `12`	Kimmeridgian `13`	Kimm. | Tithon. `14`	Tithonian `15`
Oldest bird? cf. *Archaeopteryx* sp. *Weigert 1995*	Africa's oldest pterosaur *Tendaguripterus recki* *Unwin & Heinrich 1999*	Largest Jurassic pterosaur Unnamed *Meyer & Hunt 1999*	North America's largest Jurassic theropod *Saurophaganax maximus* *Chure 1995*	Oldest theropod embryo *Lourinhanosaurus antunesi* *Mateus 1998*	Europe's largest Jurassic theropod *Torvosaurus gurneyi* *Hendrickx & Mateus 2014*	Largest carnivorous insect of the Upper Jurassic *Isophlebia aspasia* *Hagen 1866*
Tithonian `16`	Tithonian `17`	Tithonian `18`	Tithonian `19`	Tithonian `20`	Tithonian `21`	Tithonian `22`
Fastest Jurassic theropod cf. *Elaphrosaurus bambergi* *Janensch 1929*	Largest Upper Jurassic theropod "*Megalosaurus*" *ingens* *Janensch 1920*	Largest Upper Jurassic bird (estimated by footprint) Unnamed *Lange-Badre et al. 1996*	Smallest Upper Jurassic theropod *Compsognathus longipes* *Wagner 1861*	Largest Upper Jurassic insect *Kalligramma haeckeli* *Walther 1904*	Largest semi-aquatic carnivorous insect of the Jurassic *Chresmoda obscura* *Germar 1839*	South America's oldest pterosaur *Wenupteryx uzi* *Codorniú & Gasparini 2013*
Tithonian `23`	Tithonian `24`	Tithonian `25`	Tithonian `26`	Tithonian `27`	Tithonian `28`	Tithonian `29`
Smallest Upper Jurassic pterosaur *Anurognathus ammoni* *Döderlein 1923*	South America's oldest herbivorous theropod *Chilesaurus diegosuarezi* *Novas et al. 2015*	The northernmost Jurassic theropod "*Allosaurus*" sp. *Kurzanov et al. 2003*	South America's largest Jurassic theropod Unnamed *Salgado et al. 2008*	Smallest Upper Jurassic bird (estimated by footprint) *Aquatilavipes* sp. *Lockley et al. 2006*	Meteorite impact Morokweng, South Africa	**The Jurassic** ends at 12:00 a.m.

During the Lower Cretaceous, the position and separation of the continents changed greatly, as did the terrestrial plants. New species appeared. Among them birds stand out. Their presence modified the environment in a way that undoubtedly had an ecological impact. Birds must have competed with pterosaurs for food and space, as the two had similar habits and both could fly. The size of pterosaurs increased drastically during this period, an occurrence that coincides with an increase in the diversity of birds. It is suspected that both phenomena could be related. See more: *Grellet-Tinner et al. 2007*

Great theropods were still the super predators, although gigantic crocodiles disputed their domains on the banks of rivers, lakes, and seas. Some pterosaurs reached outstanding sizes, but they did not compete with the largest carnivorous dinosaurs.

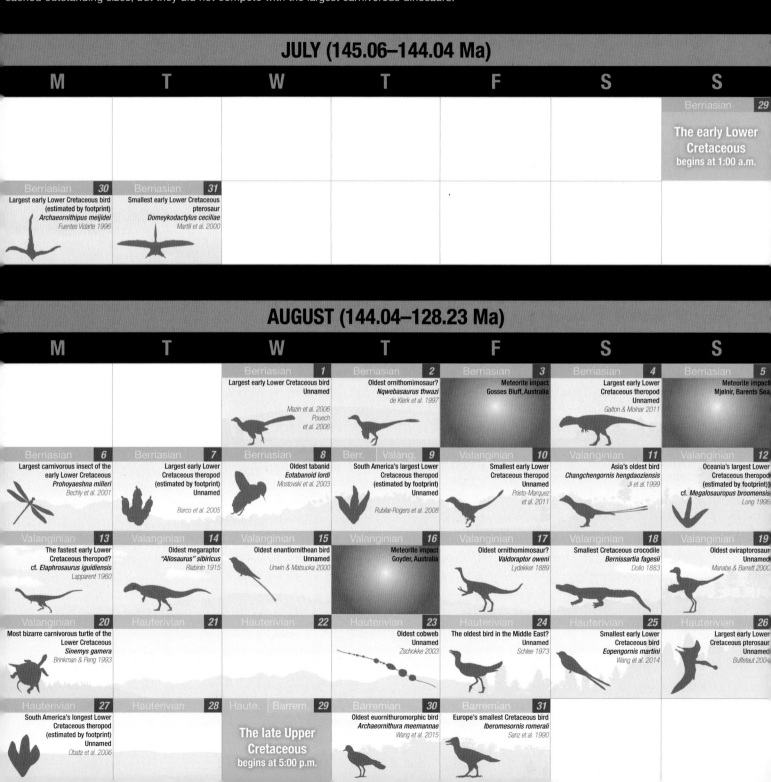

JULY (145.06–144.04 Ma)

M	T	W	T	F	S	S
						Berriasian 29 **The early Lower Cretaceous** begins at 1:00 a.m.
Berriasian 30 Largest early Lower Cretaceous bird (estimated by footprint) *Archaeornithipus meijidei* Fuentes Vidarte 1996	Berriasian 31 Smallest early Lower Cretaceous pterosaur *Domeykodactylus ceciliae* Martill et al. 2000					

AUGUST (144.04–128.23 Ma)

M	T	W	T	F	S	S
		Berriasian 1 Largest early Lower Cretaceous bird Unnamed Mazin et al. 2006 Pouech et al. 2006	Berriasian 2 Oldest ornithomimosaur? *Nqwebasaurus thwazi* de Klerk et al. 1997	Berriasian 3 Meteorite impact Gosses Bluff, Australia	Berriasian 4 Largest early Lower Cretaceous theropod Unnamed Galton & Molnar 2011	Berriasian 5 Meteorite impact Mjølnir, Barents Sea
Berriasian 6 Largest carnivorous insect of the early Lower Cretaceous *Prohoyaeshna milleri* Bechly et al. 2001	Berriasian 7 Largest early Lower Cretaceous theropod (estimated by footprint) Unnamed Barco et al. 2005	Berriasian 8 Oldest tabanid *Eotabanoid iordi* Mostovski et al. 2003	Berr. Valang. 9 South America's largest Lower Cretaceous theropod (estimated by footprint) Unnamed Rubilar-Rogers et al. 2008	Valanginian 10 Smallest early Lower Cretaceous theropod Unnamed Prieto-Marquez et al. 2011	Valanginian 11 Asia's oldest bird *Changchengornis hengdaoziensis* Ji et al.1999	Valanginian 12 Oceania's largest Lower Cretaceous theropod (estimated by footprint) cf. *Megalosauropus broomensis* Long 1998
Valanginian 13 The fastest early Lower Cretaceous theropod? cf. *Elaphrosaurus iguidiensis* Lapparent 1960	Valanginian 14 Oldest megaraptor *"Allosaurus" sibiricus* Riabinin 1915	Valanginian 15 Oldest enantiornithean bird Unnamed Unwin & Matsuoka 2000	Valanginian 16 Meteorite impact Goyder, Australia	Valanginian 17 Oldest ornithomimosaur? *Valdoraptor oweni* Lydekker 1889	Valanginian 18 Smallest Cretaceous crocodile *Bernissartia fagesii* Dollo 1883	Valanginian 19 Oldest oviraptorosaur Unnamed Manabe & Barrett 2000
Valanginian 20 Most bizarre carnivorous turtle of the Lower Cretaceous *Sinemys gamera* Brinkman & Peng 1993	Hauterivian 21	Hauterivian 22	Hauterivian 23 Oldest cobweb Unnamed Zschokke 2003	Hauterivian 24 The oldest bird in the Middle East? Unnamed Schlee 1973	Hauterivian 25 Smallest early Lower Cretaceous bird *Eopengornis martini* Wang et al. 2014	Hauterivian 26 Largest early Lower Cretaceous pterosaur Unnamed Buffetaut 2004
Hauterivian 27 South America's longest Lower Cretaceous theropod (estimated by footprint) Unnamed Obata et al. 2006	Hauterivian 28	Haute. Barrem. 29 **The late Upper Cretaceous** begins at 5:00 p.m.	Barremian 30 Oldest euornithuromorphic bird *Archaeornithura meemannae* Wang et al. 2015	Barremian 31 Europe's smallest Cretaceous bird *Iberomesornis romerali* Sanz et al. 1990		

SEPTEMBER (128.23–112.92 Ma)

M	T	W	T	F	S	S
					Barremian 1 Meteorite impact Tookoonooka, Australia	**Barremian 2** North America's oldest herbivorous theropod *Falcarius utahensis* Kirkland et al. 2005
Barremian 3 Smallest juvenile bird of the Lower Cretaceous *Dalingheornis liweii* Zhang et al. 2006	**Barremian 4** Smallest aquatic bird *Mystiornis cyrili* Kurochkin et al. 2011	**Barr. Aptian 5** Asia's largest Lower Cretaceous theropod Unnamed Buffetaut & Suteethorn 2012	**Aptian 6** Largest semi-aquatic carnivorous insect of the Cretaceous *Chresmoda shini* Zhang et al. 2009	**Aptian 7** The volcanoes of Ontong Java become active	**Aptian 8** Largest Lower Cretaceous pterosaur Unnamed Martill et al. 1996	**Aptian 9** The largest dinosaur parasite? *Pseudopulex magnus* Gao et al. 2012
Aptian 10 Largest terrestrial mammal of the Lower Cretaceous *Repenomamus giganticus* Hu et al. 2005	**Aptian 11** Africa's oldest bird Unnamed O'Connor et al. 2003	**Aptian 12** Largest Cretaceous amphibian *Koolasuchus cleelandi* Warren et al. 1997	**Aptian 13** Southernmost Cretaceous theropod Unnamed Woodward 1906 Agnolin et al. 2010	**Aptian 14** Southernmost Cretaceous bird Unnamed Close et al. 2009	**Aptian 15** The volcanoes of Ontong Java become dormant	**Aptian 16** Smallest Lower Cretaceous theropod (estimated by footprint) Unnamed Stanford et al. 2011
Aptian 17 Meteorite impacts BP Structure, Libya Mien, Sweden Oasis, Libya	**Aptian 18** South America's oldest and the smallest late Upper Cretaceous bird *Cratoavis cearensis* Carvalho et al. 2015	**Aptian 19** North America's largest Lower Cretaceous theropod *Acrocanthosaurus atokensis* Stovall & Langston 1950	**Aptian 20** Smallest late Lower Cretaceous theropod Unnamed Pittman et al. 2015	**Aptian 21** North America's oldest bird Unnamed Kirkland et al. 1997	**Aptian 22** Fastest late Lower Cretaceous theropod *Beishanlon grandis* Makovicky et al. 2009	**Aptian 23** Largest carnivorous insect of the Cretaceous *Cratostenophlebia schwickerti* Bechly 2007
Aptian 24 Smallest Lower Cretaceous pterosaur *Nemicolopterus crypticus* Wang et al. 2008	**Aptian 25** Largest Cretaceous arachnid *Britopygus weygoldti* Dunlop & Martill 2002	**Aptian 26** Meteorite impact Carswell, Canada	**Aptian 27** Smallest Lower Cretaceous bird (estimated by footprint) Unnamed Moratalla et al. 2003	**Aptian 28** Bedbug with the oldest traces of blood *Torirostratus pilosus* Yao et al. 2014	**Aptian 29** Gondwana's oldest flea *Tarwinia australis* Jell & Duncan 1986	**Aptian Albi. 30** South America's largest semi-aquatic Cretaceous carnivore *Sarcosuchus hartti* Marsh 1869

OCTOBER (112.92–100.17 Ma)

M	T	W	T	F	S	S
Albian 1 Largest Lower Cretaceous pterosaur (estimated by footprints) *Haenamichnus gainensis* Kim et al. 2012	**Albian 2**	**Albian 3**	**Albian 4**	**Albian 5**	**Albian 6** Oldest camouflaged predatory insect *Hallucinochrysa diogenesi* Pérez-de la Fuente et al. 2012	**Albian 7** Oldest web with caught prey Unnamed Peñalver et al. 2006
Albian 8 Europe's largest Lower Cretaceous theropod "*Megalosaurus*" sp. Vilanova & Piera 1873	**Albian 9** Largest semi-aquatic Cretaceous mammal *Kollikodon ritchiei* Flannery et al. 1995	**Albian 10** The last ceratosaurids *Genyodectes serus* and Unnamed Woodward 1901 Ryan 1997	**Albian 11** Largest semi-aquatic Lower Cretaceous carnivore *Sarcosuchus imperator* de Broin & Taquet 1966	**Albian 12** South America's largest Lower Cretaceous theropod *Tyrannotitan chubutensis* Novas et al. 2005	**Albian 13** Largest Lower Cretaceous bird Unnamed Molnar 1999	**Albian 14** Largest late Upper Cretaceous theropod (estimated by footprint) Unnamed Mahboubi et al. 2007
Albian 15 Largest late Upper Cretaceous theropod *Carcharodontosaurus* sp. Lapparent 1960	**Albian 16** North America's largest Lower Cretaceous theropod (estimated by footprint) *Irenichnites* sp. Lockley 1998	**Albian 17** North America's largest theropod egg *Macroelongatoolithus carlylei* Jensen 1970	**Albian 18**	**Albian 19**	**Albian 20** Largest aquatic Lower Cretaceous theropod *Spinosaurus* sp. Buffetaut & Ouaja 2002	**Albian 21** Longest Lower Cretaceous bird Unnamed Riff et al. 2002
Albian 22 Oldest spider with prey *Geratonephila burmanica* Poinar & Buckley 2012	**Albian 23** Oldest mosquito *Burmaculex antiquus* Borkent & Grimaldi 2004	**Albian 24** Longest theropod footprint *Magnoavipes* sp. Lockley, Matsukawa & Witt 2006	**Albian 25** The late Lower Cretaceous ends at 9:30 a.m.			

The largest theropods of all time lived during the Upper Cretaceous, as did giant crocodiles and pterosaurs that rivaled them in length and/or height. On the other hand, birds diversified further, and strictly aquatic or terrestrial forms appeared.

The Upper Cretaceous had two main extinctions. The first, known as the Bonarelli Event, occurred between the Cenomanian and Turonian. Several dinosaurs went extinct, causing other species to occupy the abandoned ecological niches. The extinction may have been caused by suboceanic volcanism that caused the oceans' oxygen levels to drop and the marine crust of the Pacific and the Indian Oceans to swell. There is evidence of this event all over the world. See more: *Vogt 1989; Alonso et al. 1993; Bilotte 1993; Camoin 1993; Dercourt et al. 1993; Philip 1993; Segura et al. 1993; Simó et al. 1993; Omaña et al. 2012; Larsson 1991; Larson & Olson 1991; Karakitsios et al 2007; Williamson et al. 2014; Longrich et al. 2011*

The second mass extinction occurred between the Cretaceous and Paleogene, killing 75% of terrestrial species. The survivors were so decimated that even birds and mammals were at risk of disappearing. Ammonites, belemnites, enantiornithes, ichthyosaurs, inoceramids, hesperornite birds, mosasaurs, ornithuromorphs, plesiosaurs, pliosaurs, pterosaurs, rudists, ornithischian dinosaurs, sauropod dinosaurs, non-avian theropod dinosaurs, and many microorganisms were eliminated. See more: *Longrich et al. 2011; Williamson et al. 2014*

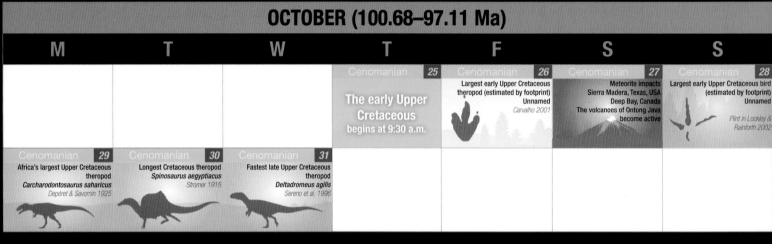

OCTOBER (100.68–97.11 Ma)

M	T	W	T	F	S	S
			Cenomanian **25** **The early Upper Cretaceous** begins at 9:30 a.m.	Cenomanian **26** Largest early Upper Cretaceous theropod (estimated by footprint) Unnamed *Carvalho 2001*	Cenomanian **27** Meteorite impacts Sierra Madera, Texas, USA Deep Bay, Canada The volcanoes of Ontong Java become active	Cenomanian **28** Largest early Upper Cretaceous bird (estimated by footprint) Unnamed *Plint in Lockley & Rainforth 2002*
Cenomanian **29** Africa's largest Upper Cretaceous theropod *Carcharodontosaurus saharicus* *Depéret & Savornin 1925*	Cenomanian **30** Longest Cretaceous theropod *Spinosaurus aegyptiacus* *Stromer 1915*	Cenomanian **31** Fastest late Upper Cretaceous theropod *Deltadromeus agilis* *Sereno et al. 1996*				

NOVEMBER (97.11–81.81 Ma)

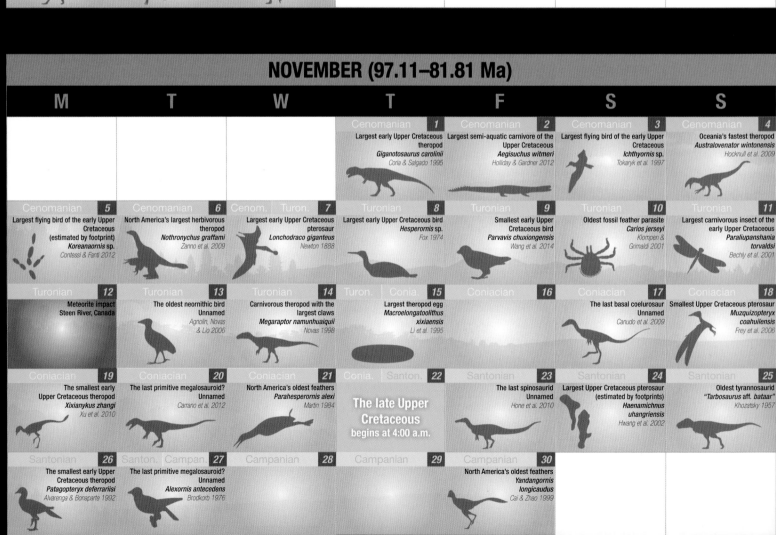

M	T	W	T	F	S	S
			Cenomanian **1** Largest early Upper Cretaceous theropod *Giganotosaurus carolinii* *Coria & Salgado 1995*	Cenomanian **2** Largest semi-aquatic carnivore of the Upper Cretaceous *Aegisuchus witmeri* *Holliday & Gardner 2012*	Cenomanian **3** Largest flying bird of the early Upper Cretaceous *Ichthyornis* sp. *Tokaryk et al. 1997*	Cenomanian **4** Oceania's fastest theropod *Australovenator wintonensis* *Hocknull et al. 2009*
Cenomanian **5** Largest flying bird of the early Upper Cretaceous (estimated by footprint) *Koreanaornis* sp. *Contessi & Fanti 2012*	Cenomanian **6** North America's largest herbivorous theropod *Nothronychus graffami* *Zanno et al. 2009*	Cenom. Turon. **7** Largest early Upper Cretaceous pterosaur *Lonchodraco giganteus* *Newton 1888*	Turonian **8** Largest early Upper Cretaceous bird *Hesperornis* sp. *Fox 1974*	Turonian **9** Smallest early Upper Cretaceous bird *Parvavis chuxiongensis* *Wang et al. 2014*	Turonian **10** Oldest fossil feather parasite *Carios jerseyi* *Klompen & Grimaldi 2001*	Turonian **11** Largest carnivorous insect of the early Upper Cretaceous *Paraliupanshania torvaldsi* *Bechly et al. 2001*
Turonian **12** Meteorite impact Steen River, Canada	Turonian **13** The oldest neornithic bird Unnamed *Agnolin, Novas & Lio 2006*	Turonian **14** Carnivorous theropod with the largest claws *Megaraptor namunhuaiquii* *Novas 1998*	Turon. Conia. **15** Largest theropod egg *Macroelongatoolithus xixiaensis* *Li et al. 1995*	Coniacian **16**	Coniacian **17** The last basal coelurosaur Unnamed *Canudo et al. 2009*	Coniacian **18** Smallest Upper Cretaceous pterosaur *Muzquizopteryx coahuilensis* *Frey et al. 2006*
Coniacian **19** The smallest early Upper Cretaceous theropod *Xixianykus zhangi* *Xu et al. 2010*	Coniacian **20** The last primitive megalosauroid? Unnamed *Carrano et al. 2012*	Coniacian **21** North America's oldest feathers *Parahesperornis alexi* *Martin 1984*	Conia. Santon. **22** **The late Upper Cretaceous** begins at 4:00 a.m.	Santonian **23** The last spinosaurid Unnamed *Hone et al. 2010*	Santonian **24** Largest Upper Cretaceous pterosaur (estimated by footprints) *Haenamichnus uhangriensis* *Hwang et al. 2002*	Santonian **25** Oldest tyrannosaurid *"Tarbosaurus aff. bataar"* *Khozatsky 1957*
Santonian **26** The smallest early Upper Cretaceous theropod *Patagopteryx deferrariisi* *Alvarenga & Bonaparte 1992*	Santon. Campan. **27** The last primitive megalosauroid? Unnamed *Alexornis antecedens* *Brodkorb 1976*	Campanian **28**	Campanian **29**	Campanian **30** North America's oldest feathers *Yandangornis longicaudus* *Cai & Zhao 1999*		

Several theories have tried to explain this massive extinction, but only three of them are supported by enough evidence. According to the theory of multiple causes, it may have even been a combination of events that caused the catastrophe. It is proposed that a decrease in ocean level and extreme volcanic activity decimated the population for quite some time, giving rise to a period of tension and stress similar to other times of crisis that dinosaurs had already endured. However, this coincided with several meteorite strikes, which caused the ecological balance to collapse. We must also consider all of the side effects, such as the release of toxic substances, atmospheric changes, large tsunamis, the blocking of sunlight, the passage of ultraviolet light, an absence of oxygen, ecological crisis, gases and volcanic dust, fires, and several others. See more: *Alvarez et al. 1980; McLean 1981; Duncan 1988; Chatterjee 1997; Sloan et al. 2002; Stewart & Allen 2002; Keller et al. 2008; Archibald & Fastovsky 2004; Jolley et al. 2010; Richard et al. 2015*

DECEMBER (81.1–66 Ma)

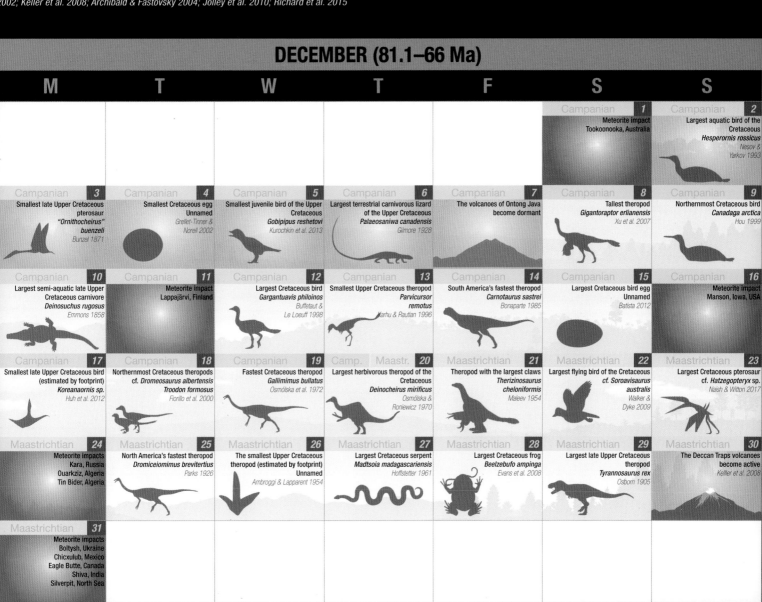

M	T	W	T	F	S	S
					Campanian **1** Meteorite impact Tookoonooka, Australia	Campanian **2** Largest aquatic bird of the Cretaceous *Hesperornis rossicus* Nesov & Yarkov 1993
Campanian **3** Smallest late Upper Cretaceous pterosaur *"Ornithocheirus" buenzeli* Bunzel 1871	Campanian **4** Smallest Cretaceous egg Unnamed Grellet-Tinner & Norell 2002	Campanian **5** Smallest juvenile bird of the Upper Cretaceous *Gobipipus reshetovi* Kurochkin et al. 2013	Campanian **6** Largest terrestrial carnivorous lizard of the Upper Cretaceous *Palaeosaniwa canadensis* Gilmore 1928	Campanian **7** The volcanoes of Ontong Java become dormant	Campanian **8** Tallest theropod *Gigantoraptor erlianensis* Xu et al. 2007	Campanian **9** Northernmost Cretaceous bird *Canadaga arctica* Hou 1999
Campanian **10** Largest semi-aquatic late Upper Cretaceous carnivore *Deinosuchus rugosus* Emmons 1858	Campanian **11** Meteorite impact Lappajärvi, Finland	Campanian **12** Largest Cretaceous bird *Gargantuavis philoinos* Buffetaut & Le Loeuff 1998	Campanian **13** Smallest Upper Cretaceous theropod *Parvicursor remotus* Karhu & Rautian 1996	Campanian **14** South America's fastest theropod *Carnotaurus sastrei* Bonaparte 1985	Campanian **15** Largest Cretaceous bird egg Unnamed Batista 2012	Campanian **16** Meteorite impact Manson, Iowa, USA
Campanian **17** Smallest late Upper Cretaceous bird (estimated by footprint) *Koreanaornis sp.* Huh et al. 2012	Campanian **18** Northernmost Cretaceous theropods cf. *Dromeosaurus albertensis* *Troodon formosus* Fiorillo et al. 2000	Campanian **19** Fastest Cretaceous theropod *Gallimimus bullatus* Osmólska et al. 1972	Camp. Maastr. **20** Largest herbivorous theropod of the Cretaceous *Deinocheirus mirificus* Osmólska & Roniewicz 1970	Maastrichtian **21** Theropod with the largest claws *Therizinosaurus cheloniformis* Maleev 1954	Maastrichtian **22** Largest flying bird of the Cretaceous cf. *Soroavisaurus australis* Walker & Dyke 2009	Maastrichtian **23** Largest Cretaceous pterosaur cf. *Hatzegopteryx sp.* Naish & Witton 2017
Maastrichtian **24** Meteorite impacts Kara, Russia Ouarkziz, Algeria Tin Bider, Algeria	Maastrichtian **25** North America's fastest theropod *Dromiceiomimus brevitertius* Parks 1926	Maastrichtian **26** The smallest Upper Cretaceous theropod (estimated by footprint) Unnamed Ambroggi & Lapparent 1954	Maastrichtian **27** Largest Cretaceous serpent *Madtsoia madagascariensis* Hoffstetter 1961	Maastrichtian **28** Largest Cretaceous frog *Beelzebufo ampinga* Evans et al. 2008	Maastrichtian **29** Largest late Upper Cretaceous theropod *Tyrannosaurus rex* Osborn 1905	Maastrichtian **30** The Deccan Traps volcanoes become active Kelller et al. 2008
Maastrichtian **31** Meteorite impacts Boltysh, Ukraine Chicxulub, Mexico Eagle Butte, Canada Shiva, India Silverpit, North Sea						

Because formations are geological events that occur over millions of years, fossils can be dated to a fairly broad period of time. This causes species to occupy several days in the Mesozoic Calendar. They were arranged by centering them between the oldest and the most recent probable time; the precise allocation of days is relative.

The World of dinosaurs
Geography of the past

Records: The largest and smallest by geographic location

In 1858, Sclater defined six biogeographic realms: Australasia, Ethiopian, Indomalaya-India, Nearctic, Neotropical, and Paleoartic. These regions are still used today, although some authors regard the insular or subcontinental regions as independent zones (Antilles, Arabian zone, Australia, Central America, Greenland, Madagascar, Southeast Asia, and New Zealand, among others).

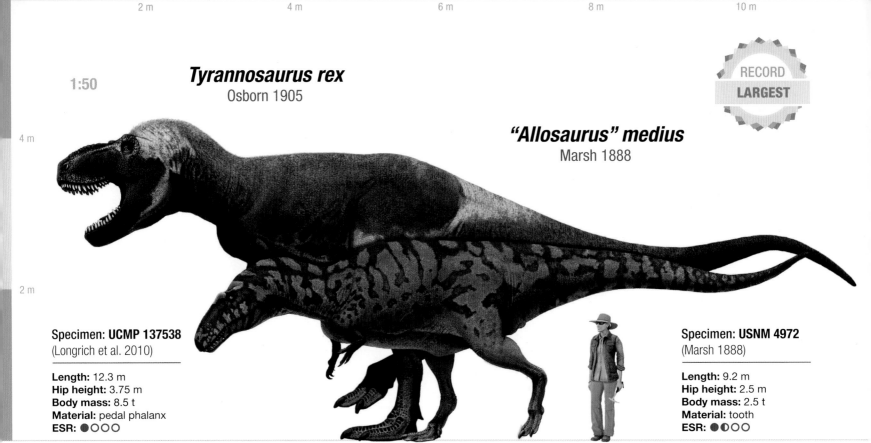

2 m 4 m 6 m 8 m 10 m

1:50

Tyrannosaurus rex
Osborn 1905

4 m

2 m

"Allosaurus" medius
Marsh 1888

Specimen: UCMP 137538
(Longrich et al. 2010)

Length: 12.3 m
Hip height: 3.75 m
Body mass: 8.5 t
Material: pedal phalanx
ESR: ●○○○

Specimen: USNM 4972
(Marsh 1888)

Length: 9.2 m
Hip height: 2.5 m
Body mass: 2.5 t
Material: tooth
ESR: ●◑○○

The largest in western North America

Theropod: *Tyrannosaurus rex* (present-day western Canada, western USA, and possibly Mexico) was the largest theropod of this continent. See more: *Longrich et al. 2010*

Mesozoic bird: The largest bird was *Hesperornis* cf. *regalis* UA 9716 (present-day Canada), which was 1.48 m long and weighed more than 25 kg. See more: *Fox 1974*

Dinosauromorph: The specimen UCMP 25793 (present-day Arizona, USA) was 2 m long and weighed 26 kg. *Atreipus* sp. (present-day Utah, USA) could have been about 2.8 m long and weighed 19 kg. See more: *Lockley & Hunt 1995; Long & Murry 1995*

Primitive saurischian: *Chindesaurus bryansmalli* and *Caseosaurus crosbyensis* (present-day Arizona, New Mexico, and Texas, USA) were the largest. Both were 2.4 m long and weighed 19 kg. See more: *Case 1927; Long & Murry 1995; Hunt et al. 1998*

The largest in eastern North America

Theropod: *"Allosaurus" medius* (present-day Maryland, USA) was of considerable size; its weight was comparable to that of a white rhinoceros. See more: *Marsh 1888*

Mesozoic bird: *Hesperornis regalis* (Kansas, USA) was 1.4 m long and weighed more than 20 kg. See more: *Marsh 1888*

Footprints

a) ⊢ 60.5 cm

b) ⊢ 86 cm

c) ⊢ 1 cm

d) ⊢ 2.5 cm

The largest

a) cf. *Eubrontes*: It is the largest theropod that left a footprint in eastern North America (present-day Virginia, USA). It was likely similar to *Sinosaurus triassicus*. See more: *Morales & Bulkey 1996*

b) *Tyrannosauripus pillmorei*: The continent's largest theropod estimated from a footprint was found in western North America (present-day New Mexico, USA). It could have been graceful or robust. See more: *Haubold 1971; Lockley & Hunt 1994*

The smallest

c) Unnamed: It is the smallest footprint that has been identified as belonging to a dinosaur. It was found in eastern North America (present-day Maryland, USA). See more: *Stanford et al. 2011*

d) *Koreanaornis* cf. *hamanensis*: This enantiornithean bird footprint is the smallest in western North America (present-day Utah, USA). See more: *Anfinson et al. 2009*

4 m

20 cm

cf. *Eubrontes*
Morales & Bulkey 1996
8.1 m / 1.1 t

Tyrannosauripus pillmorei
CU-MWC 225.1
11.4 m / 5.8-6.9 t

Unnamed footprint
Number 26
20 cm / 5 g

Koreanaornis cf. hamanensis
Anfinson et al. 2009
17.5 cm / 73 g

Theropod registry in North America

It's valid to distinguish western and eastern zones in the late Upper Cretaceous. In order to make comparisons, this will be extended throughout the Mesozoic.

In the western zone, direct remains of theropods have been located in the western United States (Idaho, Montana, Nebraska, North Dakota, Oklahoma, and Washington), and mixed remains (bones and footprints) have been found in Alaska, western Canada (Alberta and Saskatchewan), Mexico, and the western United States (Arizona, California, Colorado, New Mexico, South Dakota, Texas, Utah, and Wyoming).

In the eastern zone, there are direct remains of theropods in eastern Canada (Nova Scotia) and the eastern United States (Arkansas, Kansas, Maryland, New Jersey, Pennsylvania, and Virginia), and there are mixed remains in Greenland and the eastern United States (Alabama, North Carolina, Connecticut, the District of Columbia, Delaware, Georgia, Massachusetts, Mississippi, Missouri, New York, and Tennessee).

No Mesozoic theropod has been found in Central America.

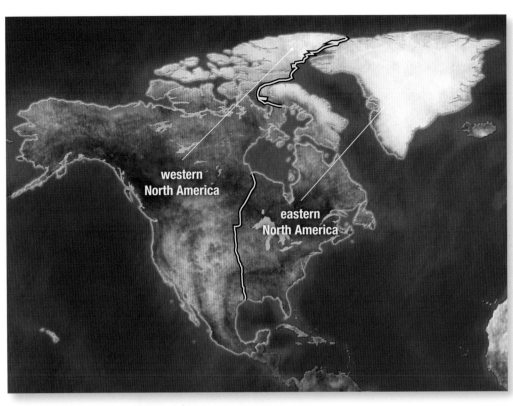

western North America

eastern North America

The smallest of western North America

Mesozoic bird: *Alexornis antecedens* (present-day Mexico) was considered a possible ancestor of coraciiform and piciform birds, but it is now known that it was an enantiornithean bird. See more: *Brodkorb 1976*

Dinosauromorph: The smallest footprints of *Rotodactylus cursorius* (present-day Arizona, USA) were 3.8 cm long, suggesting that they belonged to an animal that was about 28 cm long and weighed 60 g. *Dromomeron gregorii* (present-day Arizona and Texas, USA) was barely 85 cm long and weighed 1.2 kg, a size comparable to that of a raven. See more: *Peabody 1948; Nesbitt et al. 2009*

Theropod: The remains of juvenile coelophysoid attributed to *Protoavis texensis* TTU P 9201 (present-day Texas, USA) are the smallest. It was about 43 cm long and weighed 110 g. The adult coelophysoid cf. *Megapnosaurus* or "*Syntarsus mexicanum*" (present-day Mexico), was 1 m long and weighed 1.1 kg. See more: *Chatterjee 1991; Munter 1999*

The smallest of eastern North America

Theropod: The tiniest was the one that left the 1 cm long unnamed footprint (present-day Maryland, USA). It must have been a newborn that was approximately 22 cm long and weighed 12 g. See more: *Stanford et al. 2011*

Mesozoic bird: *Halimornis thompsoni* (present-day Alabama, USA) was a marine enantiornithean bird the size of a pigeon. See more: *Chiappe et al. 2002*

The oldest in North America

Rotodactylus cursorius (present-day Arizona, USA): They are dinosauromorph footprints that date from the Lower Triassic (Olenekian, ca. 251.2–247.2 Ma). The primitive saurischians *Caseosaurus crosbyensis* and *Chindesaurus bryansmalli* (present-day Arizona, New Mexico, and Texas, USA) date from the Upper Triassic (Norian, ca. 227–208.5 Ma). *Lepidus praecisio* (present-day Texas) is the oldest of North America's theropods (lower Norian, ca. 227–217 Ma). See more: *Case 1927; Peabody 1948; Long & Murry 1995; Sterling et al. 2015*

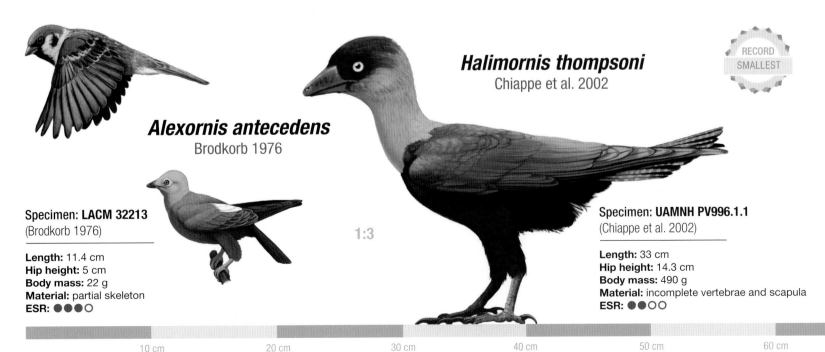

Alexornis antecedens
Brodkorb 1976

Specimen: LACM 32213
(Brodkorb 1976)

Length: 11.4 cm
Hip height: 5 cm
Body mass: 22 g
Material: partial skeleton
ESR: ●●●○

Halimornis thompsoni
Chiappe et al. 2002

RECORD SMALLEST

1:3

Specimen: UAMNH PV996.1.1
(Chiappe et al. 2002)

Length: 33 cm
Hip height: 14.3 cm
Body mass: 490 g
Material: incomplete vertebrae and scapula
ESR: ●●○○

10 cm   20 cm   30 cm   40 cm   50 cm   60 cm

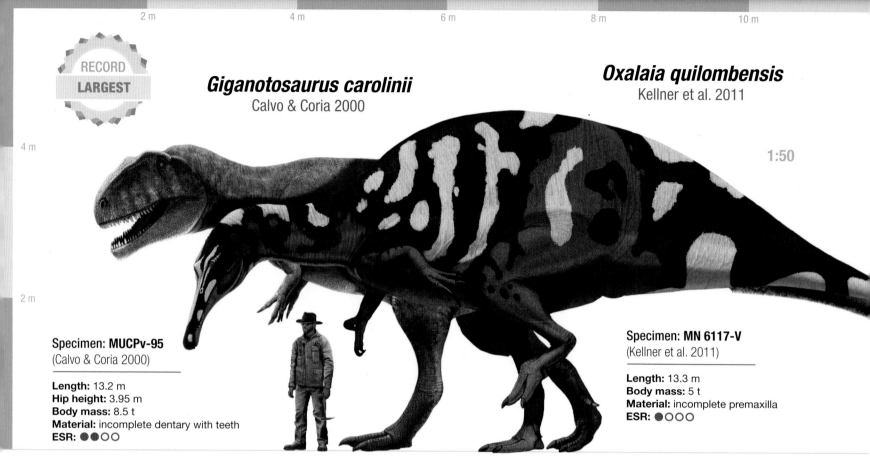

RECORD LARGEST

Giganotosaurus carolinii
Calvo & Coria 2000

Oxalaia quilombensis
Kellner et al. 2011

1:50

Specimen: MUCPv-95
(Calvo & Coria 2000)

Length: 13.2 m
Hip height: 3.95 m
Body mass: 8.5 t
Material: incomplete dentary with teeth
ESR: ●●○○

Specimen: MN 6117-V
(Kellner et al. 2011)

Length: 13.3 m
Body mass: 5 t
Material: incomplete premaxilla
ESR: ●○○○

The largest in northern South America

Theropod: *Oxalaia quilombensis* (present-day northern Brazil), a spinosaur, was about 13 m long and weighed 5 t. It was longer than a bus and almost as heavy as an African elephant. See more: *Kellner et al. 2011*

Mesozoic bird: The largest avian theropod of the Mesozoic was the enantiornithean CPP 482 (present-day northern Brazil). It was about 37 cm long and weighed about 573 g. See more: *Candeiro et al. 2012*

The largest in southern South America

Theropod: *Giganotosaurus carolinii* (present-day Argentina) was the largest terrestrial carnivore on the southern continents. See more: *Calvo & Coria 2000*

Primitive saurischians: The largest was *Frenguellisaurus ischigualastensis* (present-day Argentina), which was 5.3 m long and weighed 360 kg. It may be synonymous with *Herrerasaurus ischigualastensis*, but it was somewhat more recent. See more: *Novas 1986*

Dinosauromorph: *Teyuwasu barberenai* (present-day southern Brazil) had a massive body. It was 2.3 m long and weighed 75 kg. See more: *Kischlat 1999*

Mesozoic bird: The enantiornithean bird cf. *Soroavisaurus australis* PVL-4033 (present-day Argentina) was about 80 cm long and weighed 7.25 kg. See more: *Walker 1981; Chiappe 1993*

Footprints

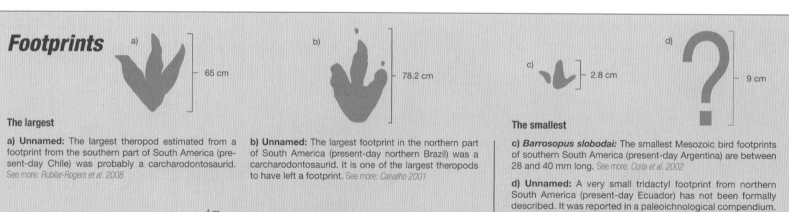

a)   65 cm

b)   78.2 cm

c)   2.8 cm

d)   9 cm

The largest

a) Unnamed: The largest theropod estimated from a footprint from the southern part of South America (present-day Chile) was probably a carcharodontosaurid. See more: *Rubilar-Rogers et al. 2008*

b) Unnamed: The largest footprint in the northern part of South America (present-day northern Brazil) was a carcharodontosaurid. It is one of the largest theropods to have left a footprint. See more: *Carvalho 2001*

The smallest

c) *Barrosopus slobodai:* The smallest Mesozoic bird footprints of southern South America (present-day Argentina) are between 28 and 40 mm long. See more: *Coria et al. 2002*

d) Unnamed: A very small tridactyl footprint from northern South America (present-day Ecuador) has not been formally described. It was reported in a paleoichnological compendium. See more: *Durán in Leonardi 1994*

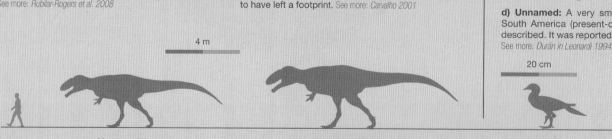

4 m

20 cm

Unnamed footprint
Trackway 22
9.1 m / 2.9 t

Unnamed footprint
SLPG - D
10.6 m / 5 t

Barrosopus slobodai
MCF-PVPH-SB 415.17 #94
15.3 cm / 280 g

Unnamed footprint
Durán in Leonardi 1994
? cm / ? g

Theropod registry in South America

To this day, discoveries of dinosaur species are less abundant in the northern area of South America than in the southern zone (also known as the Southern Cone). Because the northern area of South America and the northern part of Africa were fused for most of the Mesozoic, their fauna was very similar. Some authors call the union of South America and Africa *Samafrica*. See more: *Upchurch 2008*

In the northern area, direct remains have been located in Venezuela. Mixed remains have been found in Bolivia, northern Brazil (Amazon, Bahia, Ceará, Goiás, Maranhão, Minas Gerais, Matto Grosso, and Paraíba), Colombia, and Peru. Indirect remains (footprints) have been found in Ecuador and Guyana. In the southern zone, mixed remains were found in Argentina, southern Brazil (Arana, Rio Grande do Sul, and Sao Paulo), Chile, and Uruguay. Indirect remains have been found in Paraguay.

The smallest of northern South America

Mesozoic bird: *Cratoavis cearensis* (present-day northern Brazil) is the smallest bird of all fossil species. Its weight was less than that of the short-tailed pygmy tyrant (*Myiornis ecaudatus*), the smallest songbird, and similar to that of the smallest hummingbirds. See more: *Carvalho et al. 2015*

Theropod: The coelurosaur *Mirischia asymetrica* (present-day northern Brazil) was 1.5 m long and weighed 5 kg. A 9 cm long footprint (present-day Ecuador) was discovered in 1984 but was never formally described, so an estimate of the size of the creature that left it can't be made. See more: *Durán in Leonardi 1994; Naish, Martill & Frey 2004*

The smallest of southern South America

Mesozoic bird: *Lamarqueavis australis* (present-day Argentina) was almost twice as light as a sparrow. See more: *Agnolin 2010*

Dinosauromorph: The smallest was *Marasuchus lilloensis* (present-day Argentina), which was about 65 cm long and weighed about 320 g. *Lagosuchus talampayensis* was smaller but a juvenile. It was about 40 cm long and weighed 70 g. See more: *Romer 1971, 1972*

Theropod: The holotype of *Sarmientichnus scagliai* is a footprint of a probable abelisauroid that would have been approximately 28 cm long and weighed 285 g. See more: *Casamiquela 1964*

Primitive saurischian: *Eoraptor lunensis,* from southwestern Pangea (present-day Argentina), was 1.5 m long and weighed 5 kg. See more: *Sereno et al. 1993*

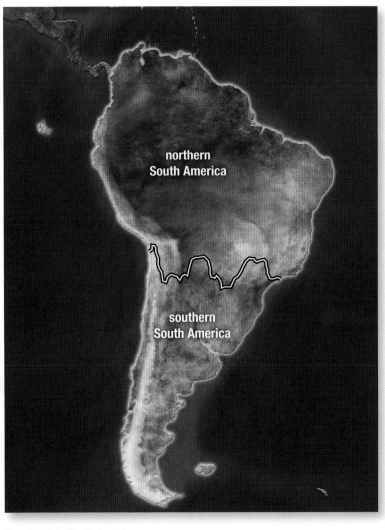

northern
South America

southern
South America

The oldest in South America

The dinosauromorphs *Lagerpeton chanarensis, Lewisuchus admixtus, Marasuchus lilloensis,* and *Pseudolagosuchus major* of the Chañares Formation (present-day Argentina) belonged to the Upper Triassic (lower Carnian, ca. 237–222 Ma). The dinosauromorph, saurischian, and theropod footprints from the Ischichuca and Los Rastros Formations are just as old. Both formations were once thought to be from the Middle Triassic, but a new revision suggests that they are more recent. See more: *Bonaparte in Leonardi 1989, 1994; Marsicano et al. 2007, 2015*

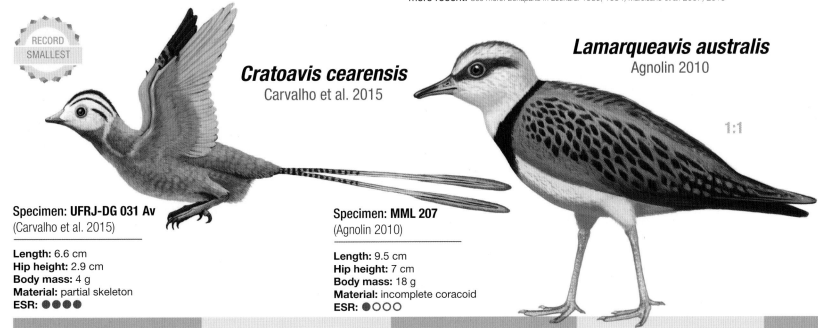

RECORD
SMALLEST

Cratoavis cearensis
Carvalho et al. 2015

Lamarqueavis australis
Agnolin 2010

1:1

Specimen: UFRJ-DG 031 Av
(Carvalho et al. 2015)

Length: 6.6 cm
Hip height: 2.9 cm
Body mass: 4 g
Material: partial skeleton
ESR: ●●●●

Specimen: MML 207
(Agnolin 2010)

Length: 9.5 cm
Hip height: 7 cm
Body mass: 18 g
Material: incomplete coracoid
ESR: ●○○○

5 cm   10 cm   15 cm   20 cm

RECORD
LARGEST

Torvosaurus gurneyi
Hendrickx & Mateus 2014

cf. *"Megalosaurus" pannoniensis*
Osi & Apesteguía 2008

1:50

4 m

2 m

Specimen: Collection V.01
(Osi & Apesteguía 2008)

Length: 6.8 m
Hip height: 1.75 m
Body mass: 750 kg
Material: tooth
ESR: ●◐○○

Specimen: ML 1100
(Hendrickx & Mateus 2014)

Length: 11.7 m
Hip height: 3.1 m
Body mass: 4 t
Material: partial skull and teeth
ESR: ●●○○

The largest in western Europe

Theropods: *Torvosaurus gurneyi* (present-day Portugal) was as long as a city bus and as heavy as an Asian elephant. See more: *Hendrickx & Mateus 2014*

Dinosauromorph: A Middle Triassic footprint (present-day France) belonged to a silesaurid that was 3.5 m long and weighed 120 kg. See more: *Demathieu 1970*

Mesozoic bird: *Gargantuavis philoinos* (present-day France) is exceptionally large for a Mesozoic bird. It was about 1.8 m tall and weighed about 120 kg. See more: *Buffetaut & Le Loeuff 1998*

The largest in eastern Europe

Theropod: cf. *"Megalosaurus" pannoniensis* was an Austrian species to which materials from western Europe (present-day France and Portugal) and eastern Europe (present-day Hungary) were assigned. It has been considered a large dromaeosaurid but is more likely a tyrannosauroid. See more: *Osi & Apesteguía 2008*

Mesozoic bird: FGGUB R.1902 (present-day Romania) was an Ornithuromorph bird that was 58 cm long and weighed 7 kg. Other reports, such as *Elopteryx*, are theropods. See more: *Wang et al. 2011*

Dinosauromorph: A footprint (present-day Czech Republic) that belonged to an animal that was 2.2 m long and weighed 30 kg. See more: *Zajic 1998*

Footprints

 a)  65 cm

b) 72 cm

c) 78 cm

d) 2 cm

e) 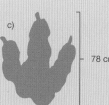 3.7 cm

The largest

a) Unnamed: The largest footprint in eastern Europe is an isolated impression (present-day Poland). The longest footprint from the Lower Jurassic, it may have belonged to a predator similar to *Sinosaurus triassicus*. See more: *Gierlinski 1991; Niedzwedzki 2006*

c) Unnamed: : The largest theropod known by footprints from western Europe (present-day Spain) may have been a carcharodontosaurid, not as long than *Torvosaurus gurneyi* but heavier. **b)** Another footprint, which may have belonged to a metriacantosaurid, could have belonged to a longer but lighter individual. See more: *Day et al. 2004; Barco et al. 2005*

The smallest

d) Unnamed: The smallest estimated by footprint of western Europe is a bird (present-day Spain). See more: *Moratalla et al. 2003*

e) Unnamed: The smallest known by footprints in eastern Europe was a dinosauromorph (present-day Poland). See more: *Niedzwiedzki et al. 2013*

4 m

20 cm

Unnamed footprint	**Unnamed footprint**	**Unnamed footprint**	**Unnamed footprint**	**Unnamed footprint**
Gierlinski 1991	T80	*Barco et al. 2005*	Los Cayos C (level C5).	*Niedzwiedzki et al. 2013*
8.6 m / 1.35 t	10.8 m / 3.2 t	10.1 m / 4 t	7.3-9.5 cm / 13-30 g	30 cm / 54 g

Theropod registry in Europe

The eastern and western zones are not divided along a geographical barrier. This division is imaginary, used in this book to help compare the faunas of both regions. In the western zone, direct remains have been located in Austria, Belgium, Denmark, Ireland, and Luxembourg. Mixed remains have been found in Germany, Spain, France, Holland, Italy, Norway, Portugal, United Kingdom, Sweden, and Switzerland. In the eastern zone, direct remains were found in Armenia, Bulgaria, Slovenia, Romania, and European Russia, while mixed remains were found in Croatia, Slovakia, Georgia, Hungary, Poland, and the Czech Republic.

The smallest of western Europe

Bird: *Iberomesornis romerali* (present-day Spain) was barely one-third the size of a sparrow. Some footprints could have belonged to a bird that was between 7.3 and 9.5 cm long and weighed between 7 and 15 g. See more: *Sanz & Bonaparte 1992*

Dinosauromorph: The old and tiny *Rotodactylus tumidus* footprints (present-day England) seem to have been produced by an animal that was about 25 cm long and weighed 12 g. See more: *Maidwell 1914*

Avemetatarsalian: *Scleromochlus taylori* (present-day Scotland) is the most primitive genus of the line that gave rise to dinosaurs and pterosaurs. They were only about 18.5 cm long and weighed 18 g, smaller than a sparrow. See more: *Woodward 1907*

Theropod: A footprint (present-day Scotland) belonged to a juvenile individual that was about 20 cm long and weighed 18 g. See more: *Clark et al. 2011*

Primitive saurischian: "Grallator pisanus" (present-day Italy) are footprints whose owner was about 80 cm long and weighed 66 g. See more: *Bianucci & Landini 2005*

The smallest of eastern Europe

Bird: The enantiornithean NVEN 1 (present-day Romania) is based on an incomplete humerus that must have belonged to a very small bird. See more: *Wang et al. 2011*

Dinosauromorph: The *Rotodactylus* sp. footprints (present-day Poland) belonged to an animal that was about 30 cm long and weighed 54 g. See more: *Niedzwiedzki et al. 2013*

Theropod: The most outstanding specimen is MTM PAL 2011.18 (present-day Hungary), a probable adult *Pneumatoraptor fodori* that was about 66 cm long and weighed 550 g. The troodontid tooth ACKK-D-8/088 from central Laurasia (present-day Slovenia) may have belonged to a young individual that was only 38 cm long and weighed 110 g. See more: *Debeljak et al. 2002; Ösi & Buffetaut 2011*

Primitive saurischians: Some footprints (present-day Poland) belonged to an animal that was 2.5 m long and weighed 20 kg. See more: *Gierslinsky et al. 2009*

The oldest in Europe

Prorotodactylus sp. (present-day Poland) were footprints of primitive avemetatarsalians that were found in the oldest geological layers of the Lower Triassic (Olenekian, ca. 251.2–247.2 Ma). See more: *See more: Brusatte et al. 2010*

RECORD SMALLEST

Unnamed species
Wang et al. 2011

1:3

Iberomesornis romerali
Sanz & Bonaparte 1992

Specimen: NVEN 1
(Wang et al. 2011)

Length: 24.4 cm
Hip height: 10.7 cm
Body mass: 200 g
Material: incomplete humerus
ESR: ●○○○

Specimen: LH-22
(Sanz & Bonaparte 1992)

Length: 8.7 cm
Hip height: 4.3 cm
Body mass: 9.5 g
Material: partial skeleton
ESR: ●●●○

10 cm   20 cm   30 cm   40 cm   50 cm   60 cm

Unnamed
Galton & Molnar 2011

1:50

Carcharodontosaurus saharicus
Sereno et al. 1996

4 m

2 m

Specimen: MB R2352
(Galton & Molnar 2011)

Length: 13 m
Hip height: 3.6 m
Body mass: 7 t
Material: tooth
ESR: ●◐○○

Specimen: SGM-Din 1
(Sereno et al. 1996)

Length: 12.7 m
Hip height: 3.75 m
Body mass: 7.8 t
Material: incomplete skull
ESR: ●●○○

The largest in North Africa

Theropod: *Carcharodontosaurus saharicus* (present-day Egypt and Morocco) was an immense theropod, heavier than an orca. It was the same size as the older *Carcharodontosaurus* sp. (present-day Niger). *Spinosaurus aegyptiacus* was longer but probably not as heavy. See more: *Stromer 1915; Lapparent 1960*

Dinosauromorph: *Diodorus scytobrachion* (present-day Morocco) was a 1.7 m long silesaurid that weighed 15.5 kg. See more: *Kammerer et al. 2012*

Mesozoic bird: Specimen CMN 50852 (present-day Morocco) was similar to *Rahonavis*, a fossil that is suspected to be an avialan or an unenlagiine theropod. If it was a bird, it would have been 1.7 m long and weighed about 8.8 kg. See more: *Riff et al. 2002, 2004*

The largest from southern Africa

Primitive saurischian: SAM-PK-K10013 (present-day South Africa) was perhaps similar to *Herrerasaurus*. Its size was comparable to that of an ostrich: It was 4.2 m long and weighed 130 kg. See more: *Kammerer et al. 2012*

Dinosauromorph: A great silesaurid (present-day Zambia), reaching a length of 3 m and a weight of 83 kg. It was a contemporary of *Lutungutali sitwensis*, so there is a possibility that they were the same species. See more: *Peecook et al. 2013*

Theropod: The carcharodontosaurid MB R2352 (present-day South Africa) is based on one of the thickest known theropod teeth, with an anteroposterior length of 52.8 mm. See more: *Galton & Molnar 2011*

Footprints

a) — 47 cm

b) — 82 cm

c) ↓— 0.8 cm

d) — 6.7 cm

The largest

a) *Megalosauripus* **sp.:** It is the largest theropod estimated by footprints in southern Africa (present-day Zimbabwe). The symmetrical shape reveals that it was an allosauroid. The location, time, and proportions indicate that the prints could have belonged to a carcharodontosaurid as old as *Veterupristisaurus*. See more: *Lingham-Soliar & Broderick 2000; Lingham-Soliar 2003*

b) **Unnamed:** The largest theropod ever to leave its footprint in the Jurassic. A similar one reached 90 cm in length, but the imprint included part of the metatarsal (present-day Morocco). See more: *Boutakiout et al. 2008, 2009*

4 m

The smallest

c) *Atreipus-Grallator*: The smallest footprint from northern Africa (present-day Morocco) belonged to a young dinosauromorph. See more: *Klein et al. 2011*

d) *Prototrisauropodiscus minimus*: The smallest footprint from southern Africa (present-day Lesotho) could have belonged to a coelophysoid. See more: *Ellenberger 1972*

15 cm

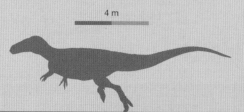

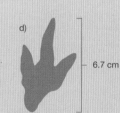

Megalosauripus sp.
Trackway 1
7.1 m / 1.4 t

Unnamed footprint
161GR1.1
12.7 m / 5.1 t

Atreipus-Grallator
CDUE 260
11 cm / 2 g

Prototrisauropodiscus minimus
Ellenberger 1972
79 cm / 530 g

Theropod registry in Africa

Here we divide Africa into two parts to compare them: the north and the south. Different distances to the equator cause differences in theropod fauna.

In the northern area, direct remains have been located in Chad, Libya, Egypt, Mauritania, Mali, Sudan, and Senegal. Mixed remains have been found in Algeria, Cameroon, Morocco, Niger, and Tunisia. In the southern area, direct remains have been located in Angola, Congo, Kenya, Malawi, Mozambique, South Africa, and Zambia, and mixed remains have been found in Ethiopia, Lesotho, Tanzania, and Zimbabwe. Indirect remains have come from Namibia.

The smallest from northern Africa

Dinosauromorph: The smallest Triassic footprints are identified as similar to *Atreipus-Grallator* of Pangea (present-day Morocco). Judging by size, their owner was a neonate that was only 11 cm long and weighed 2 g. See more: *Klein et al. 2011*

Theropod: The smallest footprint (present-day Morocco) might have been a paravian that was about 33–36 cm long and weighed 70–77 g. See more: *Belvedere et al. 2011*

Mesozoic bird There are several footprints (present-day Morocco) that may have belonged to a bird of the Middle Jurassic similar to *Archeopteryx*, which was about 59 cm long and weighed 685 g. By contrast, the only Mesozoic bird known from bone material in that area belonged to the late Lower Cretaceous and would have looked a lot like a larger *Rahonavis*. See more: *Belvedere et al. 2011*

The smallest of southern Africa

Mesozoic bird: An enantiornithean bird from the late Lower Cretaceous (present-day Tanzania) has been reported, but it has not been formally published. See more: *O'Connor et al. 2003*

Theropod: *Prototrisauropodiscus minimus* (present-day Lesotho) is a theropod footprint. Its owner could have been a coelophysoid that was 79 cm long and weighed 530 g. The ornithomimosaur *Nqwebasaurus thwazi* ("lizard runner of Nqweba") from central Gondwana (present-day South Africa) was 1.2 m long and weighed about 2.5 kg. See more: *Ellenberger 1972; de Klerk et al. 1997*

Primitive saurischian: *Saurichnium anserinum* (present-day South Africa) could be the footprint of a carnivorous dinosaur that was 88 cm long and weighed 895 g. See more: *Gurich 1926*

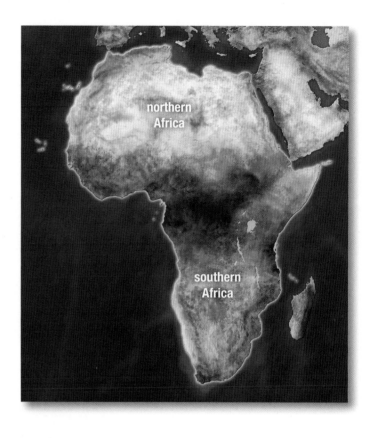

The oldest in Africa

Nyasasaurus parringtoni (present-day Tanzania) dates from the Middle Triassic (Anisian, ca. 247.2–242 Ma) and lived in south-central Pangea (present-day Tanzania). It might have been a dinosauromorph or a primitive saurischian dinosaur, but the material is too incomplete to say. The footprint of the dinosauromorph *Rotodactylus* cf. *Bessieri* from south-central Pangea (present-day Algeria) is just as old. See more: *Charig 1967; Kotanski et al. 2004; Nesbitt et al. 2013*

Nqwebasaurus thwazi
de Klerk et al. 1997

1:15

Specimen: AM 6040
(de Klerk et al. 1997)

Length: 1.2 m
Hip height: 35 cm
Body mass: 2.5 kg
Material: skull and partial skeleton
ESR: ●●●●

Unnamed
Riff et al. 2002

RECORD SMALLEST

Specimen: CMN 50852
(Riff et al. 2002)

Length: 1.7 m
Hip height: 55 cm
Body mass: 8.8 kg
Material: incomplete vertebra
ESR: ●◐○○

1 m   2 m   3 m

RECORD
LARGEST

4 m

2 m

Unnamed
Buffetaut & Suteethorn 2012

1:50

Indeterminate
Schulp & Hartman 2000

Specimen: SQU-2-7
(Schulp & Hartman 2000)

Length: 6.1 m
Hip height: 1.65 m
Body mass: 690 kg
Material: caudal vertebra
ESR: ●○○○

Specimen: PRC 61
(Buffetaut & Suteethorn 2012)

Length: 11.2 m
Hip height: 3.25 m
Body mass: 5 t
Material: maxilla and incomplete teeth
ESR: ●●○○

The largest from the Middle East

Theropod: Specimen SQU-2-7 (present-day Sultanate of Oman) was a carnivorous dinosaur that weighed more than a bull. Its affinity with other theropods is not known with certainty, although it could be a carcharodontosaurid. *See more: Schulp & Hartman 2000*

Mesozoic bird: A footprint (present-day Israel) belonged to a long-legged bird that was about 59 cm long and a weighed around 3.1 kg. *See more: Avnimelech 1966*

The largest from Cimmeria

Theropod: Specimen PRC 61 is a carcharodontosaurid from eastern Cimmeria (present-day Thailand). It was very large, twice the weight of a white rhinoceros and was almost as long as a city bus. *See more: Buffetaut & Suteethorn 2012*

Primitive saurischian: Footprints very similar in shape to those belonging to *Herrerasaurus* are the largest found in this area (present-day Thailand). It must have been a predator that was about 4.9 m long and weighed 295 kg. *See more: Le Loeuff et al. 2009*

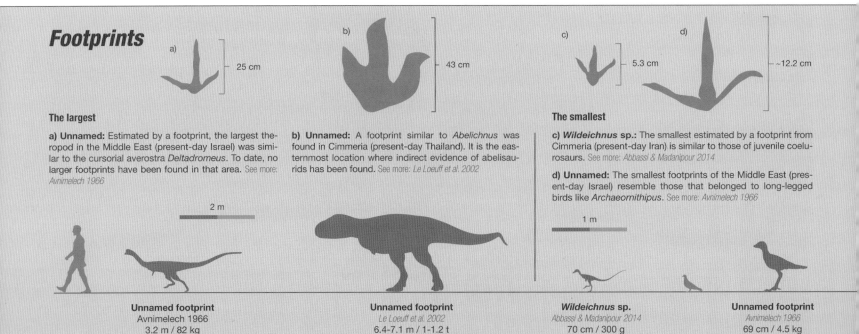

Footprints

a) — 25 cm

b) — 43 cm

c) — 5.3 cm

d) — ~12.2 cm

The largest

a) Unnamed: Estimated by a footprint, the largest theropod in the Middle East (present-day Israel) was similar to the cursorial averostra *Deltadromeus*. To date, no larger footprints have been found in that area. *See more: Avnimelech 1966*

b) Unnamed: A footprint similar to *Abelichnus* was found in Cimmeria (present-day Thailand). It is the easternmost location where indirect evidence of abelisaurids has been found. *See more: Le Loeuff et al. 2002*

The smallest

c) *Wildeichnus* sp.: The smallest estimated by a footprint from Cimmeria (present-day Iran) is similar to those of juvenile coelurosaurs. *See more: Abbassi & Madanipour 2014*

d) Unnamed: The smallest footprints of the Middle East (present-day Israel) resemble those that belonged to long-legged birds like *Archaeornithipus*. *See more: Avnimelech 1966*

2 m

1 m

Unnamed footprint
Avnimelech 1966
3.2 m / 82 kg

Unnamed footprint
Le Loeuff et al. 2002
6.4-7.1 m / 1-1.2 t

***Wildeichnus* sp.**
Abbassi & Madanipour 2014
70 cm / 300 g

Unnamed footprint
Avnimelech 1966
69 cm / 4.5 kg

MESOZOIC OF THE MIDDLE EAST and CIMMERIA

Throughout the Mesozoic, the Middle East and North Africa were joined and shared the same geographical history. Between the Triassic and the Lower Jurassic, the region formed part of south-central Pangea. In the Middle and Upper Jurassic, it was part of south-central Neopangea. At the beginning of the Cretaceous, it was part of south-central Gondwana. By the end, it was the northeastern part of Afro-Arabia.

Theropod registry in the Middle East

The dinosaur record in the Middle East is extremely poor. Direct remains have been located in Saudi Arabia, Lebanon, Oman, and Syria. Mixed remains have been found in Israel.

Cimmeria was a paleocontinent that surrounded the Paleo-Tethys Ocean. It was formed by a series of islands that extended below northeastern Pangea from the Triassic to the Lower Jurassic, was in Paleoasia from the Middle to the Upper Jurassic, and was in the eastern part of Laurasia in the Cretaceous. During the middle of the Cenozoic, Cimmeria collided with the northern continents, forming the Alpine orogeny, which includes the Alps and the Caucasus in Europe and the Himalayas, Hindu Kush, and Zagros in Asia.

Theropod registry in Cimmeria

Direct remains have been located in Armenia, Laos, Malaysia, Tibet, Turkey, and Vietnam. Mixed remains have been located in Thailand, and indirect remains (footprints) have been found in Afghanistan and Iran.

The smallest of the Middle East

Mesozoic bird: *Enantiophoenix electrophyla* ("opposite amber-eating phoenix") (present-day Lebanon) was an enantiornithean bird so small that a pigeon would look large beside it. See more: *Cau & Arduini 2008*

Theropod: The smallest were SGS 0061 and SGS 0090, which are known only by some teeth. They are similar to abelisaurids. They could be juveniles that were 1.35 m long and weighed 6 kg. See more: *Cau & Arduini 2008; Kear et al. 2013*

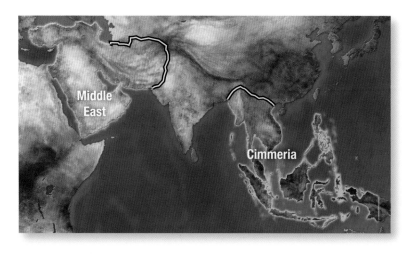

The smallest of Cimmeria

Mesozoic bird: The enantiornithean bird K3-1 from eastern Cimmeria (present-day Thailand) is one of the oldest known specimens. It was quite small, a third of a crow's weight. See more: *Buffetaut et al. 2005*

Theropod: The smallest are the specimens of coelurosaurs TF 1739-1 and TF 1739-2, from eastern Cimmeria (present-day Thailand). Taking *Compsognathus longipes* as reference, the former would be 80 cm long and weigh 340 g, while the latter would be about 1.5 m long and weigh 2.3 kg. See more: *Buffetaut & Ingavat 1984*

The oldest in the Middle East

A few feathers preserved in amber (present-day Lebanon) date from the early Lower Cretaceous (Hauterivian, ca. 132.9–129.4 Ma). They were identified as bird feathers, but it is also possible that they belonged to theropods. See more: *Schlee 1973; Schlee & Glöckner 1978*

The oldest in Cimmeria

Primitive saurischian footprints (present-day Thailand) could belong to a carnivore similar to *Herrerasaurus ischigualastensis*, which was probably 4.9 m long and weighed 295 kg. Possible theropod footprints (present-day Afghanistan) dating from the Lower Jurassic (Hettangian-Pliensbachian, ca. 201.3–182.7 Ma) have been reported. See more: *Lapparent & Stocklin1972; Le Loeuff et al. 2009*

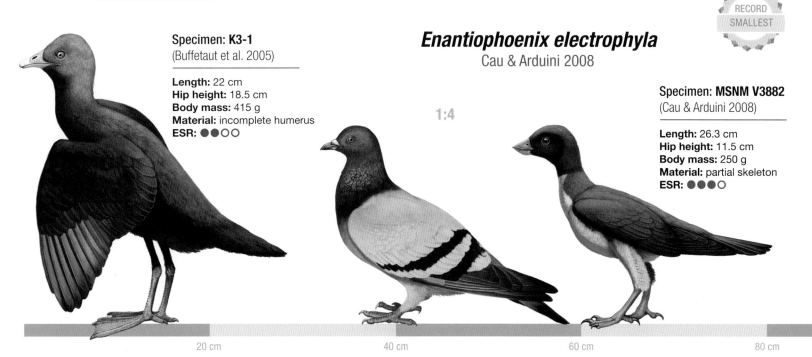

Unnamed
Buffetaut et al. 2005

Specimen: K3-1
(Buffetaut et al. 2005)

Length: 22 cm
Hip height: 18.5 cm
Body mass: 415 g
Material: incomplete humerus
ESR: ●●○○

1:4

RECORD SMALLEST

Enantiophoenix electrophyla
Cau & Arduini 2008

Specimen: MSNM V3882
(Cau & Arduini 2008)

Length: 26.3 cm
Hip height: 11.5 cm
Body mass: 250 g
Material: partial skeleton
ESR: ●●●○

20 cm   40 cm   60 cm   80 cm

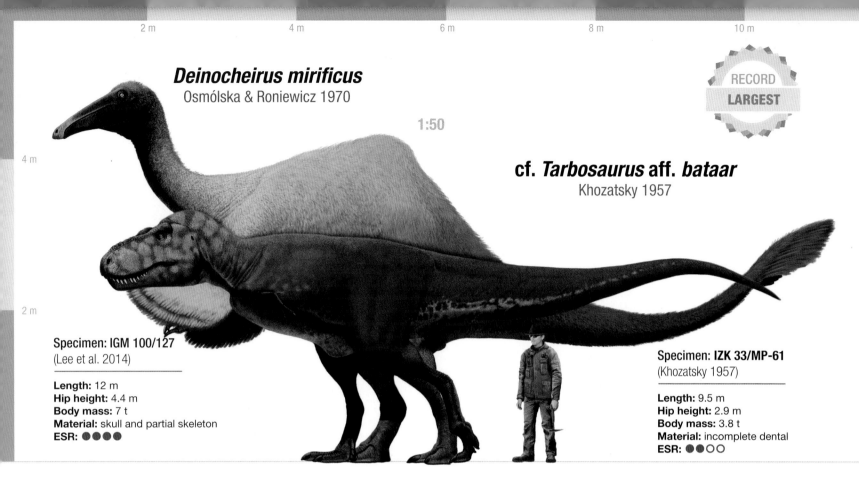

Deinocheirus mirificus
Osmólska & Roniewicz 1970

1:50

RECORD
LARGEST

cf. *Tarbosaurus* aff. *bataar*
Khozatsky 1957

Specimen: IGM 100/127
(Lee et al. 2014)

Length: 12 m
Hip height: 4.4 m
Body mass: 7 t
Material: skull and partial skeleton
ESR: ●●●●

Specimen: IZK 33/MP-61
(Khozatsky 1957)

Length: 9.5 m
Hip height: 2.9 m
Body mass: 3.8 t
Material: incomplete dental
ESR: ●●○○

The largest in western Asia

Theropods: cf. *Tarbosaurus* aff. *bataar* (present-day Kazakhstan) was almost as heavy as an Asian elephant, but it is not known if it was subadult or adult. *Turkmenosaurus kugitangensis* (present-day Turkmenistan) are footprints that belonged to a large megalosaurid that was 11.2 m long and weighed 3.4 t. See more: *Amanniazov 1985; Khozatsky 1957; Averianov et al. 2012*

Mesozoic bird: *Asiahesperornis bazhanovi* (present-day Kazakhstan) was a large hesperornithiform with aquatic habits. It was 1.25 m long and weighed 14.8 kg. Some researchers think that it is possible that it belongs to the genus *Hesperornis*. See more: *Nesov & Prizemlin 1991; Mortimer**

The largest in eastern Asia

Theropod: *Deinocheirus mirificus* (present-day Mongolia) was a huge herbivorous dinosaur with large arms, a hump, and a pygostyle tail. The largest carnivorous theropod in Asia was the specimen NHMG 8500 (present-day China), which might have been 12.9 m long and weighed 5.2 t. See more: *Lee et al. 2014; Mo & Xu 2015*

Primitive saurischian: A footprint (present-day China) is the only evidence of a carnivorous dinosaur that has been conserved from the Triassic in East Asia. It has similarities with primitive saurischians. Its owner could have been about 3 m long and weighed approximately 34 kg. See more: *Xing et al. 2014*

Mesozoic bird: *Yandangornis longicaudus* (present-day China) was a fast-running flightless bird that was about 60 cm long and weighed 2.1 kg. See more: *Cai & Zhao 1999*

Footprints

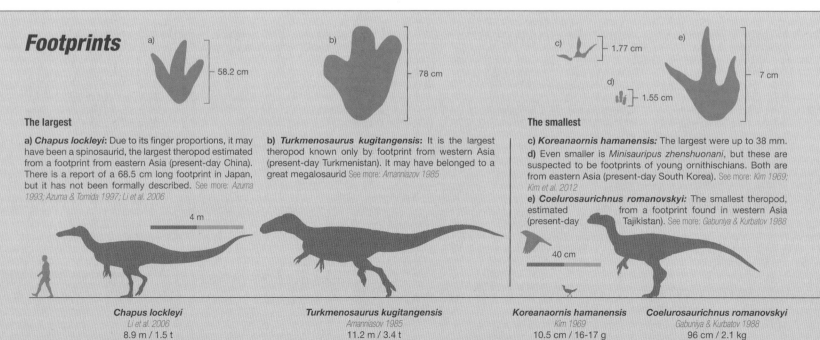

a) — 58.2 cm

b) — 78 cm

c) — 1.77 cm

d) — 1.55 cm

e) — 7 cm

The largest

a) *Chapus lockleyi*: Due to its finger proportions, it may have been a spinosaurid, the largest theropod estimated from a footprint from eastern Asia (present-day China). There is a report of a 68.5 cm long footprint in Japan, but it has not been formally described. See more: *Azuma 1993; Azuma & Tomida 1997; Li et al. 2006*

b) *Turkmenosaurus kugitangensis*: It is the largest theropod known only by footprint from western Asia (present-day Turkmenistan). It may have belonged to a great megalosaurid See more: *Amanniazov 1985*

The smallest

c) *Koreanaornis hamanensis*: The largest were up to 38 mm.
d) Even smaller is *Minisauripus zhenshuonani*, but these are suspected to be footprints of young ornithischians. Both are from eastern Asia (present-day South Korea). See more: *Kim 1969; Kim et al. 2012*
e) *Coelurosaurichnus romanovskyi*: The smallest theropod, estimated   from a footprint found in western Asia (present-day   Tajikistan). See more: *Gabuniya & Kurbatov 1988*

4 m

40 cm

Chapus lockleyi
Li et al. 2006
8.9 m / 1.5 t

Turkmenosaurus kugitangensis
Amanniasov 1985
11.2 m / 3.4 t

Koreanaornis hamanensis
Kim 1969
10.5 cm / 16-17 g

Coelurosaurichnus romanovskyi
Gabuniya & Kurbatov 1988
96 cm / 2.1 kg

ASIAN MESOZOIC

In 1942, three major geographical areas of the Mesozoic were recognized: Laurentia (North America), Angaria (Asia and Europe), and Gondwana (South America, Africa, the Middle East, Hindustan, Antarctica, and Oceania). Therefore, modern Asia is the sum of several paleocontinents that collided with each other. See more: *Jeannel 1942*

Theropod registry in Asia

Here we divide the west and east to compare them. In the western zone, direct remains have been located in Kyrgyzstan and western Russia. Mixed remains have been located in Kazakhstan, Tajikistan, Turkmenistan, and Uzbekistan. In the eastern zone, direct theropod fossils are known in North Korea, and mixed remains have been found in China, South Korea, Japan, Mongolia, and eastern Russia.

The smallest of East Asia

Mesozoic bird: *Parvavis chuxiongensis* (present-day China) was the second smallest bird of the Mesozoic (globally). It was so tiny that it would take five to equal the weight of one sparrow. Even smaller are bird embryos, including the specimen PIN 4492-4 of *Gobipipus reshetovi* (present-day Mongolia), which was 4.7 cm long and weighed 2 g.

Theropod: *Epidendrosaurus ningchengensis* and *Scansoriopteryx heilmanni* (present-day China) were juveniles that were 13 cm long and weighed 6.7 g. Some experts think that *Scansoriopteryx* dates from the Middle Jurassic, while others suggest it is from the late Lower Cretaceous. *Parvicursor remotus* (present-day Mongolia) was an adult alvarezsaurid that was 50 cm long and weighed 185 g. See more: *Karhu & Rautian 1996; Czerkas & Yuan 2002; Zhang et al. 2002; Buffetaut et al. 2005; Kurochkin, Chatterjee & Mikhailov 2013; Wang, Zhou & Xu 2014*

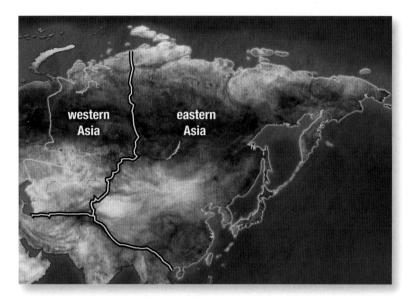

western Asia | eastern Asia

The smallest of western Asia

Mesozoic bird: *Kizylkumavis cretacea* (present-day Uzbekistan) was smaller than a sparrow. See more: *Nesov 1984*

Theropod: *"Velociraptor"* sp. ZIN PH 34/49 (present-day Kazakhstan) are remains of a probable microraptor that was 66 cm long and weighed 410 g. The therizinosaur embryo ZIN PH 35/49 (present-day Kazakhstan) was only 8.7 cm long and weighed 3.6 g. See more: *Averianov 2007; Mortimer**

The oldest in Asia

The footprint PI4 (present-day China) dates from the Upper Triassic (Norian-Rhaetian, ca. 227–201.3 Ma). It may have belonged to a primitive saurischian. *Pengxianpus cifengensis* and other footprints (present-day China), from the Upper Triassic (Rhaetian, ca. 208.5–201.3 Ma), may have belonged to theropods or primitive sauropodomorphs. See more: *Yang & Yang 1987; Xing et al. 2013, 2014*

RECORD SMALLEST

Kizylkumavis cretacea
Nesov 1984

1:1

Specimen: TsNIGRI 51/11915
(Nesov 1984)

Length: 12.5 cm
Hip height: 5.4 cm
Body mass: 25 g
Material: partial skeleton
ESR: ●●●○

Parvavis chuxiongensis
Wang et al. 2014

Specimen: IVPP V18586
(Wang et al. 2014)

Length: 7.5 cm
Hip height: 3.3 cm
Body mass: 6 g
Material: partial skeleton and feathers
ESR: ●●●●○

5 cm   10 cm   15 cm   20 cm

4 m

Rajasaurus narmadensis
Matley 1923

RECORD
LARGEST

1:40

Unnamed
Maganuco et al. 2007

2 m

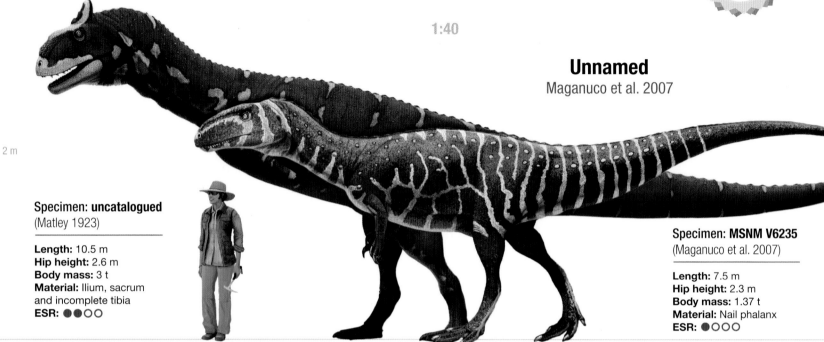

Specimen: uncatalogued
(Matley 1923)

Length: 10.5 m
Hip height: 2.6 m
Body mass: 3 t
Material: Ilium, sacrum
and incomplete tibia
ESR: ●●○○

Specimen: MSNM V6235
(Maganuco et al. 2007)

Length: 7.5 m
Hip height: 2.3 m
Body mass: 1.37 t
Material: Nail phalanx
ESR: ●○○○

The largest in Hindustan

Theropod: Some fossil bones from south-central Gondwana (present-day India) were originally assigned to the supposed armored dinosaur *Lametasaurus indicus*, but they turned out to be from a large specimen of the abelisaurid *Rajasaurus narmadensis.* See more: *Matley 1923*

The largest in Madagascar

Theropod: The ceratosaurid MSNM V6235 (present-day Madagascar) was about 7.5 m long and weighed around 1.37 t. It was shorter than *Majungasaurus crenatissimus,* but its weight was similar. At that time, Ma-

dagascar was part of south-central Neopangea. It was in contact with Africa, Hindustan, and Antarctica. *Majungasaurus crenatissimus* (formerly known as *"Megalosaurus" crenatissimus*) was an abelisaurid that inhabited Madagascar once it became an island. It was 8.1 m long and weighed 1.3 t. See more: *Sues & Taquet 1979; Maganuco et al. 2007; Goswami et al. 2012*

Mesozoic bird: The largest Mesozoic bird of Madagascar was *Vorona berivotrensis,* which was 51 cm long and weighed 4.5 kg. See more: *Forster et al. 1996*

Footprints

 a)

— 46 cm

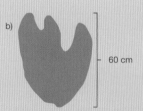 b)

— 60 cm

 c) ↓ — 5.5 cm

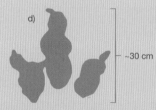 d)

— ~30 cm

The largest

a) Unnamed: The largest theropod estimated from a footprint (present-day Madagascar) is similar to *Kayentapus* and belonged to an animal that weighed over 500 kg. See more: *Wagensommer et al. 2011*

b) *Samanadrinda surghari*: It has been suggested that it could be an abelisaurid footprint, but its asymmetrical proportions correspond to those of a megalosauroid. It was found in Hindustan (present-day Pakistan). See more: *Malkani 2007*

The smallest

c) *Grallator tenuis*: The smallest footprint of Hindustan (present-day India) belonged to a coelurosaurian theropod similar to a coelophysoid. See more: *Pienskowski et al. 2015*

d) Unnamed: The smallest theropod footprints found in present-day Madagascar are likely medium-size abelisaurids. See more: *Wagensommer et al. 2011*

3 m

2 m

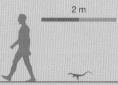

Unnamed footprint
TsI 31
6.4 m / 510 kg

Samanadrinda surghari
Malkani 2007
8.1 m / 860 kg

Grallator tenuis
Pienskowski et al. 2015
74 cm / 435 g

Unnamed footprint
TSI 8
5.6 m / 470 kg

INDO-MALGACH MESOZOIC

During the Triassic and the beginning of the Jurassic, Indo-Malgach was part of south-central Pangea. It integrated into southwestern Neopangea during the second half of the Jurassic and formed the south-central zone of Gondwana in the Lower Cretaceous until it separated from all the other continents. Finally, in the Upper Cretaceous, it divided into two regions: Madagascar and Hindustan.

The union of Hindustan and Madagascar took place 100 million–120 million years ago. It separated from Antarctica and Africa and subsequently collided with Paleoasia 55 million years ago. The separation of Hindustan and Madagascar occurred 85 million years ago. See more: *Fang et al. 2006*

Theropod registry in Hindustan

Mixed remains have been located in India and Pakistan.

Theropod registry in Madagascar

Mixed remains have been located on the island of Madagascar.

The smallest of Hindustan

Primitive saurischian: *Alwalkeria maleriensis*, from south-central Pangea (present-day India), was a basal saurischian similar to *Eoraptor lunensis*, so it could have been a primitive sauropodomorph. See more: *Chatterjee 1987*

Theropod: The coelophysoid K.33/606b was a juvenile fossil, as it was found with another individual twice as large (K.33/606a). It was 2 m long and weighed 7.5 kg. *Grallator tenuis* (present-day India) is a footprint of a coelophysoid that was about 74 cm long and weighed 435 g. See more: *Huene 1940; Pienskowski et al. 2015*

The smallest of Madagascar

Mesozoic bird: The specimen FMNH PA 747 is the smallest Mesozoic bird from Madagascar. See more: *O'Connor & Foster 2000; O'Connor 2010*

Theropod: The specimen MSNM V 5589 was a possible dromaeosaurid that was 2.1 m long and weighed 23 kg. It should be noted that some experts consider them abelisaurid teeth. *Masiakasaurus knopfleri* reached up to 4.6 m in length and 115 kg in weight. See more: *Sampson, Carrano & Forster 2001; Fanti & Therrien 2007*

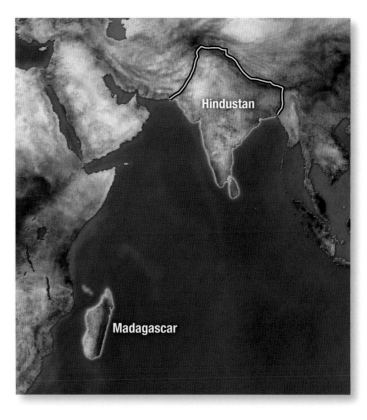

The oldest in Hindustan

The primitive saurischian *Alwalkeria maleriensis* and the coelophysoid theropods K.33/621a and K.33/606b (present-day India) date from the Upper Triassic (Carnian, ca. 237–227 Ma). See more: *Huene 1940; Chatterjee 1987*

The oldest in Madagascar

Footprints similar to those of *Kayentapus* and others similar to those of abelisaurids (present-day Madagascar) date from the Middle Jurassic (Bajocian, ca. 170.3–168.3 Ma). See more: *Wagensommer et al. 2011*

RECORD
SMALLEST

Alwalkeria maleriensis
Chatterjee 1987

Unnamed
O'Connor & Forster 2000

1:6

Specimen: FMNH PA 747
(O'Connor & Forster 2000)

Length: 20 cm
Hip height: 8.9 cm
Body mass: 119 g
Material: humerus
ESR: ●●○○

Specimen: ISI R 306
(Chatterjee 1987)

Length: 1.3 m
Hip height: 38 cm
Body mass: 2 kg
Material: femur and vertebrae
ESR: ●●○○

20 cm   40 cm   60 cm   80 cm   100 cm

4 m

2 m

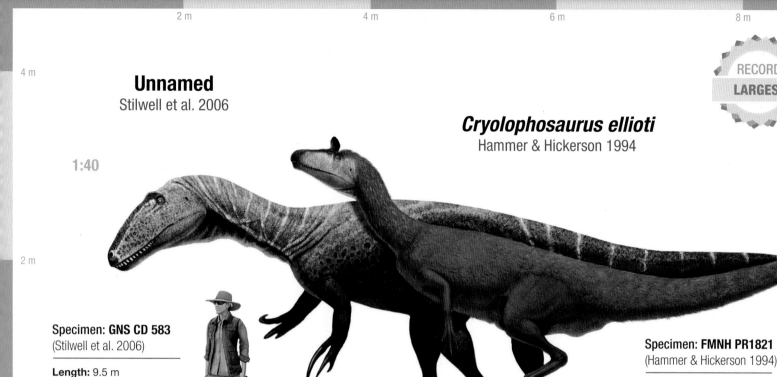

Unnamed
Stilwell et al. 2006

1:40

RECORD
LARGEST

Cryolophosaurus ellioti
Hammer & Hickerson 1994

Specimen: **GNS CD 583**
(Stilwell et al. 2006)

Length: 9.5 m
Hip height: 2.65 m
Body mass: 2 t
Material: manual phalanx
ESR: ●○○○

Specimen: **FMNH PR1821**
(Hammer & Hickerson 1994)

Length: 7.7 m
Hip height: 2 m
Body mass: 780 kg
Material: skull and partial skeleton
ESR: ●●●●

The largest in Oceania

Theropod: The specimen GNS CD 583 (present-day New Zealand) must have been a theropod heavier than a hippopotamus. Perhaps it was a megaraptor that was 9.5 m long. See more: *Molnar 1999*

Primitive saurischian: The footprints of *Eubrontes* sp. (present-day Australia) were similar to those of *Herrerasaurus*. It could have been 4.7 m long and weighed about 265 kg. See more: *Staines & Woods 1964*

Mesozoic bird: The specimen QM F37912 (present-day Australia) was similar to *Yandangornis longicaudus* but bigger. It was 1.65 m long and weighed 21 kg. See more: *Stilwell et al. 2006*

Largest in Antarctica

Theropods: *Cryolophosaurus ellioti* from the southeastern Pangea zone, it was a coelophysoid or a primitive tetanurae two car-lengths long and as heavy as the weight of a saltwater crocodile and a lion combined. It had a curious yet striking fan-shaped crest that caused it to be called "Elvisaurus" a year before receiving its official name. See more: *Hammer & Hickerson 1994*

Mesozoic bird: A femur similar to that of the present-day *Cariama* was also reported, and to date it is the largest known on that continent, with a length of 74 cm and weighing 4.1 kg. See more: *Case et al. 2006*

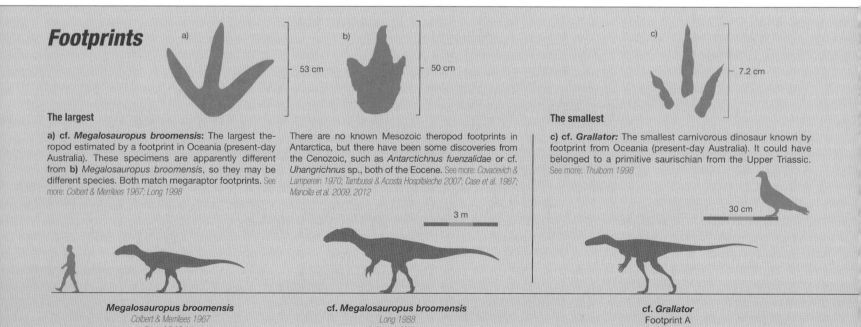

Footprints

a)   b)   c)

├ 53 cm   ├ 50 cm   ├ 7.2 cm

The largest

a) cf. *Megalosauropus broomensis*: The largest theropod estimated by a footprint in Oceania (present-day Australia). These specimens are apparently different from **b) *Megalosauropus broomensis*,** so they may be different species. Both match megaraptor footprints. See more: *Colbert & Merrilees 1967; Long 1998*

There are no known Mesozoic theropod footprints in Antarctica, but there have been some discoveries from the Cenozoic, such as *Antarctichnus fuenzalidae* or cf. *Uhangrichnus* sp., both of the Eocene. See more: *Covacevich & Lamperein 1970; Tambussi & Acosta Hospitaleche 2007; Case et al. 1987; Mancilla et al. 2009, 2012*

The smallest

c) cf. *Grallator*: The smallest carnivorous dinosaur known by footprint from Oceania (present-day Australia). It could have belonged to a primitive saurischian from the Upper Triassic. See more: *Thulborn 1998*

3 m

30 cm

Megalosauropus broomensis
Colbert & Merrilees 1967
5 m / 310 kg

cf. *Megalosauropus broomensis*
Long 1988
6.8 m / 750 kg

cf. *Grallator*
Footprint A
98 cm / 1.2 kg

OCEANIC-ANTARCTIC MESOZOIC

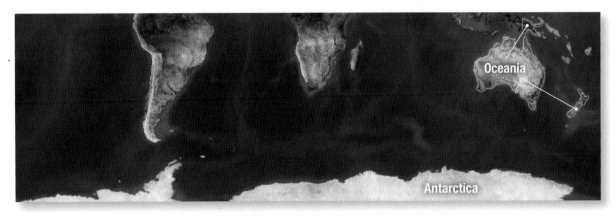

During the Triassic and the beginning of the Jurassic, this great region was part of southeastern Pangea. In the Middle and Upper Jurassic, it was part of southeastern Neopangea. During the early Cretaceous, it integrated into southern Gondwana, creating a passage between Madagascar, Hindustan, South America, and Oceania. At the end of the late Upper Cretaceous, it began to drift away from the other continents.

In the middle of the Cretaceous, some parts of what is Oceania today were farther south than parts of Antarctica at the time. Its fauna was adapted to cold environments, since temperatures remained low for long periods (up to half a year).

Theropod registry in Oceania and Antarctica

Direct remains have been located in Antarctica and New Zealand, and mixed remains have been found in Australia.

The smallest of Oceania

Mesozoic bird: *Nanantius* sp. QQM F31813 (present-day Australia) was an enantiornithean bird. See more: *Kurochkin & Molnar 1997*

Primitive saurischian: cf. *Grallator* (present-day Australia) could be a footprint whose owner would have been approximately 98 cm long and weighed 1.2 kg. See more: *Thulborn 1998*

Theropod: *Ozraptor subotaii,* from southeastern Neopangea (present-day Australia), is an enigmatic theropod that has been interpreted as a primitive abelisauroid. It could have been 2.3 m long and weighed 13 kg. On the other hand, the size of a coelophysoid footprint (present-day Australia) suggests that it belonged to a small animal that was 1.65 m long and weighed 4.9 kg. See more: *Hill et al. 1966; Long & Molnar 1998; Rauhut 2012*

The smallest of Antarctica

Mesozoic bird: A possible caradriform MLP 98-I-10-25 was the smallest. It was barely larger than a sparrow. Whether it was an adult or a juvenile is unknown. See more: *Case et al. 2007*

Theropod: The smallest of this area is an unenlagian found on James Ross Island. It was about 2.9 m long and weighed 48 kg. See more: *Case & Tambussi 1999*

The oldest in Oceania

cf. *Grallator* (present-day Australia) are footprints of probable primitive saurischians dating from the Upper Triassic (Carnian, ca. 237–227 Ma). Several footprints of coelophysoid theropods date from the Lower Jurassic (Sinemurian, ca. 199.3–190.8 Ma). See more: *Bartholomai 1966; Hill 1970; Molnar 1991; Thulborn 1998; Scanlon 2006; Cook et al. 2010*

The oldest in Antarctica

Cryolophosaurus ellioti and the teeth of an undescribed coelophysoid (present-day Antarctica) date from the Lower Jurassic (Sinemurian, 199.3–190.8 Ma). See more: *Hammer & Hickerson 1994; Ford**

RECORD SMALLEST

Nanantius sp.
Kurochkin & Molnar 1997

1:3

Unnamed
Case & Tambussi 1999

Specimen: **QM F31813**
(Kurochkin & Molnar 1997)

Length: 23.4 cm
Hip height: 10.2 cm
Body mass: 175 g
Material: partial skeleton
ESR: ●●●○

Specimen: **MLP 98-I-10–25**
(Case & Tambussi 1999)

Length: 19 cm
Hip height: 9 cm
Body mass: 40 g
Material: partial posterior limb
ESR: ●●○○

20 cm   40 cm   60 cm   80 cm   100 cm   120 cm

ca. 1000 BC—China—The first report of dinosaurs in Asia?
Among the oldest traditions that were compiled at that time was finding dragon remains in the fields. See more: *Mayor 2000*

ca. 430 BC—Egypt—The first report of dinosaurs in Africa?
On a trip to Egypt, Herodotus, the Father of Historiography, discovered vast mounds of "snake bones." It is suspected that some were the remains of theropods and other dinosaurs. See more: *Mayor 2000*

ca. 200 BC—China—The first report of dinosaurs in Asia?
Fossil bones were found in an excavation of the present-day Dragon-Head Waterway. See more: *Mayor 2007*

1671—England—The first fossil of a Jurassic theropod in western Europe
In England, the distal part of a femur of a Middle Jurassic theropod was found. Although it has been assigned to *Megalosaurus*, it could have belonged to any other carnivorous dinosaur of that time. See more: *Sanz 2007; Mortimer **

1806—USA—Western North America's first dinosaur
In an expedition to Montana directed by Clark, giant fish ribs were reported. They were probably from a dinosaur. See more: *Saindon 2003*

1828—France—The first report of a theropod in a second country
The presence of *"Megalosaurus"* was reported in Normandy, France. See more: *Caumont 1828; Spalding & Sarjeant 2012*

1834—England—The first bird fossil in western North America
"Palaeotringa" vetus was originally identified as a woodcock (*Scolopax*). For a long time, it was considered a bird of the Mesozoic, but it may belong to the Cenozoic. It is the first North American fossil bird. See more: *Morton 1834; Olson & Parrish 1987*

1854—Canada—The first mistaken report of a theropod from eastern North America
Bathygnathus borealis, from Prince Edward Island (Canada), turned out to be a plesiosaur from the Permian. See more: *Leidy 1854; Case 1905*

1856—USA—The first theropods discovered in western North America
Deinodon horridus and *Troodon formosus* were discovered in Montana. Since both are based only on teeth, the names are considered dubious. See more: *Leidy 1856*

1857—USA—The first misdiagnosis of a Mesozoic bird in eastern North America
A supposed Triassic bird from North Carolina was first named *Palaeonornis struthionoides* but turned out to be the remains of the phytosaurus *Rutiodon carolinensis*. See more: *Emmons 1857; Wetmore 1956*

1858—Germany—Western Europe's first Triassic theropod?
"Megalosaurus" cloacinus based on a tooth that could have been from a theropod or an archosaur and currently is considered dubious. See more: *Plieninger 1846; Quenstedt 1858*

1859—India—The first theropod discovered in the south of Asia (Hindustan)
"Massospondylus" rawesi is a loose tooth thought to be a prosauropod. It was discovered in 1859 in Saugor and named in 1890. See more: *Hislop 1864; Lydekker 1890; Carrano et al. 2010*

1859—South Africa—The first incorrect report of a theropod in southern Africa
Galesaurus planiceps mistaken for a dinosaur. Actually, it was a Lower Triassic synapsid. See more: *Owen 1859*

1859—Russia—The first incorrect report of a theropod in eastern Europe
Deuterosaurus biermicus was considered a dinosaur but was a synapsid from the Upper Permian of the Republic of Bashkortostan. See more: *Eichwald 1860*

1860—Brazil—The first incorrect report of a theropod in northern South America
A thoracic vertebra of *"Megalosaurus"* is mentioned, but it later turned out to be of a crocodile. See more: *Allport 1860; Campos & Kellner 1991*

1865—USA—The first theropod discovered in eastern North America
"Coelosaurus" antiquus is an ornithomimosaur tibia from Maryland. The name was already occupied by an unidentified animal that is thought to be forgotten. See more: *Leidy 1865; Owen 1854; Spamer, Daeschler & Vostreys-Shapiro 1995*

1865—India—The first incorrect report of a carnivorous dinosaur in South Asia (Hindustan)
Ankistrodon indicus is an archosauriform from the Lower Triassic. It was initially considered a dinosaur. In 1972, it was synonymized with *Proterosuchus*. See more: *Huxley 1865; Romer 1994*

1869—USA—Western North America's first Jurassic theropod
Antrodemus valens is represented by an incomplete caudal vertebra. It probably belonged to an *Allosaurus*, but this is impossible to determine. In addition, the piece was lost. See more: *Leidy 1870*

1870—USA—Western North America's first Mesozoic (and aquatic) bird
Holotype material of *Hesperornis regalis* was found two years before it was described in Kansas. See more: *Mortimer**

1872—USA—The first Mesozoic bird named in western North America
Hesperornis regalis was the first, but *Graculavus anceps*, *Colonosaurus mudgei*, and *Ichthyornis dispar* followed in the same year. They turned out to be synonymous with *Icthyornis anceps*. See more: *Marsh 1872*

1872—USA—The first Mesozoic bird named in eastern North America
Limosavis velox (now *Graculavus*) was a New Jersey bird from the border between the Mesozoic and Cenozoic. See more: *Marsh 1872; Shufeldt 1915*

1875—France—The first Cretaceous theropod of Western Europe
"Megalosaurus" superbus (now *Erectopus*), was named seven years after its discovery. See more: *Barrois 1875; Sauvage 1882; Huene 1923*

1881—Austria—Eastern Europe's first theropod
"Megalosaurus" pannoniensis is an incomplete tooth from an Upper Cretaceous theropod. See more: *Seeley 1881*

1881—USA—Western North America's first Triassic theropod
Coelurus bauri and *C. longicollis* (now *Coelophysis bauri*) are the first carnivorous Triassic dinosaurs discovered. Another discovery, made later but in the same year, is *Tanystropheus willistoni*. All were found in New Mexico. See more: *Cope 1887*

1896—Madagascar—The first theropod discovered in Madagascar
"Megalosaurus" crenatissimus was the first name given to *Majungasaurus*. See more: *Depéret 1896; Lavocat 1955*

1898-1899—Argentina—The first theropod reported in southern South America
Loncosaurus argentinus was named for the first time in a document between 1898 and 1899. It is actually a mixture of a femur of an ornithischian dinosaur and an indeterminate theropod tooth. See more: *Ameghino 1898, 1899*

1901—Argentina—The first theropod named in southern South America
Genyodectes serus was the most complete carnivorous dinosaur for sixty-two years, until *Herrerasaurus* was described. See more: *Woodward 1901*

1905—Algeria—The first Cretaceous theropod in North Africa
Two partial teeth of a probable *Spinosaurus aegyptiacus* were reported in Algeria. It was later discovered again in Egypt in 1915. See more: *Haug 1905; Stromer 1915*

1906—Australia—Oceania's first Cretaceous theropod
A nail phalanx (claw) was presented at Cape Paterson. See more: *Woodward 1906*

1910—England—The first dinosauromorph in western Europe
Saltopus elginensis was considered a dinosaur from its discovery until the year 2000. It is now known that it was a dinosauromorph. See more: *Huene 1910; Rauhut & Hungerbühler 2000*

1910—Brazil—Northern South America's first theropods
A tooth similar to *Thecodontosaurus* is mentioned in association with Cretaceous rocks. It is probably from an abelisaurid theropod. See more: *Allport 1860; Woodward 1910; Kellner & Campos 2000*

1912—Germany—Western Europe's first Triassic theropod
Procompsognathus triassicus, whose first name was "Hallopus celerrimus." *"Megalosaurus" cloacinus* precedes it, but it is not known whether it was a theropod or an archosaur. See more: *Quenstedt 1858; Fraas 1912, 1913; Jaekel 1913; Molnar 2015*

1913—Romania—The first incorrect report of a Mesozoic bird in Eastern Europe
Elopteryx nopcsai is a fragmented femur from Sinpetru, Romania. It was considered a bird until it was later identified as a paravian theropod. See more: *Andrew 1913; Harrison & Walker 1975; Csiki & Grigorescu 1998*

1915—Russia—The first Cretaceous theropod in East Asia
"Allosaurus" sibiricus was found in the Chita Oblast, Russia. It is also known as *"Chilantaisaurus" sibiricus*. See more: *Riabinin 1915*

1920—Tanzania—The first theropods discovered in southern Africa
Elaphrosaurus bambergi and *"Megalosaurus" ingens* were found in Tendaguru, Tanzania. See more: *Janensch 1920; Rauhut 2011*

1931—Chile—Southern South America's first Mesozoic bird
Neogaeornis wetzeli was a relative of modern loons. See more: *Lambrecht 1931*

1931—USA—The first Cretaceous bird footprint in western North America
A few *Ignotornis mcconnelli* footprints, belonging to semi-aquatic birds, were located in Colorado. See more: *Mehl 1931*

1932—Australia—The first theropods named in Oceania
Rapator ornitholestoides and *Walgettosuchus woodwardi* are very incomplete pieces found in New South Wales, Australia. See more: *Woodward 1906*

1932—Tanzania—The first Triassic dinosaur from southern Africa?
Thecodontosaurus alophos is a possible relative of *Nyasasaurus parringtoni*. Both are contemporary and date from the Middle Triassic. It is not known if they are primitive dinosaurs or derived dinosauromorphs, but they present intermediate characteristics between both groups. See more: *Haughton 1932; Nesbitt et al. 2013*

1933—Sweden—The first Mesozoic bird in eastern Europe
Parascaniornis stensioi was described twenty years after *Elopteryx nopscai* from Romania, but the latter is a non-avian theropod. See more: *Andrew 1913; Lambrecht 1933; Csiki & Grigorescu 1998*

1938—Brazil—First dinosauromorph discovered in southern South America
Questionable remains that had been assigned to *Hoplitosuchus aetosaurus* turned out to be a dinosauromorph whose relationship with others is currently unknown. It is much more robust than any other species. It was named *Teyuwasu barberenai* sixty-two years later. See more: *Huene 1938; Kischlat 1999, 2000; Ezcurra 2012*

1942—Brazil—The first primitive saurischian in southern South America
Spondylosoma absconditum was initially described as a prosauropod, but it has subsequently been recognized as the combined remains of a primitive dinosaur and a rauisuchid archosaur. See more: *Huene 1942; Galton 2000; Langer 2004*

1942—China—The first Jurassic theropods in East Asia
Chienkosaurus ceratosauroides, *Sinocoelurus fragilis*, and *Szechuanosaurus campi* were described in China. The first is a mixture of *Hsisosuchus* crocodile teeth and a theropod. See more: *Young 1942; Dong et al. 1983*

1948—China—East Asia's first incorrect Triassic theropod report
Sinosaurus triassicus is more recent than previously assumed. It is now known to date from the Lower Jurassic. See more: *Young 1948*

1963—Argentina—The first primitive saurischian in southern South America
Herrerasaurus ischigualastensis (suspected to be synonymous with *Ischisaurus cattoi*). See more: *Reig 1963; Langer 2004*

1967—Australia—Oceania's first Jurassic theropod
Ozraptor subotaii was found in 1967 and identified as a turtle piece. Thirty-one years later, it was recognized as a theropod. See more: *Long & Molnar 1998*

1968—Syria—The first theropod discovered in the Middle East
A tibia fragment is published in Rif Dimashqq Governorate. See more: *Hooijer 1968*

1969—Zimbabwe—The first theropod to change its nationality
Megapnosaurus rhodesiensis was described by the name of *Syntarsus* in ancient southern Rhodesia. Rhodesia has been known as the Republic of Zimbabwe since 1980. See more: *Raath 1969*

1972—Mongolia—The first Mesozoic bird in East Asia
Gobipteryx minuta is an enantiornithean bird that was named two years after its publication. See more: *Elzanowski 1974, 1976*

1979—India—The first incorrect report of a Mesozoic bird in South Asia (Hindustan)
A news report indicates that a Lower Jurassic bird was found in the Kota Formation. The remains were actually from the fish *Indocoelacanthus*. See more: *Anonymous 1979; Jain 1980; Mortimer**

1979—Argentina—Southern South America's first Jurassic theropod
Piatnitzkysaurus floresi is a fairly complete skeleton of a megalosauroid. See more: *Bonaparte 1979*

1982—India—The first Jurassic theropod in South Asia (Hindustan)
Dandakosaurus indicus is the largest of the Lower Jurassic carnivorous dinosaurs. See more: *Yadagiri 1982*

1983—Uzbekistan—The first Mesozoic bird in western Asia
Horezmavis eocretacea was an ornithuromorph of the Khdzhakul Formation. It precedes *Cretaaviculus* but could be a bird or a theropod, as it is an asymmetrical feather. See more: *Bazhanov 1969; Nesov & Borkin 1983*

1984—USA—Western North America's first dinosauromorph
Technosaurus smalli, from Texas, was considered an ornithischian. It was later discovered that it was a silesaurid. See more: *Chatterjee 1984; Nesbitt et al. 2010*

1984—Thailand—Cimmeria's first theropod found
Small, very incomplete theropods similar to *Compsognathus* are described. See more: *Buffetaut & Ingavat 1984*

1985—USA—The first primitive saurischian discovered in western North America
Chindesaurus bryansmalli is a Texas dinosaur. The name was informally mentioned ten years earlier. See more: *Long vide Murry 1985; Long & Murry 1995*

1986—Australia—Oceania's first Mesozoic bird
Nanantius eos was an enantiornithean bird discovered in Queensland. See more: *Molnar 1986*

1989—Canada—The first incorrect report of Mesozoic bird footprints from the Jurassic in western North America
Possible footprints of birds from the Upper Jurassic were mentioned, but they have recently been dated to the Lower Cretaceous. See more: *Currie 1989; McCrea et al. 2014*

1989—Antarctica—Antarctica's first Mesozoic bird
Polarornis gregorii was discovered in 1989 and was formally published in 2002. The name was unofficially mentioned in 1997. See more: *Chatterjee 1989, 1997, 2002*

1993—The theropod reported in the most countries
Megalosaurus species have been reported in Germany, Algeria, Argentina, Austria, Belgium, China, Spain, the United States, France, Holland, England, Madagascar, Morocco, Poland, Portugal, Romania, Switzerland, and Tanzania, and indeterminate reports (*Megalosaurus* sp.) have been made in Brazil, South Korea, India, and Mali with a range from the Upper Triassic to the late Upper Cretaceous. Actually, there is only one species: *Megalosaurus bucklandii* from the Middle Jurassic of England.

1994—Antarctica—Antarctica's first theropod
A year before the official publication, the popular magazine *Prehistoric Times* informally named *Cryolophosaurus ellioti* "Elvisaurus" for its strange crest. See more: *Hammer & Hickerson 1994*

1994—India—The first primitive saurischian in South Asia (Hindustan)
Alwalkeria maleriensis (formerly *Walkeria*) could be related to *Eoraptor*. See more: *Chatterjee 1987; Chatterjee & Creisler 1994*

1996—Thailand—Cimmeria's first named theropod
Siamosaurus suteethorni appeared to be akin to tyrannosaurids, but it turned out to be an allosauroid or a primitive coelurosaur. See more: *Buffetaut & Ingavat 1984; Holtz et al. 2004; Samathi 2015*

1996—Madagascar—The first Mesozoic bird in Madagascar
Vorona berivotrensis is the largest Mesozoic bird from Madagascar. See more: *Forster et al. 1996*

2002—Lebanon—The Middle East's first Mesozoic bird
An enantiornithean bird was found and named *Enantiophoenix electrophyla* six years later. See more: *Dalla-Vecchia & Chiappe 2002,*

Cau & Arduini 2008

2003—Tanzania—The first Mesozoic bird in southern Africa
An enantiornithean bird from the TNM collection. It hasn't been formally described. See more: *O'Connor et al. 2003*

2003—Argentina—Southern South America's first Triassic theropod
All reports of supposed Triassic theropods before *Zupaysaurus rougieri* were actually basal saurischians. See more: *Arcucci & Coria 1997*

2004—Morocco—The first Mesozoic bird in North Africa?
A vertebra of specimen CMN 50852 is similar to *Rahonavis*. The piece generates debate, since it is not known if it is from an unenlagian dromaeosaurid theropod or a primitive bird. See more: *Riff et al. 2004*

2005—Brazil—Northern South America's first Mesozoic bird
An enantiornithean bird (UFRJ-DG 06-Av) was reported in Sao Paulo. It hasn't been formally described. See more: *Alvarenga & Nava 2005*

2008—Lebanon—The Middle East's first named Mesozoic bird
Enantiophoenix electrophyla, an enantiornithean bird, was found six years earlier. Its name means "opposite amber-eating phoenix" because pieces of amber were found inside it. See more: *Dalla-Vecchia & Chiappe 2002, Cau & Arduini 2008*

2010—Tanzania—The first dinosauromorph in southern Africa
At least fourteen specimens of *Asilisaurus kongwe* have been reported. See more: *Nesbitt et al. 2010*

2011—Morocco—North Africa's first dinosauromorph
Diodorus scytobrachion was a phytophagous sylesaurid. Its teeth tilt forward. See more: *Kammerer et al. 2011*

RECORDS FROM LARGE GEOLOGICAL AREAS

USA—The largest Triassic geological area with dinosaurs
The geological group Dockum or Chinle is about 400 km wide and 800 km long. It has been a source of Upper Triassic fossils since 1893. It consists of the states of Arizona, Colorado, Kansas, Nevada, New Mexico, Oklahoma, Texas, and Utah. See more: *Lucas 1993; Lehman 1994; Chatterjee 1997*

Canada and the United States—The largest Jurassic geological area with dinosaurs
Morrison is a sequence of Upper Jurassic sedimentary rock covering about 1.5 million km². It includes parts of Arizona, Colorado, Idaho, Montana, Nebraska, New Mexico, Utah, and Wyoming in the United States and Alberta, Canada. Tons of dinosaur bones have been found since 1877. See more: *Trujillo et al. 2006*

China and Mongolia—The largest Cretaceous geological area with dinosaurs
The Gobi Desert covers an area of 1,295,000 km² and includes numerous sites, such as Alatan Uul, Bayanzag, Barun Goyot, Buginn Tsav, Djadochta, Gurilín Tsav, Nemegt, and Nogon Tsav, among others. It has been a source of dinosaur fossils since 1922. See more: *Lockley 1991*

China—The geological zone with the most feathered dinosaurs
The Yixian and Jiufotang Formations account for the largest number of discoveries in the world of theropods with filaments, including Mesozoic birds. The great variety of intermediary forms that have been described has revolutionized ideas regarding the origin of flight in birds. The exceptional preservation of fossils is due to phreatomagmatic volcanic eruptions and pyroclastic surges that caused massive and rapid deaths. The way the organisms were killed and interred made them resistant to decomposition. See more: *Jiang et al. 2013*

HIGH-LATITUDE DISCOVERIES OF DINOSAUROMORPHS, THEROPODS, AND MESOZOIC BIRDS

NORTHERN AND SOUTHERN RECORDS
(Based on information from Paleobiology Database)

Northernmost avemetatarsalian
Scleromochlus taylori was found at a latitude of 57.7° N, Upper Triassic paleolatitude 36.3° N (present-day Scotland). See more: *Woodward 1907; Brusate et al. 2012*

Northernmost avemetatarsalian footprints
Prorotodactylus mirus was discovered at an approximate latitude of 50.4° N, Lower Triassic paleolatitude 16.2° N (present-day Poland). See more: *Ptaszynski 2000*

Northernmost dinosauromorph
Saltopus elginensis was located at a latitude of 57.7° N, Upper Triassic paleolatitude 36.3° N, (present-day Scotland). See more: *Huene 1910*

Northernmost dinosauromorph footprints
Rotodactylus sp. were reported at a latitude of 52.1° N, Lower Triassic paleolatitude 16.3° N (present-day Germany). See more: *Fichter & Kunz 2013*

Southernmost dinosauromorphs
Ignotosaurus fragilis with a latitude of 30.1° S, Upper Triassic paleolatitude 46.7° S (present-day Argentina), is currently the southernmost discovery. *Asilisaurus kongwe* and *Lutungutali sitwensis* are currently less southerly at 10.3° S and 10.8° S, respectively, but in their time, they were very far south (Middle Triassic paleolatitude of 53.7° S, present-day Tanzania and Zambia). See more: *Nesbitt et al. 2010; Martinez et al. 2013; Peecock et al. 2013*

Southernmost dinosauromorph footprints
Dinosauromorphs footprints were located at a latitude of 29.8° S, Upper Triassic paleolatitude 49.9° S (present-day Argentina). Footprints known as *Rotadasctylopus archaeus* are sometimes mentioned as possible dinosauromorphs but in reality do not appear to be so. They were at a latitude of 28.6° S, Lower Triassic paleolatitude 65.7° S (present-day Lesotho). See more: *Ellenberger 1972; Marsicano et al 2007*

Northernmost primitive saurischians
Thecodontosaurus primus and *Zanclodon silesiacus* were located at 50.5° N, Middle Triassic paleolatitude 19.1° N (present-day western Canada). Even more extreme is *Arctosaurus osborni* with a latitude of 76.6° N, Upper Triassic paleolatitude 50.8° N (present-day western Canada), although it is very incomplete and may not have been a dinosaur. See more: *Adams 1875; Huene 1905; Jaekel 1910*

Northernmost primitive saurischian footprints
The footprints that include metatarsal markings were located at 50.9° N (Upper Triassic paleolatitude 43.4º N) (present-day Poland). See more: *Gierlinski et al. 2009*

Southernmost primitive saurischians
The basal saurischians *Eoraptor lunensis* and *Herrerasaurus ischigualasensis* were discovered at a latitude of 31.1° S, Upper Triassic paleolatitude 47.7° S (present-day Argentina). *Nyasasaurus parringtoni* and *"Thecodontosaurus" alophos* are presumably basal dinosaurs found at 10.5° S, Middle Triassic paleolatitude 53.9° S (present-day Tanzania). See more: *Haughton 1932; Sereno et al. 1993; Nesbitt et al. 2013*

Southernmost primitive saurischian footprints
Footprints of probable primitive saurischians were located at 29.8° S, Upper Triassic paleolatitude 49.9° S (present-day Argentina). *Trichristolopus dubius* is sometimes mentioned as a dinosaur footprint, although it does not seem likely. It was found at a latitude of 28.6° S, Middle Triassic paleolatitude 65.7° S (present-day Lesotho). See more: *Ellenberger 1972; Marsicano et al. 2007*

Northernmost Triassic theropod
A Greenland coelophysoid was found at a latitude of 71.8° N. In the Late Triassic, the latitude was 46° N (present-day Greenland). See more: *Nesbitt et al. 2007*

Northernmost Triassic theropod footprints
Grallator sp. and other footprints were found at a latitude of 71.5° N. and 76.6° N, Upper Triassic paleolatitude 45.8° N (present-day Greenland. See more: *Milán et al. 2004; Clemmensen et al. 2015*

Southernmost Triassic theropod
Zupaysaurus rougieri found at a latitude of 29.9° S, Upper Triassic paleolatitude 39° S (present-day Argentina). See more: *Arcucci & Coria 1997*

Southernmost theropod footprints
Eubrontes sp. and *Grallator sp.* were reported at a latitude of 27.6° S, Upper Triassic paleolatitude 59.3° S (present-day Australia). See more: *Staines & Woods 1964*

Northernmost Jurassic theropods
Allosaurus sp. PIN 4874/2 is a probable very samall ceratosaurid whose current latitude is 55.7° N, Upper Jurassic paleolatitude 58° N (present-day eastern Russia). See more: *Kurzanov et al. 2003; Carrano et al. 2012*

Northernmost Jurassic theropod footprints
Eubrontes sp., *Grallator sp.*, and *Gigandipus sp.* were found at 57.6° N, Middle Jurassic paleolatitude 47° N (present-day Scotland). Other footprints are not as far north today (44.5° N), but when impressions were made, they were farther north, at the Lower Jurassic paleolatitude of 53.4° N (present-day China). See more: *Clark & Barco-Rodríguez 1998; Matsukawa et al. 2006*

Southernmost Jurassic theropods
Cryolophosaurus ellioti and the coelophysoid FMNH PR1822 were located far south: 84.3° S, Lower Jurassic paleolatitude 57.7° S (present-day Antarctica). In addition, undetermined theropod AV 13802 was collected at a latitude of 37.4° S, Upper Jurassic paleolatitude 85.9° S (present-day New Zealand). See more: *Hammer & Hickerson 1994; Molnar et al. 1998*

Southernmost Jurassic theropod footprints
Grallator sp., *Sarmientichnus scagliai* and *Wildenichnus navesi* were discovered at a latitude of 47.6° S, Middle Jurassic paleolatitude 44.2° S (present-day Argentina). Other footprints were reported at a latitude of 27.6° S, Middle Jurassic paleolatitude 67.8° S (present-day Australia). See more: *Ball 1933; Casamiquela 1964; Rich & Vickers-Rich 2003*

Northernmost Cretaceous theropods
Teeth of undetermined tyrannosaurids cf. *Dromeosaurus albertensis* and a gigantic cf. *Troodon formosus* were found at a latitude of 70.1° N, late Upper Cretaceous paleolatitude 84.1° N (present-day Alaska, USA). See more: *Fiorillo et al. 2000, 2009; Fiorillo & Gangloff 2001*

Northernmost Cretaceous theropod footprints
Magnoavipes denaliensis and *Saurexallopus sp.* are from a latitude of 63.3° N, late Upper Cretaceous paleolatitude 72.4° N (present-day Alaska, USA). See more: *Fiorillo et al 2011, 2012*

The southernmost Cretaceous theropods
A diverse group of theropods, including megaraptors and maniraptors, were located at a position of 38.7° S, late Lower Cretaceous paleolatitude 77.8° S (present-day Australia). See more: *Woodward 1906; Agnolin et al. 2010*

Southernmost Cretaceous theropod footprints
Picunichnus benedettoi, *Bressanichnus patagonicus*, *Deferrariischnium mapuchensis* and *Abelichnus astigarrae* were found at 39.5° S, early Upper Cretaceous paleolatitude 46.6° S (present-day Argentina). *Megalosauropus broomensis* was discovered at 18° S, early Upper Cretaceous paleolatitude 51° S (present-day Australia). See more: *Colbert & Merrilees 1967; Calvo 1991*

Northernmost Mesozoic birds
Canadaga arctica is a bird fossil that was located farther north, at a latitude of 76.3° N, late Upper Cretaceous paleolatitude 72.4° N (present-day western Canada). *Hesperornis sp.* is positioned farther south, at 69.4° N, late Upper Cretaceous paleolatitude of 82.1° N (present-day Alaska, USA). See more: *Bryant 1983; Wilson et al. 2011*

Northernmost Mesozoic bird footprints
Aquatilavipes swiboldae, *Gruipeda vegrandiunus*, *Ignotornis mcconnelli*, *Uhangrichnus chuni*, and *Uhangrichnus sp.* were collected at a latitude of 63.3° N, late Upper Cretaceous paleolatitude 72.4° N (present-day Alaska, USA). See more: *Fiorillo et al. 2011*

Southernmost Mesozoic birds
Polarornis gregorii was found at latitude 64.3° S, late Upper Cretaceous paleolatitude 62.6° S (present-day Antarctica). On the other hand, the enantiornithean bird P 208183 comes from a current latitude of 38.7° S, late Lower Cretaceous paleolatitude 77.8° S (present-day Australia). See more: *Chatterjee 2002; Close et al. 2009*

Southernmost Mesozoic bird footprints
Footprints of unidentified birds were located at a latitude of 36.7° S, late Lower Cretaceous paleolatitude 74.1° S (present-day Australia). *Barrosopus slobodai*, *Ignotornis sp.*, and cf. *Aquatilavipes sp.* were found at 38.8° S, late Upper Cretaceous paleolatitude 41.7° S (present-day Argentina). See more: *Calvo 2007; Martin 2013*

The northernmost theropod oolith
Shells of a prismatoolithid were found at a latitude of 62.9º N

(paleolatitude of the Late Upper Cretaceous of 75.8° N) (present eastern Russia). See more: *Godefroit et al. 2008*

Northernmost Mesozoic bird ooliths
Dispersituberoolithus exilis and *Tristraguloolithus cracioides* were found at latitude 49.1° N, late Upper Cretaceous paleolatitude 56.8° N (present-day western Canada). See more: *Zelenitsky et al. 1996*

The southernmost theropod eggs
Arraigadoolithus patagoniensis was found at a latitude of 39.5° S, late Upper Cretaceous paleolatitude 42.3° S (present-day Argentina). Some shells of an elongatoolithid were found less to the south at 38.5° S, late Upper Cretaceous paleolatitude 44.4° S (present-day Argentina). See more: *Simon et al. 2006; Agnolin et al 2012*

Southernmost Mesozoic bird eggs
The bird eggs located farthest south, at 38.9° S, late Upper Cretaceous paleolatitude 43.5° S (present-day Argentina), are tentatively assigned to *Neuquenornis volans*. See more: *Schweitzer 2002*

The northernmost Mesozoic feather
The feather PIN 3064/10593 could be from a non-avian theropod or a bird, since it is symmetrical. It was located at 51.2° N, early Lower Cretaceous paleolatitude 52.9° N (present-day eastern Russia). See more: *Kurochkin 1985*

The southernmost Mesozoic feather
A few feathers discovered exactly one hundred years after *Archeopteryx lithographica* was found were reported at 38.6° S, late Lower Cretaceous paleolatitude 77°–79° S (present-day Australia). They may have belonged to avian theropods or birds. See more: *Talent et al. 1966*

THE MOST OR FEWEST DISCOVERIES AND DESCRIPTIONS BY COUNTRY OR CONTINENT

Scotland—The nation where the only known primitive avemetatarsalian has been named
Scleromochlus taylori is the only one known to date.

Argentina—The country where the most dinosauromorphs have been named
Ignotosaurus fragilis, Lagerpeton chanarensis, Lewisuchus admixtus, Marasuchus lilloensis and the doubtful species *Lagosuchus talampayensis* and *Pseudolagosuchus major.*

Germany, Scotland, England, Morocco, Poland, Tanzania, and Zambia—Nations where the fewest dinosauromorphs have been named
In each nation, only one species has been described.

France—The country where the most dinosaur footprints have been named
Thirteen possible ichnospecies have been named in France. A fourteenth species, *Prototodactylus lutevensis*, could be a dinosauromorph or primitive avemetatarsalian, since the footprints assigned to this species have ectaxonic leg fingers.

Canada, the Netherlands, England, and Poland—Countries where the fewest dinosauromorph footprints have been named
In each country, only one species has been named to date from a footprint. *Prorotodactylus mirus*, from Poland, could be the footprint of a dinosauromorph or a primitive avemetatarsalian.

Argentina—The country where the most primitive saurischians have been named
Herrerasaurus ischigualastensis and *Sanjuansaurus gordilloi* are well-identified species. *Eoraptor lunensis* could be a primitive sauropodomorph. The species *Ischisaurus cattoi* and *Frenguellisaurus ischigualastensis* could be synonymous with *Herrerasaurus ischigualastensis.*

India—The country where the fewest primitive saurischians have been named
Alwalkeria maleriensis is the only Indian species. It could be a primitive sauropodomorph.

Lesotho—The country where the most primitive saurischian footprints have been named
Distinguishing between primitive saurischian and theropod footprints is difficult. It seems that four species of *Qemetrisauropus* and *Prototrisauropus* can be considered possible theropod dinosaurs.

Namibia—The country where the fewest primitive saurischian footprints have been named
Saurichnium anserinum could be a possible primitive saurischian footprint, although its name is dubious.

China—The country where the most theropods have been named
Some 115 species have been described. The country that comes closest is the United States, which has 105 species

Antarctica, Algeria, Austria, Chile, South Korea, Denmark, Holland, Hungary, Italy, Kazakhstan, Pakistan, Switzerland, Venezuela, and Zimbabwe—The continents or nations where the fewest theropods have been named
Each has only one species.

United States—The country where the most theropod footprints have been named
Some thirty-four species have been named in the eastern United States. Another sixteen have been named in the west.

Croatia, Slovakia, India, Iran, Namibia, Niger, Pakistan, Peru, Sweden and Uzbekistan—Countries where the fewest theropod footprints have been named
Each has only one ichnospecies.

China—The country where the most Mesozoic birds have been named
Ninety-one species have been described. Another 107 species have been named in the rest of the world.

Germany, Australia, Brazil, Chile, Hungary, Kazakhstan, Lebanon, and Romania—Countries where the fewest Mesozoic birds have been named
Each has only one species.

China—The country where the most Mesozoic bird footprints have been named
Eleven species have been named.

Spain and Japan—The countries where the fewest Mesozoic bird footprints have been named
Each has only one ichnospecies.

The first ornithurine bird of the Mesozoic to be found in a nonmarine environment
Ambiortus dementjevi
PIN 3790-271/272
Before its discovery, most of the known Mesozoic birds came from lake, coastal, or marine deposits. Only the enantiornithean birds *Alexornis antecedens* and *Gobipteryx minuta* precede it.
See more: *Kurochkin 1982*

Prehistoric puzzle
The anatomy of theropods

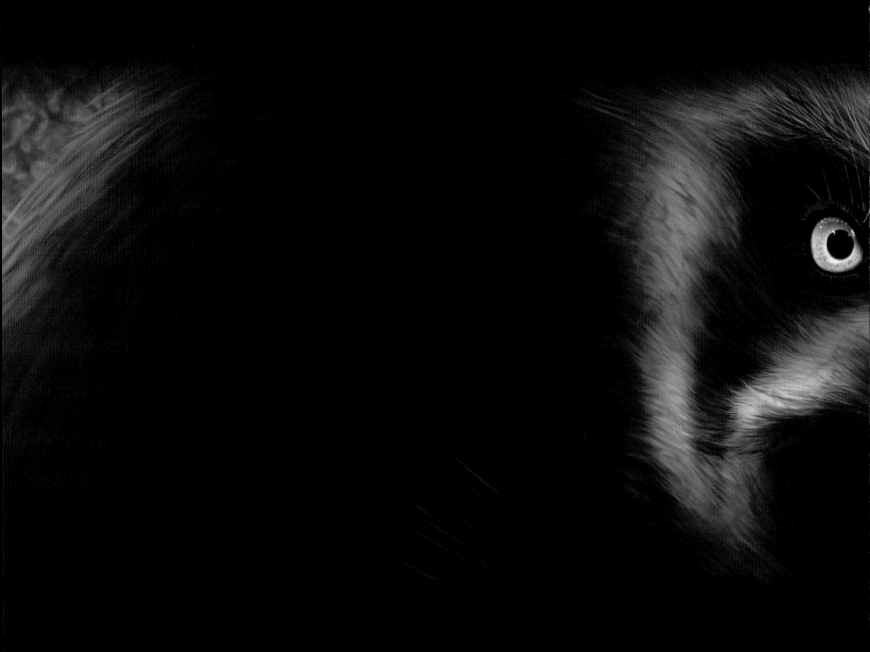

Records: The largest and smallest bones and feathers.

Most fossil remains of dinosaurs are bones and teeth. Bones are mineralized structures that are more likely to be preserved. Bones and teeth are the pieces that are most resistant to erosion and wear.

The skeleton of an adult human consists of 206 bones. Most dinosaurs, with long tails, had more bones than we do. *Tyrannosaurus rex*, for example, had about 300 bones. The exact number can not be specified, because no complete skeleton has been found.

It is fascinating to observe how the femurs of giant theropods were larger than those of an adult African elephant, while the femurs of the smaller theropods were as small as a pin. The great variety of shapes and sizes of bones in the theropod skeleton means that these animals were varied and impressive.

The smallest bones shown on the following pages correspond to the largest pieces of the smallest adult or subadult species. This is because some bones (phalanges, ribs, and minute vertebrae) are not that useful to determine their sizes.

Dinosaurs were vertebrate animals, which means that their body was supported by a bony skeleton. This is an articulated structure that serves to shape the body, anchor muscles and tendons, and protect the nervous system.

TERRIBLE HEADS

Theropod heads comprise the skull and lower jaws. Both are formed by a set of bones closely linked together, very similar to those of the "reptiles" and, in some cases, modern birds.

Measures applied to skulls:

- **OL**: Length from the snout to the occipital condyle
- **SL**: Length from the nose to the squamous
- **LQ**: Length from the snout to the quadratojugal
- **MW**: Maximum width of the skull
- **JL**: Jaw length

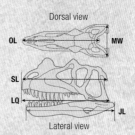

Dorsal view

Lateral view

Largest heads | 1:18

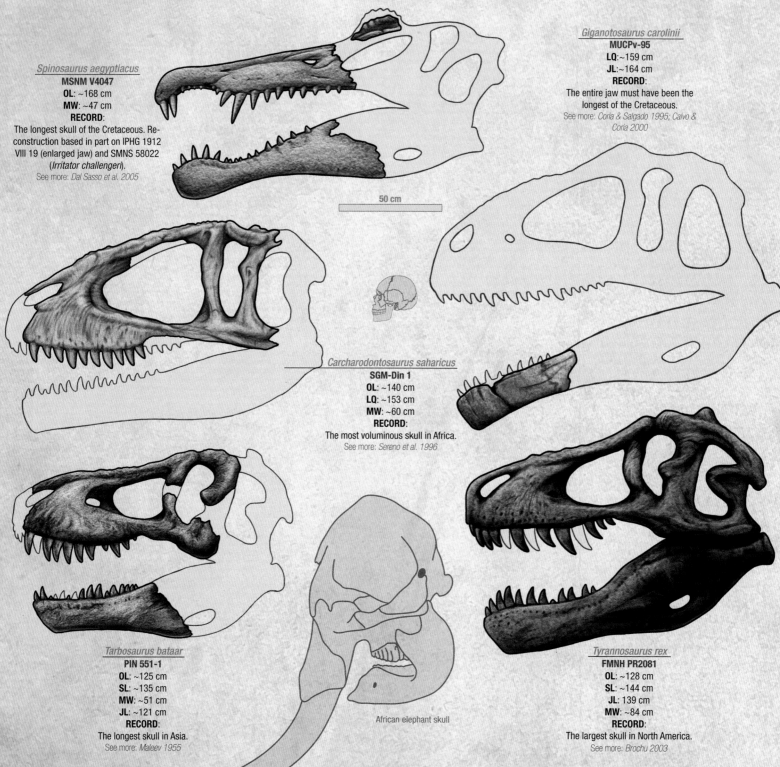

Spinosaurus aegyptiacus
MSNM V4047
OL: ~168 cm
MW: ~47 cm
RECORD:
The longest skull of the Cretaceous. Reconstruction based in part on IPHG 1912 VIII 19 (enlarged jaw) and SMNS 58022 (*Irritator challengeri*).
See more: *Dal Sasso et al. 2005*

Giganotosaurus carolinii
MUCPv-95
LQ: ~159 cm
JL: ~164 cm
RECORD:
The entire jaw must have been the longest of the Cretaceous.
See more: *Coria & Salgado 1995; Calvo & Coria 2000*

50 cm

Carcharodontosaurus saharicus
SGM-Din 1
OL: ~140 cm
LQ: ~153 cm
MW: ~60 cm
RECORD:
The most voluminous skull in Africa.
See more: *Sereno et al. 1996*

Tarbosaurus bataar
PIN 551-1
OL: ~125 cm
SL: ~135 cm
MW: ~51 cm
JL: ~121 cm
RECORD:
The longest skull in Asia.
See more: *Maleev 1955*

African elephant skull

Tyrannosaurus rex
FMNH PR2081
OL: ~128 cm
SL: ~144 cm
JL: 139 cm
MW: ~84 cm
RECORD:
The largest skull in North America.
See more: *Brochu 2003*

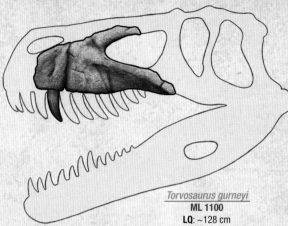

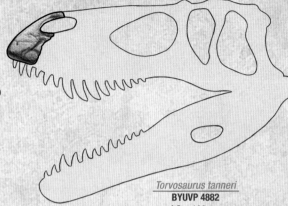

Saltwater crocodile skull

Torvosaurus gurneyi
ML 1100
LQ: ~128 cm
RECORD:
The longest skull in Europe and the
Jurassic (tied with *T. tanneri*)
See more: *Hendrickx & Mateus 2014*

Torvosaurus tanneri
BYUVP 4882
LQ: ~128 cm
RECORD:
The longest skull of the Jurassic
(tied with *T. gurneyi*)
See more: *Mateus, Walen & Antunes 2006*

Hesperornis regalis
YPM 1206
SL: 25.7 cm
JL: 25.7 cm
RECORD:
The longest skull of a Mesozoic bird.
See more: *Marsh 1875*

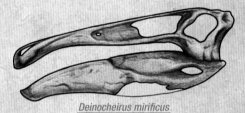

MYTH: Giant Triassic theropod
This huge maxilla (BMNH R3301), 42 cm long, was
considered to belong to a large carnivorous dinosaur, but a review has shown that it actually belonged to an archosaur. See more: *Seeley 1894; Galton & van
Heerden 1998*

Frenguellisaurus ischigualastensis
PVSJ 53
SL: ~56 cm
RECORD:
The longest skull of the Triassic. (Among
theropods, the record is held by *Zupaysau-
rus rougieri*, at 45 cm).
See more: *Novas 1993; Arcucci & Coria 2003*

Deinocheirus mirificus
IGM 100/127
OL: 102 cm
SL: 106 cm
MW: ~25 cm
JL: ~99 cm
RECORD:
The longest skull among herbivorous theropods.
See more: *Lee et al. 2014*

Smallest heads 1:1

Sparrow skull

Sinornis santensis
BPV 538
SL: ~2.65 cm
JL: ~2.4 cm
RECORD:
The shortest jaw of the Cretaceous.
See more: *Sereno & Rao 1992; Kurochkin
et al. 2013*

Gobipipus reshetovi
PIN 4492-3
SL: ~1.85 cm
JL: ~1.8 cm
RECORD:
The smallest skull of the Cretaceous
(juvenile).
See more: *Kurochkin et al. 2013*

Scleromochlus taylori
R3556
SL: 3.5 cm
JL: 3.1 cm
RECORD:
The shortest avemetatarsalian skull.
The juvenile measurements (R3146A):
SL 3.2 cm, JL 3 cm
See more: *Woodward 1907*

Cratoavis cearensis
UFRJ-DG 031 Av
SL: ~1.5 cm
RECORD:
The shortest skull of the Cretaceous.
See more: *Carvalho et al. 2015a, 2015b*

2 cm

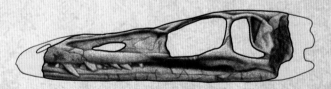

Epidexipteryx hui
IVPP V15471
SL: 4.3 cm
JL: 3.9 cm
RECORD:
The shortest Jurassic skull. The skull of *Eosinop-
teryx brevipenna* is 4.32 cm long
See more: *Zhang et al. 2008; Godefroit et al. 2013*

Epidendrosaurus ningchengensis
IVPP V12653
SL: ~2.15 cm
JL: ~1.8 cm
RECORD:
The smallest Jurassic skull (juvenile).
See more: *Lee et al. 2014*

Procompsognathus triassicus
SMNS 12591
LQ: ~6.5 cm
RECORD:
The shortest theropod skull of the Triassic. The smallest
among the early saurischians was *Agnosphitys cromhallensis*
(VMNH 1751), which was about 10 cm long
See more: *Fraas 1913; Fraser et al. 2002*

BONY RIDGES AND HORNS

Bony ridges are protuberances on the skull. They had diverse functions, either for exhibition or for other unknown functions. The "horns" were actually crests, because they were not optimal for strong impacts.

DOME

A *dome* is a very thick bone in the skull. It was used for striking, either in territorial fights or for resources, or for attracting a mate for procreation.

Crests and largest "horns"   1:5

10 cm

Ceratosaurus nasicornis
UMNH 5278
Length: ~14 x ~7 cm
RECORD:
The largest nasal crest.
See more: *Marsh 1884*

Dilophosaurus wetherilli
UCMP 37302
Length: ~43 x ~13 cm
RECORD:
The largest crest. It belonged to a subadult individual
See more: *Welles 1984*

Carnotaurus sastrei
MACN-CH 894
Length: ~8.7 x ~12.5 cm
RECORD:
Largest supraorbital horn.
See more: *Bonaparte 1985*

Largest dome   1:5

Majungasaurus crenatissimus
FMNH PR 2100
Length: ~31 cm
RECORD:
The longest dome of all belonged to a 57 cm long skull.
See more: *Sampson et al. 1998*

SCLEROTIC RING

The ring consists of small bones found inside the eyes of some "reptiles" and birds. It support the muscles of the iris.

Measures applied to the sclerotic rings:

Side view

Diameter

Largest sclerotic   1:3

5 cm

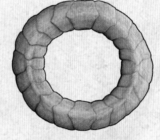

Smallest sclerotia   X10

1 mm

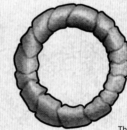

Allosaurus europaeus
ML 415
Diameter: ~6.4 cm
RECORD:
The largest sclerotic ring of the Jurassic. Based on *Allosaurus fragilis* (DINO 11541).
See more: *Mateus, et al. 2006*

Herrerasaurus ischigualastensis
PVSJ 407
Diameter: ~3.4 cm
RECORD:
The largest sclerotic ring of the Triassic.
See more: *Sereno & Novas 1992*

Deinocheirus mirificus
IGM 100/127
Diameter: ~11.6 cm
RECORD:
The largest sclerotic ring of the Cretaceous. Based on *Dromiceiomimus samueli* (ROM 840).
See more: *Lee et al. 2014*

Shanweiniao cooperorum
PVSJ 407
Diameter: 3.2 mm
RECORD:
The smallest sclerotic ring. The ring of the juvenile theropod *Epidendrosaurus ningchengensis* was about 4.5 cm in diameter.
See more: *Zhang et al. 2002; O'Connor et al. 2009*

HYOID

The tongue is supported by several bones that form the *hyoid apparatus*. These bones are little known in dinosaurs, since they tend to be cartilaginous structures; not all of them ossify.

Measures applied to the hyoid bone:

Side view

Length

Largest hyoid bones   1:5

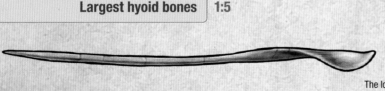

10 cm

Sinraptor dongi
IVPP 10600
Length: 50 cm
RECORD:
The longest hyoid bone. It belonged to a subadult individual.
See more: *Currie & Zhao 1994*

Carnotaurus sastrei
MACN-CH 894
Length: 35 cm
RECORD:
The longest hyoid bone of the Cretaceous.
See more: *Bonaparte 1985*

COLUMELLA (Stirrup)

Theropods, modern birds, "reptiles," and amphibians have a single bone in the middle ear known as *columella auris* or stirrup.

Measures applied to stirrups:

Side view

Length

Tyrannosaurus rex
FMNH PR2081
Length: ~14.5 cm
RECORD:
The longest stirrup.
See more: *Brochu 2003*

4 cm

Allosaurus fragilis
IVPP 10600
Length: ~10 cm
RECORD:
The longest Jurassic stirrup. It belonged to a subadult individual.
See more: *. Madsen 1976*

VERTEBRAE

The spine is made up of numerous bones that protect the spinal cord. These are generally similar to each other, except for the first and second vertebrae (*atlas* and *axis*), which support the skull and give mobility.

The neck is formed by the *cervical vertebrae*. The back, thorax, sacrum, sacral, and tail are formed by the caudal vertebrae. Some theropods and all modern birds have a *pygostyle* at the tip of the tail, formed by several fused caudal vertebrae. It serves to hold a fan of feathers.

Measures applied to the vertebrae:

- **VL**: Anterior posterior length of the vertebral body
- **VH**: Maximum height of the vertebra

Side view

VH

VL

Largest cervical vertebrae | 1:5

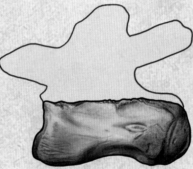

10 cm

Yangchuanosaurus shangyouensis
CV 00216
VL: 138 mm
RECORD:
The longest cervical vertebra of the Jurassic.
See more: *Dong et al. 1983*

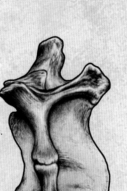

Deinocheirus mirificus
IGM 100/127
VL: 236 mm
RECORD:
The longest cervical vertebra of the Cretaceous.
See more: *Lee et al. 2014*

Sigilmassasaurus brevicollis
BSPG 2011 I 118
VL: 212 mm
RECORD:
The longest cervical vertebra of a carnivorous theropod.
See more: *Evers et al. 2015*

Lophostropheus airelensis
Collection of the University of Caen
VL: 83 mm
RECORD:
The longest cervical vertebra of the Triassic. The primitive saurischian record, 47 mm, is held by *Herrerasaurus ichigualastensis* (PVL 2566).
See more: *Reig 1963; Cuny & Galton 1993*

Smallest cervical vertebrae

2 cm

Epidexipteryx hui
IVPP V15471
VL: 5.4 mm
RECORD:
The shortest cervical vertebra of the Jurassic.
See more: *Zhang et al. 2008*

Procompsognathus triassicus
SMNS 12591
VL: ~16 mm
RECORD:
The shortest cervical vertebra of the Triassic. The smallest of the same specimen is about 14 mm long.
See more: *Fraas 1913*

X10

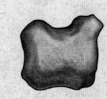

1 mm

Iberomesornis romerali
BSPG 2011 I 118
VL: 2.3 mm
RECORD:
The shortest cervical vertebra of the Cretaceous. The juveniles of this specimen are 1.7 mm long. The non-avian theropod record corresponds to specimen BEXHM: 2008.14.1, which is 7.1 mm long. Even smaller is that of the *Troodon formosus* embryo (MOR 246-1), at 4.4 mm.
See more: *Horner & Weishampel 1988; Sanz & Bonaparte 1992; Naish & Sweetman 2011*

Largest thoracic vertebrae 1:10

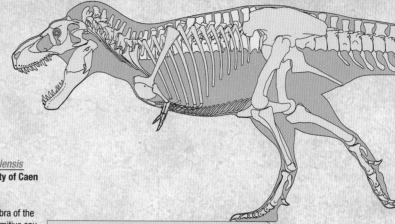

Bahariasaurus ingens
IPHG 1922 X 47
VL: 22.5 cm
RECORD:
The longest thoracic vertebra. The piece was destroyed in World War II
See more: *Stromer 1934*

Spinosaurus aegyptiacus
IPHG 1912 VIII 19
VL: 23.6 cm
VH: 165 cm
RECORD:
The highest thoracic vertebra. It belonged to a subadult individual. The piece was destroyed in World War II.
See more: *Stromer 1934*

20 cm

Unnamed
ISI R282
VL: ~7.9 cm
RECORD:
The longest sacral vertebra of a primitive saurischian.
See more: *Novas et al. 2009*

Lophostropheus airelensis
Collection of the University of Caen
VL: 7.7 cm
RECORD:
The longest thoracic vertebra of the Triassic. The record for a primitive saurischian, 5 cm, is held by *Herrerasaurus ichigualastensis* (PVL 2566). Could belong to the Lower Jurassic.
See more: *Reig 1963; Cuny & Galton 1993*

Vertebra T10 of an African elephant

"Sinosaurus shawanensis"
IPHG 1922 X 47
VL: 22.5 cm
RECORD:
The longest thoracic vertebra of the Jurassic. The vertebra of *Dandakosaurus* was equally long.
See more: *Young, 1948; Anonymous 1979; Yadagiri 1982*

Tyrannosaurus rex
FMNH PR2081
VL: ~20 cm
VH: 78 cm
RECORD:
The most massive thoracic vertebra.
See more: *Brochu 2003*

Smallest dorsal vertebrae 1:1

2 cm

Epidexipteryx hui
IVPP V15471
VL: 6.6 mm
RECORD:
The shortest thoracic vertebra of the Jurassic. Those of the juvenile *Epidendrosaurus ningchengensis* were about 1.5 mm long.
See more: *Zhang et al. 2002, 2008*

Parvicursor remotus
PIN 4487/25
VL: 7.5 mm
RECORD:
The shortest thoracic vertebra of a Cretaceous non-avian theropod. The embryo of *Troodon formosus* (MOR 993) is 2.5 mm long.
See more: *Karhu & Rautian 1996; Varricchio & Jackson 2004*

Procompsognathus triassicus
SMNS 12591
VL: 16.8 mm
RECORD:
The shortest thoracic vertebra of the Triassic. The smallest one is approximately 12.6 mm long.
See more: *Fraas 1913*

X10

1 mm

Iberomesornis romerali
BSPG 2011 I 118
VL: 1.8 mm
RECORD:
The shortest thoracic vertebra of the Cretaceous.
See more: *Sanz & Bonaparte 1992*

Largest sacral vertebrae 1:15

20 cm

Unnamed
ISI R282en
VL: ~7.9 cm
RECORD:
The longest sacral vertebra of the Triassic. It was a primitive saurischian.
See more: *Novas et al. 2009*

Tyrannosaurus rex
FMNH PR2081
VL: 28.5 cm
RECORD:
The longest sacral vertebra of the Cretaceous.
See more: *Brochu 2003*

Yangchuanosaurus shangyouensis
CV 00216
VL: 14.5 cm
RECORD:
The longest sacral vertebra of the Jurassic.
See more: *Dong et al. 1983*

Lophostropheus airelensis
Collection of the University of Caen
VL: 7.7 cm
RECORD:
The longest theropod sacral vertebra of the Triassic. It could belong to the Lower Jurassic.
See more: *Cuny & Galton 1993*

Smallest sacral vertebrae 1:1

Epidexipteryx hui
IVPP V15471
VL: ~5.7 mm
RECORD:
The shortest theropod sacral vertebra of the Jurassic.
See more: *Zhang et al. 2008*

Iberomesornis romerali
BSPG 2011 I 118
VL: 14 mm
RECORD:
The shortest sacral vertebra of the Cretaceous. Not illustrated. The smallest measure 9 mm. The shortest in a non-avian theropod is that of the *Troodon formosus* embryo (MOR 246-1), at 5.1 mm.
See more: *Sanz & Bonaparte 1992; Horner & Weishampel 1988*

Eodromaeus murphi
PVSJ 560
VL: 19 mm
RECORD:
The shortest sacral vertebra of the Triassic. Not illustrated. A smaller one is about 18 mm long. Among the primitive saurischians, the record is held by *Chindesaurus bryansmalli* (PEFO 10395), at 32 and 38 mm.
See more: *Long & Murry 1995; Martínez et al. 2011*

Largest sacra 1:15

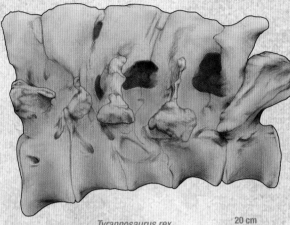

Side view

LCS

SVL: Length along the sacral bodies

Herrerasaurus ischigualastensis
PVL 2566
SVL: 16.3 cm
RECORD:
The longest Triassic sacrum.
See more: *Reig 1963*

Tyrannosaurus rex
BHI 3033
SVL: 106 cm
RECORD:
The longest sacrum of the Cretaceous.
See more: *Larsson 1992*

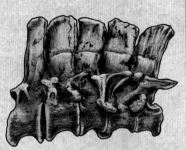

Yangchuanosaurus shangyouensis
CV 00216
SVL: 63 cm
RECORD:
The longest sacrum of the Jurassic.
See more: *Dong et al. 1983*

Smallest sacra 1:1

2 cm

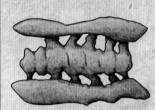

Eodromaeus murphi
PVSJ 560
SVL: ~5.5 cm
RECORD:
The shortest Triassic sacrum.
Not illustrated
See more: *Martínez et al. 2011*

Epidexipteryx hui
IVPP V15471
SVL: 30 mm
RECORD:
The shortest sacrum of the Jurassic.
See more: *Zhang et al. 2008*

Cratoavis cearensis
UFRJ-DG 031 Av
SVL: ~9.7 mm
RECORD:
The shortest sacrum of the Cretaceous. Not illustrated.
See more: *Carvalho et al. 2015a, 2015b*

Largest caudal vertebrae 1:15

Lophostropheus airelensis
Collection of the University of Caen
VL: 7 cm
RECORD:
The longest caudal vertebra of the Triassic. Among primitive saurischians, the record is held by *Frenguellisaurus ischigualastensis*, at about 6.4 cm.
See more: *Novas 1993; Cuny & Galton 1993*

Caudal vertebra of an African elephant

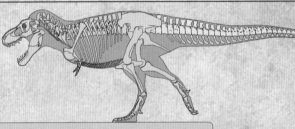

Smaller caudal vertebrae 1:1

2 cm

Allosaurus sp.
NMMNH P-26083
VL: 20 cm
RECORD:
The longest caudal vertebra of the Jurassic.
See more: *Williamson & Chure 1996*

20 cm

Deinonychus antirrhopus
YPM 5201
VL: 5.06 cm (Total length 63 cm)
RECORD:
The longest caudal vertebra of the Cretaceous. *Utahraptor ostrommaysorum* (BYU 9429) has a VL of 6.79 cm. The total length could have reached about 82 cm.
See more: *Ostrom, 1969; Kirkland et al.1993*

Tyrannosaurus rex
FMNH PR2081
VL: ~21 cm
RECORD:
The longest caudal vertebra of the Cretaceous.
See more: *Brochu 2003*

Procompsognathus triassicus
SMNS 12591
VL: 15 mm
RECORD:
The shortest caudal vertebra of the Triassic.
See more: *Fraas 1913*

Epidexipteryx hui
IVPP V15471
VL: 7.2 mm
RECORD:
The shortest caudal vertebra of the Jurassic. Those of the juvenile *Epidendrosaurus ningchengensis* were about 1.3–4.3 mm.
See more: *Zhang et al. 2002, 2008*

X10

1 mm

Cratoavis cearensis
UFRJ-DG 031 Av
VL: ~0.65 mm
RECORD:
The shortest caudal vertebra of the Cretaceous. The smallest of the specimen is about 0.49 mm long.
See more: *Carvalho et al. 2015a, 2015b*

Largest pygostyles 1:1

Nomingia gobiensis
GIN 100/119
VL: 5 cm
RECORD:
The longest pygostyle.
See more: *Barsbold et al. 2000*

Little pygostyles 1:1

2 cm

Iberomesornis romerali
BSPG 2011 I 118
VL: 9.2 mm
RECORD:
The shortest pygostyle. The *Gobipipus reshetovi* PIN 4492-1 is 8.6 mm long but belonged to a juvenile individual
See more: *Sanz & Bonaparte 1992; Kurochkin 1996*

RIBS

The ribs form the rib cage, protecting the internal organs. Their flexibility facilitates breathing. Another type of ribs are the cervicals, which are found in the neck and in some cases are fused with the vertebrae.

Larger cervical ribs   1:10

Side view

Length

20 cm

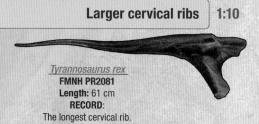

Tyrannosaurus rex
FMNH PR2081
Length: 61 cm
RECORD:
The longest cervical rib.
See more: *Brochu 2003*

Smallest cervical ribs   X5

2 mm

Epidexipteryx hui
IVPP V15471
LCV: ~4 mm
RECORD:
The shortest cervical rib.
See more: *Zhang et al. 2008*

Larger ribs   1:15

Side view

Length

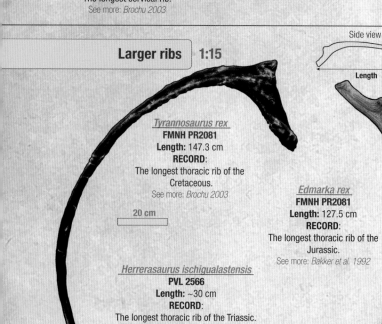

Tyrannosaurus rex
FMNH PR2081
Length: 147.3 cm
RECORD:
The longest thoracic rib of the Cretaceous.
See more: *Brochu 2003*

20 cm

Edmarka rex
FMNH PR2081
Length: 127.5 cm
RECORD:
The longest thoracic rib of the Jurassic.
See more: *Bakker et al. 1992*

Herrerasaurus ischigualastensis
PVL 2566
Length: ~30 cm
RECORD:
The longest thoracic rib of the Triassic. Not illustrated. The theropod record is held by *Liliensternus liliensterni*, although it consists of fragments
See more: *Huene 1934; Reig 1963*

Smaller ribs   1:1

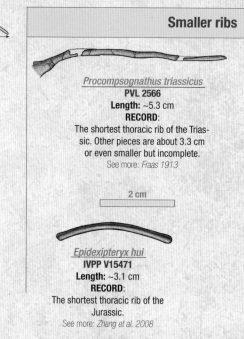

Procompsognathus triassicus
PVL 2566
Length: ~5.3 cm
RECORD:
The shortest thoracic rib of the Triassic. Other pieces are about 3.3 cm or even smaller but incomplete.
See more: *Fraas 1913*

2 cm

Epidexipteryx hui
IVPP V15471
Length: ~3.1 cm
RECORD:
The shortest thoracic rib of the Jurassic.
See more: *Zhang et al. 2008*

X5

2 mm

Iberomesornis romerali
BSPG 2011 I 118
Length: ~1.8 mm
RECORD:
The shortest thoracic rib of the Cretaceous.
See more: *Sanz & Bonaparte 1992*

CHEVRONS

Some theropods have numerous bones known as chevrons at the end of the tail. These serve to protect the blood vessels and caudal nerves and support the tail muscles.

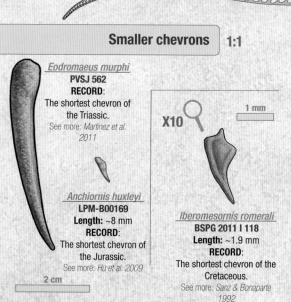

Larger chevrons   1:8

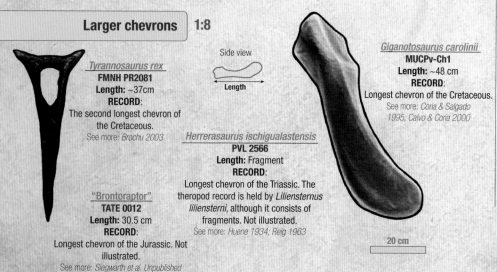

Side view

Length

Tyrannosaurus rex
FMNH PR2081
Length: ~37cm
RECORD:
The second longest chevron of the Cretaceous.
See more: *Brochu 2003*

"Brontoraptor"
TATE 0012
Length: 30.5 cm
RECORD:
Longest chevron of the Jurassic. Not illustrated.
See more: *Siegwarth et al. Unpublished*

Herrerasaurus ischigualastensis
PVL 2566
Length: Fragment
RECORD:
Longest chevron of the Triassic. The theropod record is held by *Liliensternus liliensterni*, although it consists of fragments. Not illustrated.
See more: *Huene 1934; Reig 1963*

Giganotosaurus carolinii
MUCPv-Ch1
Length: ~48 cm
RECORD:
Longest chevron of the Cretaceous.
See more: *Coria & Salgado 1995; Calvo & Coria 2000*

20 cm

Smaller chevrons   1:1

Eodromaeus murphi
PVSJ 562
RECORD:
The shortest chevron of the Triassic.
See more: *Martinez et al. 2011*

Anchiornis huxleyi
LPM-B00169
Length: ~8 mm
RECORD:
The shortest chevron of the Jurassic.
See more: *Hu et al. 2009*

2 cm

X10

1 mm

Iberomesornis romerali
BSPG 2011 I 118
Length: ~1.9 mm
RECORD:
The shortest chevron of the Cretaceous.
See more: *Sanz & Bonaparte 1992*

GASTRALIA

A series of "floating" ribs, gastralia were present in all non-avian theropods but disappeared as modern birds evolved. They are small bones that support the viscera and facilitate breathing, although in the case of birds, the sternum enlarges and apparently replaces this function.

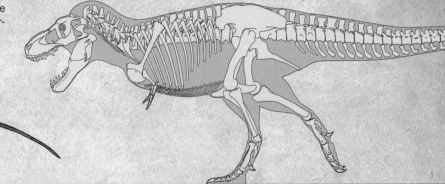

The largest gastralia   1:15

Tyrannosaurus rex
FMNH PR2081
Length: ~90 cm
RECORD:
The longest gastralium.
See more: *Brochu 2003*

20 cm

Largest scapulae and coracoids   1:15

SCAPULA AND CORACOIDS

These bones connect the arm and the clavicle or sternum with the back, supporting the muscles and tendons of arms in theropods.

Measures applied to the vertebrae:

- **CL**: Coracoid length
- **SL**: Length of the scapula
- **SCL**: Length of the scapula and coracoid

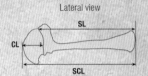

Lateral view

SL

CL

SCL

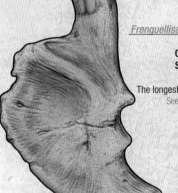

Therizinosaurus cheloniformis
IGM 100/15
CL: 36 cm
SL: ~129 cm
RECORD:
The longest scapula of the Cretaceous.
See more: *Barsbold 1976*

Frenguellisaurus ischigualastensis
PVSJ 53
CL: ~10.5 cm
SL: ~28.5 cm
RECORD:
The longest scapula of the Triassic.
See more: *Reig 1963*

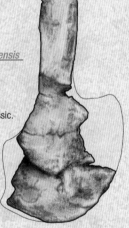

Deinocheirus mirificus
ZPAL MgD-I/6
CL: 34 cm
SL: 119 cm
SCL: 153 cm
RECORD:
The longest complete scapula-coracoid of the Cretaceous. The IGM 100/27 coracoid, 40 cm long, would be the largest of all.
See more: *Osmólska & Roniewicz 1970; Lee et al. 2014*

20 cm

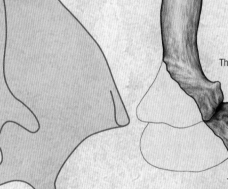

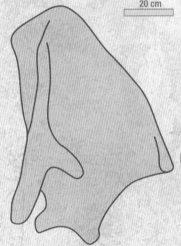

Scapula of an African elephant

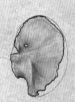

Epanterias amplexus
AMNH 5767
CL: 32.8 cm
RECORD:
The longest coracoid of the Jurassic.
See more: *Cope 1878*

Edmarka rex
CPS 1002
SL: ~95 cm
RECORD:
The longest scapula of the Jurassic
See more: *Cope 1878*

Smaller scapulae and coracoids   1:1

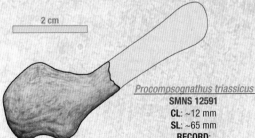

2 cm

Procompsognathus triassicus
SMNS 12591
CL: ~12 mm
SL: ~65 mm
RECORD:
The shortest coracoid and scapula of the Triassic.
See more: *Fraas 1913*

Epidexipteryx hui
IVPP V15471
CL: ~12.2 mm
RECORD:
The shortest coracoid of the Jurassic.
See more: *Zhang et al. 2008*

Iberomesornis romerali
BSPG 2011 I 118
SL: ~19.4 mm
RECORD:
The smallest scapula of the Cretaceous.
See more: *Sanz & Bonaparte 1992*

Eosinopteryx brevipenna
YFGP-T5197
SL: 23.8 mm
RECORD:
The smallest scapula of the Jurassic. The juvenile *Epidendrosaurus ningchengensis* is 11.3 mm long.
See more: *Zhang et al. 2002; Godefroit et al. 2013*

Cratoavis cearensis
UFRJ-DG 031 Av
CL: 7.7 mm
RECORD:
The shortest coracoid of the Cretaceous. The non-avian theropod record, about 10.5 mm, corresponds to *Ceratonykus oculatus* (BSPG 2011 I 118).
See more: *Alifanov & Barsbold 2009; Carvalho et al. 2015a, 2015b*

STERNAL PLATES, CLAVICLES, AND FURCULAE

Sternal plates (or, when the plates are fused, the *sternum*) protect the heart and lungs. The clavicle is present in some primitive dinosaurs, although in several theropods and birds, it is fused, forming a *furcula*, also known as the wishbone.

Measures applied to sternal plates, clavicles, and furculae:

Length

Height

Lateral view

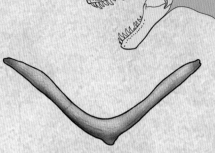

Largest sternum and furcula   1:6

Tyrannosaurus rex
FMNH PR2081
Height: ~14 cm
Length: 29 cm
RECORD:
The longest furcula of the Cretaceous.
See more: *Brochu 2003*

Hesperornis regalis
YPM 1206
Length: ~20 cm
RECORD:
The longest sternum.
See more: *Marsh 1876*

Allosaurus fragilis
UUVP 6132
Height: ~10 cm
Length: ~24 cm
RECORD:
The longest furcula of the Jurassic.
See more: *Chure & Madsen 1996*

Cristatusaurus lapparenti (Suchomimus tenerensis)
MNN GDF 500
Length: 32 cm
RECORD:
The widest furcula of the Jurassic.
See more: *Sereno et al. 1998*

Smallest sternum and furcula   1:1

2 cm

Iberomesornis romerali
BSPG 2011 I 118
Length: 7.6 mm
RECORD:
The shortest furcula.
See more: *Sanz & Bonaparte 1992*

Iberomesornis romerali
BSPG 2011 I 118
Length: 7.8 mm
RECORD:
The shortest sternum. The original fossil was not clearly appraised. This is the *Shanweiniao cooperum* model.
See more: *Sanz & Bonaparte 1992*

Coelophysis bauri
NMMNH P-42353
Length: 49 mm
RECORD:
The longest Triassic furcula.
See more: *Rinehart et al. 2007*

Epidexipteryx hui
IVPP V15471
Length: 8.8 mm
RECORD:
The shortest sternal plate of the Jurassic.
See more: *Zhang et al. 2002, 2008*

HUMERUS

This bone connects the shoulder and forearm. The shape and size of its delto-pectoral ridge helps determine if the arm was weak or strong.

Measures applied to the humeri:

Lateral view

Length

Largest humeri   1:15

20 cm

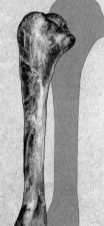
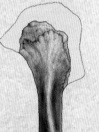

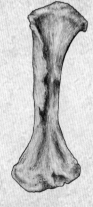

Humerus of an African elephant

Deinocheirus mirificus
ZPAL MgD-I/6
Length: 93.8 cm
RECORD:
The longest humerus of an omnivorous theropod. The one in the IGM 100/127 (shaded) was even longer; it was 100 cm long.
See more: *Lee et al. 2014*

Sigilmassasaurus brevicollis
CMN 41852
Length: ~75 cm
RECORD:
The longest humerus of a piscivorous theropod.
See more: *Russell 1996*

Saurophaganax maximus
OMNH 1935
Length: 54.5 cm
RECORD:
The longest humerus of the Jurassic.
See more: *Chure 1995*

Therizinosaurus cheloniformis
IGM 100/15
Length: 76 cm
RECORD:
The longest humerus of a herbivorous theropod.
See more: *Barsbold 1976*

Herrerasaurus ischigualastensis
MCZ 7064
Length: 26.6 cm
RECORD:
The humerus of the longest Triassic saurischian. The theropod record, 21.4 cm, is held by *Liliensternus liliensterni*.
See more: *Huene 1934; Brinkman & Sues 1987*

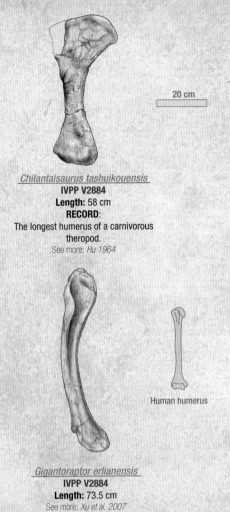

Chilantaisaurus tashuikouensis
IVPP V2884
Length: 58 cm
RECORD:
The longest humerus of a carnivorous theropod.
See more: *Hu 1964*

20 cm

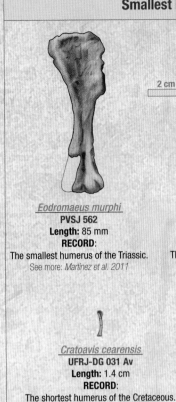

Eodromaeus murphi
PVSJ 562
Length: 85 mm
RECORD:
The smallest humerus of the Triassic.
See more: *Martinez et al. 2011*

2 cm

Epidexipteryx hui
IVPP V15471
Length: 50 mm
RECORD:
The shortest humerus of the Jurassic.
See more: *Zhang et al. 2008*

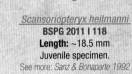

Ceratonykus oculatus
IVPP V15471
Length: 20.8 mm
RECORD:
Shortest non-avian theropod humerus of the Cretaceous.
See more: *Alifanov & Barsbold 2009*

Scansoriopteryx heilmanni
BSPG 2011 I 118
Length: ~18.5 mm
Juvenile specimen.
See more: *Sanz & Bonaparte 1992*

Cratoavis cearensis
UFRJ-DG 031 Av
Length: 1.4 cm
RECORD:
The shortest humerus of the Cretaceous. The non-avian theropod record, 20.8 mm, belongs to *Ceratonykus oculatus* (BSPG 2011 I 118). The juvenile *Scansoriopteryx heilmanni* is 18.5 mm long.
See more: *Czerkas & Yuan 2002; Alifanov & Barsbold 2009; Carvalho et al. 2015a, 2015b*

Agnosphitys cromhallensis
BSPG 2011 I 118
Length: 40 mm
RECORD:
Shortest primitive saurischian humerus.
See more: *Fraser et al. 2002*

Eosinopteryx brevipenna
YFGP-T5197
Length: 37.9 mm
RECORD:
The smallest humerus of the Jurassic. The juvenile *Epidendrosaurus ningchengensis* is 17.1 mm long. .
See more: *Zhang et al. 2002; Godefroit et al. 2013*

Gigantoraptor erlianensis
Gigantoraptor erlianensis
IVPP V2884
Length: 73.5 cm
See more: *Xu et al. 2007*

Human humerus

ULNAE AND RADII

Both bones of the forearm tend to be long and thin. In some birds, they are the longest bones in the body. Nodules called *papillae* where wing feathers are inserted appear in bird ulnae and also in some theropods.

Measures applied to ulnae and radii:

Lateral view
Length

Lateral view
Length

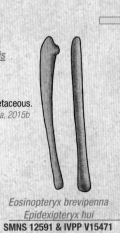

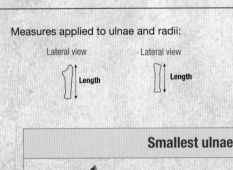

Deinocheirus mirificus
ZPAL MgD-I/6
Length: 68.8 cm
RECORD:
The longest ulna of the Cretaceous. The complete ulna of the IGM 100/127 would measure about 71.5 cm.
See more: *Osmólska & Roniewicz 1970; Lee et al. 2014*

Spinostropheus gauthieri
Collection MNHN
Length: 30 cm
RECORD:
The longest ulna of the Jurassic. Not illustrated.
See more: *Lapparent 1960*

Human ulna

Cratoavis cearensis
UFRJ-DG 031 Av
Length: 13.3 mm
RECORD:
The smallest ulna of the Cretaceous.
See more: *Carvalho et al. 2015a, 2015b*

2 cm

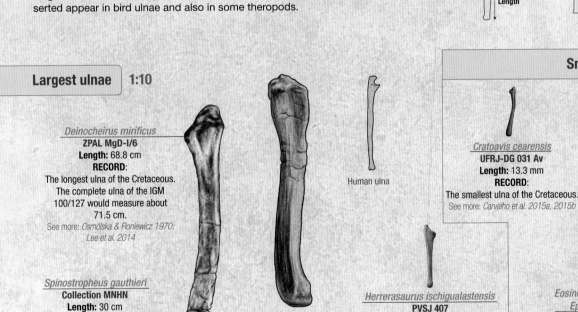

Therizinosaurus cheloniformis
IGM 100/15
Length: 62 cm
RECORD:
The second longest ulna.
See more: *Barsbold 1976*

Herrerasaurus ischigualastensis
PVSJ 407
Length: 16.8 cm
RECORD:
The longest saurischian ulna of the Triassic. The theropod record, 15.8 cm, is held by *Liliensternus liliensterni*.
See more: *Huene 1934; Sereno & Novas 1992*

Eosinopteryx brevipenna
Epidexipteryx hui
SMNS 12591 & IVPP V15471
Length: 42 mm
RECORD:
The smallest ulnae of the Jurassic. The juvenile *Epidendrosaurus ningchengensis* ulna is 15 mm long.
See more: *Zhang et al. 2002, 2008; Godefroit et al. 2013*

Procompsognathus triassicus
SMNS 12591
Length: 34.2 mm
RECORD:
The smallest ulna of the Triassic.
See more: *Fraas 1913*

20 cm

Largest radii 1:10

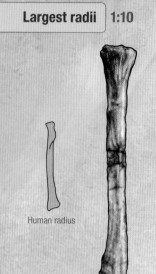

Human radius

20 cm

Allosaurus fragilis
CMN 41852
Length: 22.2 cm
RECORD:
The longest radius of the Jurassic.
It belonged to a medium-size
specimen.
See more: *Gilmore 1915*

Herrerasaurus ischigualastensis
PVSJ 373
Length: 15.3 cm
RECORD:
The longest saurischian radius
of the Triassic. The theropod
record, 15.1 cm, is held by
Liliensternus liliensterni.
See more: *Sereno & Novas 1992*

Therizinosaurus cheloniformis
IGM 100/15
Length: 55 cm
See more: *Barsbold 1976*

Deinocheirus mirificus
ZPAL MgD-I/6
Length: 63 cm
RECORD:
The longest radius of the Cretaceous. The radius of
the IGM 100/127 is about 65.5 cm long.
See more: *Osmólska & Roniewicz 1970; Lee et al. 2014*

Smallest radii 1:1

Eodromaeus murphi
PVSJ 562
Length: 6.6 cm
RECORD:
The shortest radius of the Triassic.
See more: *Martinez et al. 2011*

2 cm

Epidexipteryx hui
IVPP V15471
Length: 3.9 cm
RECORD:
The shortest radius of the Jurassic.
See more: *Zhang et al. 2008*

Iberomesornis romerali
BSPG 2011 I 118
Length: 18.2 mm
See more: *Sanz & Bonaparte 1992*

Cratoavis cearensis
UFRJ-DG 031 Av
Length: 13.3 mm
RECORD:
The shortest radius of the Cretaceous.
Mononykus olecranus (IGM N107/6)
holds the non-avian theropod record, at
about 18.2 mm.
See more: *Carvalho et al. 2015a, 2015b*

METACARPAL BONES

Metacarpals are bones of the hand. There can be up to five in the most primitive
species, three in the great majority, or none, as in the Hesperornithiformes.

Measures applied to metacarpals:

Lateral view

Length

Largest metacarpals 1:5

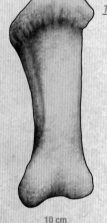

Therizinosaurus cheloniformis
IGM 100/15
Length: 28.7 cm
RECORD:
The longest metacarpal of the
Cretaceous.
See more: *Barsbold 1976*

10 cm

Metacarpal III of an African elephant

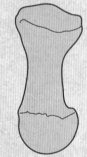

Afrovenator abakensis & Sinraptor dongi
UC UBA 1 & IVPP 10600
Length: 13.5 cm
RECORD:
The longest metacarpals of the Jurassic.
See more: *Currie & Zhao 1994; Sereno et al. 1994*

Herrerasaurus ischigualastensis
PVSJ 380
Length: 7.4 cm
RECORD:
The longest saurischian metacarpal of
the Triassic. The theropod record, 7 cm,
is held by *Liliensternus liliensterni*.
See more: *Huene 1934; Sereno 1993*

Human metacarpal III

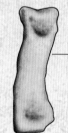

Deinocheirus mirificus
ZPAL MgD-I/6
Length: 24.6 cm
That of specimen MPC]-D 100/127
was 7.2 cm long but incomplete.
Perhaps it reached 25.6 cm in length.
See more: *Osmólska & Roniewicz
1970; Lee et al. 2014*

Megaraptor namunhuaiquii
MUCPv 341
Length: 17 cm
RECORD:
The longest metacarpal of a
carnivorous theropod.
See more: *Porfiri, Calvo & Santos 2007*

Smallest metacarpals 1:1

Eodromaeus murphi
PVSJ 562
Length: 2.8 cm
RECORD:
The shortest metacarpal of the Triassic.
See more: *Martinez et al. 2011*

2 cm

Cratoavis cearensis
UFRJ-DG 031 Av
Length: 7.2 mm
RECORD:
The second shortest metacarpal.
Those from the juvenile *Gracilornis
jiufotangensis* are between 1.9 and
12.5 mm long.
See more: *Li & Hou, 2011; Carvalho et al.
2015a, 2015b*

Linhenykus monodactylus
IVPP V17608
Length: 5.1 mm
RECORD:
The shortest non-avian theropod
metacarpal.
See more: *Xu et al. 2011*

Epidexipteryx hui
IVPP V15471
Length: 13.4 mm
RECORD:
The shortest metacarpal of the Jurassic.
The smallest of this specimen measures 5.1
mm. Those of the juvenile *Epidendrosaurus
ningchengensis* are 5.2–5.8 mm long.
See more: *Zhang et al. 2002, 2008*

STYLIFORM BONE

This bone is found only in some gliding animals, such as the theropod Yi qi.

Measures applied to the styliform bone:

Lateral view

Length

Styliform bone 1:2

2 cm

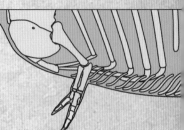

Yi qi
STM 31-2
Length: 13.3 cm
RECORD:
The only styliform bone in a theropod.
See more: *Xu et al. 2015*

HAND PHALANGES

The bones of the fingers are known as *phalanges*. The claws are formed by specialized phalanges known as *ungual phalanges*.

Measures applied to phalanges and claws:

Lateral view

Lateral view

Length

Curvature length

Straight length

Largest manual phalanges 1:5

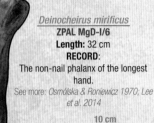

Deinocheirus mirificus
ZPAL MgD-I/6
Length: 32 cm
RECORD:
The non-nail phalanx of the longest hand.
See more: *Osmólska & Roniewicz 1970; Lee et al. 2014*

10 cm

Megaraptor namunhuaiquii
MUCPv 341
Curvature length ~37 cm
RECORD:
The claw of the longest hand of a carnivorous theropod.
See more: *Calvo et al. 2004*

Unnamed
LRF 100-106
Curvature length: ~12 cm
RECORD:
The claw of the longest hand in Oceania.
See more: *Bell et al. 2015*

Deinocheirus mirificus
ZPAL MgD-I/6
Straight length: 19.6 cm
Curvature length: 32.3 cm
RECORD:
The claw of the longest hand in Asia.
See more: *Lee et al. 2014*

Herrerasaurus ischigualastensi
MUCPv 341
Straight length: 4.7 cm
RECORD:
The claw of the longest saurischian hand of the Triassic. The theropod record, 2.4 cm, is held by *Tawa hallae* (GR 242).
See more: *Sereno 1993; Nesbitt et al. 2009*

Saurophaganax maximus
OMNH 780
Straight length: 21 cm
RECORD:
The claw of the Jurassic's longest hand.
See more: *Chure 1995*

Unnamed
BMNH R9951
Curvature length: ~34 cm
RECORD:
The claw of the longest hand in Africa. It may have belonged to a spinosaurid or carcharodontosaurid.
See more: *Ricqles 1967*

Baryonyx walkeri
-
Curvature length: 24 cm
RECORD:
The claw of the longest hand in Europe.
See more: *Charig & Milner 1986*

Therizinosaurus cheloniformis
PIN 551-483
Curvature length: ~75 cm
RECORD:
The claw of the longest Cretaceous hand.
See more: *Barsbold 1976; Zanno 2010*

Smallest hand phalanges 1:1

2 cm

Epidexipteryx hui
IVPP V15471
Straight length: 14 mm
RECORD:
The claw of the Jurassic's shortest hand. The smallest of the specimen is 7.6 mm long. Those of the juvenile *Epidendrosaurus ningchengensis* are between 5 and 7.1 mm long.
See more: *Zhang et al. 2002, 2008*

Eodromaeus murphi
PVSJ 562
Straight length: 11 mm
RECORD:
The claw of the Triassic's shortest hand.
See more: *Martinez et al. 2011*

Eosinopteryx brevipenna
IVPP V12721
Length: ~6.5 mm
RECORD:
The phalanx of the shortest foot of the Jurassic. The smallest of this specimen measures about 4.3 mm.
See more: *Godefroit et al. 2013*

X10

1 mm

Gracilornis jiufotangensis
PMOL-AB00170
Straight length: 1.5 mm
RECORD:
The claw of the shortest hand of the Cretaceous. The smallest of this specimen is 0.7 mm long.
See more: *Li & Hou 2011*

153

PELVIC WAIST

The *ilium*, *pubis*, and *ischium* form the pelvis or pelvic girdle. They are the bones of the hip that connect to the sacrum and hind legs. The pubis and ischium help give space to the viscera and the reproductive system.

Lateral view

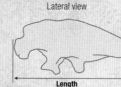

Length

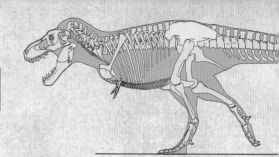

Largest ilia 1:18

50 cm

Tyrannosaurus rex
FMNH PR2081
Length: 146 cm
RECORD:
The second largest ilium. The record, 155 cm, is held by the BHI 3033.
See more: *Larson 1992; Borchu 2003*

"Brontoraptor"
TATE 0012
Length: ~97 cm
RECORD:
The longest ilium of the Cretaceous.
See more: *Siegwarth et al. Unpublished*

Lophostropheus airelensis
Collection of the University of Caen
Length: ~35 cm
RECORD:
The largest Triassic ilium. The primitive saurischian record, 24 cm, is held by *Herrerasaurus ichigualastensis* (PVL 2566).
See more: *Reig 1963; Cuny & Galton 1993*

Liliensternus liliensterni
MB.R.2175
Length: 27 cm
RECORD:
The longest complete ilium of the Triassic.
See more: *Huene 1934*

Smallest ilia 1:3   1:1

5 cm

2 cm

Procompsognathus triassicus
SMNS 12591
Length: 7 cm
RECORD:
The shortest ilium of the Triassic.
See more: *Fraas 1913*

Eoraptor lunensis
PVSJ 512
Length: 8.2 cm
RECORD:
The shortest ilium of a primitive saurischian.
See more: *Sereno et al. 1993*

Iberomesornis romerali
BSPG 2011 I 118
Length: 10.3 mm
RECORD:
The shortest ilium of the Cretaceous.
See more: *Sanz & Bonaparte 1992*

Eosinopteryx brevipenna
IVPP V12721
Length: 25 mm
RECORD:
The shortest ilium of the Jurassic.
See more: *Godefroit et al. 2013*

Largest pubes 1:18

Lateral view

Length

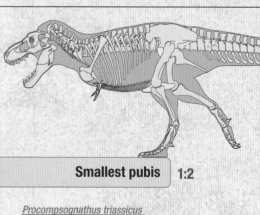

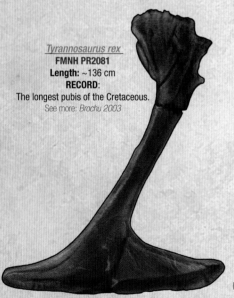

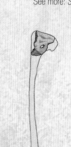

Tyrannosaurus rex
FMNH PR2081
Length: ~136 cm
RECORD:
The longest pubis of the Cretaceous.
See more: *Brochu 2003*

Bahariasaurus ingens
CPS 1010
Length: 103 cm
One of the longest pubes. It was destroyed in World War II.
See more: *Stromer 1934*

Smallest pubis 1:2

Procompsognathus triassicus
SMNS 12591
Length: 8.8 cm
RECORD:
The shortest pubis of the Triassic.
See more: *Fraas 1913*

2 cm

50 cm

Edmarka rex
CPS 1010
Length: 86.6 cm
RECORD:
The longest pubis of the Jurassic.
See more: *Bakker et al. 1992*

Lophostropheus airelensis
CPS 1010
Length: 86.6 cm
RECORD:
The longest pubis of the Triassic. The record for a primitive saurischian, 43 cm, is held by *Herrerasaurus ichigualastensis* (PVL 2566).
See more: *Reig 1963; Cuny & Galton 1993*

Iberomesornis romerali
BSPG 2011 I 118
Length: ~1.4 cm
RECORD:
The shortest pubis of the Cretaceous. The record for a non-avian theropod of the Cretaceous, 7.4 cm, is held by *Sinosauropteryx prima* (NIGP 127587)
See more: *Sanz & Bonaparte 1992; Chen et al. 1998*

Epidexipteryx hui
BSPG 2011 I 118
Length: 2.8 cm
RECORD:
The shortest pubis of the Jurassic.
See more: *Zhang et al. 2008*

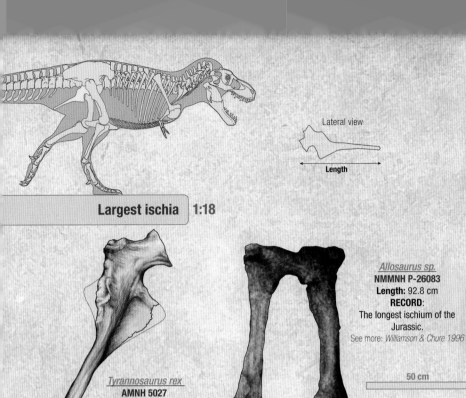

Largest ischia 1:18

Lateral view

Length

Smallest ischia

1:3 5 cm **1:1** 1 cm

Iberomesornis romerali
PVSJ 562
Length: 9 mm
RECORD:
The shortest ischium of the Cretaceous.
See more: *Sanz & Bonaparte 1992*

Eodromaeus murphi
PVSJ 562
Length: 11.6 cm
RECORD:
The shortest Triassic ischium.
See more: *Martinez et al. 2011*

Anchiornis huxleyi
LPM-B00169
Length: 22.4 mm
RECORD:
The shortest ischium of the Jurassic.
See more: *Hu et al. 2009*

Allosaurus sp.
NMMNH P-26083
Length: 92.8 cm
RECORD:
The longest ischium of the Jurassic.
See more: *Williamson & Chure 1996*

50 cm

Tyrannosaurus rex
AMNH 5027
Length: ~124 cm
RECORD:
The longest ischium of the Cretaceous.
See more: *Osborn 1912*

FEMURS

Femurs form the thigh. Sometimes this bone's minimum circumference and length are used to estimate a dinosaur's weight, but this method is no longer considered reliable.

Measures applied to femurs:

Lateral view

Length

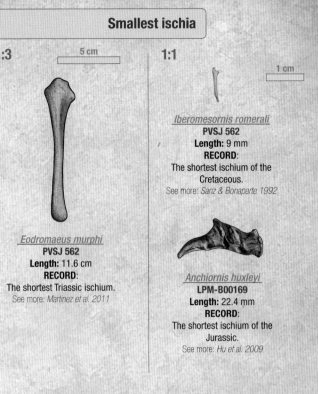

Larger femurs 1:15

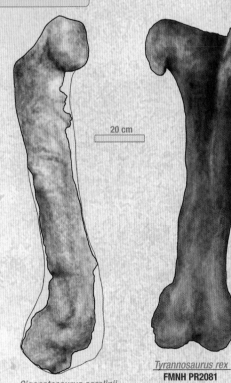

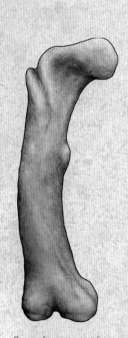

20 cm

Human femur

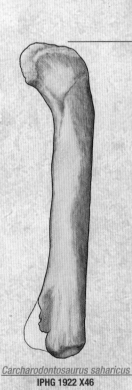

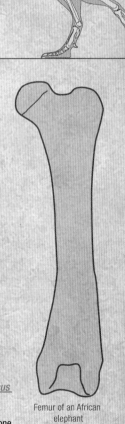

Femur of an African elephant

Giganotosaurus carolinii
MUCPv-Ch1
Length: 143 cm
RECORD:
The longest femur of the Cretaceous. The largest specimen has a 153 cm long femur inferred.
See more: *Coria & Salgado 1995*

Tyrannosaurus rex
FMNH PR2081
Length: 132 cm
(Minimum circumference: 58 cm)
RECORD:
The most massive femur.
See more: *Brochu 2003*

Saurophaganax maximus
OMNH 01708
Length: 113.5 cm
RECORD:
The longest femur of the Jurassic.
See more: *Chure 1995*

Unnamed
SMNS 51958
Length: 60 cm
RECORD:
The longest femur of the Triassic. The primitive saurischian record, about 52 cm, is held by *Herrerasaurus ichigualastensis* (MCZ 7064).
See more: *Reig 1963; Galton 1985*

Carcharodontosaurus saharicus
IPHG 1922 X46
Length: 126 cm
The longest femur in Africa, and one of the largest found. This material was destroyed in World War II.
See more: *Stromer 1934; Sereno et al. 1996*

Largest femurs 1:15

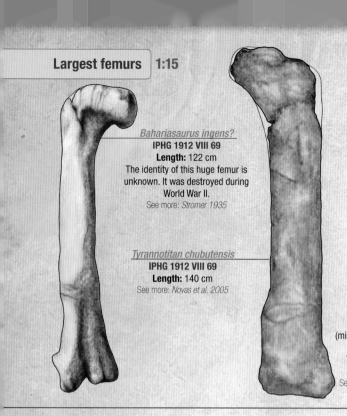

Bahariasaurus ingens?
IPHG 1912 VIII 69
Length: 122 cm
The identity of this huge femur is unknown. It was destroyed during World War II.
See more: *Stromer 1935*

Tyrannotitan chubutensis
IPHG 1912 VIII 69
Length: 140 cm
See more: *Novas et al. 2005*

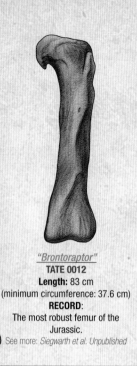

"Brontoraptor"
TATE 0012
Length: 83 cm
(minimum circumference: 37.6 cm)
RECORD:
The most robust femur of the Jurassic.
See more: *Siegwarth et al. Unpublished*

Smallest femurs

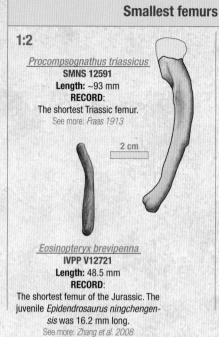

Procompsognathus triassicus
SMNS 12591
Length: ~93 mm
RECORD:
The shortest Triassic femur.
See more: *Fraas 1913*

2 cm

Eosinopteryx brevipenna
IVPP V12721
Length: 48.5 mm
RECORD:
The shortest femur of the Jurassic. The juvenile *Epidendrosaurus ningchengensis* was 16.2 mm long.
See more: *Zhang et al. 2008*

1:1

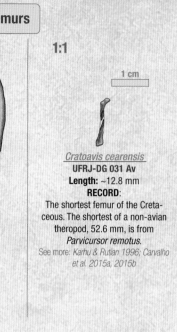

1 cm

Cratoavis cearensis
UFRJ-DG 031 Av
Length: ~12.8 mm
RECORD:
The shortest femur of the Cretaceous. The shortest of a non-avian theropod, 52.6 mm, is from *Parvicursor remotus*.
See more: *Karhu & Rutian 1996; Carvalho et al. 2015a, 2015b*

TIBIAS AND FIBULAS

In the fastest animals, these two leg bones are usually longer than the femur.

Measures applied to the tibia and fibula:

Lateral view | Lateral view

Length | Length

Largest tibias 1:15

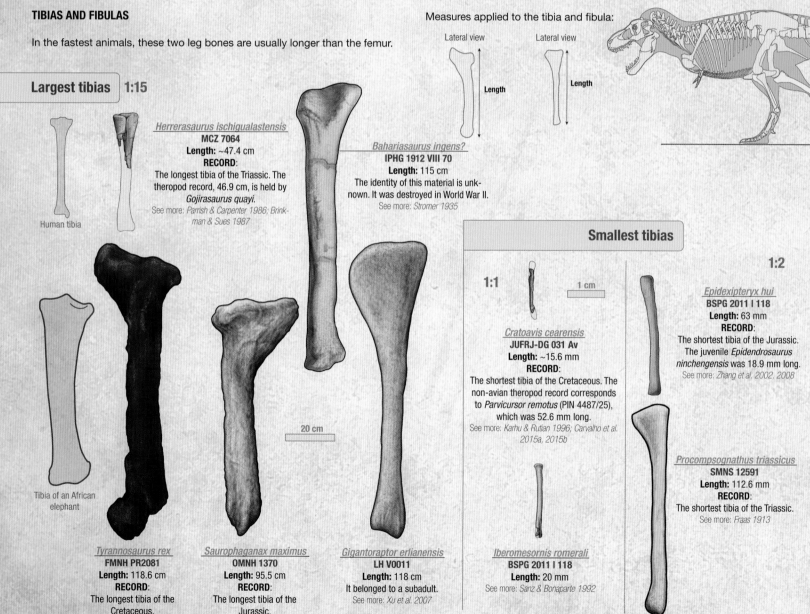

Human tibia

Herrerasaurus ischigualastensis
MCZ 7064
Length: ~47.4 cm
RECORD:
The longest tibia of the Triassic. The theropod record, 46.9 cm, is held by *Gojirasaurus quayi*.
See more: *Parrish & Carpenter 1986; Brinkman & Sues 1987*

Bahariasaurus ingens?
IPHG 1912 VIII 70
Length: 115 cm
The identity of this material is unknown. It was destroyed in World War II.
See more: *Stromer 1935*

Tibia of an African elephant

20 cm

Tyrannosaurus rex
FMNH PR2081
Length: 118.6 cm
RECORD:
The longest tibia of the Cretaceous.
See more: *Brochu 2003*

Saurophaganax maximus
OMNH 1370
Length: 95.5 cm
RECORD:
The longest tibia of the Jurassic.
See more: *Chure 1995*

Gigantoraptor erlianensis
LH V0011
Length: 118 cm
It belonged to a subadult.
See more: *Xu et al. 2007*

Smallest tibias

1:1

1 cm

Cratoavis cearensis
JUFRJ-DG 031 Av
Length: ~15.6 mm
RECORD:
The shortest tibia of the Cretaceous. The non-avian theropod record corresponds to *Parvicursor remotus* (PIN 4487/25), which was 52.6 mm long.
See more: *Karhu & Rutian 1996; Carvalho et al. 2015a, 2015b*

Iberomesornis romerali
BSPG 2011 I 118
Length: 20 mm
See more: *Sanz & Bonaparte 1992*

1:2

Epidexipteryx hui
BSPG 2011 I 118
Length: 63 mm
RECORD:
The shortest tibia of the Jurassic. The juvenile *Epidendrosaurus ninchengensis* was 18.9 mm long.
See more: *Zhang et al. 2002, 2008*

Procompsognathus triassicus
SMNS 12591
Length: 112.6 mm
RECORD:
The shortest tibia of the Triassic.
See more: *Fraas 1913*

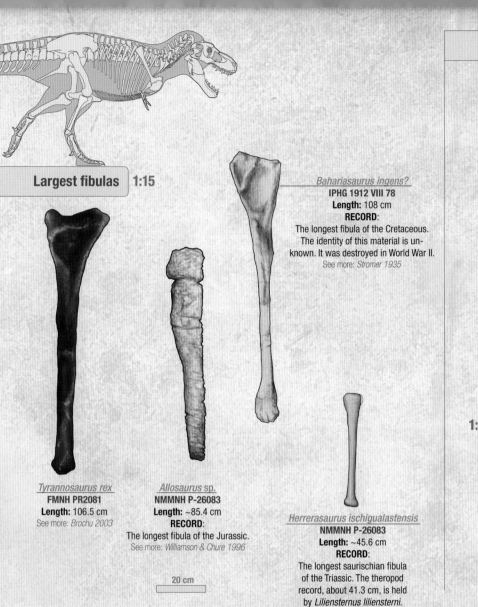

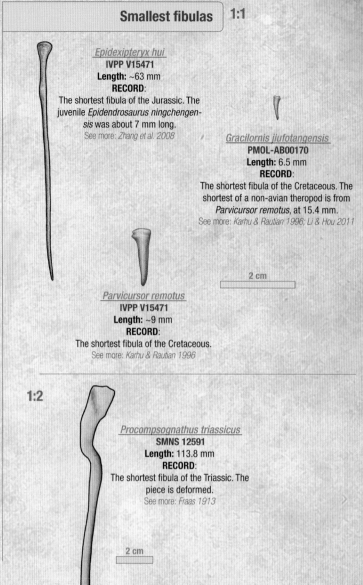

Largest fibulas 1:15

Bahariasaurus ingens?
IPHG 1912 VIII 78
Length: 108 cm
RECORD:
The longest fibula of the Cretaceous.
The identity of this material is un-
known. It was destroyed in World War II.
See more: *Stromer 1935*

Smallest fibulas 1:1

Epidexipteryx hui
IVPP V15471
Length: ~63 mm
RECORD:
The shortest fibula of the Jurassic. The
juvenile *Epidendrosaurus ningchengen-
sis* was about 7 mm long.
See more: *Zhang et al. 2008*

Gracilornis jiufotangensis
PMOL-AB00170
Length: 6.5 mm
RECORD:
The shortest fibula of the Cretaceous. The
shortest of a non-avian theropod is from
Parvicursor remotus, at 15.4 mm.
See more: *Karhu & Rautian 1996; Li & Hou 2011*

Parvicursor remotus
IVPP V15471
Length: ~9 mm
RECORD:
The shortest fibula of the Cretaceous.
See more: *Karhu & Rautian 1996*

2 cm

Tyrannosaurus rex
FMNH PR2081
Length: 106.5 cm
See more: *Brochu 2003*

Allosaurus sp.
NMMNH P-26083
Length: ~85.4 cm
RECORD:
The longest fibula of the Jurassic.
See more: *Williamson & Chure 1996*

20 cm

Herrerasaurus ischigualastensis
NMMNH P-26083
Length: ~45.6 cm
RECORD:
The longest saurischian fibula
of the Triassic. The theropod
record, about 41.3 cm, is held
by *Liliensternus liliensterni*.
See more: *Huene 1934; Brink-
man & Sues 1987*

1:2

Procompsognathus triassicus
SMNS 12591
Length: 113.8 mm
RECORD:
The shortest fibula of the Triassic. The
piece is deformed.
See more: *Fraas 1913*

2 cm

PATELLA (kneecap)

The patella forms the knee that is present in some birds. It works in a similar way
to the patella of mammals.

Measurements applied to the patella:

Lateral view

Length

Largest patella 1:2

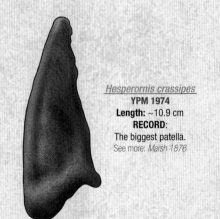

Hesperornis crassipes
YPM 1974
Length: ~10.9 cm
RECORD:
The biggest patella.
See more: *Marsh 1876*

MYTH: Reconstructed dinosaur skeletons in museums reflect the exact dimensions of the animals in life

This is usually not the case. Dinosaurs are often known only from several broken
bones. Reconstructing a skeleton requires a lot of speculation about dimensions, so the margin of error is usually high. On the other hand, science advances. Today we know that dinosaurs had more developed articular cartilage than
mammals, so in many cases, the separation between bones must be greater.

TARSI

Usually, tarsi are the bones between the tibia and metatarsals. They form our ankles. In theropods, the largest are the talus and the *calcaneus*. The tarsus is known as the *tibiotarsus* when a bird's tibia fuses with the talocalcaneal joint.

Measures applied to the tarsi:

Lateral view

Length

Larger tarsi 1:10

20 cm

Tyrannosaurus rex
AMNH 5027
Length: ~42 cm
RECORD:
The largest talocalcaneal. It is 37.2 cm wide.
See more: *Welles & Long 1974*

Allosaurus fragilis
Uncatalogued
Length: 20.8 cm
RECORD:
The largest talocalcaneal joint of the Jurassic.
See more: *Welles & Long 1974*

Herrerasaurus ischigualastensis
PVL 2566
Length: ~14 cm
RECORD:
The largest talocalcaneal joint of the Triassic. The theropod record, 7.9 cm, is held by *Liliensternus liliensterni*.
See more: *Huene 1934; Brinkman & Sues 1987; Reig 1963*

Smallest tarsi X5 🔍

2 mm

Shanweiniao cooperorum
Uncatalogued
Length: ~3 mm
RECORD:
The smallest talocalcaneal joint of the Cretaceous.
See more: *O'Connor et al. 2009*

Procompsognathus triassicus
SMNS 12591
Length: ~ 1 cm
RECORD:
The smallest talocalcaneal joint of the Triassic. Badly preserved.
See more: *Fraas 1913*

Epidexipteryx hui
Uncatalogued
Length: 7 mm
RECORD:
The smallest talocalcaneal joint of the Jurassic. The juvenile *Epidendrosaurus ninchengensis* is about 2 mm long.
See more: *Zhang et al. 2002, 2008*

METATARSALS

Long bones that form the upper part of the foot. Theropods usually have four on each foot. In cursorial animals they are often very long and narrow, while in graviportal animals they are short and broad. Some theropods are "arctometatarsals," i.e., having the middle metatarsal pinched between the two that flank it.

Measures applied to the metatarsals:

Lateral view

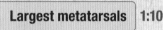

Length

Largest metatarsals 1:10

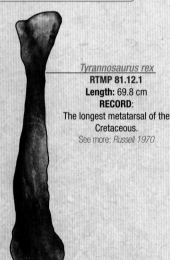

Tyrannosaurus rex
RTMP 81.12.1
Length: 69.8 cm
RECORD:
The longest metatarsal of the Cretaceous.
See more: *Russell 1970*

20 cm

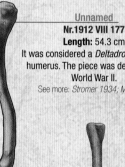

Saurophaganax maximus
OMNH 01338
Length: 47 cm
RECORD:
The longest metatarsal of the Jurassic.
See more: *Chure 1995*

Liliensternus liliensterni
MB.R.2175
Length: 23 cm
RECORD:
The longest metatarsal of the Triassic.
See more: *Huene 1934*

Unnamed
Nr.1912 VIII 177
Length: 54.3 cm
It was considered a *Deltadromeus agilis* humerus. The piece was destroyed in World War II.
See more: *Stromer 1934; Mortimer**

Herrerasaurus ischigualastensis
PVL 2566
Length: 22.3 cm
RECORD:
The longest primitive saurischian metatarsal of the Triassic.
See more: *Reig 1963*

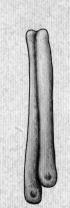

Gigantoraptor erlianensis
LH V0011
Length: 58.3 cm
See more: *Xu et al. 2007*

Smallest metatarsals 1:1

Epidexipteryx hui
IVPP V15471
Length: 31 mm
RECORD:
The shortest metatarsal of the Jurassic.
See more: *Zhang et al. 2008*

Procompsognathus triassicus
SMNS 12591
Length: 69.4 mm
RECORD:
The shortest metatarsal of the Triassic.
See more: *Fraas 1913*

2 cm

Cratoavis cearensis
JUFRJ-DG 031 Av
Length: 8.9 mm
RECORD:
The shortest metatarsal of the Cretaceous.
See more: *Carvalho et al. 2015a, 2015b*

Parvicursor remotus
IGM 100/99
Length: 14 mm
RECORD:
The shortest metatarsal of non-avian theropod. The tarsometatarsal was 58.1 mm long, but the center piece was much shorter.
See more: *Karhu & Rautian 1996*

Iberomesornis romerali
BSPG 2011 I 118
Length: 11.8 mm
See more: *Sanz & Bonaparte 1992*

PEDAL PHALANGES

The toe bones are known as *phalanges*. Claws are formed by a special phalanx known as the *ungual phalanx*.

Measures applied to phalanges and claws:

Lateral view
Length

Lateral view
Curvature length
Straight length

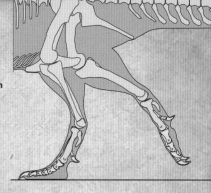

Largest pedal phalanges   1:3

5 cm

Herrerasaurus ischigualastensis
NMMNH P-26083
Straight length: 7.5 cm
RECORD:
The phalanx of the longest foot of a primitive saurischian.
See more: *Reig 1963*

Utahraptor ostrommaysorum
BYU 9429
Straight length: 20.6 cm
Curvature length: 24 cm
RECORD:
The longest foot claw of the Cretaceous.
See more: *Kirkland et al. 1993*

Sinraptor dongi
IVPP 10600
Straight length: 11.1 cm
RECORD:
The second-longest foot claw of the Jurassic.
See more: *Currie & Zhao 1994*

Tyrannosaurus rex
FMNH PR2081
Straight length: 20.4 cm
RECORD:
The most massive foot claw.
See more: *Brochu 2003*

Liliensternus liliensterni
MB.R.2175
Straight length: 5.5 cm
RECORD:
The longest foot claw of the Triassic. The record for a primitive saurischian, 7.5 cm, is held by *Herrerasaurus ischigualastensis* (PVL 2566).
See more: *Huene 1934; Reig 1963*

Allosaurus sp.
NMMNH P-26083
Straight length: ~12 cm
RECORD:
The longest foot claw of the Jurassic.
See more: *Williamson & Chure 1996*

Smallest pedal phalanges   1:1

2 cm

X5 2 mm

Anchiornis huxleyi
LPM-B00169
Straight length: 14.9 mm
RECORD:
The shortest foot claw of the Jurassic.
See more: *Hu et al. 2009*

Procompsognathus triassicus
SMNS 12591
Straight length: 10.6 mm
RECORD:
The shortest foot claw of the Triassic. The smallest of this specimen is 7.2 mm long.
See more: *Fraas 1913*

Procompsognathus triassicus
SMNS 12591
Length: 17.1 mm
RECORD:
The shortest pedal phalanx of the Triassic. The smallest of this specimen is 5.2 mm long.
See more: *Fraas 1913*

Iberomesornis romerali
BSPG 2011 I 118
Straight length: 2.5 mm
RECORD:
The shortest foot claw of the Cretaceous. The smallest of this individual is 1.8 mm long.
See more: *Sanz & Bonaparte 1992*

Cratoavis cearensis
UFRJ-DG 031 Av
Length: ~3.2 mm
RECORD:
The shortest pedal phalanx of the Cretaceous. The smallest of this individual is about 1.4 mm long.
See more: *Sanz & Bonaparte 1992*

Pedopenna daohugouensis
IVPP V12721
Straight length: ~13 mm
See more: *Xu & Zhang 2005*

Albinykus baatar
IGM 100/3004
Straight length: 11.5 mm
See more: *Nesbitt et al. 2011*

Eosinopteryx brevipenna
IVPP V12721
Straight length: ~7.7 mm
RECORD:
The shortest foot claw of the Jurassic. The smallest of this specimen is about 5.2 mm long.
See more: *Godefroit et al. 2013*

Eosinopteryx brevipenna
IVPP V12721
Length: ~6.5 mm
RECORD:
The shortest pedal phalanx of the Jurassic. The smallest of this individual is about 4.3 mm long.
See more: *Godefroit et al. 2013*

TEETH

The teeth are organs that are housed in the premaxillary, maxillary (bones of the upper jaw) and in the dentary (bones of the lower jaw), although in various species these are absent.

Measurements applied to teeth:

- **CH**: Crown height
- **APW**: Anteroposterior width
- **LMT**: Lateromedial thickness

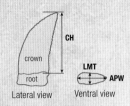

CH

crown

root

LMT

APW

Lateral view   Ventral view

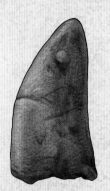

The largest teeth 1:2

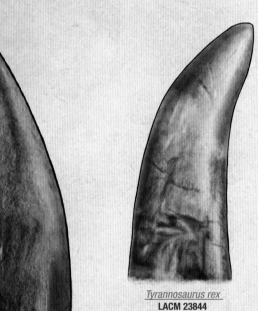

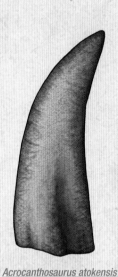

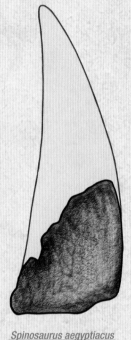

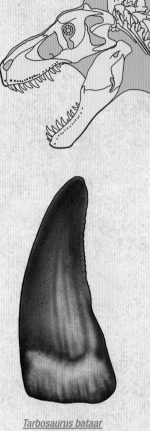

Tyrannosaurus rex
LACM 23844
CH: 117 mm
APW: 54.5 mm
LMT: 34.4 mm
RECORD:
The widest tooth. The largest measurements of the FMNH specimen PR2081 ("Sue") are CH 108.8 mm, APW 52.1 mm, and LMT 37.2 mm, so they were thicker.
See more: *Smith, Vann & Dodson 2005*

Acrocanthosaurus atokensis
NCSM 14345
CH: 93 mm
APW: 42 mm
RECORD:
The largest of the late Lower Cretaceous.
See more: *Currie & Carpenter 2000*

Spinosaurus aegyptiacus
MSNM V4047
APW: ~54 mm
LMT: ~43 mm
RECORD:
The largest conodont tooth.
See more: *Dal Sasso et al. 2005*

Tarbosaurus bataar
MSNM V4047
CH: ~120 mm
APW: ~50 mm
RECORD:
The largest tooth in Asia.
See more: *Maleev 1955*

10 cm

Tyrannosaurus rex
MOR 1125
CH: 100 mm
CH+ root: 305 mm
RECORD:
The longest complete piece. The total length, including the root, is about 30.5 cm.

Tarbosaurus luanchuanensis
IVPP V4733
CH: 110 mm
APW: 47 mm
RECORD:
The second-largest tooth in Asia.
See more: *Dong 1979*

Carcharodontosaurus sp.
MNNHN coll
CH: 125 mm
APW: 47 mm
RECORD:
The largest of the early Upper Cretaceous.
See more: *Lapparent 1960*

Unnamed
MB R2352
APW: 52.8 mm
LMT:~24.9 mm
RECORD:
The largest tooth in Africa.
See more: *Galton & Molnar 2011; Smith, Vann & Dodson 2005*

Giganotosaurus carolinii
MUCPv-52
CH: ~90 mm
APW: ~45 mm
LMT: 21 mm
RECORD:
The largest tooth in South America. It belonged to an animal smaller than the holotype (MUCPv-Ch1).
See more: *Calvo 1999*

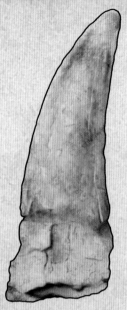

"*Megalosaurus*" *ingens*
MB R 1050
CH: 120 mm
CH + root: 155 mm
APW: 48 mm
RECORD:
The longest tooth of the Jurassic.
See more: *Janensch 1920*

"*Megalosaurus*" *ingens*
MNHUK R6758
CH: ~145 mm
APW: 50 mm
RECORD:
The widest tooth of the Jurassic.
See more: *Charig 1979; Charig in Galton & Molnar 2011*

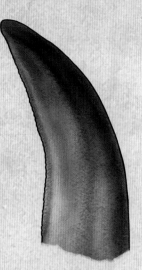

Torvosaurus gurneyi
ML 1100
CH: ~120 mm
APW: 48 mm
RECORD:
The largest tooth in Europe. The complete tooth, including a portion of the root, is 12.7 cm long.
See more: *Mateus et al. 2006; Hendrickx & Mateus 2014*

Carcharodontosaurus saharicus
SGM-Din 1
CH: ~130 mm
APW: 46.7 mm
See more: *Sereno et al. 1996*

Spinosaurus
IPHG 1912 VIII 19
CH: 85 mm
APW: 34 mm
CH + root: 230 mm
(The full root is not shown here.)
See more: *Stromer 1915*

10 cm

Megalosaurus bucklandii
OUM J13505
CH:~46 mm
APW: 20 mm
RECORD:
The first tooth discovered.
See more: *Buckland 1824*

Frenguellisaurus ischigualastensis
PVSJ 53
CH:~60 mm
APW: ~24 mm
RECORD:
The largest of the Triassic (saurischian). Among the theropods, the record is held by *Procompsognathus* sp., measuring 2.7 x ca. 1.17 cm.
See more: *Cuny & Ramboer 1991; Novas 1993; Godefroit & Cuny 1997*

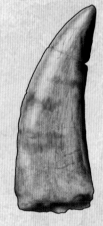

Troodon sp.
AK498-V-001
CH: 14.3 mm
APW: 9 mm
RECORD:
The largest serrated tooth of a larger theropod.
See more: *Fiorillo 2008*

Megalosaurus insignis
Collection Museum du Havre
CH: 120 mm
APW: 42 mm
See more: *Valenciennes 1863; Eudes-Delongchamps & Lennier in Lennier 1870*

1:1

Procompsognathus triassicus
SMNS 12591
CH: ~27.5 mm
APW: ~2 mm
RECORD:
The smallest tooth of the Triassic. Dimensions of the teeth of the ave-metatarsalian *Scleromochlus taylori*: CH 0.6–1.2 mm, APW 0.5 mm.
See more: *Woodward 1907; Fraas 1913*

Smallest teeth | **X10** 🔍

Euronychodon portucalensis
TV 18–20
CH: 2.05 mm
APW: ~1.1 mm
RECORD:
The Cretaceous species based on the tiniest piece. Measurements of the smallest pieces: CH 1.01–1.45 mm, APW 0.65–0.74 mm.
See more: *Antunes & Sigogneau-Russell 1991; Rauhut 2002*

Unnamed
IPFUB GUI 94
CH: 1.02 mm
APW: 0.92 mm
RECORD:
The smallest tooth of the Jurassic.
See more: *Zinke 1998*

Archaeornithoides deinosauriscus
ZPAL MgD-II/29
CH: ~0.17 mm
APW: 0.24 mm
RECORD:
The smallest tooth of the Cretaceous. (It belonged to a juvenile.) Measurements of the largest: CH 2.2 mm, APW 0.78 mm.
See more: *Elzanowski & Wellnhofer 1992*

Koparion douglassi
DINO 3353
CH: 2 mm
APW: 2 mm
RECORD:
The Jurassic species based on the tiniest piece.
See more: *Chure 1994*

Archaeopteryx lithographica
BMNH 37001
CH: 1.5 mm
APW: 9 mm
RECORD:
The largest Jurassic bird tooth.
See more: *Howgate 1984*

Mononykus olecranus
IGM N107/6
CH: 1 mm
APW: ~0.3 mm
RECORD:
The smallest tooth of the Cretaceous.
See more: *Perle et al. 1993*

Camptodontus yangi
IGM N107/6
CH: 11 mm
APW: 7.7 mm
RECORD:
The largest Cretaceous bird tooth.
See more: *O'Connor & Chiappe 2011*

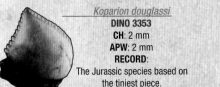

INTEGUMENT (The skin and its specializations)

The integumentary system covers an animal completely. It includes the skin, dermal plates, osteoderms, rhamphotheca (beak), horns, nails, filaments, and advanced feathers.

RHAMPHOTHECA

Horny cover that protects the snout and helps with feeding. It forms the beak. It was present in herbivorous theropods, in some Mesozoic birds, and in other beak-faced dinosaurs.

The largest beak 1:2

2 cm

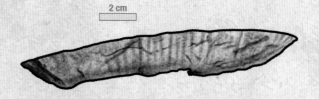

Gallimimus bullatus
IGM 100/11
Length: ~15.4 cm
RECORD:
The longest rhamphotheca.
See more: *Sampson et al. 1998*

OSTEODERM

Ossification of the skin, common in crocodiles, armadillos, and ankylosaurs. *Ceratosaurus* was the only theropod with an osteoderm.

Largest osteoderm 1:1

2 cm

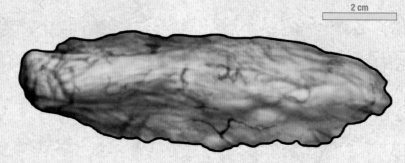

Ceratosaurus dentisulcatus
UMNH 5278
Length: 10 cm
(3.3 cm wide)
RECORD:
The largest osteoderm.
See more: *Madsen & Welles 2000*

FEATHERS

Largest feathers 1:2

1:3

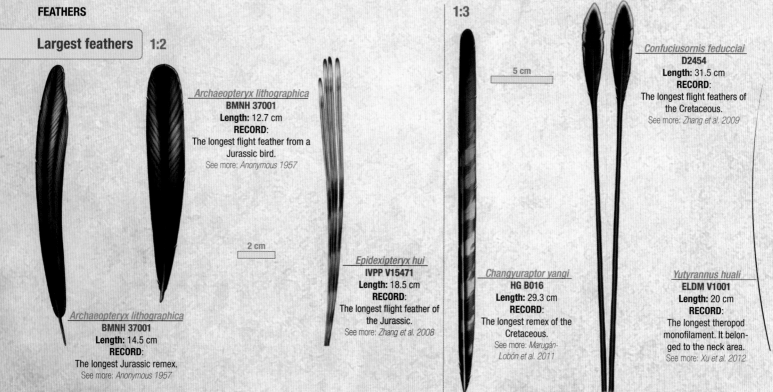

Archaeopteryx lithographica
BMNH 37001
Length: 12.7 cm
RECORD:
The longest flight feather from a Jurassic bird.
See more: *Anonymous 1957*

2 cm

Archaeopteryx lithographica
BMNH 37001
Length: 14.5 cm
RECORD:
The longest Jurassic remex.
See more: *Anonymous 1957*

Epidexipteryx hui
IVPP V15471
Length: 18.5 cm
RECORD:
The longest flight feather of the Jurassic.
See more: *Zhang et al. 2008*

5 cm

Changyuraptor yangi
HG B016
Length: 29.3 cm
RECORD:
The longest remex of the Cretaceous.
See more: *Marugán-Lobón et al. 2011*

Confuciusornis feducciai
D2454
Length: 31.5 cm
RECORD:
The longest flight feathers of the Cretaceous.
See more: *Zhang et al. 2009*

Yutyrannus huali
ELDM V1001
Length: 20 cm
RECORD:
The longest theropod monofilament. It belonged to the neck area.
See more: *Xu et al. 2012*

1699—England—The first theropod tooth
A theropod tooth catalogued with the number 1328 in a book by naturalist Edward Lhuyd. See more: *Lhuyd 1699; Delair & Sarjeant 2002*

1728—England—The first complete theropod bone
A femur identified as "specimen A1" is the oldest piece still preserved today. It is in the Woodwardian Museum, Cambridge.

1755—England—The first dinosaur vertebrae
Joshua Platt discovered three large vertebrae in Stonesfield, England. Unfortunately, the pieces were lost and not identified. Perhaps they were of a theropod. Their membership in *Megalosaurus* is not secure. See more: *Platt 1758*

1797—England—The first theropod jaw
The *Megalosaurus bucklandii* holotype was discovered in 1797 in the Stonesfield quarry. In 1818, it was identified as a giant marine reptile. It was finally published in 1824 along with fragments that may belong to other individuals. See more: *Gunter 1925; Buffetaut 1991*

1815—England—The first theropod skull
The skull of *Megalosaurus bucklandii* was discovered in 1815. At first, it was identified as a giant monitor lizard. It was mentioned in 1822, and the pieces were formally published two years later. See more: *Parkinson 1822; Buckland 1824*

1855—Germany—The first small theropod
"*Pterodactylus*" *crassipes* was the first Mesozoic theropod discovered whose size would have been smaller than that of a human. It is the first known *Archeopteryx* specimen, and at that time, it was considered a pterosaur. See more: *Meyer 1857*

1855—England—The first theropod with high vertebrae
The fossil collector Samuel Husband Beckles sent Richard Owen some vertebrae with high neural spines. These were referred to *Megalosaurus bucklandii*, causing the first reconstruction of the theropod to be a humpback lizard. Later, the vertebrae were assigned to a different species, "*Megalosaurus*" *dunkeri*. Eventually, the species was assigned its own genus, *Becklespinax altispinax*. See more: *Owen 1856; Huene 1923; Olshevsky 1991*

1859—Germany—The first theropod skull
The skeleton of the holotype specimen of *Compsognathus longipes* was quite complete. It included the skull. See more: *Wagner 1859*

1859—Germany—The first almost-complete dinosaur
The holotype of *Compsognathus longipes* was discovered in Solhofen. Its skeleton was found almost completely articulated and with evidence of its last dinner. Even so, it was not recognized as a dinosaur until 1896, although that possibility was raised in 1868. See more: *Wagner 1859; Huxley 1868; Marsh 1896*

1865—Kansas, USA—The Mesozoic bird with the most teeth
Hesperornis regalis has 28 maxillary teeth and 66 in the dentary, so it had about 94 teeth in total. See more: *Marsh 1875*

1870—Kansas, USA—The first toothed bird discovered
Ichthyornis dispar was discovered in 1870 and described two years later. This discovery preceded the first specimen of *Archaeopteryx* to be preserved with teeth, which was discovered between 1874 and 1875 and not described until 1884. See more: *Marsh 1872, 1873; Dames 1884*

1876—France—The holotype specimen of the Triassic theropod known by the tiniest fragment of material
"*Megalosaurus*" *obtusus* is a tooth 2.6 cm long. Its anteroposterior length is 0.8 cm. It could be from a theropod or a sauropodomorph. See more: *Henry 1876*

1877—USA—The Mesozoic bird with the fewest phalanges in the hands
The anterior limbs of *Hesperornis regalis* were so atrophied that it lacked forearms and fingers. See more: *Marsh 1877*

1880—USA—The Mesozoic bird with the shortest tail in proportion to its size
The tail of *Hesperornis regalis* was only 11% of its total length. See more: *Marsh 1880*

1883—Germany—Theropod species based on the most incomplete material
Several theropods were identified and named from fossil bone fragments. Perhaps the most extreme case is "*Ornithocheirus*" *hilsensis*, which is based on an incomplete II-2 phalanx that was thought to be a metacarpal IV of a pterosaur. It is impossible to name a new dinosaur from this bone, as it does not differ much from other theropods. See more: *Koken 1883; Mortimer**

1884—USA—The first theropod with "horns"
Ceratosaurus nasicornis was the first dinosaur with ridges in the shape of antlers. They were considered attack weapons but may have been used for display. See more: *Marsh 1884; Rowe & Gauthier 1990*

1884—Germany—The first Mesozoic bird skull
Archaeopteryx lithographica HMN 1880 was the first specimen with a skull. It was discovered in 1874 and described ten years later. See more: *Dames 1884*

1898—USA—The first fossil scales in a Mesozoic bird
Scales and feathers were found in a specimen of *Parahesperornis alexi* in Kansas. See more: *Williston 1898*

1901—Germany—The first skin impression of a theropod
Polygonal forms in the abdominal part of the type specimen of *Compsognathus longipes* were mentioned. See more: *Huene 1901; Ostrom 1978*

1901—USA—The first and only theropod osteoderm
A row of pointed osteodermic plates on the caudal column of *Ceratosaurus nasicornis* was reported. See more: *Marsh 1884, 1892; Gilmore 1920*

1903—USA—The Mesozoic bird with the most sacral vertebrae
Hesperornis regalis had a very long sacrum consisting of fourteen vertebrae. See more: *Lucas 1903*

1905—USA—The theropod with the shortest neck in proportion to its size
The neck of *Tyrannosaurus rex* is about 10% of its total length. See more: *Osborn 1905; Larson & Carpenter 2008*

1907—Scotland—The oldest skin print on an avemetatarsalian
The impression left by skin in the ventral region of *Scleromochlus taylori* includes small rectangular scales 1 mm in diameter. It dates from the Upper Triassic (Carnian, ca. 237–227 Ma). See more: *Woodward 1907*

1915—Egypt—The theropod with the highest thoracic vertebrae
Spinosaurus aegyptiacus had extremely developed vertebrae whose huge spines were 8.1 times as tall as they were long. See more: *Stromer 1915*

1916—Canada—The first toothless theropod discovered
The first specimen of *Struthiomimus altus* with a preserved skull reveals that the theropod had no teeth. It was found in Alberta, Canada. See more: *Osborn 1916*

1917—USA—The Mesozoic bird with the most cervical vertebrae
Hesperornis had seventeen vertebrae in its neck, which must have been very flexible. See more: *Sternberg 1917*

1917—USA—Mesozoic birds with the fewest thoracic vertebrae
Hesperornithian birds such as *Baptornis advenus* or *Hesperornis regalis* had only seven vertebrae in the thorax. See more: *Sternberg 1917*

1933—USA—The first clavicle discovered in a theropod
Segisaurus was found in Arizona in 1933 and described three years later. Because the clavicle had not been conserved in any other dinosaur, primitive arcosaurs were thought to be the origin of birds. See more: *Camp 1936*

1950—USA—The theropod with the smallest legs in proportion to its size
The very short hind legs of *Acrocanthosaurus atokensis* were only 24% of its total length. See more: *Stovall & Langston 1950*

1955—Mongolia—The theropod with the smallest arms in proportion to its size
Tarbosaurus bataar owns the record for the smallest arm in proportion to total length: 8%. See more: *Maleev 1955*

1962—USA—The first estimates of body mass in theropods
The probable weights of *Allosaurus fragilis* and *Tyrannosaurus rex* were estimated for the first time: 2.09 t and 6.89 t, respectively. See more: *Colbert 1962*

1963—Argentina—The most robust Triassic basal saurischian femur
The femur of *Herrerasaurus ischigualastensis* PVL 2566 is the thickest. Its minimum circumference is 16.95 cm, or 35.1% of the femur's length. See more: *Reig 1963*

1963—USA—The Mesozoic bird species known by the most incomplete material
"*Lonchodytes*" *pterygius* (UCMP 53961), from Wyoming, is the distal part of a metacarpal that is only 6.6 mm long, so it is very difficult to verify it as a species. See more: *Brodkorb 1963*

1969—USA—The theropod with the most elongated caudal vertebrae
The caudal vertebrae of *Deinonychus antirrhopus* were up to 13.3 times longer than they were wide. This is because the vertebrae have very developed processes. See more: *Ostrom 1969*

1970—Brazil—The saurischian with the fewest cervical vertebrae
Among dinosaurs, primitive saurischians such as *Eoraptor*, *Herrerasaurus*, or *Staurikosaurus* had the lowest number of vertebrae in the neck: nine pieces. See more: *Colbert 1970; Sereno & Novas 1993; Sereno, Martínez & Alcober 2013*

1970—Brazil—The saurischians with the fewest sacral vertebrae
The sacrums of *Eoraptor*, *Herrerasaurus*, *Sanjuansaurus*, and *Staurikosaurus* had only three vertebrae. See more: *Colbert 1970; Alcober & Martínez 2010; Sereno, Martínez & Alcober 2013*

1970—Brazil—The saurischians with the most thoracic vertebrae
Basal saurischians had a greater number of vertebral parts. *Staurikosaurus* or *Eoraptor* had fifteen. See more: *Colbert 1970; Sereno, Martínez & Alcober 2013*

1971—Argentina—The dinosauromorph with the most phalanges in the feet
It is possible that *Lagosuchus talampayensis* had sixteen phalanges in each foot. The norm for dinosauromorphs and basal saurischians was fifteen phalanges distributed in five fingers. Theropods and birds usually have fourteen phalanges in four fingers. See more: *Romer 1971*

1972—Mongolia—The theropod with the smallest head in proportion to its size

Among theropods, *Gallimimus bullatus* had the proportionally shortest skull: 6% of the animal's total length. See more: *Osmólska, Roniewicz & Barsbold 1972*

1972—France—The theropod with the longest tail in proportion to its size

MNHN CNJ 79, an adult *Compsognathus longipes*, had an incomplete tail, but it is estimated that it would have been 59% of the animal's total length. See more: *Bidar, Demay & Thomel 1972*

1972—Brazil—The first gastrointestinal impression

The specimen of *Santanaraptor placidus* (MN/UFRJ MN 4802-V) is the first dinosaur to present digestive tract marks. Twenty-eight years later, a similar fossil was found (SMNK 2349), but it is unknown if it belongs to the same species. See more: *Kellner 1995; Martill et al. 2000; Peyer 2006*

1974—Mongolia—The first toothless Mesozoic bird

The first bird of the Mesozoic with a toothless beak was *Gobipteryx*, but it is certain that *Cimolopteryx rara* (originally *rarus*) and *C. retusus*, discovered eighty-five years before, were edentulous birds. See more: *Marsh 1889; Elzanowski 1974*

1981—Mongolia—Theropods with the most sacral vertebrae

Among non-avian theropods, "*Ingenia*" and *Nemegtomaia* had the largest known number of sacral vertebral pieces: eight. See more: *Barsbold 1981; Lu et al. 2004*

1981—Mongolia—The most extreme arctometatarsal condition

When metatarsal III is tight between metatarsals II and IV, the condition is known as arctometatarsal. It was a very common trait caused by convergence in at least five groups of theropods. The most extreme cases are found in *Avimimus* oviraptorosaurs and advanced alvarezsaurids. In the latter, metatarsal III was so reduced that it lost its importance for locomotion. See more: *Kurzanov 1981; Perle et al. 1994; Snively, Russell & Powell 2004*

1985—Argentina—The theropod with shortest arms in relation to its legs

The arms of *Carnotaurus sastrei* were so short that they were barely 14% the length of its hind legs. See more: *Bonaparte 1985*

1985—Argentina—The theropod predator with the smallest head in proportion to its size

Carnotaurus sastrei stands out. Its skull was only 13% of its total length See more: *Bonaparte 1985*

1987—USA—Strangest denticles

Most denticles in theropods point upward slightly and are very small. The denticles in troodontids were the exception. They were positioned at an angle of ninety degrees and usually were very large in proportion to the size of the tooth. See more: *Currie 1987*

1991—Germany—The holotype specimen of a dinosauromorph known by the tiniest fragment of material

Avipes dillstedtianus retains only three incomplete metatarsals 35 mm long. See more: *Huene 1932; Rauhut & Hungerbuhler 2000*

1991—Portugal—The Cretaceous theropod holotype known by the tiniest fragment of material

Euronychodon portucalensis is known from only two isolated teeth. The specimen type (CEPUNL TV 20) is 1.8 mm high and 0.65 mm long (anteroposterior). See more: *Antunes & Sigogneau–Russell 1991*

1992—Argentina—The saurischian with the most phalanges in the hands

Basal saurischians had more phalanges than theropods because they were shrinking as they evolved. *Herrerasaurus* had ten in each claw, primitive theropod forms conserved only nine, and several of

them fused in the birds. See more: *Sereno & Novas 1992*

1993—China—The oldest theropod with high thoracic vertebrae

Yangchuanosaurus zigongensis ((formerly known as *Szechuanosaurus*) dates from the Middle Jurassic (Callovian, ca. 166.1–163.5 Ma). Metriacanthosaurids were the first theropods to present elongated neural spines. See more: *Gao 1993; Bailey 1997*

1994—Spain—The theropod with the most teeth

Pelecanimimus polyodon had 14 premaxillary teeth, 60 maxillary teeth, and 150 dentary teeth: a total of 224 teeth. See more: *Pérez-Moreno et al. 1994*

1994—USA—The Jurassic theropod holotype specimen known by the tiniest fragment of material

The troodontid *Koparion douglassi* (DINO 3353), from Utah, is known by only a small tooth 2 mm long. Its anteroposterior length is 2 mm. See more: *Chure 1994*

1994—Spain—The first nonskeletal ridge on a theropod

In the fossil of *Pelecanimimus polyodon*, a soft cephalic crest can be seen. It is preserved thanks to bacterias. See more: *Perez-Moreno et al. 1994*

1995—Spain—The theropod tooth with the smallest denticles of the Cretaceous

The dromaeosaurid tooth MPZ 96/97 (1.5 mm height, 1.07 mm anteroposterior length, and 0.67 mm lingual lip length) contains up to 26.6 denticles per millimeter in the mesial margin. They are very tiny, only 0.0375 x 0.005 mm in size See more: *Ruiz–Omeñaca et al. 1995*

1995—Portugal—The theropod tooth with the smallest denticles of the Jurassic

The teeth of cf. *Archeopteryx* sp. have up to 24 denticles per millimeter. See more: *Weigert 1995*

1995—Germany—The strongest theropod femur of the Triassic

The minimum circumference of the femur of *Liliensternus liliensterni* MB.R.2175 is 11 cm (25% of its total length). See more: *Huene 1934*

1996—Germany, USA—The oldest skin print of a theropod

Conserved in the juvenile specimens of *Allosaurus* sp. CJW 295 (present-day Wyoming, USA) and in the tail of *Juravenator starki* (present-day Germany), the skin print dates from the Upper Jurassic (Kimmeridgian, ca. 157.3–152.1 Ma). See more: *Weege & Schumde 1996; Pinegar et al. 2003; Gohlich & Chiappe 2006*

1996—USA—The slowest tooth replacement

After a tooth had been used for some time, its root was absorbed and the crown fell off, and it was replaced by a new one. *Tyrannosaurus rex* used a tooth for about 777 days, on average. See more: *Erickson 1996*

1996—USA—The strongest theropod femur of the Jurassic

The minimum circumference of the femur of *Allosaurus* sp. NMMNH P-26083 is 46 cm (41.8% of the length of the femur). In proportion to size, the most robust is "Brontoraptor" TATE 0012. Its minimum perimeter corresponds to 45.3% of the femur's length. See more: *Siegwarth et al. 1994; Williamson & Chure 1996*

1996—China—The theropod with the most caudal vertebrae

Sinosauropteryx prima had a very long and articulated tail, although it is incomplete. It is estimated to have fifty-nine to sixty-four vertebrae. See more: *Ji & Ji 1996*

1996—USA—The fastest tooth replacement

Deinonychus antirrhopus replaced each of its numerous teeth approximately every 290 days. See more: *Erickson 1996*

1996—Mongolia—The theropod with the most leg bone fusions

In some adult theropods, the leg bones fused. In *Bagaraatan ostromi*, it happened in such a way that the talus, calcaneus, tibia, and fibula are a single rigid piece, giving the leg greater strength when running. See more: *Osmólska 1996*

1997—China—The Mesozoic bird with the most premaxillary teeth

Largirostrornis sexdentoris had six teeth in each premaxilla, a total of a dozen. It was most common in Mesozoic birds to have zero to five teeth maximum on each side of the premaxilla. See more: *Hou 1997*

1998—France—The most robust Mesozoic bird femur

The minimum circumference of the femur of *Gargantuavis philoinos* MDE-A08 was very robust, 14.8 cm (41.7% of the estimated length of the femur). See more: *Buffetaut & Le Loeuff 1998*

1999—China—The Mesozoic bird with the fewest cervical vertebrae

Zhongornis haoae or *Yandangornis longicaudus* have only nine pieces. See more: *Cai & Zhao 1999; Gao et al. 2008*

1999—China—The theropod with the most fused caudal vertebrae

The pygostyle of *Beipiaosaurus inexpectus* was formed of seven caudal vertebrae. See more: *Xu, Tang & Wang 1999*

2000—China—The theropod with the fewest thoracic vertebrae

Caudipteryx holds the record with only nine thoracic vertebrae. See more: *Zhou, Wang, Zhang & Xu 2000*

2000—Mongolia—The first theropod with a pygostyle

A pygostyle is typical of birds but has been documented in some theropods. *Nomingia gobiensis* was the first time it was identified in a non-avian theropod. The structure emerged in various groups by convergence. The pygostyle consists of fused final caudal vertebrae. Its function is to support and give mobility to tail feathers. See more: *Barsbold et al. 2000*

2001—Canada, Mongolia—The first beak discovered in a theropod

A horny beak was reported for the first time in *Gallimimus* IGM 100/1133 and *Ornithomimus edmontonicus* RTMP 95.110.11. See more: *Norell et al. 2001*

2001—China—The Mesozoic bird with the largest head in proportion to its size

Longipteryx chaoyangensis had a very long snout, 28% of its total length. See more: *Zhang, Zhou, Hou & Gu 2001*

2002—China—The theropod with the most cervical vertebrae

Heyuannia huangi has thirteen neck vertebrae. Most theropods had about ten vertebral pieces. See more: *Lu 2002*

2003—China—The Mesozoic bird with the most caudal vertebrae

Shenzhouraptor sinensis (*Jeholornis prima*) had up to twenty-seven pieces. See more: *Ji & Ji 1996; Zhou & Zhang 2003*

2003—China—The theropod with the longest neck in proportion to its size

The neck of *Sapeornis chaoyangensis* was so long that it was 30% of its total length. See more: *Zhou & Zhang 2003*

2003—USA—The most robust theropod femur of the Cretaceous

The minimum perimeter of the central part of the femur of the *Tyrannosaurus rex* specimens FMNH PR2801 and MOR 1128 is 58 cm (44.5% and 46% of the length of the femurs, respectively). Among herbivores, *Deinocheirus mirifcus* IGM 100/127 stands out:

56 cm (42.4% of the length of the femur). See more: *Brochu 2003; Lee et al. 2014*

2003—China—The Mesozoic bird with the largest arms in proportion to its size
Sapeornis chaoyangensis had very large arms for its size. Each was 87% of its total length. See more: *Zhou & Zhang 2003*

2004—USA—The most recent skin impression in a theropod
Skin prints of *Tyrannosaurus rex* have been found in specimens BHI 6230 and MOR 1125, which date from the late Upper Cretaceous (Maastrichtian, ca. 69–66 Ma). See more: *Schweitzer et al. 2004; Larson 2008*

2005—Argentina—The theropod with the largest head in proportion to its size
The skull of *Buitreraptor gonzalezorum* was so long that it was 15.4% of its total length. See more: *Makovicky, Apesteguia & Agnolin 2005*

2006—Germany—The theropod with the fewest sacral vertebrae
Among early theropods, the most common number is five vertebrae, but the juvenile *Juravenator starki* had only three vertebral parts. See more: *Gohlich & Chiappe 2006*

2006—USA—The theropod with the greatest capacity for arm rotation
It has been proven that *Allosaurus* could rotate his hands about 140 degrees from their original position. See more: *Senter 2006*

2006—China—The Mesozoic bird that could rotate its arms the most
The adaptation for flight has caused birds to lose the ability to rotate their arms. For example, the crow barely reaches a rotation of 20 degrees. *Sapeornis chaoyangensis* could rotate its arms 112 degrees, far surpassing current birds. See more: *Senter 2006*

2006—Spain—The Mesozoic bird that could rotate its arms the least
Eoalulavis hoyasi could barely rotate its arms 29%. See more: *Senter 2006*

2006—China—The strangest legs of a Mesozoic bird
Dalingheornis liweii is the only Mesozoic bird whose feet are zygodactyl—that is, finger II (internal) and III (middle) point forward, while finger IV (external) and I (thumb) point back. This arrangement allows modern birds to hold onto branches or climb trunks. Only *Shandongornipes muxiai* footprints present this condition. Maybe they were left by the same type of enantiornithean bird. See more: *Li et al. 2005; Zhang et al. 2006*

2006—USA—The theropod that could rotate its arms the least
Bambiraptor could rotate its hands only 75% maximum. See more: *Senter 2006*

2008—China—The Mesozoic bird with the fewest sacral vertebrae
Archaeopteryx had five sacral vertebrae, as did most non-avian theropods. Among Mesozoic birds, the juvenile *Zhongornis* had the fewest (four). See more: *Gao et al 2008*

2008—China—The Mesozoic bird with the fewest sacral vertebrae
Among theropods, *Epidexipteryx* stands out. It had fourteen vertebral pieces. See more: *Zhang et al. 2008*

2008—China—Theropods with the fewest caudal vertebrae
Epidexipteryx had an extraordinarily short tail, containing only sixteen vertebrae. The juvenile *Zhongornis haoae* had only thirteen vertebral parts in its tail. See more: *Zhang et al. 2008*

2008—China—The theropod with the largest legs in proportion to its size
The hind legs of *Epidexipteryx hui* were 77% of the animal's total length. See more: *Zhang et al. 2008*

2008—China—The Mesozoic bird with the fewest caudal vertebrae
Specimens of *Confuciusornis* have between seven and ten vertebrae in the tail, but it was accompanied by a pygostyle (fused terminal vertebrae); so, in reality, the number is greater. The juvenile *Zhongornis haoae* had only thirteen vertebral parts in the tail. See more: *Hou 1997; Gao et al. 2008*

2009—Argentina—The most pneumatized non-avian theropods
Megaraptors had pneumatic gaps not only in the vertebrae, like many other theropods, but also in the furcula, ilium, and gastralia, so they may have had a system of air sacs similar to that of birds. Their thermoregulation and respiration must have been very efficient. See more: *Sereno et al. 2009*

2009—China—The first toothless theropod of the Jurassic
Before the discovery of *Limusaurus inextricabilis*, all toothless theropods were from the Cretaceous. See more: *Xu et al. 2009*

2010—Romania—The Mesozoic bird with the most sickle claws
The discovery of *Balaur bondoc* was surprising, partly because it had two claws on its hind legs, similar to dromaeosaurs (raptors). See more: *Csiki et al. 2010*

2011—Italy—The theropod with the most fossilized organs
Scipionyx samniticus is exceptional, because the fossil has three-dimensional permineralizations in several soft tissues: trachea, intestines, liver, and muscles. See more: *dal Sasso & Maganuco 2011*

2011—China—The Mesozoic bird with the most pointed nails
The toenails of *Gansus yumenensis* had two tips. Perhaps they helped the bird avoid slipping in the mud or were used for grooming, as some birds do today. See more: *Li et al. 2011*

2011—China—The theropod with the fewest phalanges in the hands
Linhenykus monodactylus is the only non-avian theropod that has a single finger with two phalanges. Previously, it was thought that *Mononykus olecranus* would also have just one finger, but it apparently had two small atrophied fingers. See more: *Xu et al. 2011*

2011—Canada—The theropod with the fewest phalanges in the legs
Avimimus portentosus and ornithomimosaurs had only three toes on each foot. Each had twelve phalanges. See more: *Lambe 1902; Kurzanov 1981*

2012—The theropod with the largest tooth in proportion to its size
The theropod with the largest tooth in proportion to its size is *Daemonosaurus chauliodus*. Its premaxillary tooth may have reached between 10% and 12.8% of its skull's length. See more: *Sues et al. 2012; Headden**

2013—Argentina—The most incomplete dinosauromorph species
Ignotosaurus fragilis is known only by an ilium. An unnamed lagerpetid similar to *Dromomeron* is known only by a femur (present-day Colorado, USA). See more: *Small 2009; Martínez et al. 2013*

2015—USA—The theropod that could open its jaws the widest
A study of crocodiles, birds, and theropods revealed that *Allosaurus fragilis* could open its jaws up to 79 degrees. The paleontologist Bob Bakker made a similar estimate seventeen years earlier. See more: *Lautenschlager 2015*

2015—Canada—Skull of the most incomplete theropod species
The specimen type of *Boreonykus certekorum* (RTMP 89.55.47) is based only on an incomplete frontal bone. It is possible that the other bones assigned to the species do not belong. See more: *Bell & Currie 2015; Cau 2015; Mortimer**

2015—Brazil—The smallest Mesozoic dinosaur
The enantiornithean bird *Cratoavis cearensis* is the smallest known Mesozoic dinosaur. It was smaller than even the smallest extinct bird fossil. It was approximately 6.6 cm long (almost 15 cm with the tail feathers) and weighed a little more than 4 g. See more: *Carvalho, et al. 2015a, 2015b*

2015—Spain—The first theropod podotheca
Skin impressions and hind leg claws of *Concavenator corcovatus* were preserved. Its morphology, in terms of the arrangement and shape of the scales, was very similar to that of current birds. See more: *Cuesta et al. 2015*

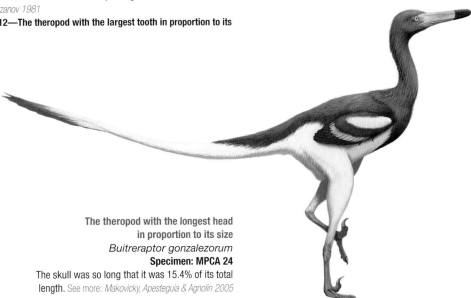

The theropod with the longest head in proportion to its size
Buitreraptor gonzalezorum
Specimen: MPCA 24
The skull was so long that it was 15.4% of its total length. See more: *Makovicky, Apesteguia & Agnolin 2005*

The origin of feathers was unrelated to flight. They existed many millions of years before it. This is suggested by the many morphological changes that the integumentary system underwent—from a filament to several filaments to true feathers. It is possible that the function of monofilament feathers was to help with thermoregulation and intraspecies communication (sexual attraction, social hierarchy, indication of maturity, etc.) or interspecies communication (camouflage, threat or warning signals, mimicry, etc.). Flight was apparently made possible by a combination of evolutionary developments in internal and external anatomy that were further defined when the need arose to reach new sources of food or to escape from predators. See more: *Lingham-Soliar 2009*

Symmetry and asymmetry

The flight feathers of birds are asymmetrical—that is, the rachis is not exactly in the center of the barbs. Some theropod gliders, such as *Microraptor*, and older birds, such as *Archeopteryx* or *Jeholornis*, already had markedly asymmetrical feathers. This characteristic was eventually present in all flying birds.

On the other hand, it is important to note that *Archeopteryx*, unlike *Anchiornis*, *Xiaotingia*, and other very similar forms, is the oldest and most primitive theropod whose body was covered with bilaterally symmetrical feathers. This book uses this difference as a characteristic feature to define the ambiguous term *bird*, since Carl Linnaeus founded the clade considering that they had bodies covered with feathers, not simple monofilaments. See more: *Linnaeus 1758; Dyke et al. 2013; Foth et al. 2014*

Color

The color of modern bird feathers is influenced both by the shape and arrangement of melanosomes and by the feathers' physical structure, which reflects part of the visible light waves. Of these two characteristics, only melanosomes have been preserved in dinosaurs. If they are long and narrow, the color is black or gray. If they are short and wide, they are reddish. Feathers that lack them are white. If the melanosomes are aligned in the same direction, they result in a type of iridescence (like, for example, the metallic blue in the feathers of some ducks). See more: *Li et al. 2010; Zhang et al. 2010*

Whiskers

Face whiskers are a type of feather or rigid hair whose function is tactile, because they are very sensitive. They are a very common structure in mammals and birds, so perhaps they were also present in dinosaurs and pterosaurs. See more: *Persons & Currie 2015*

1836—USA—The first theropod footprints with feather impressions
Ornithichnites diversus clarus, O. diversus platydactylus and *O. ingens* (now *Steropoides*) from the Lower Jurassic of Massachusetts, have markings that show tufts of feathers emerging from their heels. See more: *Hitchcock 1836*

1855—Germany—The first fossil feather of the Jurassic
The first specimen of *Archeopteryx lithographica* was discovered in Riedenbur, Germany. The fossil is very incomplete. It was interpreted as a pterosaur ("*Pterodactylus*" *crassipes*) in 1857, and its true nature was not discovered until 113 years later. The fossil shows traces of feathers. See more: *Meyer 1857, Ostrom 1970*

1857—Germany—The first name for a feathered dinosaur
Pterodactylus crassipes was the name given to the first specimen of *Archeopteryx*. At that time, it was considered a pterosaur. It was not given the name combination "Archeopteryx crassipes" because the name of *Archeopteryx lithographica* is historically protected against any name that precedes it. See more: *Meyer 1855; Ostrom 1970*

1865—Massachusetts, USA—Oldest theropod footprints with feathers
It is thought that the footprints *Eubrontes minusculus* PMNH-EHC 1/7 belonged to a resting coelophysoid. The impression shows two hind legs with metatarsals, the ischium, and the belly. The most unusual thing is the presence of a coat of monofilaments. The fossil was known as early as 1926, but it took seventy-one years for its true nature to be recognized.

Steropoides diversus and *S. ingens* had strange feathers in their heels. *Plesiornichnus mirabilis* AC 51/16 also has featherlike impressions near the tread, but experts question it. They date from the Lower Jurassic (Hettangian-Pliensbachian, ca. 201.3–199.3 Ma). See more: *Hitchcock 1836, 1865; Lull 1904, 1926, 1952; Olsen & Baird 1986; Gierslinsky 1994; Gierslinsky 1997; Rainforth 2005*

1955—Spain—The first Cretaceous and European feather discovered
A loose feather was reported in El Montsec, Spain, exactly one hundred years after the first feather from the Jurassic. See more: *Meyer 1855; Ferrer-Condal 1955; Kellner 2002*

1966—Australia—The first Cretaceous feather in Oceania
Several isolated feathers are reported in Koonwarra. See more: *Talent et al. 1966; Kellner 2002*

1969—Kazakhstan—The theropod species described only by a feather
Cretaaviculus sarysuensis is known only by an isolated feather that is 1.7 cm long. Its shape is asymmetrical, so it was identified as a possible paravian four years before *Sinosauropteryx* was described. See more: *Bazhanov 1969; Nesov 1992*

1970—Kyrgyzstan—The first mistaken report of fossilized Triassic feathers
Longisquama insignis has strange appendages that were once considered feathers, but they apparently constitute a different structure with no relation to feathers. See more: *Sharov 1970; Senter 2004*

1971—Kazakhstan—The first Jurassic feather of western Asia
Remains of feathers that were later named *Praeornis sharovi* were discovered. These were later interpreted as cycad leaves (*Cycadites saportae*), until they were again recognized as authentic feathers, thanks to new specimens and chemical studies. They may have belonged to any type of dinosaur. See more: *Rautian 1978; Nesov 1992; Dzik et al. 2010; Mortimer**

1973—Lebanon—The first Cretaceous feather in the Middle East
Several isolated feathers of a bird possibly older than *Enantiophoenix electrophyla* were reported. See more: *Schlee 1973; Kellner 2002; Cau & Arduini 2008*

1973—Lebanon—The oldest feather preserved in amber
The oldest specimen was found in the city of Jezzine. It dates from the early Lower Cretaceous (Hauterivian, ca. 132.9–129.4 Ma) of south-central Gondwana (present-day Lebanon). See more: *Schlee 1973*

1978—Kazakhstan—The indeterminate dinosaur known only by feathers
Praeornis sharovi is known only for two strange feathers that were reinterpreted as cycad leaves, although later studies revalidated that they were authentic feathers. Considering the diversity of feathers in ornithischian dinosaurs, they can be assigned only to indeterminate dinosaurs. See more: *Rautian 1978; Glazunova et al. 1991; Nesov 1992; Dzik, Sulej & Niedzwiedzki 2010; Mortimer**

1985—Mongolia and Russia—The first Cretaceous feather in East Asia
Several feathers were reported in Gurvan Eren, Shenn Khuduk, and Transbaikalia. In the latter, there is a color pattern. See more: *Kurochkin 1985; Kellner 2002*

1985—Kazakhstan—The Mesozoic bird species based only on feathers
Ilerdopteryx viai is known for nine feathers, but thanks to their asymmetrical nature, we know that they belonged to bird. See more: *Lacasa-Ruiz 1985*

1987—Mongolia—The first theropod discovered with complex feathers
A prediction of feathers on forearms predates the discovery of *Protarchaeopteryx* by ten years. The study of the forearm of *Avimimus portentosus* already indicated the presence of feathers. See more: *Kurzanov 1987; Norman 1988*

1988—Brazil—The first Mesozoic feather discovered in northern South America
A fossil feather from the Santana Formation was thought to be evidence of a bird. Nowadays, it is known that it could also belong to an indeterminate theropod. See more: *Martins-Neto & Kellner 1988; Kellner 2002*

1995—Canada—The first evidence of dinosaur feathers in North America
Filament tags were found on the arms of an adult specimen of *Ornithomimus edmontonicus* but were not identified until seventeen years later. See more: *Zelenitsky et al. 2012*

1995—Canada—The first Cretaceous feather in western North America
Fragments of feathers preserved in amber were found in Alberta. See more: *Davis & Briggs 1995, Kellner 2002*

1995—USA—The first Cretaceous feather in eastern North America
Remains of feathers preserved in amber were discovered in New Jersey. See more: *Grimaldi & Case 1995, Kellner 2002*

1996—China—The first theropod with direct evidence of feathers (filaments)
The presence of feathery filaments in *Sinosauropteryx prima* was so surprising that it was considered a probable primitive bird. This was because feathers were thought to be a unique characteristic of birds. Nowadays, it is known that these structures were from the basal trunk of dinosaurs. See more: *Ji & Ji 1996*

1997—China—The first theropod with complex feathers
Despite being a terrestrial dinosaur, *Protarchaeopteryx robusta* had flight feathers. See more: *Ji & Ji 1997*

1997—USA—The first controversy over the evidence of feathers in non-avian theropods
Researchers suspected that the monofilaments preserved in *Sinosauropteryx prima* were composite collagen fibers, but new analyses showed that they were really a "protoplume" cover. See more: *Ji & Ji 1996; Morell 1997; Currie & Chen 2001*

1998—Madagascar—The first feathered theropod in Africa (inferred)
From a study of their bones, the presence of feathers in *Rahonavis ostromi* was demonstrated. See more: *Chiappe 2007; Forster et al. 1998*

1999—Mongolia—The first feathered theropod shown through biochemical analysis
The year after the genus *Shuvvuia deserti* was described by means of electron microscopy, it was found that its body must have been

covered with feathers. This was demonstrated by the presence of products of the decay of the protein beta-keratin. See more: *Schweitzer et al. 1999*

1999—China—The first herbivorous theropod with feathers
Beipiaosaurus inexpectus is the first herbivorous dinosaur in which body filaments were preserved. See more: *Xu et al. 1999*

2002—China—The first non-avian theropod with "four wings"
Microraptor zhaoianus (Cryptovolans pauli) is the first non-avian theropod with asymmetrical feathers on both its hind and lower limbs. See more: *Czerkas et al. 2002*

2002—China—The Mesozoic bird with the longest flight feathers of the Cretaceous
The largest arm feathers of *Shenzhouraptor sinensis* were 21 cm long, longer than the arm and forearm feathers combined. A feather from the specimen of *Confuciusornis sanctus* (BSP 1999 I 15) was 20.7 cm long but was not an adult. See more: *Ji et al. 2002; Nudds & Dyke 2010*

2002—China—The theropod with the longest flight feathers of the Cretaceous
The oviraptorosaur *Caudipteryx zhoui* had feathers that were up to 20 cm long. A series of structures in *Dromiceiomimus brevitertius* (= *Ornithomimus edmontonicus*) specimens TMP 2009.110.1 and TMP 2008.70.1 formed a large filamentous covering on the arms. Its appearance is not entirely understood. See more: *Currie et al. 1998; Zelenitsky et al. 2012*

2002—Canada—The largest theropod with direct evidence of complex feathers
Two specimens of *Dromiceiomimus brevitertius* (= *Ornithomimus edmontonicus*), a juvenile (TMP 2009.110.1) and an adult (TMP 2008.70.1) that were about 3.5 m long and weighed 81 kg, had structures in their arms similar to bird wings. See more: *Zelenitsky et al. 2012*

2006—China—The Mesozoic bird with the largest feather in proportion to its size
The two caudal feathers of *Dapingfangornis sentisorhinus* were 13 cm long and exceeded the length of its skeleton (126% of the total). See more: *Li et al. 2006*

2007—China—The non-avian theropod with the finest filaments
Monofilaments of *Sinosauropteryx prima* were 0.08–0.3 mm in diameter See more: *Ji & Ji 1996; Lingham-Soliar, Feduccia & Wang 2007*

2009—China—The theropod with the thickest filaments
Monofilaments of *Beipiaosaurus inexpectus* STM31-1 were up to 3 mm thick, twice as wide as those of the type IVPP 11559.

Peculiarly, unlike monofilaments of other theropods, they have a flattened shape, while others in the neck area are rigid and hollow. See more: *Lee et al. 2014*

2009—China—The Mesozoic bird with the longest rectrices of the Cretaceous
A caudal feather in the *Confuciusornis feducciai* D2454 is approximately 31.5 cm long. Some adult specimens of *Confuciusornis sanctus* are 40 cm long, but they did not retain their rectrices, which perhaps were as long as their body. See more: *Zhang et al. 2009*

2009—China—The theropod with the longest flight feathers of the Cretaceous
Tail feathers of the *Changyuraptor yangi* parave were 29.27 cm long. See more: *Marugán-Lobón et al. 2011*

2010—Germany—Longest remiges of the Jurassic
Archaeopteryx lithographica HMN MB. 1880/81 has feathers up to 14.5 cm long, although the largest may have reached more than 15 cm. The largest specimen of *Archeopteryx* did not conserve feathers, so their maximum size is unknown. See more: *Anonymous 1957*

2010—China—Longest flight feather of the Jurassic
The longest caudal feather of *Epidexipteryx hui* was 18.5 cm long. It is incomplete. See more: *Zhang et al. 2008*

2010—China—The proportionally longest feather of the Jurassic
Epidexipteryx hui had very long feathers compared to its body, which was about 30 cm long. Its feathers were over 18.5 cm long. See more: *Zhang et al. 2008*

2010—China—The Mesozoic bird with the most colorful plumage
According to a study of the melanosomes of *Confuciusornis sanctus* (IVPP V13171), the theropod had a reddish tone accompanied by gray, red-brown, and black shades. The tones are similar to those of the zebra finch (*Taeniopygia guttata*). See more: *Zhang et al. 2010*

2010—Argentina—The first non-avian feathered theropod in western Europe
Although *Juravenator starki* was discovered in 1998, it wasn't proven to have feathers until a later study using ultraviolet rays. See more: *Göhlich & Chiappe 2010*

2011—Alabama, USA—Longest isolated feather
The fossil feather KIS-706 was the largest of the Mesozoic. It was 16.5 cm long and could have been from a bird or non-avian theropod. See more: *Knight et al. 2011*

2011—Canada—The most recent feather conserved in Mesozoic amber
Feathers similar to modern aquatic birds date from the late Upper Cretaceous (Campanian, ca. 83.6–72.1 Ma) and belonged to the western area of Laurasia. See more: *Ryan, Chatterton, Wolfe & Currie 2011*

2012—China—The longest theropod filament
The largest monofilaments were from the neck of *Yutyrannus huali*. They were 20 cm long. See more: *Xu et al. 2012*

2012—China—The theropod with the brightest tone
Microraptor zhaoianus (= *M. gui*) BMNHC PH881 had melanosomes aligned in the same direction as iridescent birds. It may have been black or dark blue, with a lustrous touch. See more: *Li et al. 2012*

2012—China—The largest feathered dinosaur
The type specimen of *Yutyrannus huali* ZCDM V5000 was a large carnivore that was approximately 8.4 m long and weighed 1.5 t. This makes it the largest feathered fossil. See more: *Xu et al. 2012*

2014—Mongolia—The largest feathered theropod (inferred)
IGM 100/127 of *Deinocheirus mirificus* has a pygostyle, a structure that gives support to and enables the movement of a crest of feathers at the tip of the tail. It was 12 m long and weighed about 7 t. See more: *Lee et al. 2014*

2015—Brazil—The first Mesozoic bird with preserved feathers in southern South America
Cratoavis cearensis is a tiny bird found in the state of Cerará, Brazil. It was preserved with evidence of feathers. Standing out among them are rectrices that are longer than the bird's skeleton. See more: *Carvalho et al. 2015a, 2015b*

Feathered theropod
Avimimus portentosus
A forearm bone (ulna) of *Avimimus* has structures similar to the anchorage points of rectrices, so the presence of feathers was accepted even in the absence of further evidence. Sixteen years later came the discovery of *Protarchaeopteryx*, a close relative of *Avimimus* that supported the interpretation of wings on the forearm. See more: *Kurzanov 1981; Ji & Ji 1987*

Theropod life
The biology of the theropods

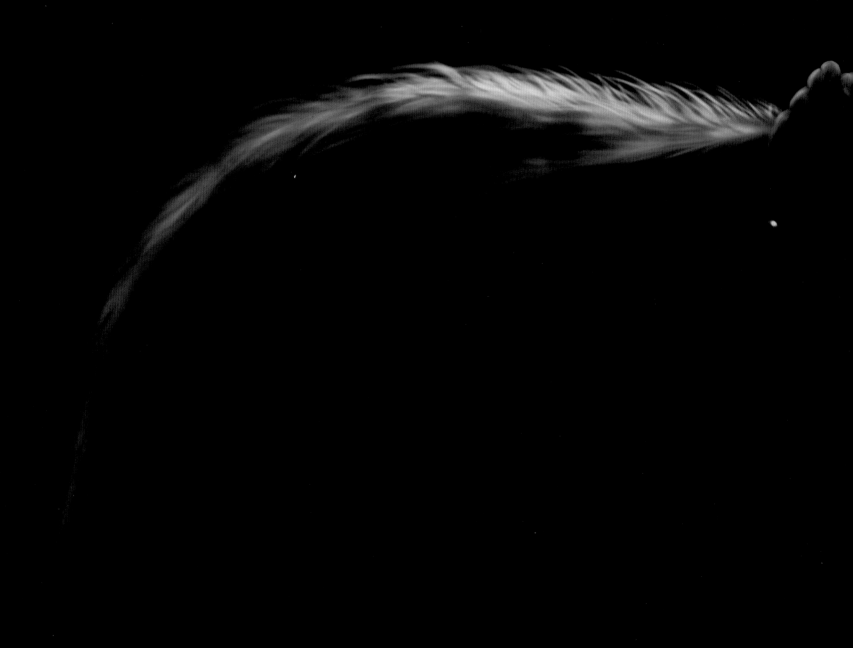

Records: The fastest, the largest and smallest fliers, and the biggest and smallest eggs, among others.

Contrary to the popular image of dinosaurs as violent and voracious monsters, their behavior was probably similar to that of present-day animals: getting food and water, reproducing, and protecting themselves from other animals, climate, diseases, and parasites.

Theropods lived in diverse habitats, such as aquatic, aerial, arboreal, semi-aquatic, and terrestrial. They could be classified as runners, fossorials (diggers), generalists (mixed capacities), swimmers, climbers, fliers, or jumpers, according to their habitat.

TERRESTRIALS

Almost all theropods and dinosauromorphs were subcursorial to a high degree. It is possible that the avemetatarsalians and the lagerpetids moved by jumping, whereas the later forms began to adapt to locomotion from steps.

FLIERS

The first avian theropods had to be full-time or occasional climbers, and although some could glide, none managed controlled flight. It is possible that the ability to flap was developed as a way to help climb inclined planes in order to take advantage of new food sources or to escape from predators. True flight began with confuciusornithine birds, although limited in form, since they lacked an alula (a structure that allows controlled flight and landing) and the highly developed sternum of modern birds. Enantiornithean birds developed flight independent of ornithuromorph birds, which flew like present-day birds. *See more: Chatterjee & Templin 2007; Parsons & Parsons 2007; Alexander et al. 2010; Boris 2014*

FOSSORIALS OR EXCAVATORS

Theropods have not been found in caves or underground nests, but there is evidence that dromaeosaurs or troodontids dug to catch mammals hidden in burrows. *See more: Simpson et al. 2010*

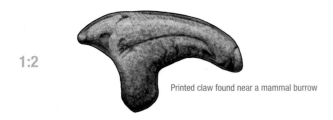

1:2

Printed claw found near a mammal burrow

The two specimens of *Mei long* IVPP V12733 and DNHM D2154 were found coiled, so it has been suggested that they died in a burrow. They may have dug it themselves or occupied another animal's den. *See more: Gao et al. 2012*

SWIMMERS

Interdigital membranes can be observed in some theropod and Mesozoic bird footprints, suggesting that they must have had an underwater or aquatic life-style, although it is believed that, in some cases, soft sediment pressure formed the marks. Direct evidence of fins on hands was reported in a *Compsognathus* specimen, but this was shown to be an error six years later. In contrast, this has been proven in *Yanornis martini*; its feet were lobed, as seen in loons.

In any case, we know that some theropods could move across the sea to take advantage of natural resources offered by this habitat, or to access new territories. Among those that were better adapted to aquatic life are spinosaurids and hesperornithines. It has also been proposed that Ceratosaurus was a good swimmer because of its long body and flattened tail. It has been found associated with fish and does not show much wear on its teeth. *See more: Gilmore 1920; Bidar, Demay & Thomel 1972; Halstead 1975; Osborn 1978; Bakker 2004; Falkingham et al. 2009; Amiot et al. 2010; Ibrahim et al. 2014; Bell & Chiappe 2015*

Mesozoic birds' adaptation to swimming can be compared between the diverse present genera. *See more: Enciclopedia Salvat*

Levels of adaptation to swimming

Level 1 Albatross

Fast and shallow dives, light bones, flying. *Volgavis* was similar to frigatebirds (*Fregata*), which almost never land on water.

Level 2 Gannet

Fast and semideep dives, light bones, flying. *Cimolopteryx*, *Lamarqueavis*, and other primitive charadriiformes lasted a short time in the water.

Level 3 Cormorant

Fast or long-lasting deep dives, light bones, generally flying, and good swimmers. The lifestyle of *Gansus* and *Mystiornis* was similar to that of these birds

Level 4 Diving petrel

Deep and long-lasting dives, some solid bones, generally flying, and good swimmers. *Neogaeonis gaviiform* would be on this level

Level 5 Razorbill

Deep and lasting dives, some solid bones, able to exclude air from its feathers, generally flying, and very good swimmers. *Brodavis* was possibly a hesperornithiform flying bird.

Level 6 Penguin

Deep and very durable dives, solid bones, able to exclude air from its feathers, lacking air sacs, able to pursue prey while diving with great skill, flightless but excellent swimmers. Advanced hesperornithiform birds such as Baptornithidae and Hesperornithidae

The smallest Mesozoic waterbird
Mystiornis cyrili
PM TSU 16/5-45
Due to the shape of the metatarsals, it is believed that this was a diving bird that was only 25 cm long and weighed about 200 g. *See more: Kurochkin et al. 2011*

1870—The first flying bird of the Mesozoic to receive a valid name
Ichthyornis anceps
Specimen: YPM 1208
Because the mandible of *Ichthyornis anceps* has teeth, it was thought to belong to a marine reptile and was named after *Colonosaurus mudgei* and *Graculavus anceps*. *Enaliornis* was named six years earlier, in 1864, but not formalized until 1876. This fact and fifteen preceding reports (which turned out to be remnants of pterosaurs, Cenozoic birds, or flightless avians) make *I. anceps* the first valid name. See more: *Seeley 1864, 1876; Marsh 1872, 1873, 1880; Galton & Martin 2002*

The oldest long-legged bird
Archaeornithipus meijidei
Specimen: SRT-6
Archaeornithipus meijidei footprints were so ancient that they fall between the Jurassic boundary and the early Lower Cretaceous (Barremian, ca. 145–139.8 Ma). Due to their proportions, we know the birds were long-legged. See more: *Fuentes-Vidarte 1996*

The most complete theropod impression
Eubrontes
Specimen: SGDS.18.T1
Although some had very long arms, most theropods could not walk on four limbs because of the position of their hands, with palms pointed toward the chest. For this reason, theropod handprints are very scarce, except in resting positions. An impression of *Eubrontes* includes the marking of both legs with metatarsals, both hands, belly marks, and ischiatic callus, as well as rearrangement imprints. See more: *Milner et al. 2009; Wilson, Marsicano & Smith 2009*

RUNNING THEROPODS

Some theropods rank among the fastest animals, as the fastest could reach speeds of 80 km/h (22 m/s). Some swifts exceed 170 km/h in direct flight. Eagles, hawks, and swifts even exceed 300 km/h when they dive.

Almost all theropods but those too adapted to aquatic life (loons, grebes, hesperornithiforms, or some spinosaurs) or aerial life (hummingbirds and swifts, which are quite clumsy on land) have the ability to run fast.

Due to giant theropods' enormous weight (more than 2.5 t), they probably could not run faster than 45 km/h. It should be noted that young Tyrannosaurus may have run at speeds close to 60 km/h, approaching the speed of the ostrich, the fastest running bird of today.

Among the fastest dinosaurs are ornithomimosaurs and other groups adapted for running, such as *Anzu, Dakotaraptor, Elaphrosaurus,* and others. Their anatomy is adapted for running at high speeds. They have very long legs with short femurs in proportion to the tibia and metatarsals.

MYTH: All non-avian theropods were runners

Until 2014, it was thought that they all could run, but the discovery of a partial skeleton of *Spinosaurus*, an animal adapted to aquatic life with hind legs too short to run, shattered this belief. *Spinosaurus* could probably not travel faster than 15 km/h on land. See more: *Ibrahim et al. 2014*

1 meter

HESPERORNITHES
? km/h
Any Hesperornithes would be very agile in water but would crawl on land.

Epidendrosaurus ningchengensis
6 km/h
Some young dinosaurs were so tiny that their running speed was very low.

10 km/h

1 meter

Liliensternus liliensterni
47 km/h
The fastest Triassic theropod.

Usain Bolt
44.7 km/h
The fastest human in history.

Ostrich
65 km/h
Can run for a long time without fatigue.

Herrerasaurus ischigualastensis
42 km/h
The fastest primitive saurischian. *Frenguellisaurus* may have reached 46 km/h.

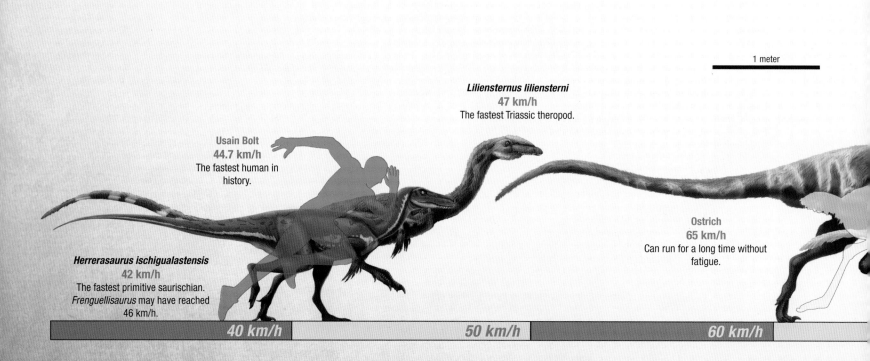

40 km/h   50 km/h   60 km/h

These characteristics are present in the fastest modern land animals. These theropods could reach speeds comparable to those of a racehorse but in no case could they approach the top speed of cheetahs and antelopes, which can surpass 100 km/h and would easily leave them behind. Humans are faster than is commonly believed, as athletic people can reach speeds between 30 and 39 km/h (8.3–10.8 m/s), while a professional sprinter exceeds 43 km/h (12 m/s). These speeds are comparable to those of many theropods, although we can't maintain the same pace in a prolonged race.

Ironically, some of the slowest theropods were sometimes considered very fast. This is not due to their running ability, but to their tiny size, which prevented them from developing very high speeds despite a high frequency of steps. This can be seen in current birds such as the roadrunner (*Geococcyx californianus*), which has the amazing ability to take up to twelve steps per second, almost three times more than the ostrich. Due to its tiny stride, however, it hits a top speed less than half of what the bigger bird accomplishes.

MYTH: Compsognathus ran at 64 km/h

Using computer programs, a study suggested that the small *Compsognathus* would be one of the fastest dinosaurs, as fast as an ostrich. To achieve such a speed, the theropod would have had to take about twenty steps per second, practically twice as many as a roadrunner, which would have been quite unlikely. See more: *Thulborn 1990; Seller & Manning 2007*

1:10

Silesaurus opolensis
28 km/h
The fastest dinosauromorph.

Epidexipteryx hui
18 km/h
The slowest theropod of the Jurassic.

Saltopus elginensis
20 km/h
The slowest dinosauromorph.

Roadrunner
27 km/h
The modern bird with the fastest pace.

Procompsognathus triassicus
28 km/h
The slowest theropod of the Triassic.

20 km/h

30 km/h

Elaphrosaurus bambergi
68 km/h
The fastest of the Jurassic. The adult of this species may have exceeded 70 km/h.

Ornithomimus edmontonicus
(Dromiceiomimus brevitertius)
72 km/h
The fastest dinosaur of the Mesozoic.

1:35

Dinornis giganteus
81 km/h
The fastest bird of all time.

Cheetah
105 km/h
Today's fastest terrestrial animal, due in part to its flexible spinal column.

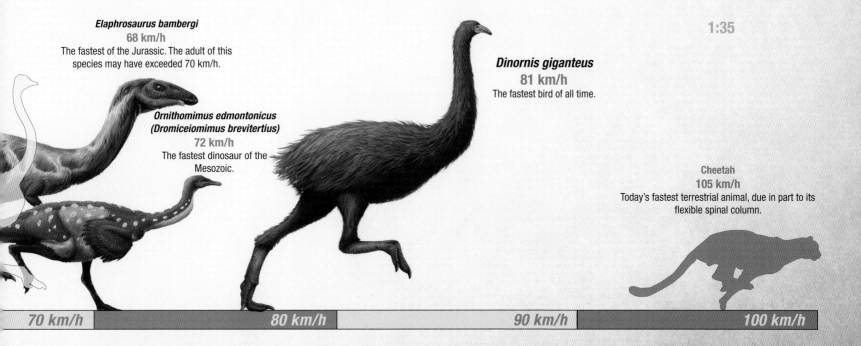

70 km/h

80 km/h

90 km/h

100 km/h

HOW TO CALCULATE THE SPEED OF A THEROPOD

At present there are some studies focused on calculating the speed of dinosaurs, especially theropods. The speeds of extinct animals are complicated to calculate, but they can be estimated based on the stride length, the probable number of steps per second (cadence), the mass of the organism, and the proportions of the leg bones. Most investigations today are based on fossil footprints or on the proportions of the appendicular skeletons (extremities and the corresponding waists that anchor them to the body), which are compared with current animals that are biomechanically similar. In the case of theropod dinosaurs, the studies usually use terrestrial birds.

Terrestrial animals are usually classified into four different categories according to their locomotion abilities (Coombs 1978):

Graviportal: These animals are quadrupeds whose legs allow them to carry large weights. They are not able to gallop or trot, so their speed is modest. In the case of mammals and dinosaurs, the legs are columnar (elephants and sauropods).

Mediportal: The legs are mostly adapted to carry weight, but with some cursorial adaptations, such as flexed legs, digitigrade posture, etc. All animals of this group are able to trot, and some even gallop (hippos, rhinos, ceratopsids, nodosaurids, ankylosaurids, etc.).

Subcursorial: This category includes practically all theropods. This group has moderately cursorial characteristics and little or no modification in the limbs for supporting weight. Some of these animals are very good runners. The fastest animals are included—felines, canids, theropods, bipedal sauropodomorphs, basal ornithopods, etc.

Cursorial: The appendicular skeleton is designed for running with extensive cursorial adaptations. (This group includes many ungulate mammals, ratites, and ornithomimosaurs).

The speed that each animal type can reach is limited by different characteristics and factors. Weight plays an important role. Different studies suggest that the optimal weight for an animal runner is between 50 and 500 kg. Above 0.5 t and below 5 kg, animals' top speed decreases considerably. Another important characteristic, especially in bipeds, is the proportion between the femur, tibia, and metatarsal. In general, the shorter the femur is in relation to the tibia, the greater the ability to move fast becomes. Also important is possessing relatively long metatarsals, as long as they are within some limits.

Maximum hypothetical speeds of theropods, and of humans, have been estimated for this book. This was achieved by comparing modern running birds ranging from the smallest roadrunner (300 g), to the large rhea and to the heaviest and fastest ostrich (*Struthio camelus*). Steps per second (cadence), stride length, body mass, and the relationship between the length of the femur and the sum of the lengths of the tibia and third metatarsal were carefully studied. The most important factors influencing the rate and stride ratio (ratio between stride length and leg height) are body mass and the proportion of the femur versus the tibia-metatarsal. Testing the methodology in present-day animals produced excellent results. Applying it to extinct theropods yielded reasonably logical results, comparable to those of current studies.

As might be expected, the fastest theropods are ratite birds, some averostra similar to *Elaphrosaurus*, and ornithomimosaurs, which include animals that weighed between 50 and 500 kg. The slowest, on the other hand, are the smallest theropods. Despite a very high cadence, leg length prevented great speeds. It is necessary to consider that these estimates are approximate, since in studies of athletes with the same stride length, speed is not always the same. This suggests that this type of calculation should be regarded with caution. Other factors that alter the animal's speed are muscular capacity and tail muscles, which, in some cases, would allow greater traction. See more: *Charig 1972; Russell 1972; Alexander 1976; Coombs 1978; Thulborn 1982, 1990; Demathieu 1984, 1986; Paul 1988; Lockley 1991; Christiansen 1998; Ruiz 2003; Peterson & Currie 2011; Sellers & Manning 2012*

CURSORIALITY

Body mass (kg)	Optimal cadence	Optimal stride ratio
0-0.5	12	3
+0.5-1	11	3
+1-2	10	3
+2-4	9	3
+4-10	8	3
+10-20	7	3
+20-50	6	3
+50-100	5	3
+100-250	4.5	2.8
+250-500	4	2.6
+500-1000	3.5	2.4
+1000-2000	3	2.2
+2000-4000	2.5	2
+4000-8000	2	1.8
+8000	1.5	1.6

X

Tibia + Met/ femur	x Optimal cadence	x Optimal stride ratio
0.9-1	0.89	0.85
+1-1.1	0.9	0.86
+1.1-1.2	0.91	0.87
+1.2-1.3	0.92	0.88
+1.3-1.4	0.93	0.89
+1.4-1.5	0.94	0.9
+1.5-1.6	0.95	0.91
+1.6-1.7	0.96	0.93
+1.7-1.8	0.97	0.95
+1.8-1.9	0.98	0.97
+1.9-2.0	0.99	0.99
+2.0-3.8	1	1

Speed calculation. The two tables can be used to calculate the velocity of any bipedal animal. The table on the left shows estimated optimal cursorial values for an animal of a specified body mass. The table on the right presents the values by which the optimal parameters of the table on the left have to be multiplied based on the proportion obtained by dividing the sum of the length of the tibia and third metatarsal by the femur length, always taking into account the body mass. Once the cadence (C) and stride ratio (SR) are obtained, they are multiplied by the leg length (LL) in millimeters (length of femur + tibia + metatarsal III). To obtain the speed in km/h, the space-time variable (ST) must be assigned a value equal to 0.0036. To obtain the speed in m/s, VE = 0.001.

$$V = (LL \times C \times SR) \times ST$$

For example, the femur of *Sinraptor dongi* IVPP 10600 is 876 mm long, the tibia is 776 mm long, and metatarsal III is 410 mm long. This gives a sum of 2,062 mm for the leg and a length ratio of 1.35 (tibia + metatarsal III/femur). The animal's estimated body mass is 1,400 kg. The optimal cadence for an animal of this size would be 3. The stride ratio would be 2.2. The result obtained from the tibia-metatarsal III/femur ratio, however, indicates that the cadence (C) must be multiplied by 0.93, and the stride ratio (SR) by 0.89 (right table), yielding a result of 2.79 and 1.96, respectively. The end result is 40.6 km/h or 11.3 m/ss: $V_{(km/h)}$= (2062 x 2.79 x 1.96) x 0.00036 I $V_{(m/s)}$= (2062 x 2.79 x 1.96) x 0.001

Theropod velocity table. Specimens that have full legs or that have reliable records of their proportions are included. Small jumpers like *Scleromochlus* or *Lagerpeton*, aquatic *Spinosaurus*, and the hesperornithiforms were omitted because their way of moving did not correspond with the formula used here. Flying birds were not included, as they would fly up before being able to run. (Enantiornithes may have been able to travel at a speed of only 1–2 km/h on land). They are added with caution to *Microraptor* and other climbing birds since their long leg feathers and curved nails would affect their ability to run.

Theropod velocity

Genus and species	Specimen	Leg length (mm)	Tibia + Met/femur	Mass (kg)	Cadence	Stride ratio	Speed km/h	Speed m/s	Record
Dinosauromorphs									
Lagosuchus talampayensis (juvenile)	UPLR 09	112	1.72	0.08	11.64	2.85	13.3	3.7	Slowest young dinosauromorph
Saltopus elginensis	NHMUK R3915	157	2.34	0.11	12	3	20.3	5.7	Slowest dinosauromorph
Lewisuchus admixtus	PULR 01	239	1.27	1.4	9.2	2.64	20.9	5.8	-
Marasuchus lilloensis	PVL 3871	169	1.93	0.225	11.88	2.97	21.4	5.9	-
Silesaurus opolensis	ZPAL Ab III/361	452	1.26	13	6.44	2.64	27.7	7.7	Fastest primitive dinosauriform
Primitive saurischians									
Staurikosaurus pricei	MCZ 1669	584	1.54	12.7	6.65	2.73	38.2	10.6	-
Ischisaurus cattoi (juvenile)	MACN 18.060	700	1.45	32	5.64	2.7	38.4	10.7	-
Herrerasaurus ischigualastensis	PVL 2566	1107	1.34	145	4.19	2.49	41.6	11.6	Fastest primitive saurischian
Coelophysoid therapod									
Procompsognathus triassicus	SMNS 12591	278	1.9	1.3	9.8	2.91	28.5	7.9	Slowest Triassic theropod
Podokesaurus holyokensis (juvenile)	-	255	1.97	0.9	10.89	2.97	29.7	8.2	-
Coelophysis bauri (juvenile)	AMNH 7246	337	1.76	2.7	8.73	2.85	30.2	8.4	-
Coelophysis bauri	UCMP 129618	653	1.67	32	5.76	2.79	37.8	10.5	-
Megapnosaurus rhodesiensis	QG 1	563	1.71	13	6.79	2.85	39.2	10.9	Slowest Lower Jurassic theropod
Liliensternus liliensterni (subadult)	MB.R.2175	1079	1.45	110	4.23	2.52	41.4	11.5	Fastest Triassic theropod
Megapnosaurus kayentakatae	MNA V2623	741	1.68	30	5.76	2.79	42.9	11.9	-
Dilophosaurus wetherilli	UCMP 77270	1448	1.54	350	3.76	2.34	45.9	12.7	-
Segisaurus halli	UCMP 32101	408	1.81	4.4	10.78	2.91	46.1	12.8	Fastest Lower Jurassic theropod
Indeterminate theropod									
Chilesaurus diegosuarezi	SNGM-1935	358	1.51	4.9	7.6	2.73	26.7	7.4	-
Avetheropods									
Limusaurus inextricabilis	IVPP V15923	614	1.92	10	7.92	2.97	52	14.4	-
Elaphrosaurus bambergi	HMN Gr.S. 38-44	1518	1.92	210	4.46	2.77	67.5	18.8	Fastest early Upper Jurassic and fastest African theropod
Deltadromeus agilis	SGM-Din 2	1891	1.55	960	3.33	2.18	49.4	13.7	Fastest early Upper Cretaceous theropod
Ceratosaurids									
Ceratosaurus magnicornis	MWC 1	1384	1.2	560	3.19	2.09	33.2	9.2	-
Ceratosaurus nasicornis	USNM 4735	1431	1.31	550	3.26	2.14	35.9	10	-
Noasaurids									
Velocisaurus unicus	MUCPv 41	365	1.7	4.7	7.68	2.79	28.1	7.8	-
Masiakasaurus knopfleri	FMNH PR2481	503	1.6	14	6.65	2.73	32.8	9.1	Fastest Madagascar theropod
Eoabelisaurus mefi	MPEF PV 3990	1507	1.35	445	3.72	2.32	46.8	13	-
Abelisaurids									
Indosuchus raptorius	-	2000	1.29	2000	2.3	1.76	29.1	8.1	-
Majungasaurus crenatissimus	FMNH PR 2778	1351	1.14	715	3.18	2.01	31.1	8.6	-
Ekrixinatosaurus novasi	MUCPv-294	1775	1.29	1400	2.76	1.94	34.2	9.5	-
Aucasaurus garridoi	MCF-PVPH-236	1665	1.34	600	3.26	2.14	41.8	11.6	-
Carnotaurus sastrei	MACN-CH 894	2511	1.44	1850	2.82	1.98	50.5	14	Fastest South American theropod
Megalosauroids									
Sciurumimus albersdoerferi (juvenile)	BMMS BK	137	1.71	0.225	11.64	2.85	16.3	4.5	-
Chuandongocoelurus primitivus (juvenie)	CCG 20010	554	1.76	14	6.79	2.85	38.6	10.7	-
Piatnitzkysaurus floresi	PVL 4073	1326	1.4	278	3.72	2.31	41	11.4	-
Megalosaurus bucklandii	-	1730	1.4	950	3.26	2.14	43.4	12.1	-
Afrovenator abakensis	UC UBA 1	1791	1.36	790	3.26	2.14	45	12.5	-
Eustreptospondylus oxoniensis (subadult)	-	1255	1.41	230	4.23	2.52	48.2	13.4	Fastest Middle Jurassic and fastest European theropod

Spinosaurids									
Genus and species	**Specimen**	**Leg length (mm)**	**Tibia + Met/femur**	**Mass (kg)**	**Cadence**	**Stride ratio**	**Speed km/h**	**Speed m/s**	**Record**
Cristatusaurus lapparenti (Suchomimus)	MNN GDF500	2466	1.29	2500	2.3	1.76	**35.9**	**10**	-
Baryonyx sp.	La Rioja	2020	1.34	1300	2.79	1.96	**39.8**	**11.1**	-
Allosauroids									
Allosaurus fragilis (subadult)	Uncatalogued	2141	1.2	2050	2.28	1.74	**30.6**	**8.5**	-
Allosaurus sp.	AMNH 290	2220	1.25	2100	2.3	1.76	**32.4**	**9**	-
Saurophaganax maximus	OMNH	2512	1.21	3200	2.3	1.76	**36.6**	**10.2**	-
Sinraptor dongi (subadult)	IVPP 10600	2062	1.35	1400	2.79	1.96	**40.6**	**11.3**	-
Allosaurus fragilis (juvenile)	USNM 4734	1787	1.32	1050	3.26	1.96	**41.1**	**11.4**	-
Yangchuanosaurus yandonensis	CV 00214	1365	1.33	433	3.72	2.31	**42.2**	**11.7**	-
Erectopus sauvagei	MNHN 2001-4	1137	1.37	225	4.19	2.49	**42.7**	**11.9**	-
Carcharodontosaurids									
Acrocanthosaurus atokensis	OMNH 10168	2690	1.11	4900	1.82	1.57	**27.7**	**7.7**	-
Giganotosaurus carolinii	MUCPv-Ch1	3180	1.22	7000	1.84	1.58	**33.3**	**9.2**	-
Concavenator concorvatus	MCCM-LH 6666	1355	1.38	410	3.72	2.41	**43.1**	**12**	-
Neovenator salerii	MIWG 6348	1755	1.4	850	3.26	2.14	**44.1**	**12.2**	-
Coelurosaurs									-
Juravenator starki (juvenile)	JME Sch 200	144	1.77	0.25	11.64	2.85	**17.2**	**4.8**	-
Sinosauropteryx prima (juvenile)	GMV 2123	154	1.9	0.22	11.76	2.91	**19**	**5.3**	-
Sinosauropteryx prima (subadult)	NIGP 127587	248	1.88	1.1	9	3	**24.1**	**6.7**	Slowest early Lower Cretaceous theropod
Ornitholestes hermanni	AMNH 619	488	1.36	12	6.51	2.64	**30.2**	**8.4**	-
Compsognathus longipes	MNHN CNJ 79	323	1.93	2.3	8.91	2.97	**30.8**	**8.6**	-
Sinosauropteryx sp.	NGMC 2124	349	2.23	2.2	9	3	**33.9**	**9.4**	-
Huaxiagnathus orientalis	CAGS-IG02-301	449	1.74	6.4	7.76	2.85	**35.7**	**9.9**	-
Gasosaurus constructus	IVPP V7264	1045	1.46	150	4.23	2.52	**40.1**	**11.1**	-
Nedcolbertia justinhofmanni (juvenile)	TMP 96.90.2	451	2.12	3.4	9	3	**43.8**	**12.2**	-
Proceratosaurs									
Dilong paradoxus	IVPP V14243	501	1.77	26	5.82	2.85	**29.9**	**8.3**	-
Tanycolagreus topwilsoni (subadult)	TPII 2000-09-29	959	1.69	114	4.32	2.6	**38.8**	**10.8**	-
Megaraptors									
Chilantaisaurus tashuikouensis	IVPP V2884.2	2614	1.2	4100	1.82	1.57	**26.9**	**7.5**	Slowest early Upper Cretaceous theropod
Fukuiraptor kitadaniensis	FPDM 9712201	1312	1.59	320	3.8	2.37	**42.5**	**11.8**	-
Australovenator wintonensis	AODF 604	1469	1.54	450	3.8	2.37	**47.6**	**13.2**	-
Tyrannosauroids									
Yutyrannus huali	ZCDMV5000	1925	1.26	1500	2.76	1.76	**33.7**	**9.4**	-
Bistahieversor sealeyi	OMNH 10131	2407	1.33	2800	2.33	1.78	**35.9**	**10**	-
Dryptosaurus aquilunguis	ANSP 9995	1983	1.48	1100	2.82	1.98	**39.9**	**11.1**	-
Alectrosaurus olseni	AMNH 6554	1865	1.88	560	3.43	2.33	**53.7**	**14.9**	-
Appalachiosaurus montgomeriensis	RMM 6670	2049	1.61	1000	3.36	2.23	**55.3**	**15.4**	-
Tyrannosaurids									
Tyrannosaurus rex (robust morph)	FMNH PR2081	3161	1.38	8265	1.41	1.44	**23.1**	**6.4**	Slowest theropod of the late Upper Cretaceous
Tyrannosaurus rex (gracile morph)	MOR 555	3122	1.43	6500	1.88	1.62	**34.2**	**9.5**	-
Daspletosaurus torosus	AMNH 5438	2330	1.33	2700	2.33	1.78	**34.8**	**9.7**	-
Tarbosaurus efremovi (subadult)	PIN 552-1	2455	1.53	2100	2.38	1.82	**38.3**	**10.6**	-
Albertosaurus arctunguis	ROM 807	2636	1.58	2550	2.38	1.82	**41.1**	**11.4**	-
Lythronax argestes	UMNH VP 20200	2032	1.54	1400	2.85	2	**41.7**	**11.6**	-
Gorgosaurus libratus	AMNH 5458	2707	1.54	2900	2.38	1.82	**42.2**	**11.7**	-
Tarbosaurus efremovi (juvenile)	PIN 552-2	1851	1.85	715	3.43	2.33	**53.3**	**14.8**	-
Gorgosaurus libratus (juvenile)	ROM 1247	2037	1.8	520	3.43	2.33	**58.6**	**16.3**	Fastest juvenile theropod

Ornithomimosaurs

Genus and species	Specimen	Leg length (mm)	Tibia + MTs/femur	Mass (kg)	Cadence	Stride ratio	Speed km/h	Speed m/s	Record
Nqwebasaurus thwazi (subadult)	AM 6040	322	1.81	2.5	8.82	2.91	30.6	8.5	-
Deinocheirus mirificus	MPC-D 100/127	3100	1.35	7000	1.86	1.6	33.2	9.2	-
Hexing qingyi	JLUM-JZ07b1	404	1.99	4.5	7.92	2.97	34.2	9.5	-
"Grusimimus" (cf. Harpymimus)	GIN 960910KD	768	1.81	35	5.88	2.91	47.3	13.1	-
Garudimimus brevipes	GIN 100/13	988	1.66	90	4.8	2.79	47.6	13.2	-
Gallimimus bullatus (juvenile)	MgD-1/94	789	1.96	26	5,94	2.97	50.1	13.9	-
Anserimimus planinychus	IGM 100/300	1216	1.8	115	4.37	2.66	50.9	14.1	-
Gallimimus bullatus (juvenile)		1030	1.86	64	4.9	2.91	52.9	14.7	-
Struthiomimus currellii	ROM 851	1227	1.82	111	4.41	2.72	53	14.7	-
Beishanlong grandis (subadult)	FRDC-GS GJ (06)01-18	1723	1.61	500	3.84	2.42	57.6	16	Fastest late Lower Cretaceous theropod
Struthiomimus altus	AMNH 5339	1380	1.88	185	4.41	2.72	59.6	16.6	-
Struthiomimus sp.	AMNH 5257	1458	1.84	220	4.41	2.72	63	17.5	-
Struthiomimus sedens	USNM 4736	1807	1.86	410	3.92	2.66	67.8	18.8	-
Gallimimus bullatus	IGM 100/11	1925	1.89	500	3.92	2.52	68.5	19	Fastest Asian theropod
Dromiceiomimus brevitertius	CMN 12228	1443	2.08	140	4.5	3	70.1	19.5	-
Dromiceiomimus brevitertius (subadult)	CMN 12068	1326	2.18	98	5	3	71.6	19.9	Fastest Mesozoic and fastest North American theropod

Alvarezsaurs

Genus and species	Specimen	Leg length (mm)	Tibia + MTs/femur	Mass (kg)	Cadence	Stride ratio	Speed km/h	Speed m/s	Record
Parvicursor remotus	PIN 4487/25	187	2.55	0.19	12	3	24.2	6.7	Slowest Asian theropod
Parvicursor sp.	IGM 100/120	230	2.59	0.34	12	3	29.8	8.3	-
Xixianykus zhangi (subadult)	XMDFEC V0011	231	2.3	0.4	12	3	29.9	8.3	-
Shuvuuia deserti	MPD 100/120	256	2.36	0.6	11	3	30.4	8.4	-
Ceratonykus oculatus	MPC 100/24	293	2.45	0.8	11	3	34.8	9.7	-
Mononykus olecranus	IGM N107/6	431	2.12	3.4	9	3	41.9	11.6	-
Haplocheirus sollers	IVPP V15988	628	1.93	15	6.93	2.97	46.5	12.9	-

Therizinosaurs

Genus and species	Specimen	Leg length (mm)	Tibia + MTs/femur	Mass (kg)	Cadence	Stride ratio	Speed km/h	Speed m/s	Record
Neimongosaurus yangi	LH V0001	796	1.17	115	4.1	2.44	28.7	8	-
Segnosaurus galbinensis	IGM 100/82	2093	1.18	2050	2.23	1.74	29.2	8.1	-
Beipiaosaurus inexpectus	IVPP 11559	647	1.44	43	4.7	2.7	29.6	8.2	-
Nothronychus graffami	UMNH VP 16420	1567	1.25	785	3.22	2.11	38.3	10.6	-
Jianchangosaurus yixianensis (juvenile)	41HIII-0308A	694	2.36	21	6	3	44.9	12.5	-

Oviraptorosaurs

Genus and species	Specimen	Leg length (mm)	Tibia + MTs/femur	Mass (kg)	Cadence	Stride ratio	Speed km/h	Speed m/s	Record
Yulong mini (juvenil)	HGM 41HIII-0107	204	1.83	0.525	10.78	2.91	23	6.4	-
Protarchaeopteryx robusta	NGMC 2125	371	1.97	1.8	9.9	2.97	39.3	10.9	-
Ajancingenia yanshini	MPC-D100/32	680	1.68	+20	5.76	2.79	39.3	10.9	-
Heiyuannia huangi	HYMV1-1	710	1.78	+20	5.76	2.79	41.1	11.4	-
Khaan mckennai	MPC D-100/1002	536	1.73	9.2	7.76	2.85	42.7	11.9	-
Conchoraptor gracilis	MPC-D102/03	638	1.66	17	6.72	2.79	43.1	12	-
cf. Ajancingenia sp.	MPC no catalogado 1	641	1.69	16.5	6.72	2.79	43.2	12	-
Caudipteryx zhoui	NGMC 97-4-A	450	2.06	2.2	9	3	43.7	12.2	-
Wulatelong gobiensis	IVPP V 18409	723	1.84	27	5.88	2.91	44.5	12.4	-
Chirostenotes pergracilis	TMP.79.30.1	890	1.86	46	4.9	2.91	45.7	12.7	-
Caudipteryx dongi	IVPP V 12344	472	2.11	2.3	9	3	45.9	12.8	-
Citipati osmolskae	MPC-D100/978	934	1.71	51	4.85	2.85	46.5	12.9	-
Avimimus portentosus	PIN 3907/1	593	2.16	9	8	3	51.3	14.3	-
Citipati sp.	MPC-D100/42	865	1.84	36	5.88	2.91	53.3	14.8	-
Similicaudipteryx yixianensis	IVPP V12556	639	1.96	6.7	7.92	2.97	54.1	15	-
Gigantoraptor erlianensis	LH V0011	2863	1.6	2000<	2.85	2	58.7	16.3	Fastest theropod weighing more than a ton
Anzu wyliei	CM 78001	1444	1.86	195	4.41	2.72	62.4	17.3	-

Paraves

Genus and species	Specimen	Leg length (mm)	Tibia + Met/femur	Mass (kg)	Cadence	Stride ratio	Speed km/h	Speed m/s	Record
Epidendrosaurus ningchengensis (juvenile)	IVPP V12653	47.2	1.88	0.067	11.76	2.91	5.8	1.6	Slowest young Jurassic theropod
Scansoriopteryx heilmanni (juvenile)	CAGS02-1/DM 607	47.8	1.9	0.07	11.76	2.91	5.9	1.6	Slowest young Cretaceous theropod
Buitreraptor gonzalezorum	MPCA 24.5	396	1.69	5.5	7.68	2.79	30.5	8.5	-
Microraptor gui	IVPP V13352	333	1.84	1.45	9.8	2.91	34.2	9.5	-
Microraptor zhaoianus (juvenile)	NGMC 00-12-A	322	2.07	1.1	10	3	34.8	9.7	-
Mahakala omnogovae	IGM 100/1033	271	2.43	0.45	12	3	35.1	9.8	-
Sinornithosaurus millenii	IVPP V12811	417	1.81	2.9	8.82	2.91	38.5	10.7	-
Zhenyuanlong suni	JPM-0008	529	1.74	6.4	7.76	2.85	42.1	11.7	-○
Bambiraptor feinbergi	FIP 002-136	500	1.94	6	7.92	2.97	42.3	11.8	-
Austroraptor cabazai	MML-195	1455	1.6	340	3.8	2.37	47.2	13.1	-
Tianyuraptor ostromi (subadult)	STM1-3	601	2.01	7.1	8	3	51.9	14.4	-

Raptors

Genus and species	Specimen	Leg length (mm)	Tibia + Met/femur	Mass (kg)	Cadence	Stride ratio	Speed km/h	Speed m/s	Record
Velociraptor mongoliensis	IGM 100/986	592	1.49	24	5.64	2.7	32.5	9	-
Tsagaan mangas	IVPP V16923	610	2.65	23	5.76	2.79	35.3	9.8	-
Deinonychus antirrhopus	MCZ 4371	870	1.6	76	4.75	2.73	40.6	11.3	-
Saurornitholestes langstoni	TMP 88.121.39	614	1.9	11.7	6.86	2.91	44.1	12.3	-
Achillobator gigantus	FR.MNUFR-15	1230	1.44	165	4.23	2.42	45.3	12.6	-
Dakotaraptor steini (subadult)	STM1-3	1551	1.78	220	4.37	2.6	63.4	17.6	Fastest predator theropod

Troodontids

Genus and species	Specimen	Leg length (mm)	Tibia + Met/femur	Mass (kg)	Cadence	Stride ratio	Speed km/h	Speed m/s	Record
Eosinopteryx brevipenna	YFGP-T5197	154	2.16	0.1	12	3	19.9	5.5	-
Aurornis xui	YFGP-T5198	200	2.04	0.26	12	3	26	7.2	-
Jinfengopteryx elegans	CAGS-IG-04-0801	225	2.2	0.37	12	3	29.1	8.1	-
Anchiornis huxleyi	LPM-B00169	228	2.44	0.26	12	3	29.5	8.2	-
Mei long	IVPP V12733	245	2.02	0.43	12	3	31.8	8.8	-
Philovenator curriei	IVPP V 10597	299	2.45	0.98	11	3	35.5	9.9	-
Sinovenator changii	IVPP V12615	357	2.04	1.3	10	3	38.6	10.7	-
Sinusonasus magnodens	IVPP V 11527	435	2.09	2.3	9	3	42.3	11.8	-
Saurornithoides mongoliensis	AMNH 6516	582	1.91	12	6.93	2.97	43.1	12	-
Sinornithoides youngi	IVPP V9612	449	2.2	2.5	9	3	43.7	12.1	-
Talos sampsoni	UMNH VP 19479	578	2.07	+10	7	3	43.7	12.1	-
Troodon formosus	MOR 748	1113	1.3	170	4.14	2.64	43.8	12.2	-

Avialans

Genus and species	Specimen	Leg length (mm)	Tibia + Met/femur	Mass (kg)	Cadence	Stride ratio	Speed km/h	Speed m/s	Record
Dalianraptor cuhe	D2139	160	2.27	0.2	12	3	20.7	5.8	-
Jixiangornis orientalis (juvenile)	CDPC-02-04-001	196	1.73	0.65	10.67	2.85	21.5	6	-
Jeholornis palmapenis (subadult)	SDM 20090109	174	2.01	0.35	12	3	22.6	6.3	-
Archaeopteryx lithographica	-	208	1.97	0.42	11.88	2.97	26.4	7.3	Fastest Jurassic running bird
Balaur bondoc (juvenile)	EME PV.313	348	1.6	3.6	8.55	2.73	29.2	8.1	-
Rahonavis ostromi	UA 8656	256	1.91	0.95	10.89	2.97	29.8	8.3	-
Yandangornis longicaudus	M1326	308	1.91	0.75	10.89	2.97	35.9	10	Fastest Mesozoic and fastest Asian running bird

Ornithuromorphs

Genus and species	Specimen	Leg length (mm)	Tibia + Met/femur	Mass (kg)	Cadence	Stride ratio	Speed km/h	Speed m/s	Record
Patagopteryx deferrariisi	MACN-N-11	282	1.82	2	8.82	2.91	23.9	7.2	-

Ratites

Genus and species	Specimen	Leg length (mm)	Tibia + Met/femur	Mass (kg)	Cadence	Stride ratio	Speed km/h	Speed m/s	Record
Pachyornis elephantopus	-	1145	2.48	130	4.5	2.8	51.9	14.4	-
Palaeotis weigelti	-	613	3.23	7	8	3	53	14.7	-
Cassowary (Casuarius casuarius)	-	1030	2.9	60	5	3	55.6	15.4	-
Aepyornis hildebrandti	-	1563	2.61	300	4	2.6	58.5	16.3	-
Greater Rhea (Rhea americana)	-	904	2.93	25	6	3	58.6	16.3	-
Emu (Dromaius novaehollandiae)	-	1099	3.56	45	5	3	59.3	16.5	-
Ostrich (Struthio camelus)	-	1377	3.34	115	4.5	2.8	62.5	17.4	Fastest present-day and fastest African running bird
Aepyornis maximus	-	1755	2.77	370	4	2.6	65.7	18.3	Fastest Madagascar running bird

Genus and species	Specimen	Leg length (mm)	Tibia + Met/femur	Mass (kg)	Cadence	Stride ratio	Speed km/h	Speed m/s	Record
Dinornis maximus	-	1978	3.21	380	4	2.6	**74.1**	20.6	-
Dinornis giganteus	-	1790	3.41	245	4.5	2.8	**81.2**	22.6	**Fastest running bird of all time** Lived in Oceania during the Cenozoic Era
Neognaths									
Roadrunner (*Geoccocyx californianus*)	MVZ 176050	210	2.73	0.3	12	3	**27.2**	7.5	-
Psilopterus lemoinei	-	555	2.96	15	7	3	**42**	11.7	-
Mesembriornis milneedwardsi	MMP-S1551	1110	3.01	106	4.5	2.8	**50.3**	14	-
Paraphysornis brasiliensis	DGM-1418-R	1215	2.47	240	4.5	2.8	**55.1**	15.3	-
Patagornis marshi	BMNH-A516	906	2.99	38	6	3	**58.7**	16.3	-
Anomalopteryx didiformis	-	914	2.32	40	6	3	**59.2**	16.5	Fastest South American running bird
Ibandornis lawsoni	-	1340	3.12	110	4.5	2.8	**60.8**	16.9	-
Ilbandornis woodburnei	-	1220	3.21	80	5	3	**65.9**	18.3	-
Dromornis stirtorni	-	1765	2.76	460	4	2.6	**66.1**	18.4	-
Other present-day animals									
Human	-	950	0.94	75	4.55	2.55	**38.8**	10.8	-

Slowest herbivorous theropod
Chilesaurus diegosuarezi
Specimen: SNGM-1935
Among all herbivorous theropods, *Chilesaurus* was the slowest.
See more: *Novas et al. 2015*

Fastest juvenile theropod
Gorgosaurus libratus
Specimen: ROM 1247
The juveniles of this large predator were so fast that they may have been about 50% faster than adults.
See more: *Russell 1970*

MESOZOIC MICROFLIERS

How small were the flying dinosaurs of the Mesozoic? We know that trochilids (hummingbirds) are incredibly tiny, so small that they can be mistaken for bumblebees or beetles. *Cratoavis cearensis*, the smallest of all enantiornithean birds, did not achieve this extreme specialization. It must have weighed as little as the smallest songbird of today: the short-tailed pygmy tyrant (*Myiornis ecaudatus*). The adult weighs only 4.2 g.

If we compare the size of this bird with that of other Mesozoic flying beings, we can see that, as of now, there are insects that can overcome it by far. (A goliath beetle weighs more than fifty bee hummingbirds, *Mellisuga helenae*.) See more: *Guinness 1995; Perrins 2006; Carvalho et al. 2015a, 2015b*

Apparently, there are many animals that can "fly." Among them are fish, amphibians, lizards, snakes, marsupials, rodents, and a large number of other animals that take advantage of gravity and the wind to move through the air. Insects, birds, and bats, however, are the only living animals capable of active flight. They can ascend or descend freely, change direction, and perform complex maneuvers. This group is joined by the extinct pterosaurs. See more: *Fichter 1972; Wellnhofer 1991*

Parvavis chuxiongensis
Specimen: **IVPP V18586**
Wingspan: ~19 cm
Length: 7.5 cm
Weight: 6 g

RECORD:
The smallest northern dinosaur
of the Mesozoic.
See more: *Wang et al. 2014*

1:2

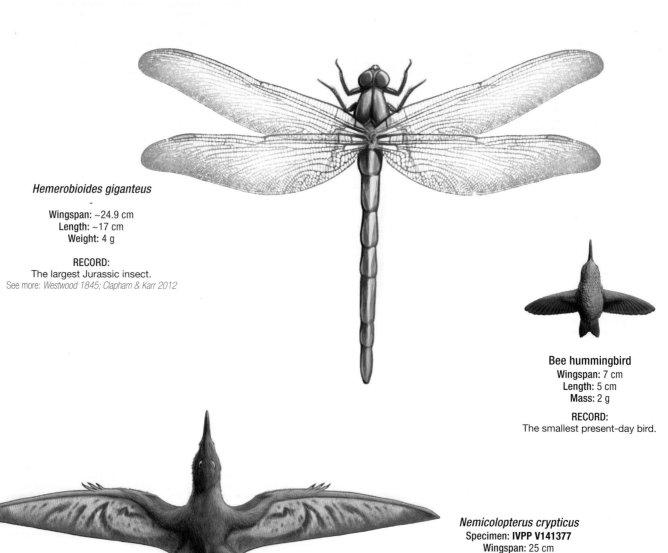

Hemerobioides giganteus
-
Wingspan: ~24.9 cm
Length: ~17 cm
Weight: 4 g

RECORD:
The largest Jurassic insect.
See more: *Westwood 1845; Clapham & Karr 2012*

Bee hummingbird
Wingspan: 7 cm
Length: 5 cm
Mass: 2 g

RECORD:
The smallest present-day bird.

Nemicolopterus crypticus
Specimen: **IVPP V141377**
Wingspan: 25 cm
Length: 9 cm

RECORD:
The smallest pterosaur.
See more: *Wang et al. 2008*

Prohoyaeshna milleri
Specimen: **MNEMG 1996.220**
Wingspan: ~19.8 cm
Length: ~12.5 cm
Weight: -

RECORD:
The largest insect of the Cretaceous.
See more: *Bechly et al. 2001; Clapham & Karr 2012*

Cratoavis cearensis
Specimen: **UFRJ-DG 031 Av**
Wingspan: ~10 cm
Length: 6.6 cm
Weight: 4 g

RECORD:
Smallest dinosaur of the Mesozoic.
See more: *Carvalho et al. 2015*

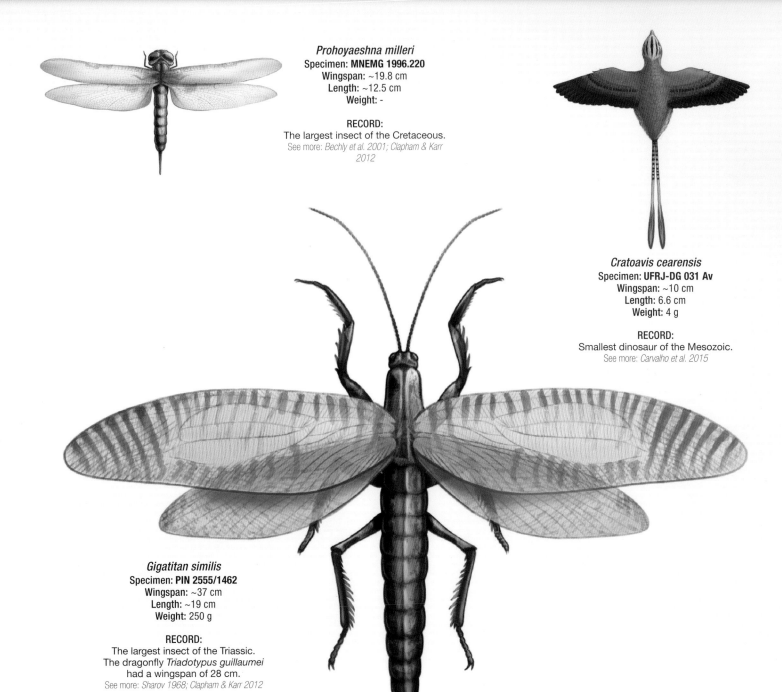

Gigatitan similis
Specimen: **PIN 2555/1462**
Wingspan: ~37 cm
Length: ~19 cm
Weight: 250 g

RECORD:
The largest insect of the Triassic.
The dragonfly *Triadotypus guillaumei*
had a wingspan of 28 cm.
See more: *Sharov 1968; Clapham & Karr 2012*

Kalligramma haeckeli
Specimen: **BSPGM Munich**
Wingspan: 24 cm
Length: 7 cm
Weight: -
See more: *Walther 1904*

Curiosity: Eyes of Sauron

Two theories attempt to explain why spots in the form of "eyes," also known as ocelli, are present in many lepidopterans. The first is that they frighten predators, as can be clearly seen in the owl butterfly (*Caligo idomeneus*). Other experts suggest that the spots attract attention, since very attractive prey are usually toxic. In spite of their appearance, kalligrammatids were not butterflies; they were insects similar to lacewings, ant lions, and other neuropterans, which may have imitated the eyes of predators of their time: the dinosaurs. See more: *Stevens 2005; Stevens et al. 2008*

The largest flying animals of all time were the pterosaurs, which weighed over 250 kg. In contrast, the largest birds of the Mesozoic were thirty-four times lighter. See more: *Paul 2002*

1:10

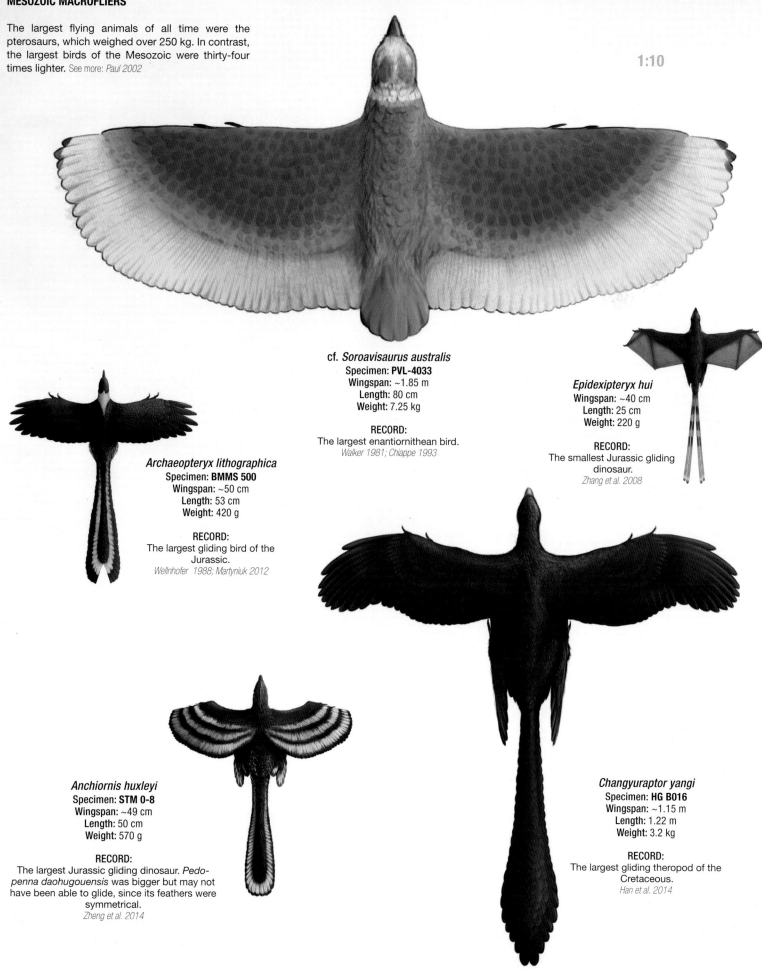

cf. *Soroavisaurus australis*
Specimen: **PVL-4033**
Wingspan: ~1.85 m
Length: 80 cm
Weight: 7.25 kg

RECORD:
The largest enantiornithean bird.
Walker 1981; Chiappe 1993

Epidexipteryx hui
Wingspan: ~40 cm
Length: 25 cm
Weight: 220 g

RECORD:
The smallest Jurassic gliding dinosaur.
Zhang et al. 2008

Archaeopteryx lithographica
Specimen: **BMMS 500**
Wingspan: ~50 cm
Length: 53 cm
Weight: 420 g

RECORD:
The largest gliding bird of the Jurassic.
Wellnhofer 1988; Martyniuk 2012

Anchiornis huxleyi
Specimen: **STM 0-8**
Wingspan: ~49 cm
Length: 50 cm
Weight: 570 g

RECORD:
The largest Jurassic gliding dinosaur. *Pedopenna daohugouensis* was bigger but may not have been able to glide, since its feathers were symmetrical.
Zheng et al. 2014

Changyuraptor yangi
Specimen: **HG B016**
Wingspan: ~1.15 m
Length: 1.22 m
Weight: 3.2 kg

RECORD:
The largest gliding theropod of the Cretaceous.
Han et al. 2014

1858—The first discovered flying and aquatic bird
Bones were discovered that six years later were called "Palaeo-colyntus barretti" and "Pelargornis sedgwicki." These names were invalid until the bones were formally described and illustrated in 1876. See more: *Seeley 1864, 1876; Galton & Martin 2002*

1872—Mesozoic birds best adapted to aquatic life
The legs of Hesperornithiformes were very similar to those of modern aquatic birds. They had very long and asymmetrical fingers and may have been covered with a membrane that would serve as a fin. They lacked forearms, which reduced water resistance, and their atrophied arms were long, which may have helped them maintain balance or swim. Their teeth, being located longitudinally, were very similar to those of the mosasaurs. Some species are found only in sediments that were part of the sea. See more: *Marsh 1872; Gregory 1951; Martin & Cordes-Person 2007*

1910—Scotland—The oldest dinosauromorph in resting position
Saltopus elginensis was found preserved in a resting position similar to that adopted by warm-blooded animals. It dates from the Upper Triassic (Carnian, ca. 237–227 Ma) See more: *Huene 1910*

1933—Arizona, USA—The oldest theropod in resting position
Segisaurus halli se encontrówas found protecting itself from a sandstorm. It dates from the Lower Jurassic (Norian, ca. 227–208.5 Ma). See more: *Anonymous 1933; Camp 1936*

1981—Argentina—The largest flying theropods
PVL-4033, attributed to cf. *Soroavisaurus australis*, is based on a tibiotarsus of an enantiornithean bird, whose owner must have been 80 cm long and weighed 7.25 kg. An even larger bird was *Bauxitornis mindszentyae*, with a tibiotarsus that could belong to an enantiornithean bird (88.4 cm long and weighing 9.5 kg) or to a nonflying animal similar to Balaur bondoc, which is more likely. See more: *Walker 1981; Chiappe 1993; Walker & Dyke 2009; Dyke & Osi 2010; Mortimer**

1992—Argentina—The smallest land birds of the Mesozoic
Patagopteryx deferrasii was a nonflying bird that was 40 cm long and weighed 2 kg, making it somewhat larger than a tinamou. The specimens of *Alamitornis minutus* were much smaller, only 6.3 cm long and weighing 8 g. It is possible that they were all juveniles. See more: *Alvarenga & Bonaparte 1992; Agnolin & Martinelli 2009*

1993—Sweden and Russia—The largest aquatic Mesozoic bird
Among aquatic Mesozoic birds, the largest is *Hesperornis rossicus*, which was 1.6 m long and weighed 30 kg. See more: *Nesov & Yarkov 1993, Rees & Lindgren 2005*

1995—Mongolia—The most recent theropod in resting position.
Some specimens of *Citipati osmolskae* have been found in nesting positions, protecting their eggs with their arms like modern birds. They date from the late Late Cretaceous (Campanian, ca. 83.6–72.1 Ma). See more: *Clark 1995; Norell et al. 1995; Clark, Norell & Chiappe 1999; Clark, Norell & Barsbold 2001*

2000—USA—The largest land bird in the Mesozoic
Among Mesozoic birds, *Gargantuavis philoinos* was gigantic compared to its close relatives. It had an imposing height of about 1.8 m and a weight of 120 kg. See more: *Buffetaut & Le Loeuff 1998*

2000—Japan—The oldest flying bird
The enantiornithean bird SBEI 307, from Japan, is the oldest. It dates from the early Lower Cretaceous (Valanginian, ca. 139.8–132.9 Ma), although the fingerprints of *Pullornipes aureus*, from China, are even older: Upper Jurassic (Tithonian, ca. 152.1–145 Ma). See more: *Unwin & Matsuoka 2000, Lockley et al. 2006*

2000—China—The oldest gliding theropods
The paravian theropods *Epidexipteryx hui* and possibly *Pedopenna daohugouensis* were gliders. The former had some membranes, and the latter had "rear wings" with symmetrical feathers (unlike *Microraptor*), so its gliding would have been very rudimentary. It dates from the Middle Jurassic (Callovian, ca. 166.1–163.5 Ma). See more: *Xu & Zhang 2005*

2000—USA—The least cursorial theropod
Acrocanthosaurus atokensis (OMNH 10168) had a very long femur in relation to the sum of the tibia and metatarsal, with a stride ratio of 1.1. It is possible that its prey were not very fast animals, such as small or juvenile sauropods. See more: *Currie & Carpenter 2000*

2001—Mongolia—The most cursorial theropod
Parvicursor sp. IGM 100/120 had a very short femur in relation to the sum of the tibia and metatarsal, with a stride ratio of 2.58. It must have been very fast for its tiny size. See more: *Suzuki et al. 2001, 2002*

2007—Brazil—The oldest saurischian in a resting position
Tucking the feet into the body serves to conserve heat. The oldest dinosaur found in this pose is *Guaibasaurus candelariensis* UFRGS PV 0725T (Norian, ca. 227–208.5 Ma). See more: *Bonaparte et al. 2007*

2008—China The smallest gliding theropod of the Mesozoic
The paravian *Epidexipteryx hui* could have been a glider, as there is evidence of the remnants of membranous wings similar to those of *Yi qi*. Not counting its long caudal feathers, it was about 30 cm long and weighed about 220 g. See more: *Zhang et al. 2008, Carvalho et al. 2015; Xu et al. 2015*

2011—Western Russia—The smallest Mesozoic waterbird
The enantiornithean bird *Mystiornis cyrili* had very short and flat metatarsals II, a typical feature of diving waterfowl. It was around 25 cm long and weighed 165 g. See more: *Kurochkin et al. 2011*

2011—Morocco—The largest aquatic theropod
Spinosaurus aegyptiacus, which was 16 m long and weighed 7.5 t, had very short legs. It moved better in water than on land. See more: *Dal Sasso et al. 2005; Ibrahim et al. 2014*

2014—Non-avian theropods best adapted to aquatic life
The most advanced spinosaurids were so adapted to aquatic life that, like crocodiles, they had sensory structures that made it possible to detect movement in water. They also had short hind legs and solid bones with a reduced medullary cavity, so they were better able to sink in water. It is possible that they spent most of their time in water. See more: *Ibrahim et al. 2014*

2014—China—The largest theropod glider
The largest may be *Changyuraptor yangi*, which was 1.25 m long and weighed 3.2 kg. Other older microraptors, such as *Zhenyuanlong* or *Tianyuraptor*, had short wings, so it is thought they were terrestrial. See more: *Zheng et al. 2010; Han et al. 2014; Lu & Brusatte 2015*

2015—China—The smallest terrestrial theropod
The smallest non-avian theropod was IVPP parasite V22530, which was 46 cm long and weighed 175 g. See more: *Pittman et al. 2015*

2015—Brazil—The smallest flying bird of the Mesozoic
Cratoavis cearensis, which was 6.6 cm long and weighed 4 g, was smaller than most modern birds. It was apparently an almost fully grown individual. See more: *Carvalho et al. 2015*

1:2

Mongolarachne jurassica
Specimen: CNU-ARA-NN2010008
Body length: 2.46 cm
Total length: 15 cm
The largest spider of the Jurassic was similar to the golden silk orb-weaver (*Nephila*), which weaves very tough, large cobwebs in which it catches large flying insects and birds. It may have had a similar way of life, hunting even small theropod gliders. See more: *Selden, Shih & Ren 2011, 2013*

1:10

Xiaotingia zhengi
Specimen: STM 27-2
Length: 50 cm
Weight: 530 g
No primitive glider achieved controlled flight, because active flight requires a broad sternum with a keel to support the strong muscles needed to stay in the air, an alula to control landings, and other structures. The discovery of *Xiaotingia* has complicated the concept of *bird*, which currently has different definitions. See more: *Xu et al. 2006, 2009*

SUPERBITES

Theropods are among the dinosaurs with the most powerful bites. Crocodiles are today's most powerful biters, followed by hippos, hyenas, turtles, sharks, felines, and others.

To calculate bite force, numerous studies have been conducted on both current and extinct animals. Different methods have been used, ranging from mechanical devices, such as gnatodynamometers, to engineering techniques such as Finite Element Analysis using 3-D technology.

The results obtained so far differ markedly, due to the fact that measurements are usually obtained from different positions in the mouth, making results often difficult to compare. The measurements are presented in units of force: newtons (N). These units can easily be converted to kilogram-force by dividing the result by 9.8.

Most common measurements:

- **Unilateral:** On one side of the mouth, usually covering two or more teeth on the back.
- **Bilateral:** On both sides of the mouth at the same time. This measurement usually gives the highest results, between 30% and 100% higher than those obtained unilaterally.
- **Tooth:** On a single tooth, usually posterior or an isolated tooth.
- **In the canines:** On the fangs of carnivores. It is usually applied to felines.
- **Tip of the beak:** On the front of the beak. Applied to birds.

57,158 N / **5,832 Kg**
Measure: Posterior tooth

10,2550 N / **10,464 Kg**
Measure: Unilateral

The theropod with the most powerful bite

Tyrannosaurus rex had a jaw designed to break bones, more so than any other theropod. The results calculated by different experts vary from between 13,400 and 57,158 N in a single posterior tooth up to the incredible figure of 235,000 N bilaterally. Not all scientists agree with this data. See more: *Erickson et al. 1996; Meers 2002; Bates & Falkingham 2012*

The most powerful of the Mesozoic

Deinosuchus rugosus, which weighed about 8 t and was 11 m long, was one of the largest crocodiles that inhabited Earth. Its huge 1.8 m long jaws could exert a pressure ten times higher than most of its current relatives. See more: *Schwimmer 2002; Erickson, et al. 2012*

134 N / **14 Kg**
Measure: Posterior tooth

1,3172 N / **1,344 Kg**
Measure: Unilateral

The weakest bite calculated for a theropod

Erlikosaurus andrewsi possessed the weakest bite of all measured theropods. The low strength of their jaws indicates that their diet was vegetarian, based mainly on soft buds and leaves. The cranial morphology of the *Erlikosaurus* suggests that bite force decreased in the more advanced tericinosaurs. See more: *Lautenschlager 2012*

The most powerful bite

Crocodiles have more powerful bites than any other terrestrial or aquatic animal today. The most outstanding species are the American alligator and the huge saltwater crocodile. An alligator weighing just 240 kg produced a bite force above 13,000 N. On the other hand, a 4.6 m long saltwater crocodile weighing 530 kg exerted a force of 16,400 N. The largest measured specimen was 6.7 m long. It is estimated to have a bite force of over 34,000 N. See more: *Erickson et al. 2003, 2012*

8,200 N / **837 Kg**
Measure: Unilateral

Raptors: Even more dangerous

Deinonychus bite marks on *Tenontosaurus* bones reveal that these raptors possessed a bite force comparable to that of modern crocodiles of similar size, although its relative, *Dromaeosaurus*, may have had even greater jaw strength. See more: *Therrien et al. 2005; Gignac et al. 2010*

+310 N / **+32 Kg**
Measure: Tip of the beak

The tiny big biter

The tiny hawfinch is 18 cm long and weighs 50 g and is capable of breaking cherry seeds with its beak. This requires a bite force between 310 and 700 N, while most birds of a similar size produce barely 10 N of bite force. It may have a more powerful bite, in relation to its size, than any other vertebrate. For its part, a parrot can bite with a force of 755 N (77 kg). See more: *Sims 1955; Asimov 1979; Van der Meij 2004*

4,500 N / **459 Kg**
Measure: Unilateral

The most powerful bite (current terrestrial carnivore)

Although the results obtained unilaterally on spotted hyenas (*Crocuta crocuta*) show a bite force of up to 4,500 N, it is believed that they can exert more than 9,000 N bilaterally. This is the force necessary to break the femurs of hippos and giraffes, observed behavior in these animals. See more: *Binder & van Valkenburgh 2000; Mason 2002*

1,185 N / **121 Kg**
Measure: Unilateral

Humans are not so weak

Most studies indicate that humans have powerful bites that vary between 965 and 1,500 N, figures comparable to those obtained in wolves, but the record belongs to a man descended from the Inuit who raised the figure to 4,437 N! See more: *Pruim et al. 1980; van Eijden 1991; Guinness Record 1992; Zhao & Ye 1994*

3,441 N / **341 Kg**
Measure: Bilateral

MYTH: *Carnotaurus* was a super predator

For a long time, *Carnotaurus* was considered a hunter of large prey, but a study revealed that its bite was too weak for it to prey on large animals. See more: *Mazzeta et al. 2009*

Bite force

Theropods

Genus and species	Force (N)	Force (Kg)	Bite type	Mass (kg)	Reference
Albertosaurus sarcophagus	3413	348	Back teeth	2500	Bates & Falkingham 2012
Allosaurus fragilis	8724	890	Back teeth	1200	Bates & Falkingham 2012
Sinocalliopteryx sp.	8880	906	Back teeth	100	Xing et al. 2012
Giganotosaurus carolinii	13258	1353	Back teeth	7000	Mazzetta et al. 2004
Gorgosaurus libratus	6053	618	Isolated tooth	~2500	Bell & Currie 2010

Other extinct animals

Genus and species	Force (N)	Force (Kg)	Bite type	Mass (kg)	Reference
Dunkleosteus terrelli	7400	755	Bilateral	~1000	Philip et al. 2009
Kronosaurus queenslandicus	27716	2828	Bilateral	6000	McHenry 2009
Pliosaurus kevani	48278	4926	Bilateral	6000	Foffa et al. 2014
Homotherium serum	787	80	Canines	187	Christiansen 2007
Smilodon fatalis	1926	197	Canines	218	Christiansen 2007
Smilodon populator	2258	230	Canines	278	Christiansen 2007
Paranthropus boisei	3471	354	Unilateral	50	Constantino et al. 2010
Australopithecus africanus	2598	265	Unilateral	40	Constantino et al. 2010
Homo erectus	2075	212	Unilateral	80	Constantino et al. 2010

Present-day animals

Genus and species	Force (N)	Force (Kg)	Bite type	Mass (kg)	Reference
Nile crocodile	3172-22000	324-2245	Unilateral	87 ~ 500	Erickson et al. 2012; National Geographic
Gavial crocodile	2006	205	Unilateral	121	Erickson et al. 2012
Komodo dragon	149	15	Unilateral?	70	Domenic et al. 2011
Gorilla	2865	292	Unilateral	150	Constantino et al. 2010
Orangutan	3424	349	Unilateral	57	Lucas et al. 1994
Lion	4168	425	Bilateral	163	Thomason 1991
Tiger	3007	307	Unilateral	200	Christiansen & Adolfssen 2005
Hippopotamus	8100	827	Unilateral	~1300	National Geographic
Labrador retriever	1100	112	Unilateral	31	Strom & Holm 1992
Wolf	1412	144	Bilateral	32	Thomason 1991
Polar bear	2404	245	Unilateral	400	Christiansen & Adolfssen 2005
Peccary	2280	233	Unilateral	40	Constantino et al. 2010
Great white shark	4577	467	Bilateral	423	Wroe et al. 2008

THE INTELLIGENCE OF THEROPODS

Intelligence is difficult to measure in humans, not to mention animals, and even more so in extinct species. In 1973, Jerison invented the *encephalization quotient* or EQ method to measure the "cognitive abilities" of animals. The EQ is the ratio between the mass of the real brain and the mass of the brain that is predicted for an animal of a given size. In this system, an EQ with a value equal to 1.0 is considered normal or pro-medium, so a higher or lower value could be identified as more or less "intelligent" than the average. Humans, for example, have an EQ of around 7.0, while some dolphins can have an EQ of up to 4.0, chimps have an EQ close to 2.5, elephants above 2.0, and dogs 1.2.

Grant Hurlburt (1996) perfected the Jerison method by proposing three new equations, one each for mammals (MEQ), birds (BEQ), and "reptiles" (REQ). The sparrow has a BEQ close to 1.35, the pigeon has a BEQ around 0.85, and the ostrich has a BEQ of 0.35, while the crow, one of the smartest birds, has a BEQ greater than 2.5. Possessing a REQ around 1.0, the iguana would be an example of medium "intelligence." In this last group, there are no great variations; all reptiles have similar cognitive abilities.

Today, we estimate dinosaur "intelligence" by measuring endocranial cavities. So if you want to estimate the "cognitive abilities" of dinosaurs, you must first make a mold of the endocranium and then measure its volume. To estimate the actual mass that the brain may have had, a percentage is usually applied to the volume of the intracranial cavity, since the space occupied by the dura mater and endocranial fluids can represent a very high percentage. In the case of "reptiles," the brain represents about 37% of the total volume of the endocranium. In birds and most mammals, the brain occupies practically the entire cavity, so a percentage of 95% is recommended for these groups. This means that the applied percentages will be different depending on the type of dinosaur. To obtain the approximate mass of the brain in the least-derived dinosaurs,* a factor of 0.37 should be applied. For non-avian dinosaurs with feathers,** however, the recommended factor is 0.95

* Example 1: Primitive saurischians, non-celurosaurs theropods, basal coelurosaurs (tyrannosaurids and the like).

** Example 2: ornithomimosaurs, alvarezsaurs, and maniraptors.

Animal intelligence

Theropods

Genus and species (individual)	Endocranial volume (ml)	Brain mass (g)	Body mass (kg)	REQ	BEQ	Endocranial volume reference
Acrocanthosaurus atokensis (OMNH 10146)	190	70.3	2000	1.49	0.12	Franzosa 2004
Allosaurus fragillis (UUVP 294)	188	69.6	2000	1.47	0.11	Franzosa 2004
Allosaurus fragillis (UUVP 5961)	169	62.5	2000	1.32	0.10	Franzosa 2004
Archaeopteryx lithographica (BMNH 37001)	1.6	1.5	0.31	4.11	0.44	Hurlburt et al 2013
Bambiraptor feinbergi (KUPV 129737)	14	13.3	2.75	10.75	1.06	Hurlburt et al 2013
Byronosaurus jaffei (IGM 100/983)	4.6	4.4	7.2	2.08	0.20	Franzosa 2004
Carcharodontosaurus saharicus (SGM-Din 1)	264	97.7	7800	0.97	0.07	Hurlburt et al 2013
Ceratosaurus magnicornis (MWC 1)	88	32.6	560	1.39	0.11	Franzosa 2004
Citipati osmolskae (IGM 100/978)	25	23.8	50	3.86	0.34	Franzosa 2004
Dromiceiomimus brevitertius (CMN 12228)	88	83.6	140	7.69	0.66	Franzosa 2004
Giganotosaurus carolinii (MUCPV-CH 1)	250	92.5	7000	0.98	0.07	Paulina 2012
"Nanotyrannus lancensis" (ROM 1247)	129	47.7	540	2.08	0.169	Hurlburt et al 2013
"Nanotyrannus lancensis" (CMN 7541)	111	41.1	295	2.5	0.208	Hurlburt et al 2013
Majungatholus atopus (FMNH PR. 2100)	168	62.2	465	2.94	0.24	Franzosa 2004
Sinraptor dongi (IVPP 10600)	95	35.2	1400	0.91	0.07	Paulina 2012
Troodon formosus (RTMP 86.36.457)	41	39	20	10.51	0.97	Franzosa 2004
Tyrannosaurus rex (AMNH 5117)	314	116.2	6500	1.28	0.095	Hurlburt et al 2013
Tyrannosaurus rex (FMNH PR2081)	414	153.2	8265	1.48	0.109	Hurlburt et al 2013
Tyrannosaurus rex (AMNH 5029)	382	141.3	6500	1.56	0.115	Hurlburt et al 2013

Present-day birds

Mallard duck (TMM collection)	6.8	6.5	0.72	-	1.14	Franzosa 2004
Masai ostrich	-	42.1	123	-	0.36	Crile et al. 1940
Indian bustard	-	12.9	5.5	-	0.68	Crile et al. 1940
Ground hornbill	-	26.3	2.15	-	2.43	Crile et al. 1940
Eurasian eagle-owl	-	13.7	1.18	-	1.8	Crile et al. 1940
Northern screamer (KU 81969)	8	7.6	2.2	-	0.69	Franzosa 2004
Crow	-	9.3	0.337	-	2.56	Crile et al. 1940
Red junglefowl	-	3.55	2.2	-	0.32	Crile et al. 1940
Sparrow	-	1.03	0.024	-	1.35	Crile et al. 1940
Kiwi bird (ANMH 18456)	9.4	8.9	0.88	-	1.4	Franzosa 2004
Rock dove	-	2.7	0.282	-	0.83	Crile et al. 1940
Thicket tinamou (KU 34658)	-	1.4	0.44	-	0.34	Franzosa 2004

Present-day "reptiles"

American alligator (ROM 8328)	27	10	238	0.69	-	Hurlburt et al. 2013
American alligator (ROM 8333)	33	12.2	277	0.77	-	Hurlburt et al. 2013
American alligator	-	14.4	205	1.05	-	Crile et al. 1940
Common boa constrictor	-	0.44	1.83	0.45	-	Crile et al. 1940
American crocodile	-	15.6	134	1.47	-	Crile et al. 1940
Iguana	-	1.44	4.2	0.92	-	Crile et al. 1940
Python	-	1.13	6.1	0.59	-	Crile et al. 1940
Green sea turtle	-	8.6	114	0.89	-	Crile et al. 1940

BRAINS AND THE SENSES IN THEROPODS AND MESOZOIC BIRDS

1928—Canada—The theropod with the largest eyes in relation to its size

Dromiceiomimus samueli had the largest eyes in relation to its size among dinosaurs, although they do not surpass those of a present-day ostrich (*Struthio camelus*). See more: *Park 1928*

1982—USA, Canada—The theropod with binocular vision

Troodon eyes face forward, so it had a great ability to perceive the depth and distance of objects. Other theropods that also had this capacity, though less developed, are *Bambiraptor, Carnotaurus, Tyrannosaurus*, and other troodontids. See more: *Russell & Séguin 1982; Jaffe 2006; Stevens 2006; Persons & Currie 2011*

1996—Canada—Theropods with the highest "cognitive abilities" (data updated in the table on the previous page)

Calculations of the brain ratios of dinosaurs have been made. The highest values are usually obtained by ornithomimosaurs, such as *Dromiceiomimus brevitertius* and *Troodon formosus* (REQ = 8.6 and REQ = 7.1, respectively) . See more: *Russell 1972; Hurlburt 1996; Franzosa 2004; Evans 2005*

1996—Montana, USA—The young theropod with the largest proportional brain (data updated in the table on the previous page)

Young dinosaurs had higher brain proportions than adults. The brain of *Bambiraptor feinbergi* AMNH 30556 is far superior to that of an adult. It had a REQ of 13.1. See more: *Hurlburt 1996; Wharton 2002; Burnham 2004; Franzosa 2004*

1996—USA—The largest theropod brain of the Jurassic

Allosaurus fragilis UUVP 3304, which has a 18.2 cm long brain. Unfortunately, as it is only a fragment, it has not been possible to estimate the animal's full size. See more: *Chure & Madsen 1996; Wharton 2002*

2002—USA—The theropod with the smallest eyes in relation to its size

The orbit in *Tyrannosaurus rex* was only 4.1% of the size of its skull. This suggests that it would have had proportionally very small eyes. In young individuals, this ratio rises to 8.35%. See more: *Henderson 2002*

2002—Argentina—The largest primitive saurischian brain of the Triassic

The brain of the juvenile *Herrerasaurus ischigualastensis* PVSJ 407 was approximately 10 cm long. See more: *Sereno & Novas 1992; Wharton 2002*

2007—Theropod vision

Birds' eyes excel at detecting fast and very slow movements. Their vision is 70% more efficient than human eyes. It is possible that some theropod dinosaurs had similar characteristics. See more: *Jones et al. 2007; Sampson & Witmer 2007*

2007—Madagascar—The theropod with the least motor coordination

According to a study of the region of the cerebellum that coordinates the movements of the ocular and locomotor muscles, *Majungasaurus crenatissimus* would have had little capacity to capture fast and small prey and may have hunted medium or large animals See more: *Sampson & Witmer 2007*

2011—Mongolia—Theropods with the most developed sense of smell

According to a study of current birds based on the olfactory bulb, *Tsaagan mangas* had a better sense of smell than the vast majority of theropods. It was similar to the turkey vulture (*Cathartes aura*), which has the best sense of smell of modern birds. See more: *Norell et al. 2006; Zelenitsky et al. 2011*

2011—Argentina—The most developed sense of smell of the Mesozoic

Lithornithid birds had one of the best senses of smell among extinct birds. They are known from the Paleocene, although it is possible that *Limenavis patagonica*, a bird of the Upper Cretaceous, was related to them See more: *Clarke & Chiappe 2001; Zelenitsky et al. 2011; Mortimer**

2011—USA—The largest theropod brain of the Cretaceous

The brain of *Tyrannosaurus rex* specimen FMNH PR2081 had a volume of 414 cc, suggesting that the brain weighed about 153.2 g. The human brain has an average volume of 1,400 cc. See more: *Osborn 1905; Hans et al. 2000; Larsson 2001; Cosgrove et al. 2007; Hurlburt et al. 2013*

2011—Mongolia—Theropods with the most developed hearing

The acoustic lobes of the alvarezsaurid *Ceratonykus oculatus* were exceptionally large, so it may have had a very developed ear. Troodontids had asymmetrical ears, just like present-day owls; one was positioned higher than the other. This characteristic suggests that they used their hearing to locate and capture hidden prey. See more: *Mateus 2006; Alifanov & Barsbold 2009; Alifanov & Saveliev 2011*

2011—Canada—The theropod with the least developed sense of smell

Dromiceiomimus brevitertius would have had the worst sense of smell among dinosaurs. This is similar to the ostrich (*Struthio camelus*), whose sense of smell is the worst among most of today's birds. See more: *Parks 1926; Zelenitsky et al. 2011*

2011—USA—The Mesozoic bird with the least developed sense of smell

The sense of smell of *Hesperornis regalis* was so reduced that it was the worst of all non-avian theropods. It was very similar to the Adélie penguin (*Pygoscelis adeliae*). Both birds are aquatic. See more: *Marsh 1872; Zelenitsky et al. 2011*

2011—Morocco—The theropod with the worst "cognitive abilities"

The *Carcharodontosaurus saharicus* SGM-Din 1 specimen has a REQ of 2.3. Its brain was 26 cm long and weighed 224.4 g, and it had a body mass of 7.4 t. Considering that its body would be somewhat larger (7.8 t) and that the real brain mass accounts for about 37% of the volume of the endocranium in nonderived and non-avian theropods, the REQ should have been only 0.83, a smaller proportion than that of the desert monitor (*Varanus griseus*) but larger than that of the American crocodile (*Crocodylus acutus*). See more: *Hurlburt 1996; Larson 2001; Franzosa 2004*

2014—Mongolia—The theropod with the largest eyes

The eyes of *Deinocheirus mirifcus* were 8 cm in diameter, about 3.3 times the size of human eyes. It is possible that cf. *Troodon formosus* AK498-V-001 from Alaska will overtake it, but it is too fragmentary to know. Among the modern birds, the ostrich (*Struthio camelus*) has the largest eyes, 5 cm in diameter. See more: *Güntürkün 1998; Fiorillo 2008; Schmitz 2009; Lee et al. 2014*

The theropod with the largest nose
Monolophosaurus jiangi
Specimen: IVPP 84019
It's likely it had a good sense of smell because it had a very developed nose. It also had a deep groove that passed under the nostril, a unique feature whose function is unknown.
See more: *Currie & Zhao 1994; Brusatte et al. 2010*

5 cm

10 cm

15 cm

20 cm

25 cm

Vervain hummingbird
Mellisuga minima
Khanna 2005
1 x 0.8 cm

cf. *Martinavis*
Dyke et al. 2012
4 x 2.5 cm

1:1

**Enantiornithean bird
IGM 100/1027**
Grellet-Tinner & Norell 2002
2.58 x 1.58 cm

Jinfengopteryx elegans
Ji et al. 2005
1 x 0.7 cm

LPRP-USP 0359
Marsola et al. 2014
31.4 x 1.95 cm

House sparrow
Passer domesticus
BTO Bird facts
2.27 x 1.55 cm

20 cm

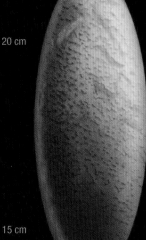

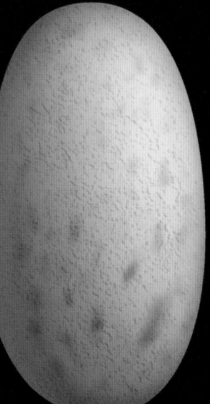

Sankofa pyrenaica
UM1
López-Martínez
& Vicens 2012
7 x 4 cm

Neuquenornis volans
Schweitzer et al. 2002
MUCPv-284
4.7 x 2.9 cm

15 cm

*Triprismatoolithus
stephesi*
ES 101
Jackson & Varricchio 2010
7 x 3 cm

Confuciusornis sanctus
B072
Kaiser 2007
2.5 x 1.7 cm

Preprismatoolithus coloradensis
MWC 122.3.1/HEC 457
11 x 6 cm

**Enantiornithean bird
STM29-8**
Ji et al. 2005
0.58 cm

10 cm

Sinosauropteryx prima
NIGP 127587
Chen et al. 1998
3.6 x 2.6 cm

5 cm

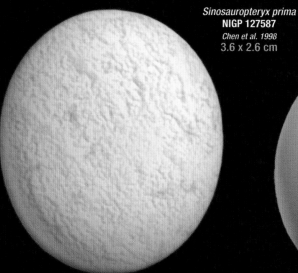

Rock pigeon
Columba livia
Ibrahim & Sani 2010
4 x 2.9 cm

Indeterminate theropod
Campos & Bertini 1985
9.1 x 5.8 cm

Dendroolithus microporosus
IVPP V 16857.1
Mikhailov 1994
7 x 6 cm

Domestic chicken
Gallus domesticus
UNECE 2010
5.9 x 4.4 cm

Most non-avian theropod eggs were cylindrical like those of crocodiles, spherical like those of turtles, or ovoid like those of chickens. Mesozoic bird eggs were ovoid or cylindrical.

THE SMALLEST

Late Upper Cretaceous (IUC) Mongolia—The smallest Mesozoic bird egg in Asia (Paleoasia)
The enantiornithean bird egg IGM 100/1027 was only 2.58 x 1.59 cm. See more: *Grellet-Tinner & Norell 2002*

IUC China—The smallest unlaid egg from Asia (Paleoasia)
Tiny eggs were found developing inside a *Sinosauropteryx prima* (GMV 2123 or NIGP 127586). See more: *Chen et al. 1998*

IUC Romania—The smallest Mesozoic bird egg in Europe
The enantiornithean nesting colony (possibly similar to *Martinavis*) had several 4 x 2.5 cm eggs. See more: *Dyke et al. 2012*

IUC Spain—The smallest theropod egg in Europe
Sankofa pirenaica eggs were very similar in shape to those of modern birds, but they had a two-layered shell, similar to non-avian theropods. The eggs measure 7 x 4 cm. See more: *López–Martínez & Vicens 2012*

IUC Brazil—The smallest Mesozoic bird egg in South America
A 3.14 x 1.95 cm egg found in northern South America was the first in this area. See more: *Marsola et al. 2014*

IUC Brazil—The smallest theropod egg in South America
A 9.1 x 5.8 cm egg was considered an ornithischian dinosaur until a study of its shell showed that it was from a theropod. See more: *Campos & Bertini 1985; Kellner, Campos, Azevedo & Carvalho 1998*

IUC Montana, USA—The smallest theropod egg in North America
The egg of the alvarezsaurid *Triprismatoolithus stephesi*, belonged to one of the smallest theropods of North America. It measures 7 x 3 cm. See more: *Jackson & Varricchio 2010*

IUC India—The largest theropod egg from Hindustan
The eggs of *Subtiliolithus kachchhensis* are known only from pieces of shells. They are 11% thicker than those of *S. microtuberculatus* (present-day Mongolia), which measures 72 x 30 mm and are similar. There is no correlation of thickness and volume in dinosaur eggs, so it is impossible to estimate their size from this data. See more: *Loyal et al. 2012*

IUC Mongolia—The smallest spherical theropod egg
The smallest therizinosaur egg is *Dendroolithus microporosus*, which measures 7 x 6 cm in diameter. It is spherical, unlike the subcylindrical form common in other theropods. See more: *Mikhailov 1994*

Upper Jurassic (UJ) Colorado, USA—The smallest theropod egg of the Jurassic
The *Preprismatoolithus coloradensis* egg measures 11 x 6 cm. It has been associated with *Allosaurus*. See more: *Hirsch 1994*

OTHER RECORDS

1923—IUC Mongolia—The first theropod nest
The first nests were thought to be *Protoceratops*, but later it was discovered that they were of an oviraptorosaurid. The nests of several theropods had eggs arranged like a ring, with a hole in the center for hatching and to protect them with their arms. They were partially buried in the sediment, unlike more advanced birds such as ornithuromorphs and the modern birds. See more: *Carpenter 1999; Tanaka et al. 2014*

1972—IUC Mongolia—The first theropod embryo
An incomplete metatarsal inside a shell fragment offers the first evidence of an unhatched dinosaur. See more: *Sochava 1972*

1991—IUC Mongolia—The largest Mesozoic bird egg in Asia (Paleoasia)
ZPAL MgOv-II/9d ovule (present-day Mongolia) measures 1.6 x 1.3 cm. Other pieces measure 2.7 x 1.5 cm but are of dubious identification. See more: *Sabath 1991*

1993—UJ Germany—The smallest theropod egg in Europe
In the holotype specimen of *Compsognathus longipes*, 1 cm spheres were found. They have been interpreted as dermal ossifications or eggs. See more: *Huene 1901; Griffiths 1993*

1995—IUC China—Largest theropod nest
It is estimated that the nest of *Macroelongatoolithus xixianensis* was about 2.1 m in diameter and contained about twenty-six eggs. See more: *Li et al. 1995*

1996—IUC China—The most bountiful dinosaur egg field
Numerous places in Spain, France, and China have provided hundreds or thousands of eggs. The most outstanding is in Heyuan, China, where more than 17,000 eggs have been unearthed since 1996. An example of its productivity is that 10,008 eggs were found in 2004. See more: *Fang et al. 2005; Tanaka et al. 2012; Guinness Record*

1996—IUC Canada and Montana, USA—The strangest theropod laying pattern
The laying patterns of theropods were usually circular, forming a ring. An exceptional case is three pairs of *Continuoolithus canadensis* eggs, which were arranged in two parallel rows. Other eggs of *C. canadensis* (present-day Montana, USA) have been found in parallel. The authors suggest that this pattern was not due to the dinosaur. See more: *Zelenitsky et al. 1996; Jackson et al. 2015*

1996—Mongolia—First relationship between pupae and dinosaurs
Some insect pupae (*Fictovichnus gobiensis*) were found inside eggs. They were 15–18 mm x 8–9 mm in size and originally interpreted as beetles but are currently considered wasp pupae. See more: *Johnston et al. 1996; Molina 2015*

2000—Late Lower Cretaceous (ILC) China—The thickest theropod shell
The egg *Macroelongatoolithus xixiaensis* had a shell that was 4.75 mm thick. As a comparison, the eggshell of an ostrich (*Struthio camelus*) is 1.6–2.2 mm thick. See more: *Zelenitsky, Carpenter, Currie 2000; Senut 2000*

2005—IUC China—The smallest possible theropod egg from Asia (Paleoasia)
Egg structures within *Jinfengopteryx elegans* are considered possible eggs. Measuring only 1 cm in diameter, they may be ovules or immature eggs. See more: *Ji et al. 2005*

2006—IUC Mongolia—The most porous theropod eggs
The egg *Protoceratopsirovum*, 12 x 5 cm, had about 2,600 pores. Despite its name, it was a theropod egg. See more: *Deeming 2006*

2007—ILC China—The oldest Mesozoic bird egg
The egg of *Confuciusornis sanctus* B072 dates from the early Lower Cretaceous (Barremian, ca. 129.4–125 Ma) See more: *Kaiser 2007; Dyke & Kaiser 2010*

2008—IUC Mongolia—The theropod nest with the most eggs
A nest of *Citipati osmolskae* contained up to thirty eggs, an amount similar to the number found in ostrich nests (*Struthio camelus*), which are communal. See more: *Varricchio et al. 2008*

2009—IUC France, Uruguay—The thickest Mesozoic bird shell
Except for the eggshells of ratite land birds, those of birds are usually very thin. *Lanceoolithus junggarensis* shells were very thick, 2.6 mm, but they may have been a non-avian theropod. The shell of the bird egg *Ageroolithus fontllongensis* was between 0.25 and 0.36 mm thick, a little thinner than that of the domestic chicken. See more: *Vianey-Liaud & López-Martínez 1997; Fang 2009*

2009—ILC Spain—Eggs eroded by snails
The egg *Prismatoolithus* sp. MPZ 2000/3558 and an elongatoolitid have marks of *Radulichnus*, a bioerosion caused by aquatic gastropod mollusks. The marks were located mainly in the interior, precisely where more protein would be. See more: *Gámez-Vintane et al. 2009*

2010—IUC—The non-avian theropod egg with more layers in the shell
Alvarezsaurids had numerous anatomical features that resembled birds. They even laid eggs with three-layered shells, similar to those of birds. See more: *Jackson & Varricchio 2010; Agnolin et al. 2012*

2012—IUC Spain—The thinnest theropod shell
The egg *Sankofa pyrenaica* had a shell whose thickness oscillated between 0.19 and 0.34 mm. The egg *Pseudogeckoolithus nodosus* had an even thicker shell, 0.22–0.36 mm. Not considering the texture, the latter is even thinner (0.13–0.29 mm). See more: *Vianey-Liaud & López-Martínez 1997; López-Martínez & Vicens 2012*

2012—IUC Mongolia—The thinnest Mesozoic bird shell
The egg *Parvoolithus tortuosus* had a shell thickness of only 0.1 mm. See more: *Mikhailov 1996*

2012—IUC Romania—The nesting colony with the highest concentration of eggs
An area of 80 x 50 cm and 20 cm high contained forty-six eggs, of which only seven are almost complete, and thousands of broken shells. It is suggested that the nest was reused for many seasons. See more: *Dyke et al. 2012*

2013—IUC Argentina—The most extensive Mesozoic nesting colony
The most extensive nesting colony is enantiornithean (tentatively assigned to *Neuquenornis australis*) and covered an area of 2.25 x 1.35 m (1,718 cm^2). The eggs were buried by a dune. Sixty-five complete, partial, or broken eggs have been recovered, making it the most abundant as well. See more: *Chiappe & Calvo 1994; Schweitzer et al. 2002; Fernández et al. 2013*

20 cm   40 cm   60 cm   80 cm

100 cm

Domestic chicken
Gallus domesticus
UNECE 2010
5.9 x 4.4 cm

*Subtilioolithus
microturberculatus*
PIN 4230-3
Mikhailov 1991
7.2 x 3.2 cm

**North Island brown
kiwi**
Apteryx mantelli
Coulborne 2002
11.6 x 7 cm

*Ellipsoolithus
khedanensis*
Loyal et al. 2012
11 x 8 cm

**Eurornithean bird
Nº 6**
Batista 2012
11 x 8.1 cm

1:4

**Indeterminate theropod
Nº 5**
Batista 2012
11.3 x 8.5 cm

80 cm

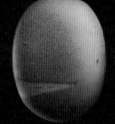

**Enantiornithean bird
ZPAL MgOv-II/9d**
Sabath 1991
1.6 x 1.3 cm

*Lourinhanosaurus antunesi
Preprismatoolithus sp.*
Mateus et al. 1997
13.7 x 10 cm

*Arriagadoolithus
patagonicus*
Agnolin et al. 2012
17.5 x 7 cm

60 cm

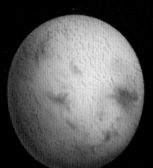

40 cm

*Macroelongatoolithus
carlylei*
IMNH 2428\49608-2
Jensen 1970
39.8 x 10.2 cm

*Dendroolithus
fengguangcunensis*
Fang et al. 2005
17 x 15 cm

*Macroelongatoolithus
xixiaensis*
Jin et al. 2007
61 x 17.9 cm

20 cm

Ostrich
Struthio camelus
Khanna 2005
17 x 13.6 cm

Macroolithus yaotunensis
Sato et al. 2005
17.5 x 8 cm

Mute swan
Cygnus olor
Czapulak 2002
12.4 x 8 cm

Vorompatra or elephant bird
Aepyornis maximus
Murray et al. 2004
35.1 x 25.9 cm

THE LARGEST

ILC China—The largest theropod egg in Asia (Paleoasia)
The largest eggs of *Macroelongatoolithus xixiaensis* were spectacular, up to 61 x 17.9 cm in size. It must have been an animal similar in size and shape to Gigantoraptor, although older. See more: *Jin et al. 2007*

IUC Mongoliaz—The largest Mesozoic bird egg in Asia (Paleoasia)
The egg *Subtilioolithus microtuberculatus* measured 7.2 x 3 cm and was very elongated compared to other bird eggs. See more: *Sabath 1991*

IUC China—The largest unlaid egg
Two eggs of *Macroolithus yaotunensis* were found within the pelvic area of a partial skeleton of *Heiyuannia huangi*. Thanks to this discovery, it is suggested that theropods had two functional oviducts. Modern birds have only one, an adaption that made them lighter and improved their flight capacity. See more: *Varricchio et al. 1997; Sato et al. 2005*

UJ Portugal—The largest European theropod egg of the Jurassic
The eggs *Preprismatoolithus* sp., assigned to the theropod *Lourinhanosaurus antunesi*, were 11.7–13.7 cm x 8–10 cm in size. See more: *Mateus et al. 1997*

IUC Uruguay—The largest Mesozoic bird egg in South America
An 11 x 8.1 cm egg is the largest of all Mesozoic birds. It was similar in size to that of the kiwi (*Apteryx mantelli*). See more: *Alvarenga & Bonaparte 1992; Batista 2012*

IUC Argentina, Uruguay—The largest theropod egg in South America
The eggs of the *Bonapartenykus ultimus* alvarezsaurid, named *Arriagadoolithus patagonicus*, were 7 cm wide. (In comparison, the eggs of *Triprismatoolithus stephensi* could reach 17.5 cm in length.) Another theropod egg found in Uruguay is shorter but wider, 11.3 x 8.5 cm. See more: *Mones 1980; Agnolin et al. 2012; Batista 2012*

eUC Idaho, USA—The largest theropod egg in North America
An egg of *Macroelongatoolithus carlylei* (formerly known as *Boletuoolithus*) measuring 39.8 x 10.2 cm is very similar to *Macroelongatoolithus xixiaensis* from East Asia, suggesting that there were large oviraptorosaurs in western North America. See more: *Simon 2014*

IUC India—The largest theropod egg from Hindustan
The only complete theropod egg from Hindustan is *Ellipsoolithus khedanensis*, which is 11 x 8 cm in size. See more: *Loyal et al. 2012*

IUC China—The largest spherical theropod egg
Dendroolithus fengguangcunensis were eggs of tericinosaurs that ranged between 15 and 17 cm in diameter. See more: *Fang et al. 2005*

1:3

The theropod egg with the most confusing name
Protoceratopsidovum
Size (mm): 150 x 57
Although the name means "egg of *Protoceratops*," they are oviraptorosaur eggs (a theropod). They were orignally confused for protoceratopsid eggs. A name change to *"Oviraptoroolithus"* has been proposed but has not been successful. See more: *Mikhailov 1994; Fang, Yu & Ling 2009*

Egg/theropod ratio

Weights are real or estimated using the Dickison 2007 method

Oogenus and species	Largest egg	Largest egg Weight– egg/ adult theropod	Bibliography
Confuciusornis sanctus 362 g female 600 g male	~2.5 x1.7 cm (estimated) 4.1 g	1.14-0.68% or 88-146 times as large	Kaiser 2007; Dike & Kaiser 2010; Marugán-Lobón et al. 2011
Sinosauropteryx prima 1.1 kg	3.6 x 2.6 cm 13.7 g	1.24% or 80.3 times as large	Chen et al. 1998
cf. *Martinavis* 730 g	4 x 2.5 cm 13.8 g	1.89% or 52.9 times as large	Dong & Currie 1995; Varricchio & Barta 2015
Neuquenornis volans 280 g	4.7 x 2.9 cm 19 g	6.78% or 14.7 times as large	Dong & Currie 1995; Varricchio & Barta 2015
cf. *Machairasaurus* 32 kg	15 x 5.5 cm 245 g	0.76% or 131 times as large	Dong & Currie 1995; Varricchio & Barta 2015
Gobioolithus major *Gobipteryx minuta* 140 g	5.35 × 3.2 cm 31 g	22% or 45 times as large	Dong & Currie 1995; Varricchio & Barta 2015
Troodon sp. nov. *Prismatoolithus levis* 39 kg	16 x 7 cm 331 g	0.85% or 118 times as large	Horner & Weishampel 1988; Zelenitsky & Hills 2011
Bonapartenykus ultimus *Arraigadoolithus patagonicus* 34 kg	17.5 x 7 cm 484 g	1.42% or 70 times as large	Agnolin et al. 2012
Citipati osmolskae 83 kg	19 x 7.2 cm 566 g	0.68% or 147 times as large	Clark et al. 1999
Macroolithus yaotunensis *Heiyuannia huangi* 22.5 kg	17.5 x 8 mm 633 g	0.28% or 35 times as large	Sato et al. 2005
Preprismatoolithus sp. *Lourinhanosaurus antunesi* 310 kg (subadult)	13.7 x 10 cm 774 g	0.25% or 400 times as large	Mateus & Mateus 1997; Araujo et al. 2012

OTHER RECORDS

2013—ILC China—The smallest Mesozoic bird ovule from Asia (Paleoasia)
Ovules (unfertilized eggs) are usually smaller in size than mature eggs. In the enantiornithean birds STM29-8 and STM10-45 6.72, ovules were found that were between 5.83 and 8.83 mm in size. See more: *Zheng et al. 2013*

2014—ILC Japan—The last Mesozoic bird egg named
Plagioolithus fukuiensis is known only by 0.44 mm thick shells. See more: *Imai & Azuma 2014*

2015—ILC Japan—The last theropod egg named
In 2015, *Nipponoolithus ramosus* was named. It was found together with *Elongatoolithus* sp., *Prismatoolithus* sp., and an indeterminate prismatoolithid. They were contemporaries. See more: *Tanaka, Zelenitsky, Saegusa, Ikeda, Debuhr & Terrien 2015*

2015—ILC Thailand—The small egg that was not a bird
The SK1-1 egg (1.8 x 1.1 cm) appeared to be from a bird because of its three-layer shell, but they were shown to be anguimorph eggs (glass lizards) when embryos were found inside. See more: *Buffetaut 2005; Fernandez et al. 2015*

2015—IUC China—The first cuticle in a theropod oolite
The eggs *Macroolithus yaotunensis* were found to have their original blue-green pigment, similar to the color of emu eggs (*Dromaius novaehollandiae*). It is known that this pigmentation is useful for camouflage. This finding provided the first evidence of the preservation of the pigment in a dinosaur egg. See more: *Wiemann et al. 1993*

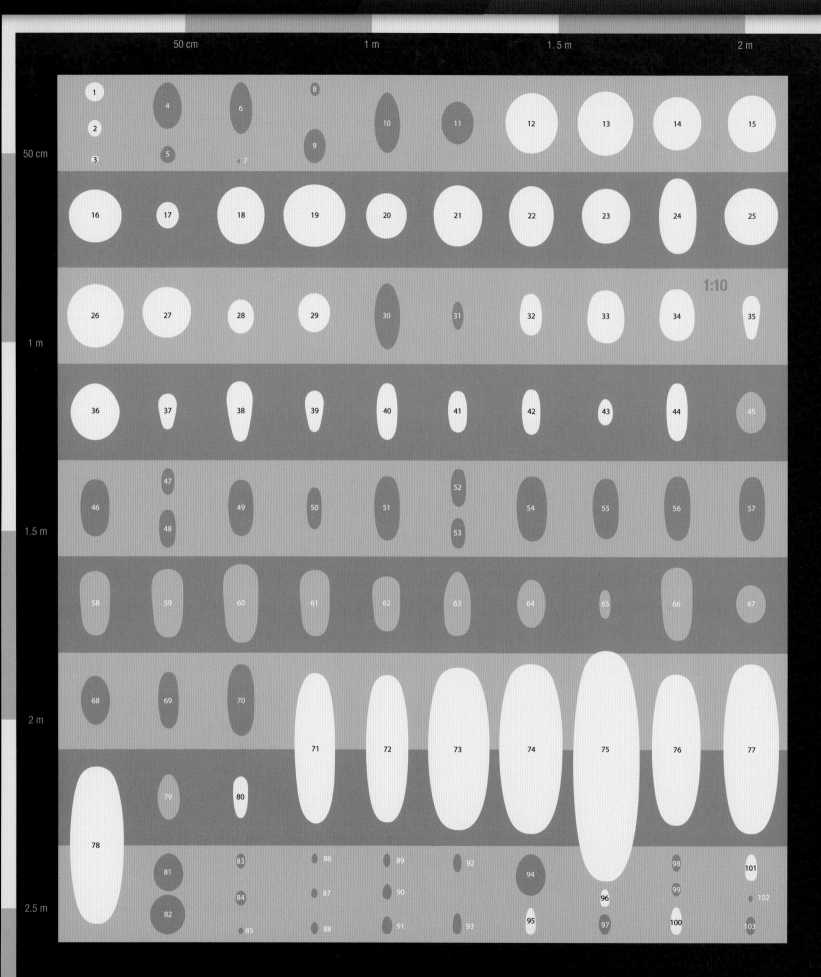

FOSSIL EGGS (OOLITES) IN THEROPODS AND MESOZOIC BIRDS

ca. 10,000 BC—Mongolia—Object with the oldest dinosaur oolites
Late Paleolithic or Early Neolithic people pierced dinosaur eggshells for use as necklace ornaments. See more: *Andrews 1943; Pauc & Buffetaut 1994*

1859—France—The first dinosaur shells in western Europe
The first fragments of dinosaur eggs were interpreted as giant bird eggs. Their age is unknown. See more: *Pouech 1859; Taquet 2001*

1859—England—The first incorrect report of a dinosaur egg in western Europe
"Oolithes" bathonicae (now *Testudo flexoolithus*) is a complete egg found in England. It was attributed to dinosaurs or pterosaurs, but it was a turtle. See more: *Buckman 1859; Hirsch 1996*

1877—France—The first microstructural study of a dinosaur egg
Before the egg was analyzed under a microscope, it was coated with epoxy resin and then sawed into very thin sections. Created almost a century and a half ago, this methodology is still used today. See more: *Gervais 1877*

1901—Germany—The first theropod ovules
A set of spheres, measuring 1 cm in diameter, were discovered in *Compsognathus longipes* (BSP AS I 536) and interpreted as dermal ossifications. Some time later, it was suggested that they could be eggs, but they are too small and it is not an adult specimen. They may be ovules. See more: *Huene 1901; Griffiths 1993*

1904—Portugal—The first theropod egg of the Jurassic in western Europe?
An egg was related to multiple theropod skeletons, but this was not proven, so the record is maintained by another more recent discovery. See more: *Lapparent & Zbyszewski 1957; Mateus et al. 1997*

1913—USA—The first shell in western North America
Several shells were located in the Blackfeet Indian Reservation, Montana. See more: *Gilmore 1913; Carpenter 2000*

1922—Mongolia—The first theropod egg in East Asia
The first egg of a theropod dinosaur was located in the Gobi Desert. At first, it was attributed to *Protoceratops andrewsi*, being a theropod oviraptorosaur. See more: *Andrews 1932; Carpenter et al. 1996*

1925—Mongolia—The first detailed study of Mesozoic bird shells
A study of two shells from Mongolia determined their thickness and microstructure. See more: *Straelen 1925*

1954—China—The first theropod egg named in Asia
Fossil eggs found in 1923 were named *Oolithus elongathus*. See more: *Young 1954*

1958—France—The first oolite of a dinosaur with pathologies
Shells with multiple layers are reported. The anomaly occurs in birds and reptiles that are under stress. See more: *Dughi & Sirugue 1958; Sochava 1971*

1961—Kyrgyzstan—The first theropod shells in western Asia
Some shells were reported in the former Soviet Union. See more: *Bazhanov 1961; Nesov & Kaznyshkin 1986*

1966—USA—The first theropod egg in western North America
Oolithes carlylensis (now *Macroelongatoolithus carlylei*) from Utah, was also called *Boletuoolithus*. See more: *Jensen 1966, 1970; Bray 1998; Zelenitsky, Carpenter & Currie 2000*

1977—Uzbekistan—The first complete Mesozoic bird egg from East Asia
In West Asia, only fragments had been found until they were able to find whole eggs. See more: *Nesov & Kaznyshkin 1977*

1980—Uruguay—The first theropod egg from southern South America
Due to the ellipsoid form of *Tacuarembovum oblongum*, it was attributed to an ornithischian, but it is now considered a theropod. An attempt was made to change the name to *Tacuaremboolithus* without success. See more: *Mones 1980; Carpenter & Alf 1994; Faccio 1994; Batista & Perea 2010*

Theropod oolites (Their silhouettes are illustrated on the previous page)

Oolites, probably of indeterminate dinosaurs
(Masses estimated as per Dickinson 2007)

N.°	Oogenus and species	Size (mm)	Mass (g)	Bibliography
1	Oolithes obtusatus UJ, England	5 x 5	0.1	Carruthers 1871
2	Oolithes nanshiungensis IUC, China	44 x 36	32	Young 1965
3	Oolithes sphaericus UJ, England	20 x 20	4.5	Carruthers 1871

Theropod oolites, indeterminate or without family

N.°	Oogenus and species	Size (mm)	Mass (g)	Bibliography
4	Continuoolithus canadensis IUC, Canada, Montana-USA	123 x 77	413	Zhao et al. 2009
5	Parvoblongoolithus jinguoensis IUC, China	45.5 x 40.4	42	Zhang et al. 2014
6	Reticuloolithus hirschi IUC, Canada	~137 x 60	279	Zelenitsky & Sloboda 2005
7	Unnamed ILC, China	10 x 7	0.3	Ji et al. 2005
8	Sinosauropteryx prima ILC, China	36 x 26	13.7	Chen et al. 1998
9	Unnamed IUC, Brazil	91 x 58	173	Campos & Bertini 1985
10	Unnamed ILC, Montana-USA	~160 x 70	443	Makovicky & Grellet-Tinner 2000
11	Unnamed IUC, Uruguay	113 x 85	462	Batista 2012

Phaceloolithidae

N.°	Oogenus and species	Size (mm)	Mass (g)	Bibliography
12	Dendroolithus dendritics IUC, China	160 x 140	1772	Fang et al. 1998
13	Dendroolithus fengguangcunensis IUC, China	170 x 150	2295	Fang et al. 2000
14	Dendroolithus furcatus IUC, China	141 x 128	1305	Xiaosi et al. 1998
15	Dendroolithus guoqingsiensis IUC, China	150 x 130	1433	Fang et al. 2000
16	Dendroolithus hongzhaiziensis IUC, China	140 x 140	1550	Zhou et al. 1998
17	Dendroolithus microporosus IUC, Mongolia	70 x 60	142	Mikhailov 1994
18	Dendroolithus sanlimiaoensis IUC, China	152 x 126	1363	Fang et al.1998
19	Dendroolithus tumiaolingensis IUC, China	165 x 165	2695	Zhou et al. 1998
20	Dendroolithus verrucarius IUC, Mongolia	120 x 110	820	Mikhailov 1994
21	Dendroolithus xichuanensis IUC, China	160 x 120	1302	Zhao & Zhao 1998
22	Dendroolithus wangdianensis IUC, China	162 x 130	1547	Zhao & Li 1988
23	Dendroolithus zhaoyingensis IUC, China	145 x 130	1385	Fang et al. 1998
24	Dendroolithus sp. UJ, Portugal	200 x 100	1130	Zelenitsky 1996
25	Paradendroolithus qinglongshanensis China	150 x 145	1782	Zhou et al. 1998
26	Phaceloolithus huanensis IUC, China	168 x 150	2268	Zeng & Zhang 1979
27	Placoolithus taohensis IUC, China	134 x 130	1280	Zhao & Zhao 1998
28	Unnamed IUC, Mongolia	90 x 70	242	Manning et al. 2000
29	Unnamed IUC, South Korea	103 x 85	421	Deeming 2006

Arriagadoolithidae

N.°	Oogenus and species	Size (mm)	Mass (g)	Bibliography
30	Arriagadoolithus patagoniensis IUC, Argentina	~175 x 70	485	Agnolin et al. 2012
31	Triprismatoolithus stephensi IUC, Canada	75 x 30	38	-

Prismatoolithidae

N.°	Oogenus and species	Size (mm)	Mass (g)	Bibliography
32	Preprismatoolithus coloradensis UJ, Colorado-USA	110 x 60	224	Hirsch 1994
33	Preprismatoolithus sp. UJ, Portugal	140 x 100	791	Ribeiro et al. 2014
34	Preprismatoolithus sp. UJ, Portugal	137 x 94	684	Antunes et al. 1998

1984—Peru—The first theropod egg in northern South America
Theropod or bird shells were found in Peru. See more: *Kerourio & Sigé 1984*

1985—Brazil—The first theropod egg in northern South America
An egg was assigned to a probable ceratopsid that was later recognized as a theropod. See more: *Campos & Bertini 1985; Kellner et al. 1998*

1989—USA—The first theropod egg of the Jurassic in North America
Prismatoolithus coloradensis (now *Preprismatoolithus*), from Utah, has been linked to *Allosaurus*. See more: *Hirsch et al. 1989; Hirsch 1994*

1991—Mongolia—The first Mesozoic bird eggs in East Asia
Laevisoolithus sochavai and *Subtilioolithus microtuberculatus* were identified as enantiornithean birds. See more: *Mikhailov 1991*

1992—Portugal—The first theropod egg in western Europe
Prismatoolithus sp. is reported. There was a previous report that was not reviewed. See more: *Lapparent & Zbyszewski 1957; Mateus et al. 1997; Dantas et al. 1992*

1995—India—The first Mesozoic bird egg in South Asia (Hindustan)
Subtiliolithus kachchhensis are bird eggshells. See more: *Khosla & Sahni 1995*

1997—Portugal—The first theropod egg of the Jurassic in Europe
Preprismatoolithus sp. are fossil eggs that were assigned to *Lourinhanosaurus antunesi*. There is a possibility that another oolite previously found belongs to a theropod, but this is currently in doubt. See more: *Lapparent & Zbyszewski 1957; Mateus et al. 1997, 2001*

1998—Spain—The first bird egg in Europe?
Ageroolithus fontllongensis may be the shells of a large bird, very similar to those of modern ratites. See more: *Vianey-Liaud & López-Martínez 1997*

1998—Brazil—The first theropod shells of northern South America
Fragmented eggs were located in northern Brazil. See more: *Kellner et al. 2014*

1998—India—The first theropod egg in South Asia (Hindustan)
Ellipsoolithus khedaensis are complete eggs. Up to thirteen of them were found in some nests. It is speculated that they may be abelisaurid eggs. See more: *Loyal et al. 1998; Mohabey 1998*

2001—Argentina—The first Mesozoic bird egg in southern South America
Complete enantiornithean bird eggs and eggshells were found, some containing embryos. See more: *Schweitzer et al. 2001, 2002*

2003—Morocco—The first theropod shells in North Africa
Three types of shells were reported: *Rodolphoolithus arioul, Pseudogeckoolithus tirboulensis* and *Tipoolithus achloujensis*. See more: *Vianey-Liaud & García 2003*

2012—Argentina—The first theropod associated with eggs in southern South America
For the first time, a genus of theropod and its fossil eggs are named in the same publication. The alvarezsaurid *Bonapartenykus ultimus* was found with two associated eggs, called *Arraigadoolithus patagoniensis*. See more: *Agnolin et al. 2012*

2012—Romania—The first Mesozoic bird egg of western Europe
Several eggs of enantiornithean birds similar to *Martinavis* are discovered See more: *Dyke et al. 2012*

2012—Romania—The first nesting colony in western Europe
A colony associated with multiple enantiornithean shells, eggs, and bones of embryos, juveniles, and adults was found. See more: *Dyke et al. 2012*

2013—China—The first case of sex identification in a Mesozoic bird
An international team that reviewed more than one hundred specimens of *Confuciusornis sanctus* revealed the existence of sexual dimorphism in this bird. An easy feature to observe is long rectrices, which are unique in males. See more: *Chinsamy et al. 2013*

2014—Brazil—The first bird egg of northern South America
A well-preserved egg of an enantiornithean bird was found in Sao Paulo. See more: *Marsola et al. 2014*

2015—China—The first cuticle in a theropod oolite
Macroolithus yaotunensis eggs were pigmented greenish blue, which camouflaged them. This finding provides the first evidence of preservation of the outer shell in a fossil shell. See more: *Wiemann et al. 1993*

Nº	Oogenus and species	Size (mm)	Mass (g)	Bibliography
35	*Prismatoolithus gebiensis* IUC, Mongolia	116 x 48	151	Zhao & Li 1993
36	*Prismatoolithus hanshuiensis* IUC, China	150 x 130	1433	Zhou et al. 1998
37	*Prismatoolithus heyuanensis* IUC, China	95 x 50	134	Lu 2006
38	*Prismatoolithus levis* IUC, Canada, Montana-USA	160 x 70	443	Zelenitsky & Hills 1996
39	*Prismatoolithus tiantaiensis* IUC, China	110 x 50	155	Fang et al. 2000
40	*Protoceratopsidovum fluxuosum* IUC, Mongolia	150 x 57	275	Mikhailov 1994
41	*Protoceratopsidovum minimum* IUC, Mongolia	110 x 50	155	Mikhailov 1994
42	*Protoceratopsidovum sincerum* IUC, Mongolia	120 x 50	170	Mikhailov 1994
43	*Sankofa pyrenaica* IUC, Spain	70 x 40	64	López-Martínez & Vicens 2012
44	Unnamed IUC, Mongolia	153 x 56.5	276	Deeming 2006
Elongatoolithidae				
45	*Ellipsoolithus khedaensis* IUC, India	110 x 80	398	Mohabey 1998
46	*Elongatoolithus andrewsi* IUC, China	151 x 77	506	Zhao 1975
47	*Elongatoolithus chichengshanensis* ILC, China	70 x 35	48	Fang et al. 2003
48	*Elongatoolithus chimeiensis* IUC, China	100 x 45	114	Fang 2007
49	*Elongatoolithus elongatus* IUC, China	149 x 67	378	Young 1954
50	*Elongatoolithus excellens* IUC, Mongolia	110 x 40	99	Mikhailov 1994
51	*Elongatoolithus frustrabilis* IUC, Mongolia	~170 x ~70	471	Mikhailov 1994
52	*Elongatoolithus jianchangensis* IUC, China	100 x 40	90	Fang 2007
53	*Elongatoolithus laijiaensis* IUC, China	80 x 38	65	Fang et al. 2003
54	*Elongatoolithus magnus* IUC, China	172 x 82	653	Zeng & Zhang 1979
55	*Elongatoolithus sigillarius* IUC, Mongolia	160 x 70	443	Mikhailov 1994
56	*Elongatoolithus subtitectorius* IUC, Mongolia	~170 x ~70	471	Mikhailov 1994
57	*Elongatoolithus taipinghuensis* IUC, China	170 x 70	471	Yu 1998
58	*Macroolithus mutabilis* IUC, Mongolia	~170 x ~80	615	Mikhailov 1994
59	*Macroolithus rugustus* IUC, China	181 x 85	739	Young 1965
60	*Macroolithus yaotunensis* IUC, China, Mongolia	208 x 94	1038	Zhao 1975
61	*Macroolithus* sp. IUC, China	175 x 80	633	Sato et al. 2005
62	*Nanshiungoolithus chuetienensis* IUC, China	145 x 76	473	Zhao 1975
63	*Paraelongatoolithus reticulatus* IUC, China	170 x ~72	498	Wang et al. 2010
64	*Tacuaremborum oblongum* IUC, Uruguay	118 x 76	388	Mones 1980
65	*Trachoolithus faticanus* IUC, Mongolia	~75 x 30	38	Mikhailov 1994
66	*Undulatoolithus pengi* IUC, China	194 x 83.5	763	Wang et al. 2013
67	Unnamed IUC, Uruguay	100 x 80	362	Batista 2012
68	Unnamed IUC, Uruguay	130 x 77	436	Batista 2012
69	Unnamed IUC, China	155 x 55	256	Varrichio & Barta 2015
70	Unnamed IUC, China	190 x 72	567	Varrichio & Barta 2015
Macroelongatoolithidae				
71	*Macroelongatoolithus carlylei* ILC, Idaho, Montana, Utah-USA	398 x 108	2785	Jensen 1970
72	*Macroelongatoolithus goseongensis* IUC, South Korea	390 x 115	3095	Kim et al. 2011

DEVELOPMENT AND GROWTH IN DINOSAUROMORPHS, THEROPODS, AND MESOZOIC BIRDS

1868—USA—The first discovered juvenile theropod
Aublysodon mirandus is an indeterminate juvenile tyrannosaurid from Montana. It is too old to be a breeding *Tyrannosaurus* sp., as was originally suggested. See more: *Leidy 1868; Carpenter et al. 1994*

1932—England—The first juvenile theropod of the Jurassic
"Megalosaurus" incognitus (now *Iliosuchus*) was considered a juvenile similar to the tyrannosaurid *Stokesosaurus*, but it is likely an indeterminate theropod. See more: *Huene 1932; Carrano et al. 2012*

1947—USA—The first baby theropod of the Triassic
Several neonates of *Coelophysis* were reported at Ghost Ranch, New Mexico, United States. See more: *Colbert 1989; Carpenter & Alf 1994*

1972—Mongolia—The first dinosaur embryo
An embryo was identified in the Gobi Desert eggs. Although it is not specified to what type of dinosaur it belonged, it is probably a theropod. See more: *Sochava 1972*

1981—Mongolia—The first Mesozoic bird embryo in East Asia
Numerous enantiornithean embryos in different stages of development, *Gobipteryx minuta* and *Gobipipus reshetovi*, were found in the Gobi Desert. See more: *Elzanowski 1981; Kurochkin et al. 2013*

1988—USA—The first non-avian theropod embryos in western North America
After extensive analysis, several Montana embryos that were originally thought to be the ornithopod *Orodromeus makelai* turned out to be the theropod *Troodon* cf. *formosus*. See more: *Horner & Weishampel 1988; Varricchio et al. 2002*

1993—Mongolia—The first non-avian theropod embryo in East Asia
An oviraptorosaur embryo was attributed to *Oviraptor philoceratops* but was later identified as *Citipati osmolskae*. See more: *Norell et al. 1994, 2001*

1993—Mongolia—The non-avian theropod with the fastest development
The small theropods reached adulthood in a short time. For example, alvarezsaurids *Albynikus baatar* (355 g) and *Shuvuuia deserti* (2.7 kg) reached their adult stage in two and three years, respectively. See more: *Nesbitt et al. 2011*

1993—USA—The oldest theropod of the Jurassic
The oldest individuals of *Allosaurus fragilis* were between twenty-two and twenty-eight years old. They became sexually mature between six and eight years. They grew at a rate of about 200 kg per year. See more: *Bybee et al. 2006*

1995—China—The first herbivorous theropod embryos
The first embryo assigned to a therizinosaur was scientifically described in 1997, although it was already known in popular publications two years before. Another egg discovered at the same time, *Macroelongatoolithus*, had an oviraptorosaur embryo. See more: *Cohen et al. 1995; Manning et al. 1997; Kundrat et al. 2007*

1995—Morocco—The first theropod embryos in North Africa
Tiny teeth have been interpreted as belonging to embryos of therizinosaurs and dromaeosaurs, and a third is thought to belong to a troodontid or an ornithopod. See more: *Sigogneau-Russell et al. 1998*

1997—Portugal—The first non-avian theropod embryo in Europe (and the first Jurassic embryo)
Some fossil eggs of *Preprismatoolithus* sp. (present-day Portugal) had fragments of theropod embryos that belong to the primitive coelurosaur *Lourinhanosaurus antunesi*. See more: *Mateus et al. 1997, 2001*

2001—Argentina—The first Mesozoic bird embryo in southern South America
An Argentine egg contained the enantiornithean embryo (MUCPv-284) that was tentatively assigned to *Neuquenornis volans*. See more: *Schweitzer et al. 2001, 2002*

2003—China—The fastest developing Mesozoic bird
Confuciusornis sanctus reached adulthood in just five months. Although it seems like a long time compared to modern birds, it was much shorter than the time required by several non-avian theropods. See more: *de Ricqlès et al. 2003*

Nº	Oogenus and species	Size (mm)	Mass (g)	Bibliography
73	*Macroelongatoolithus xixiaensis* IUC, South Korea	430 x 165	7024	-
74	*Macroelongatoolithus xixiaensis* (*Longiteresoolithus*) IUC, China	450 x 170	7803	Li et al. 1995
75	*Macroelongatoolithus xixiaensis* eUC, China	610 x 179	11727	Jin et al. 2007
76	*Macroelongatoolithus qiaoxianensis* (*Megafusoolithus*) eUC, China	400 x 130	4056	Wang et al. 2010
77	*Macroelongathoolithus zhangi* eUC, China	450 x 150	6075	Fang et al. 2000
78	Unnamed eUC, China	417 x 143	5096	Simon 2014
Montanoolithidae				
79	*Montanoolithus strongorum* eUC, Montana-USA	120 x 60	244	Zelenitsky & Therrien 2008
Oblongoolithidae				
80	*Oblongoolithus glaber* eUC, Mongolia	~110 x 40	99	Mikhailov 1996
Pinnatoolithidae				
81	*Lanceoloolithus xiapingensis* IUC, China	100 x 80	362	Fang et al. 2009
82	*Pinnatoolithus shitangensis* IUC, China	105 x 95	535	Fang et al. 2009
Oolites of Mesozoic birds, indeterminate or without family				
83	*Parvoolithus tortuosus* IUC, Mongolia	40 x 25	14	Mikhailov 1996
84	*Protornithoolithus tumendongensis* IUC, China	40 x ~29	19	Fang 2007
85	Unnamed IUC, Mongolia	16 x 13	1.5	Sabath 1991
86	Unnamed IUC, Mongolia	25.8 x 1.6	3.7	Grellet-Tinner & Norell 2002
87	*Confuciusornis sanctus* ILC, China	25 x 17	4.1	Kaiser 2007; Dyke & Kaiser 2010
88	Unnamed IUC, Brazil	31.4 x 1.95	6.7	Marsola et al. 2014
89	Unnamed ILC, China	35 x 20	7.9	Zhang & Zhou 2004
90	Unnamed IUC, Rumania	40 x 25	14	Varricchio & Barta 2015
91	Unnamed IUC, Argentina	47 x 29	22.3	Schweitzer et al. 2002
92	Unnamed IUC, China	47.5 x 22.3	13.3	Balanoff et al. 2008
93	Unnamed Uzbekistán	~55 x ~23	13.3	Nesov & Kaznyshkin 1986
94	Unnamed IUC, Uruguay	110 x 81	408	Batista 2012
Laevisoolithidae				
95	*Laevisoolithus sochavai* IUC, Mongolia	~70 x 27	28.8	Mikhailov 1991
96	Unnamed Uzbekistan	46 x 24	15	-
Gobioolithidae				
97	*Gobioolithus major* IUC, Mongolia	53.5 x 32	31	Mikhailov 1996
98	*Gobioolithus minor* IUC, Mongolia	46 x 24	15	Mikhailov 1996
99	*Gobipipus reshetovi* IUC, Mongolia	36 x 24	11.7	Kurochkin et al. 2013
Subtilioolithidae				
100	*Subtiliolithus microtuberculatus* IUC, Mongolia	72 x 30	36.6	Mikhailov 1991
101	*Styloolithus sabathi* IUC, Mongolia	70 x 32	40.5	Varrichio & Barta 2015
Not theropod eggs				
102	Unnamed ILC, Thailand	18 x 11	1.2	Buffetaut 2005
103	*Testudoflexolithus bathonicae* MJ, England	46 x 26	18.3	Buckman 1859

2004—Canada and the USA—The oldest Cretaceous theropods
The *Tyrannosaurus rex* FMNH PR2801 is the oldest. It was calculated to be twenty-eight years old. It is known that *Tyrannosaurus rex* became adult at the age of sixteen and stopped growing at nineteen. A specimen of *Albertosaurus sarcophagus* also reached twenty-eight years of age. See more: *Erickson et al. 2004, 2006*

2009—USA—The oldest theropod of the Triassic
Coelophysis bauri weighed 3 kg when it was one year old. At age four, it reached sexual maturity. The largest specimens were around seven years old. See more: *Rinehart et al. 2009*

2013—USA—The first theropod embryo of the Jurassic in western North America
Allosaurus embryos have been found in Wyoming. See more: *Carrano, Mateus & Mitchell 2013*

The smallest Triassic juvenile theropod
Unnamed
Specimen: TTU P 9201
Material: Metatarsal
Northwestern Pangea (present-day Texas, USA)
It was considered part of the remains of the supposed bird *Protoavis texensis* but in reality was a coelophysoid that was 43 cm long and weighed 110 g.

See more: *Chatterjee 1991*

The smallest juvenile Cretaceous bird
Gobipipus reshetovi
Specimen: PIN 4492-4
Material: Partial skeleton and shells
Eastern Laurasia (Mongolia)
An enantiornithean that was 4.7 cm long and weighed 2 g. The name of this embryo has existed informally since 1995.
See more: *Mourer-Chauvire 1995; Kurochkin et al. 2013*

The smallest juvenile theropod of the Jurassic
Epidendrosaurus ningchengensis
Specimen: IVPP V12653
Material: Partial skeleton and filaments
Northeastern Pangea (China)
A paravian that was 13 cm long, weighed 6.7 g, and had very big hands.
See more: *Zhang et al. 2002*

The youngest juvenile bird of the Jurassic
Archaeopteryx lithographica
Specimen: MJ SoS 2257
Material: Partial skeleton and feathers
Northeastern Pangea (Germany)
An avialan that was 28 cm long and weighed 60 g. It was known as *Jurapteryx recurva*.
See more: *Wellnhofer 1974*

The bird that was described twice at the same time
Liaoxiornis delicatus
Lingyuanornis parvus
Specimen: GMV 2156
IVPP V 130723
Material: Partial skeleton and feathers
Eastern Laurasia (China)
Both halves of the skeleton of a juvenile enantiornithean were described, named, and catalogued separately, because they were housed in two different museums. See more: *Hou & Chen 1999; Ji & Ji 1999*

Smallest juvenile Cretaceous theropod
Unnamed
Uncatalogued
Material: Incomplete pelvis
Eastern Gondwana (New Zealand)
It was an abelisauroid 20 cm long and weighing 5.3 g.
See more: *Molnar et al. 2006*

Eggshells

Oogenus and species	Shell thickness (mm)	Period and location	Bibliography
Oolites of indeterminate dinosaurs			
Dinosauriovum Disparovum compositum (n.d.)	-	IUC, Kazakhstan	Vialov 1986
Dinosauriovum Grumuliovum asperum (n.d.)	-	IUC, Kazakhstan	Vialov 1986
Dinosauriovum Grumuliovum punctatum (n.d.)	-	IUC, Kazakhstan	Vialov 1986
Dinosauriovum Grumuliovum tuberculatum (n.d.)	-	IUC, Kazakhstan	Vialov 1986
Dinosauriovum Ornatiovum multifarium (n.d.)	-	IUC, Kazakhstan	Vialov 1986
Dinosauriovum Ornatiovum partentosum (n.d.)	-	IUC, Kazakhstan	Vialov 1986
Dinosauriovum Ornatiovum subtilis Vialov 1986 (n.d.)	-	IUC, Kazakhstan	Vialov 1986
"Dongyangoolithus nanmaensis" Jin 2008 (n.n.)	-	IUC, China	Jin 2008
Theropod oolites, indeterminate or without family			
Apheloolithus shuinanensis (n.n.)	-	IUC, China	Zhao et al. 2009; Glut 2013
Continuoolithus sp.	1.2-1.28	IUC, New Mexico-USA	Tanaka et al. 2011
cf. Continuoolithus	0.51-0.81	IUC, Texas-USA	Sankey 2010
Nipponoolithus ramosus	0.37-0.53	ILC, Japan	Tanaka et al. 2015
Unnamed	-	ILC, Montana-USA	Janke 1996
Phaceloolithidae			
"Dendroolithus siliculose" (n.n.)	-	IUC, China	-
"Placoolithus tiantaiensis" (n.n.)	-	IUC, China	Jin 2008
Unnamed	0.7-0.9	IUC, Mongolia	Manning et al. 2000
Prismatoolithidae			
Preprismatoolithus coloradensis	0.7-1.14	UJ, Colorado-USA	Hirsch 1994
Preprismatoolithus sp.	0.7-0.99	UJ, Portugal	Ribeiro et al. 2014
Preprismatoolithus sp.	1.2	UJ, Portugal	Antunes et al. 1998
Prismatoolithus caboti	0.5-0.6	IUC, France	García et al. 2000
Prismatoolithus hukouensis	0.7-1	IUC, China	Zhao 1999
Prismatoolithus jenseni	0.83-1.16	IUC, Utah-USA	Bray 1999
Prismatoolithus levis	0.72-0.98	IUC, Canada, Montana-USA	Zelenitsky & Hills 1996
Prismatoolithus matellensis	1.06-1.22	IUC, France	Vianey-Liaud & Crochet 1993
Prismatoolithus tenuis	0.24-0.6	IUC, France	Vianey-Liaud & Crochet 1993
Prismatoolithus trempii	0.25-0.53	IUC, Spain	Vianey-Liaud & Crochet 1993
Prismatoolithus sp.	1	IUC, Morocco	García et al. 2003
Pseudogeckoolithus nodosus	0.3-0.35	IUC, Spain	Vianey-Liaud & López-Martínez 1997
Pseudogeckoolithus tirboulensis	0.22-0.36	IUC, Morocco	Vianey-Liaud & García 2003
Spheruprismatoolithus condensus	0.66-0.94	IUC, Utah-USA	Bray 1999
"Spheruprismatoolithus reticulata" (n.n.)	0.66-0.94	IUC, Canada	Zelenitsky 1995
Trigonoolithus amoae	0.33-1.04	ILC, Spain	Moreno-Azanza et al. 2013
Elongatoolithidae			
Elongatoolithus tiantiaensis	1.55	eUC, China	Fang et al. 2000
Elongatoolithus yangjiagouensis	0.6	IUC, China	Fang 2007
Heishanoolithus changii	0.3-1.4	ILC, China	Zhao & Zhao 1999
Lepidotoolithus guofenglouensis	0.6-1.6	IUC, China	Xue et al. 1996
Macroolithus lashuyuanensis	2.3-2.7	IUC, China	Fang et al. 2009
Oolithus laminadermus	-	IUC, China	Chao & Chang 1974
Porituberoolithus warnerensis	0.5-0.65	IUC, Canada	Zelenitsky, Hills & Currie 1996

Oogenus and species	Shell thickness (mm)	Period and location	Bibliography
Porituberoolithus sp.	0.42-0.5	IUC, New Mexico-USA	Tanaka et al. 2011
Porituberoolithus sp.	-	IUC, New Mexico-USA	Aguillon-Martinez et al. 2004
Porituberoolithus sp.	-	IUC, Texas-USA	Sankey 2010
Rodolphoolithus arioul	0.23-0.55	IUC, Morocco	Vianey-Liaud & García 2003
"Spongioolithus aenigmaticus"	0.94-1.24	IUC, Utah-USA	Zelenitskyi 1995
Spongioolithus hirschi	1.2-1.55	IUC, Utah-USA	Bray 1999
Trachoolithus sp.	-	IUC, India	Mohabey 2001
Unnamed	1.255	eUC, Brazil	Campos & Bertini 1985
Macroelongatoolithidae			
Unnamed	2.14-2.4	eLC, South Korea	Hun 2014
Pinnatoolithidae			
Lanceoolithus huangtangensis	0.3-0.8	IUC, China	Fang 2009
Lanceoolithus junggarensis	2.6	IUC, China	Fang 2009
Pinnatoolithus nanxiongensis	1.7	IUC, China	Fang 2009
Pinnatoolithus sangequanensis	3	IUC, China	Fang 2009
Oolites of Mesozoic birds, indeterminate or without family			
Ageroolithus fontllongensis	0.25-0.36	eUC, Spain	Vianey-Liaud & López-Martínez 1997
Ageroolithus sp.	0.278	eUC, France	Garcia et al. 1999
Dispersituberoolithus exilis	0.26-0.28	IUC, Canada	Zelenitsky et al. 1996
Plagioolithus fukuiensis	0.44	ILC, Japan	Imai & Azuma 2014
"Rarituberoolithus warnerensis"	0.7-0.9	IUC, Canada	Zelenitsky 1995
Tristraguloolithus cracioides	0.15	IUC, Canada	Zelenitsky et al. 1996
Tubercuoolithus tetonensis	0.831-1.186	IUC, Montana-USA	Jackson & Varricchio 2010
Dinosauriovum Laeviovum laevis (n.d.)	-	IUC, Kazakhstan	Vialov 1975; Vialov 1986
Unnamed	0.166	IUC, Mongolia	Grellet-Tinner & Norell 2002
Laevisoolithidae			
Unnamed	-	IUC, France	Vianey-Liaud, Khosla & Geraldine-García 2003
Subtilioolithidae			
Subtiliolithus kachchhensis	0.35-0.45	IUC, India	Khosla & Sahni 1995
Tipoolithus achloujensis	0.43-0.73	IUC, Morocco	García et al. 2003
Styloolithus sabathi	0.25	IUC, Mongolia	Varricchio & Barta 2015

The following abbreviations were used for the epochs:

UJ = Upper Jurassic
eLC = Early Lower Cretaceous
ILC = Late Lower Cretaceous
eUC = Early Upper Cretaceous
IUC = Late Upper Cretaceous

GROWTH

The theropods had diverse ontogeny patterns (development from birth to maturity). Some pups were very similar to adults, while others were different in appearance and underwent noticeable changes. Some parts of the body tend to grow more quickly than others, which is known as *allometry*. When an extinct animal is reconstructed from fragments, this causes inaccuracies, since all bones don't grow the same way. For example, the femur of the subadult of *Yangchuanosaurus shangyouensis* CV 00215 was 9% longer than the skull, while in the adult ZDM 0024, it was 13% shorter. Dinosaur growth was not linear, since the pace of growth was very rapid in the early stages of life and continuous throughout their lives (in some species). See more: *Erickson et al. 2004, 2006*

Curiosities

Theropods were precocious. They began to reproduce before they reached their maximum size. The opposite happens in birds: They delay sexual maturity until they finish growing completely. See more: *Erickson et al. 2007*

Shift change

Before dermestid beetles, which did not appear until the Cretaceous, the carcasses of dinosaurs and their waste were cleaned by cockroaches or termites. The bones of an *Allosaurus* has marks suggesting that the corpse had been cleaned by saprophagous insects, which feed on decomposing matter. See more: *Hasiotis et al. 1999, Chin & Bishop 2004, Hasiotis 2004, Bader 2005; Britt et al. 2008, Háva 2015*

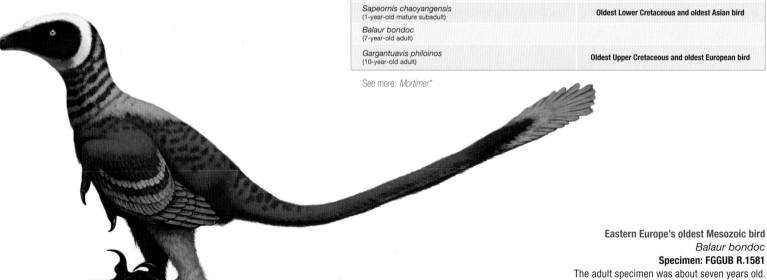

Adult and subadult theropods	
Triassic theropods	
Species	**Record**
Coelophysis bauri (7-year-old adult)	Oldest Triassic theropod
Jurassic theropods	
Limusaurus inextricabilis (5-year-old subadult)	
Allosaurus fragilis (6- to 28-year-old adult)	Oldest Upper Jurassic and oldest North American theropod
Guanlong wucaii (12-year-old adult)	Oldest Middle Triassic theropod
Megapnosaurus rhodesiensis (7- to 13-year-old adult)	Oldest Lower Triassic theropod
Jurassic Birds	
Archaeopteryx lithographica (15- to 17-month-old mature subadult)	Oldest Jurassic bird
Cretaceous theropods	
Albinykus baatar (2-year-old adult)	Oldest Jurassic bird
Shuvuuia desert (3-year-old adult)	
Dromiceiomimus brevitertius (O. edmontonicus) (2- to 3-year-old mature subadult)	
Buitreraptor (4-year-old mature subadult)	
Talos sampsoni (4-year-old mature subadult)	
Changyuraptor yangi (5-year-old adult)	
Troodon sp. (5- to 12-year old adult)	
Deinonychus antirrhopus (6- to 14-year-old adult)	Youngest adult theropod
Sinornithomimus dongi (7-year-old mature subadult)	
Megaraptor namunhuaiquii (7- to 12-year-old adult)	
Gigantoraptor erlianensis (11-year-old adult)	
Citipati osmolskae (13-year-old adult)	
Abelisauridae MMCh-PV 69 (14-year-old adul)	Oldest South American theropod
Tyrannosaurus rex (16- to 28-year-old adul)	Oldest late Upper Cretaceous and oldest North American theropod
Daspletosaurus sp. AMNH 5438 (17- to 26-year-old adul)	
Spinosaurus aegyptiacus (17-year-old mature subadult)	Oldest early Upper Cretaceous and oldest African theropod
Albertosaurus sacophagus (17- to 26-year old adult)	
Gorgosaurus libratus (14- to 22-year-old adult)	
Cretaceous Birds	
Confuciusornis sanctus (5-month old adult)	Youngest adult bird
Sapeornis chaoyangensis (1-year-old mature subadult)	Oldest Lower Cretaceous and oldest Asian bird
Balaur bondoc (7-year-old adult)	
Gargantuavis philoinos (10-year-old adult)	Oldest Upper Cretaceous and oldest European bird

See more: *Mortimer**

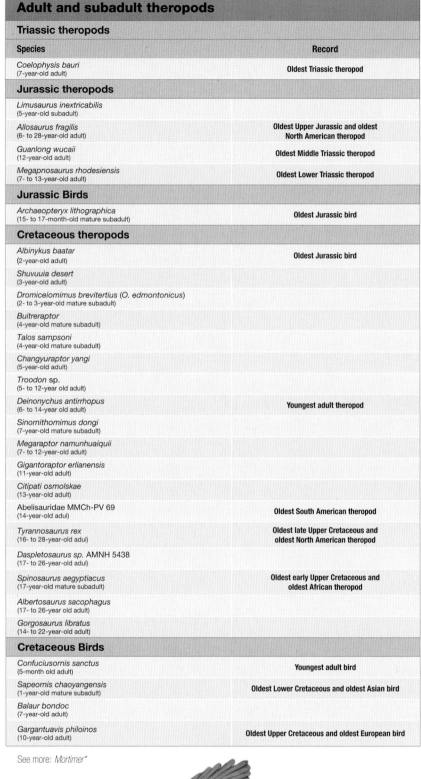

Eastern Europe's oldest Mesozoic bird
Balaur bondoc
Specimen: FGGUB R.1581
The adult specimen was about seven years old.
See more: *Csiki, Vremir, Brusatte & Norell 2010*

Theropods cover the widest range among all dinosaurs in terms of what they ate:

ZOOPHAGOUS (consumers of animals)

Carnivorous: They feed on any soft animal body part—whether meat, skin, viscera, etc. The teeth are usually knife-shaped (ziphodonts), with or without serrated edges. Theropods and the primitive saurischians were mostly carnivorous, as were some dinosauromorphs.

> **Biophagous:** Carnivores that eat only live prey. Hunting included the capture of small prey (small game) or large prey (big game). There is evidence that theropods attacked live prey, thanks to the fact that they survived and their wounds healed.
> **Scavengers or necrophagous:** Carnivores that do not hunt or kill their prey but take advantage of the carrion they encounter. Some theropods may have specialized in this type of diet, but this is difficult to determine.
> **"Crustaceavores":** (from the Latin words *crusta*, "crust," and *vorus*, "devourer"): This term can be used to designate carnivorous animals that prey on crustaceans (ostracods, shrimps, or crabs). There are examples among the enantiornitheans and several present-day birds.
> **Insectivores or entomophagous:** Carnivores that specialize in eating insects and other arthropods. Their teeth are usually bulbous, tiny, and numerous, if they have any. It is likely that some alvarezsaurs and Mesozoic birds hunted these small animals.
>
> > **Apivores:** Insectivores that eat bees and similar insects. There are present-day birds that are apivores.
> > **Culicivores:** Insectivores that consume mosquitoes. No evidence of this regime is known outside of modern birds.
> > **Myrmecophagous:** Insectivores that eat social insects. Ant or termite eaters usually have long snouts, sticky tongues, and strong arms to break into mounds. It is believed that certain alvarezsaurs specialized in myrmecophagy.
>
> **Malacophagous or molluscivores:** Carnivores that eat mollusks with or without their shells. Present-day birds specialize in collecting snails or clams and have very long or curved beaks in order to reach their food. Others consume them with the shell included.
> **Ophiophagous:** Carnivores that consume mainly snakes. This habit is known only in modern birds.
> **Ornithophagous:** Carnivores that prefer to eat birds. A pellet that included three enantiornithean birds of different species is known.
> **Paedophagous:** Carnivores that feed on the offspring of other species. Some theropods have been associated with nests and offspring of other dinosaurs, so this regime could be common in certain species.
> **Piscivores or ichthyophagous:** Carnivores that usually eat fish. They have long and conical teeth to catch slippery prey and long jaws that can be closed quickly. Some theropods and birds of the Mesozoic were clearly piscivorous.
> **Vermivores or scolecophagous:** Carnivores that depend exclusively on worms. Some modern birds take advantage of this natural resource.

Coprophagous: They feed on excrement. Some present-day birds commonly eat it.
Lactophagous or galactophagous: They feed on milk. Some present-day birds produce a substance rich in protein and fat to feed their young.
Lithophagous: They eat stones to aid digestion or to obtain minerals. The practice has been documented in theropods and present-day birds.
Sanguivores or hematophagous: They lick or suck blood. The habit of feeding only on blood is not proven in dinosaurs, but it is known that some modern birds occasionally consume this food.
Oophagous or ovivores: They consume eggs. It is suspected that some theropods consumed the eggs of other dinosaurs.
Osteophagous: They ingest bones. It is known that some carnivorous dinosaurs swallowed bones. In addition, dinosaurs with very fat teeth with deep roots and strong masseter muscles could also take advantage of this resource.
Pterophagous: They eat feathers. Some present-day birds ingest their own feathers or offer them to their young. This is a reinforcement habit that does not provide complete nutrition.

PHYTOPHAGUS (consumers of plants)

Anthophagous: They ingest flowers. Anthophagy is rare but practiced by some present-day birds.
Herbivores: They eat herbs. Graminoids (grasses) and forbs (broad leaves) were rare plants in the Mesozoic.
Foliovores or foliophagous: They consume leaves, and some also eat the branches. Leaf eaters usually have teeth with or without denticles. Some had a beak to cut the leaves and resist the continuous wear produced by the collection of this food. Several theropods were folivores.
Frugivores or carpophagous: They eat fruit. Some Mesozoic birds were found with seeds inside them; perhaps they got them by eating fruit.
Graminivores: They feed on leguminous seeds. The first leguminous plants appeared at the beginning of the Cretaceous in Hindustan.
Granivores: They collect seeds to eat. Some Mesozoic birds were found with seeds developed inside them, so they consumed seeds not contained in fruit. Birds usually have very hard beaks and require gastroliths or strong acids to grind their food.
Meliphagous: They eat honey. Some present-day birds steal honey from bees and other insects or take it directly from the resin that is released from trunks. Although eating honey is a rare habit, it is likely that some enantiornithean birds practiced it.
Oligophagous: They consume only one type of plant. No known theropod shows evidence of consuming a single type of plant.
Nectarivorous or nectivorous: They consume the nectar of flowers and are important pollinators. Some present-day birds feed exclusively from this source.
Polinivores: They are phytophages that eat pollen. Like nectarivores, they can be pollinators.
Rhizophagous: They eat roots. Whether theropods were rhizophagous can not be checked.
Xylophagous: They ingest wood. No theropod or bird consumes only wood, but it is known that other types of dinosaurs included wood in their diet at times.

OTHERS

"Waxvores" (from the Latin words cera, "wax," and vorus, "devourer"): Animals that eat wax. This term can be used to describe modern birds that eat the wax produced by bees. Eating wax is exceptional among vertebrates and has not been reported in fossil species.
Durophagous: Animals that eat hard materials. They have teeth or very strong spikes that are not too sharp, and powerful jaw muscles. They usually take advantage of what other animals leave, such as shells, bones, plants, or seeds. Some theropods were able to break the hardest parts of their victims' bodies. The enantiornithean birds of the Boahiornitidae family had rough teeth, which is interpreted as allowing the consumption of hard material.
Microphagous: They consume particles so tiny that they do not make a precise selection.

> **Filter feeders:** Microphages that absorb water in which organisms are floating. Filter feeding may have been practiced by some Mesozoic birds.
> **Limnivores:** Microphages that dig into the mud to collect food. It is possible that some birds of the Mesozoic probed the mud using long and thin beaks.

Mycophagous, micovores or fungivores: They consume fungi. No bird feeds exclusively on this source.
Monophagous: They eat only one type of food. There are several cases in modern birds, but no known fossil species provides evidence of it.
Omnivores or polyphagous: They take advantage of various sources of food. Polyphagy is common in opportunistic species that take advantage of diverse sources of energy. Since stomachs have been discovered containing very different prey, it has been speculated that some theropods would have this eating habit.
Saprophagous: They take advantage of decomposing organic matter. Scavengers are saprophagous, but the practice extends to other sources, such as plants.

Saurornitholestes cf. *langstoni*
Specimen: ALMNH 2001.1   *Lambeosaurus lambei*
Specimen: ROM 1218

It is difficult to determine whether a dinosaur skeleton with tooth marks was hunted by a theropod, a group of theropods, or if it was already dead when it was eaten. If we check the differences between modern predators and their prey, either mammals or birds, we can recognize a limit that suggests some popular scenes are myths.

Lonely hunters

Felines, mongooses, raccoons, bears, dogs, and other carnivorous mammals usually kill prey smaller than themselves. Some felines or weasels can take down prey up to almost five times their own weight, but the largest predators usually do not hunt prey that is more than three times their own weight. This suggests that it would be difficult for a large theropod to kill prey that was three times its own weight. Subduing such prey would be dangerous. In contrast, very small carnivores tend to deal with organisms smaller than themselves, although there are always exceptions. We must also consider that large modern pachyderms (elephants, hippos, and rhinoceroses) are very dangerous and difficult to overcome, due to their size, strength, and the hardness of their skin. We can suggest that the ancient image of a 3 t *Allosaurus* killing a 20 t *Apatosaurus* is not viable. The false killer whale (*Pseudorca crassidens*) has a specialized way of feeding on some giant whales; it bites and tears a piece of meat from the victim without killing it, a practice known as "flesh grazing." It is possible that some theropods fed on huge sauropods this way, although it has not been proven. See more: *Brinkman et al. 1998; Paul 1988, 1998; Hone & Benton 2005; Tucker & Rogers 2014*

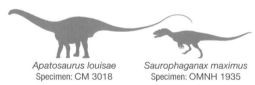

Apatosaurus louisae
Specimen: CM 3018   *Saurophaganax maximus*
Specimen: OMNH 1935

Apatosaurus louisae (CM 3018) 20 t vs. *Saurophaganax maximus* (OMNH 1935) 4.5 t: **A difference of 4.4 times.**

Apatosaurus louisae (CM 3018) 20 t vs. *Allosaurus fragilis* (NMMNH 26083) 2.9 t: **A difference of 7 times.**

Pack hunters

Chances of success improve significantly when hunting is done in a group. Hyenas and felines usually kill prey up to almost five times their own weight or even more. Dogs exceed this limit, killing prey eight times as heavy and, in the case of a pack of wolves (*Canis lupus*), killing a bison (40 kg vs. 500 kg), up to thirteen times as heavy. Comparing this data, we can suggest that it is not impossible that a group of *Deinonychus* found associated with the skeleton of a 1 t *Tenontosaurus tilletti* (YPM 5466) could hunt it. In this case, the difference in weight is about fifteen times, practically the same as that observed between groups of lions and young elephants or adult giraffes. Another popular image, that of a group of *Saurornitholestes* subduing a medium-size *Lambeosaurus,* is unlikely because the weight difference is quite extreme: about thirty times. On the other hand, the discovery of an enormous titanosaur along with the teeth of several *Tyrannotitan* may be a case of a group hunt, although a dangerous one for the carnivores. See more: *Maxwell & Ostrom 1995; Canale et al. 2014*

Indeterminate titanosaur   *Tyrannotitan chubutensis*
Specimen: MPEF-PV

Indeterminate titanosaur 28 t vs. *Tyrannotitan chubutensis* (MPEF-PV) 7 t: **A difference of 4 times.**

Deinonychus antirrhopus
Specimens: YPM 5201, 5202, 5203, 5205 & 5206

Tenontosaurus tilletti
Specimen: YPM 5466

Tenontosaurus tilletti (YPM 5466) 1 t vs. *Deinonychus antirhopus* (YPM 5201, 5202, 5203, 5205 & 5206) (68 + 57 + 48 + 48 + 43 kg) 264 kg: **A difference of 15 to 18 times.**

Saurornitholestes cf. *langstoni* (ALMNH 2001.1) 87 kg vs. *Lambeosaurus lambei* (ROM 1218) 2.5 t: **A difference of 29 times**

Share or fight for food

Very few birds or "reptiles" coordinate a group hunt and then distribute the remains peacefully. Normally, the law of the strongest and the most astute reigns. This would indicate why some theropods have deep wounds similar to those left by their congeners, or why some juveniles and subadults appear with wound marks caused by carnivores. This occurs in *Allosaurus fragilis, Daspletosaurus torosus, Deinonychus antirrhophus, Herrerasaurus ischigualastensis, Sinraptor dongi,* and *Tyrannosaurus rex.* Though uncooperative when distributing food, footprints provide evidence of gregarious behavior. Perhaps for this reason, juveniles of some tyrannosaurids were faster than adults, and avoided becoming victims of the adults. See more: *Tanke & Rochschild 2002; Roach & Brinkman 2007; Li et al. 2007; Hone & Tanke 2015*

Apex predators

Apex predators are the dominant species in an ecosystem. For example, the killer whale is capable of imposing itself on the biggest sharks, lions push away hyenas, and a bear can steal wolf kills. It is uncommon for more than one dominant predator to occupy the same territory unless the predators specialize in different prey. Among theropods that have occupied the upper part of the food chain in different places and times are some abelisaurids, primitive megalosauroids, megalosaurids, metriacanthosaurids, allosaurids, carcharodontosaurids, primitive megaraptors, megaraptorids, tyrannosauroids, tyrannosaurids, and, rarely, some dromaeosaurids. They were responsible for regulating a number of herbivores, so their disappearance always affected the balance of an ecosystem. See more: *Bakker et al. 1992*

Diplodocus carnegii vs. Allosaurus fragilis
Specimens: CM 84 vs. AMNH 680
The difference in weight between the two was 5.2 times (11.5 t vs. 2.2 t), so it is likely that this predator required help to subdue a prey of that size.
See more: *Chure 2000; Rothschild, Tanke & Ford 2001; Bakker, Robert & Gary 2004; Roach & Brinkman 2007*

THE DIET OF DINOSAUROMORPHS, THEROPODS, AND MESOZOIC BIRDS

1797—England—The first carnivorous theropod
The holotype of *Megalosaurus bucklandii* was discovered in 1797 and was already appreciated as a large and carnivorous animal. See more: *Gunter 1925; Buffetaut 1991*

1838—France—The first gastrolith in a theropod
It is believed that *Poekilopleuron* had gastroliths, but the material was destroyed in World War II, so it is no longer possible to verify. See more: *Eudes-Deslongchamps 1838*

1841—England—The first piscivorous theropod
Suchosaurus cultridens was mistaken for a crocodile due to its incompleteness and its piscivorous adaptation. It may be a synonym of *Baryonyx*, since they were contemporary. See more: *Owen 1841*

1844—USA—The first dinosaur coprolite
A few coprolites from the Lower Jurassic of the Connecticut River valley were identified as birds. Due to their age, they may have been produced by dinosaurs. See more: *Hitchcock 1844*

1881—Germany—The first stomach content in a theropod
A small skeleton was inside the type specimen of *Compsognathus longipes*. It was suggested that it may have been an embryo, but it was a lizard, *Bavarisaurus* cf. *macrodactylus*. See more: *Longrich et al. 2010*

1888—England—The first herbivorous theropod
There is doubt whether *Thecospondylus daviesi* is an ornithomimosaur or a therizinosaur. See more: *Seeley 1888; Allain et al. 2014; Mortimer**

1893—England—The first incorrect report of a scavenger theropod
A fragment of *Sarcolestes leedsi* is an ankylosaur that was considered a carnivorous dinosaur because of its sawlike teeth, a typical feature in many phytophagous animals. See more: *Lydekker 1893; Galton 1983*

1903—USA—The first evidence of theropod bites
Caudal vertebrae of a *Brontosaurus* showed tooth marks, presumably left by *Allosaurus*. It is not known if *Allosaurus* left them before or after the *Brontosaurus* died. See more: *Riggs 1903; Fastovsky & Smith 2004*

1911—USA—The oldest theropod with gastroliths
There were gastroliths in the type specimen of *Podokesaurus holyokensis* dating from the Early Jurassic (Hettangian, 201.3–199.3 Ma). See more: *Talbot 1911; Wings 2004*

1984—USA—The first herbivorous dinosauromorph
Technosaurus smalli was considered a sauropodomorph or an ornithischian because its teeth were adapted for a plant-based diet. It is now known that it was a silesaurid dinosauromorph. See more: *Chatterjee 1984*

1989—USA—The oldest cannibal theropod
The remains found within *Coelophysis bauri* AMNH 7224, from New Mexico, were considered to be a cannibalized hatchling, but it was found some time later to be a small crocodilomorph. *Coelophysis* bones present in coprolites and in a regurgitalith attributed to the species now reaffirms its cannibalism. From the Upper Triassic (Rhaetian, 208.5–201.3 Ma). See more: *Colbert 1989; Rinehart et al. 2005; Nesbitt et al. 2006*

1995—Canada, USA—Most recent non-avian theropod coprolite
Some large coprolites from the late Upper Cretaceous (upper Maastrichtian, ca. 69–66 Ma) are assigned to *Tyrannosaurus rex*. See more: *Anonymous 1995; Chin et al. 1998; Tokaryk & Byrant 2004*

1996—USA—The deepest theropod bite marks
The specimen of *Triceratops* (MOR 799) from Montana has between 58 and 70 tooth marks in the pelvis and sacrum. The most notorious mark is 2.5 cm wide and 3.7 cm deep, which has been compared to the tooth of an adult *Tyrannosaurus rex*, its probable author. See more: *Erickson & Olson 1996*

1996—Mongolia—The smallest insectivorous theropod
The smallest alvarezsaurid was *Parvicursor* remotus, which was also one of the tiniest non-avian theropods. See more: *Karhu & Rautian 1996*

1998—Italy—The theropod with the most prey inside
A high diversity of remains has been found in a juvenile *Scipionyx*: two species of teleost fish, two lizards of different sizes, and other undetermined remains.
See more: *Sasso & Signore 1998*

1999—Canada—The largest theropod coprolite
TMP 98.102.7 measures 64 x 17 x 11 cm, with a volume of about 6 L. It is suggested that it was produced by a *Gorgosaurus libratus*, a *Daspletosaurus* sp., or another contemporary tyrannosaurid. See more: *Eberth & Sloboda 1999; Chin et al. 2003*

2000—Argentina, Canada, USA, Mexico—The largest terrestrial carnivorous theropods
Gigantosaurus carolinii and *Tyrannosaurus rex* were the largest predatory dinosaurs. They were 13.2 m and 12.3 m long, respectively. Each weighed about 8.5 t. See more: *Calvo & Coria 2000; Longrich et al. 2010; Hartman**

2000—New Mexico, USA—The oldest theropod with gastroliths
Tentative gastroliths are known in a specimen of *Megapnosaurus kayentakatae*. See more: *Matora vide Whittle & Everhart 2000*

2000—USA—The most recent non-avian theropod with gastroliths
Gastroliths were discovered in *Tyrannosaurus rex*. The species dates from the late Late Cretaceous (upper Maastrichtian, 72.1–66 Ma). See more: *Currie 1997; Wings 2004*

2000—China—The non-avian theropod with the smallest gastroliths
Caudipteryx zoui gastroliths were from 4 to 4.5 mm long. See more: *Ji et al. 1998*

2001—China—The theropod with the most seeds inside it
Over fifty *Carpolithus* seeds were found in a specimen of *Shenzouraptor sinensis*. They could have been eaten in fruit or on their own. See more: *Zhou & Zhang 2002*

2001—China—The first and only theropod with a swallowed poisonous prey
Zhangeotherium remains were found inside a *Sinosauropteryx prima*. It is known that this symmetrodont had poisonous spines on its hind legs. See more: *Currie & Chen 2001*

2001—Spain—The first theropod regurgitalith with the greatest diversity of species
Pellets (balls formed by the remains of undigested and regurgitated food) from a probable theropod or bird were found. They contained three skeletons of juvenile birds of different species and sizes. See more: *Sanz et al. 2001*

2001—China—The first herbivorous theropod of the Jurassic
Eshanosaurus deguchianus is an enigmatic theropod or sauropodomorph, similar to therizinosaurs, although it is too old to be one. A similar case occurs with *Chilesaurus diegosuarezi* (present-day Chile), which is difficult to classify due to convergent characters. See more: *Xu et al. 2001; Barrett 2009; Novas et al. 2015; Mortimer**

2003—China—The first Mesozoic bird with gastroliths
A specimen of *Sapeornis chaoyangensis* contained numerous gastroliths that were 2–2.5 mm in diameter. See more: *Zhou & Zhang 2003*

2003—Poland—The first herbivorous dinosauromorph in western Europe
Twenty specimens of *Silesaurus opolensis* were found together. Teeth and the pre-bone reveal that it ate plants. See more: *Dzik 2003*

2003—Canada—The first coprolite with muscle markings
Within RTMP 98,102.7, a tyrannosaurid coprolith, the muscle fibers of its prey were preserved. See more: *Chin et al. 1999, 2003*

2004—China—The Mesozoic bird with half-eaten prey
Few fossils show half-consumed prey. The first known case among the birds is in *Yanornis martini*, which clutched the fish *Jinanichthys longicephalus* in its beak. Similar specimens were discovered nine years later. See more: *Yuan 2004; Zheng et al. 2013*

2004—China—The most common prey in Mesozoic birds
The osteoglossiform fish *Jinanichthys* has been found as food in several specimens of *"Confuciusornis" jianchangensis*, *Confuciusornis sanctus*, and *Yanornis martini*. See more: *Yuan 2004; Dalsätt et al. 2006; Zheng et al. 2013*

The omnivorous theropod with the largest crest
Anzu wyliei
Specimen: ROM 1247
Their beaks were not adapted for crushing food, so it is more likely that oviraptorosaurs consumed plants, small animals, and even eggs. *Anzu* had the largest ridge known in a Cretaceous theropod. It was about 30 cm long and 21 cm high. See more: *Funston et al. 2013; Lamanna et al. 2014*

2004—Brazil—Theropod or ornithopod fossil urine?

Only one dinosaur urolith has been documented. These are strange marks very similar to those produced by ostriches when urinating. This fact suggests that its author must have had a penis similar to that of only 3% of current birds. See more: *Fernandes et al. 2004; Brennan & Prum 2011; Souto & Fernandes 2015*

2004—Brazil—The first dinosaur urolith

Urine marks similar to those left by an ostrich were found. They are associated with traces of an indeterminate bipedal dinosaur. See more: *Adorna-Fernandes et al. 2004*

2005—Argentina—The oldest primitive saurischian coprolite

An analysis of fossil feces assigned to *Herrerasaurus ischigualastensis* showed that this predator had the ability to digest bones. They date from the Upper Triassic (lower Carnian, 237–222 Ma). See more: *Hollocher et al. 2005*

2005—Algeria, Egypt, and Morocco—The largest piscivorous theropod

Spinosaurus aegyptiacus had difficulty walking on land because it was so adapted to aquatic life. The largest specimen (MSNMV4047) was the longest theropod (16 m long). It weighed only 7.5 t because its body was fairly light for its size. It is suggested that it was mainly piscivorous because of its conical teeth and elongated face. See more: *Dal Sasso et al. 2005*

2005—New Mexico, USA—The oldest stomach contents

The NMMNH cololite P-42352, attributed to *Coelophysis bauri* from the end of the Upper Triassic (Rhaetian, ca. 208.5–201.3 Ma), has the remains of a presumably cannibalized juvenile. See more: *Rinehart et al. 2005*

2005—New Mexico, USA—The oldest theropod coprolite

Some fossil feces are assigned to *Coelophysis bauri* due to their proximity to the abundant fossils of this predator from the Upper Triassic (Rhaetian, ca. 208.5–201.3 Ma). See more: *Rinehart et al. 2005*

2005—China—The smallest herbivorous theropod

Hongshanornis longicresta was found with seeds inside it, suggesting that it was a granivorous bird (grain feeder). It was one of the smallest ornithuromorphs. It was about 12 cm long and weighed 58 g. See more: *Zhou & Zhang 2005; Chiappe et al. 2014*

2006—China—The Mesozoic bird with the smallest gastroliths

Yanornis martini gastroliths were quite small. They were between 0.2 and 2.7 mm in diameter. Gastroliths in *Hongshanornis longicresta*, a smaller bird, were 1 mm long, on average. See more: *Zhou & Zhang 2006; Zheng et al. 2011*

2007—China—The theropod species with the greatest diversity of prey

Remains of *Sinornithosaurus millenii* (dromeosaurid) and *Confuciusornis* were found within a specimen of *Sinocalliopteryx gigas*. In another specimen of the same species, ornithischian remains were found. Furthermore, a tooth of this species was found embedded in a rib of *Dongbeititan dongi*, suggesting that *Sinocalliopteryx gigas* was an opportunistic predator. A similar case occurred with *Microraptor zhaoianus*: The remains of a mammal, an enantiornithean bird, and a fish were found in three different specimens.

Tyrannosaurus rex is a complicated case, since it can not be confirmed that all the bones that bear the great predator's tooth marks were hunted by it. These marks appear on the bones of *Edmontosaurus*, *Thescelosaurus*, *Triceratops*, even other *Tyrannosaurus*. See more: *Erickson & Olson 1996; Ji et al. 2007; Gong et al. 2010; Larsson et al. 2010; Longrich et al. 2010; O'Connor, Zhou & Xu 2011; Xing et al. 2012, 2013*

2008—Lebanon—The most unexpected gastric contents

Amber pieces between 0.5 and 1.8 mm long were found inside *Enantiophoenix electrophyla*, suggesting that it consumed sap or honey. The remains hardened during fossilization. There is also the possibility that amber was used as gastroliths. Very few modern vertebrates eat plants with resin. Some that do are the dusky grouse (*Dendragapus obscurus*) and the western capercaillie (*Tetrao urogallus*), although they consume it only if necessary. See more: *Cau & Arduini 2008; del Hoyo et al. 1994*

2008—USA—The first evidence of predation by a theropod

Many theropod bite marks have been interpreted as the results of being eaten as carrion, accidents, or an attack. The circumstances under which they were created are unknown. Healed wounds on a young *Triceratops* specimen are evidence of an attack while it was alive. See more: *Happ 2008*

2009—Canada—The first ichnospecies based on a theropod bite

Ichnites are based mainly on footprints, although in some cases they can be marks of such activities as resting, scratching, or biting. *Linichnus serratus* was created to identify a particular form of bite mark in bones. See more: *Jacobsen & Bromley 2009*

2009—China—The oldest herbivorous theropod

Limusaurus inextricabilis dates from the Upper Jurassic (Oxfordian, ca. 163.5–157.3 Ma). The oldest is the dubious theropod *Eshanosaurus deguchiianus* of the Lower Jurassic (Hettangian, ca. 201.3–199.3 Ma), which could have been a sauropodomorph with some convergence with tericinosaurs. See more: *Xu et al. 2009*

2010—China—The oldest insectivorous dinosaur

Because alvarezsaurs have fine faces, tiny teeth, specialized arms, and developed chests and they are small in size, they may have fed on social insects. The most primitive species, *Haplocheirus sollers*, had sturdy arms and a significantly enlarged thumb; perhaps it consumed termites. It dates from the Upper Jurassic (Oxfordian, 163.5–157.3 Ma). See more: *Choiniere et al. 2010*

2010—USA—The most recent cannibalism in non-avian theropods

There is evidence of cannibalism in at least four specimens of *Tyrannosaurus rex* from the late Upper Cretaceous (upper Maastrichtian, 72.1–66 Ma). See more: *Longrich et al. 2010*

2011—China—The smallest piscivorous theropod

It is suggested that long-lived enantiornithean birds fed on aquatic animals, including fish and crustaceans. The smallest is *Shanweiniao cooperorum*, which was 9 cm long and weighed about 10.5 g. See more: *Zhang et al. 2001; O'Connor et al. 2009*

2011—China—The first venomous theropod?

According to researchers, the unusually long fangs of *Sinornithosaurus millenni* have a fine groove that may have been used to inject a toxic substance into prey, but this groove is common in all theropod teeth. Another unusual feature in *S. millenni* is a rough structure in its jaws where a gland could have been. Some authors mention that several individuals of the species lack this structure. So was it poisonous or not? We will probably never know. See more: *Gianechini et al. 2011; Gong et al. 2011; Bergillos & Rivas 2013*

2011—South Korea—The largest theropod bite marks

Predation marks were found on the caudal vertebrae of the sauropod *Pukyongosaurus millenniumi*. They were 17 cm long, 2 cm wide, and 1.5 cm deep See more: *Paik et al. 2011*

2012—Argentina—The largest insectivorous theropod

Advanced alvarezsaurids have anatomical characteristics that suggest an insectivorous diet. Even the largest of all, *Bonapartenykus ultimus*, which was 3 m long and weighed 35 kg, was smaller than the giant anteater (*Myrmecophaga tridactyla*), which consumes between 30,000 and 35,000 ants daily. See more: *McGhee 2011; Agnolin et al. 2012*

2014—China—The smallest durophagous theropod

Bohaiornithid enantiornithean birds stood out for having teeth adapted for hard food. The smallest of them was *Longusunguis kurochkini*, which was 18 cm long and weighed 88 g. See more: *Wang et al. 2014*

2014—Argentina—The largest dinosaur consumed by theropods

The largest fossil that has been linked to theropod feeding activity is a titanosaur that was at least 24 m long and weighed about 28 t. Five other disarticulated skeletons were found along with fifty-seven teeth of *Tyrannotitan chubutensis*, abelisaurids, and other indeterminates with teeth similar to dromaeosaurs. See more: *Canale et al. 2014*

2014—Canada, USA, Mexico—The largest durophagous theropod

Tyrannosaurs had the ability to break bones with their thick teeth and powerful jaws. The largest of all was *Tyrannosaurus rex*, which was 12.3 m long and weighed 8.5 t. Four of this species' teeth were identified in Mexico, but some authors question it. See more: *Erickson et al. 1996; Meers 2002; Longrich et al. 2010; Bates & Falkingham 2012; Brañas et al. 2014; Ramírez-Velasco & Hernández-Rivera*

2014—Mongolia—The largest herbivorous theropod

The largest theropods that had anatomical characteristics of a herbivore is *Deinocheirus mirificus*. The largest individual (IGM 100/127) was up to 12 m long and weighed 7 t. See more: *Lee et al. 2014*

2014—Mongolia—The theropod with the most gastroliths

Deinocheirus mirificus.

2014—Mongolia—The largest gastrolith in a theropod

Deinocheirus mirificus gastroliths were 8–87 mm in diameter. In ostriches, they can be up to 100 mm and do not usually exceed 1% of their body weight. See more: *Wings 2002, 2004; Lee et al. 2014*

2015—Mongolia—The largest omnivorous theropod

Direct evidence has been found that oviraptorosaurs hunted the young of other theropods. Their jaws were also adapted for a phytophagous diet. The largest of all was a huge specimen assigned to *Gigantoraptor erlianensis* (MPC-D 107/17). It was 8.9 m long and weighed 2.7 t. See more: *Norell et al. 1994; Norell et al. 2000; Funton, Currie & Murray 2013*

2015—China—The smallest carnivorous theropod

A primitive dromaeosaurid similar to *Sinornithosaurus* was an adult that was 46 cm long and weighed 175 g. Some enantiornithean birds had ziphodonts (knife-like teeth) that could have been used to hunt invertebrates. See more: *Pittman et al. 2015*

The largest theropod coprolite
Specimen: TMP 98.102.7
This piece from the late Upper Cretaceous of western Laurasia (present-day Canada) is 64 cm long, 17 cm wide, and 11 cm high. Its total volume is about 6 L. See more: *Eberth & Sloboda 1999; Chin et al. 2003*

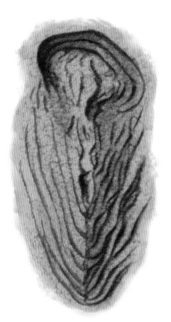

The only dinosaur fossil urine
Specimen: MPA-001
Three strange marks similar to those produced by urinating ostriches were found associated with ornithopod and theropod footprints. They date from the late Lower Cretaceous and were from western Gondwana (present-day Brazil). The largest is 75 cm long and 45 cm wide. See more: *Fernandes et al. 2004; Brennan & Prum 2011; Souto & Fernandes 2015*

The theropod with the most curved jaw
Noasaurus leali
Specimen: MACN-PV 622
Perhaps all the noasaurids had curved jaws with front teeth that projected out- and downward, suggesting they had a specialized diet. This feature is also seen in *Masiakasaurus*, although it is not as pronounced as in *Noasaurus*. See more: *Bonaparte & Powell 1980; Sampson et al. 2001; Headden**

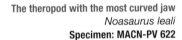

The theropod with the largest teeth in proportion to its size
Daemonosaurus chauliodus
Specimen: CM 76821
Its name means "lizard demon of big teeth." Its short head accentuates the length of its teeth. This is a unique combination in primitive theropods. See more: *Sues et al. 2011; Headden**

When we talk about dinosaur pathology, we refer to the evidence of diseases or injuries that they suffered and affected their bones. In some cases, pathologies can be identified in footprints, teeth, skin, and eggs. Here we divide paleopathologies into two parts: (1) physical-traumatic disorders or traumas incurred by accidental damage or aggression from another organism (fractures, bites, scratches, blows, etc.) and (2) nontraumatic osteopathies (or osteopathies only), which include diseases that do not involve physical impact damage (such as deformations, bone fusion, malformations, tumors, and infections).

RECORDS IN PRIMITIVE SAURISCHIANS, THEROPODS, AND MESOZOIC BIRDS

1838—France—The first theropod pathology
An inflammation was detected in the chevron of *Poekilopleuron*. It also had a phalanx with an exostosis (benign tumor) and a third unknown pathology. See more: *Eudes-Deslongchamps 1838*

1879—USA—The first traumatic theropod pathology
Labrosaurus ferox was based on a specimen damaged by an *Allosaurus fragilis*. See more: *Marsh 1879*

1884—USA—The first traumatic theropod leg pathology
Three metatarsals of the only conserved leg of the *Ceratosaurus nasicornis* type specimen had strange fusions reminiscent of birds. They were later recognized as a healed fracture. See more: *Marsh 1884; Molnar 2001*

1896—Wyoming, USA—The first theropod that was diagnosed by traumatic pathological pieces
Labrosaurus ferox (a probable synonym of *Allosaurus fragilis*) was considered a strange predator until it was recognized that it had a deformation in the lower jaw due to trauma. See more: *Marsh 1896; Gilmore 1920*

1915—Egypt, USA—The first fractures discovered in theropods
Healed fractures were found in an *Allosaurus fragilis* scapula. In the same year, ruptures were reported in the neural spines of *Spinosaurus aegyptiacus*. See more: *Gilmore 1915; Stromer 1915*

1919—USA—The first theropod pathologies of the Cretaceous
A pathological fusion is reported in the vertebrae of a *Tyrannosaurus rex*. See more: *Moodie 1919*

1933—USA—The first dinosaur hemangioma
An unidentified dinosaur from Utah had an abnormal buildup of blood vessels. See more: *Anonymous 1933; Tanke & Rothschild 2002*

1935—USA—The first report of a traumatic pathology in theropod footprints
A possible finger amputation in *Eubrontes* and an apparent pattern of limping in Connecticut were identified. A report of a *Sauroidichnites abnormis* footprint precedes it by ninety-one years. Although sometimes interpreted as such, it does not seem to belong to a theropod. See more: *Hitchcock 1844; Abel 1935; Lockley 1991; Lockley et al. 1994*

1947—USA—The oldest mass death in theropods
Hundreds of skeletons of *Coelophysis bauri* were found in the town of Ghost Ranch, New Mexico. Of different sizes and ages, they died because of a river flood. They date from the Upper Triassic (Rhaetian, ca. 208.5–201.3 Ma). See more: *Colbert 1947; Schwartz & Gillette 1994*

1966—Romania—The first possible osteopetrosis in a dinosaur
A rare disease known as "marble bones" due to its extreme density was present in an unidentified dinosaur (specimen BMNH R. 5505-06). See more: *Campbell 1966*

1969—Mongolia—The first dinosaur egg with pathologies in Asia
A review of the microstructure in shells reveals multiple layers, an anomaly that occurs in birds and reptiles under stress. See more: *Sochava 1969, 1971*

1971—Mongolia—The first theropod found in battle
A *Velociraptor mongoliensis* was found intertwined with a *Protoceratops andrewsi*. Each intended to hurt the other. The *Velociraptor* had its sickle-shaped claw directed toward the Protoceratops's neck and was biting its arm. The two must have died simultaneously. They may have been buried by a sandstorm. See more: *Kielan-Jaworowska & Barsbold 1972; Carpenter 1998*

1971—Texas, USA—The first tracks interpreted as evidence of a "battle" between dinosaurs
Tracks left by a large theropod run parallel to the trail of a sauropod, suggesting that the theropod was pursuing the sauropod, but it is now thought that there is not enough evidence that this really happened. Other tracks have inspired similar interpretations, but they, too, lack sufficient support. See more: *Bird 1939; Farlow et al. 2012*

1972—USA—The first degenerative disease in a theropod
Allosaurus chevrons have a rough texture on their surface, in addition to being strangely short. See more: *Petersen, Isakson & Madsen 1972*

1972—USA—The first amputation discovered in a theropod
A specimen of *Allosaurus fragilis* provided evidence of an amputated and infected toe (osteomyelitis), but some paleontologists suggest that they are not true pathologies, as they may have emerged after death. See more: *Petersen, Isakson & Madsen 1972; Tanke & Rothschild 2002*

1976—USA—The first evidence of cannibalism in theropods
Unidentified tyrannosaurid bones show signs of being bitten by individuals of the same species. See more: *Tracy 1976*

1977—Wales—The oldest theropod footprints with pathologies
A possible malformation is mentioned in digit III of *Anchisauripus thomasi*. See more: *Tucker & Burchette 1977*

1982—Wyoming, USA—The youngest theropod bone pathology
The hatchling UCM 41666 of *"Pectinodon" bakkeri* had a congenital tooth defect. See more: *Carpenter 1982*

1982—Canada—The first injury in a juvenile theropod
A cyst appears in the skull of specimen TMP79.8.1, a juvenile of *Troodon formosus*. See more: *Currie 1985*

1987—The first traumatic pathology in theropod teeth
Examined theropod teeth revealed cracks and enamel fractures. See more: *Rensberaer 1987*

1988—USA—The first bite marks of a theropod?
Edmontosaurus caudal vertebrae show damage to the tail that has been attributed to a *Tyrannosaurus* attack, but some authors suspect the injury could have been caused when another hadrosaur stepped on the tail. See more: *Carpenter 1988; Tanke & Rothschild 2014*

1991—Wyoming, USA—The theropod specimens with the most trauma
Allosaurus fragilis MOR 693, nicknamed "Big Al," died young. He was a subadult that had nineteen bone fractures (ribs, vertebrae, and phalanges) and signs of a toe infection (osteomyelitis). Another more recent specimen, SMA 0005 or "Big Al 2," has thirteen damaged bones, among which are the dentary, cervical, and thoracic vertebrae, ribs, scapula, humerus, ischium, and phalanges of the foot. See more: *Breithaupt 1996; Foth et al. 2015*

1991—USA—The first avulsion reported in a theropod
A damaged tendon in the arm of a *Tyrannosaurus rex* was reported. See more: *Larson 1991*

1991—Mongolia—The first pathologies discovered in theropod oolites
Some specimens of *Elongatoolithus andrewsi*, *E. elongatus*, and *Macroolithus rugustus* have anomalies in their shells. See more: *Zhao et al. 1991*

1992—USA—The first recognized cancer in a theropod
A malignant tumor known as chondrosarcoma can be seen in the humerus of an *Allosaurus fragilis*. See more: *Graham 1992; Greenberg 1992; Taylor 1992; Wycombe 1992*

1992—Germany—The first traumatic pathology in a theropod of the Triassic
A fracture in the fibula of *Procompsognathus triassicus* was identified. See more: *Sereno & Wild 1992*

1993—Argentina—The first traumatic pathology in a primitive saurischian
A specimen of *Herrerasaurus* has bite marks that could have been left by another individual of the same species or by a carnivorous archosaur. See more: *Sereno & Novas 1993*

1993—Argentina—The oldest intraspecies predation marks in primitive saurischians
It is suspected that the wounds found in a specimen of *Herrerasaurus ischigualastensis* were made by another individual of the same species. They date from the Upper Triassic (lower Carnian, ca. 237.0–232.0 Ma). See more: *Sereno & Novas 1993*

1993—USA—The most recent intraspecies predation marks in theropods
There is evidence of *Tyrannosaurus rex* attacks on other individuals of the same species in the late Upper Cretaceous (upper Maastrichtian, ca. 69–66 Ma). See more: *Longrich et al. 2010*

1993—Canada and USA—The most recent intraspecies theropod predation marks
A juvenile *Triceratops* specimen (SUP 9713) shows healed wounds in the scaly bone and a fractured horn that was bitten off. Since the injuries are partially healed, we know that the juvenile survived a *Tyrannosaurus rex* attack. Lesions with bite marks, including one where the tip of a tooth remains embedded in the bone, have been found in specimens of *Edmontosaurus*. See more: *Carpenter 1998; Happ & Carpenter 2008; DePalma II et al. 2013; Bell & Campione 2014*

1993—Argentina—The oldest traumatic bone pathology in a primitive saurischian
Three bite marks left by an unknown predator were found in a *Herrerasaurus* specimen. It dates from the Upper Triassic (lower Carnian, ca. 237–232 Ma). See more: *Sereno & Novas 1993*

1993—Mongolia—The first discovered remains of a theropod consumed by a mammal
Archaeornithoides deinosauriscus was a baby dromaeosaurid

that has mammalian bite marks and passed through the digestive tract of a predator before fossilizing. See more: *Elzanowski & Wellnhofer 1992,1993; Clark et al. 2002*

1994—Portugal—The largest theropod footprints with pathologies

Footprints 67 cm long from the Upper Jurassic (present-day Portugal) left a track containing a few disparate steps, as if the individual was limping. See more: *Dantas et al. 1994; Lockley et al. 1994*

1995—China—The first evidence of an intraspecific fight between Jurassic theropods

Tooth marks on the head of *Sinraptor dongi* are interpreted as evidence of interaction with another *Sinraptor* over territory or food. Similar behavior has been observed in modern crocodiles. See more: *Farlow & Molnar 1995; Tanke & Currie 1995*

1995—Canada—The first evidence of an intraspecies fight between Cretaceous theropods

Tooth marks are reported on the skull of cf. *Albertosaurus*, produced by another specimen of the same species. See more: *Tanke & Currie 1995*

1996—USA—The first evidence of infection in a theropod

Possible marks of infection were identified in a *Tyrannosaurus rex* skull. See more: *Anonymous 1996*

1996—Colorado, USA—The oldest theropod oolite with pathologies

Preprismatoolithus coloradensis MWC 122.2/HEC 418-R2, from the Upper Jurassic (Kimmeridgian, ca. 157.3–152.1 Ma), has various anomalies in its shell. One fragment shows an irregular structure, while another has several units of overgrowth. See more: *Hirsch 1994*

1996—Mongolia—The most recent non-avian theropod oolite with pathologies

Some irregularities have been reported in the shells of *Elongatoolithus andrewsi*, *E. elongatus*, and *Macroolithus yaotunensis*. All date from the late Upper Cretaceous (Maastrichtian, 72.1–66 Ma). See more: *Zhao et al. 1991*

1996—China—The oldest possible dinosaur parasite

Hadropsylla sinica was an ancestral flea from the middle of the Jurassic (Callovian, ca. 166.1–163.5 Ma). Because of its large size, it has been interpreted as a large pterosaur and dinosaur parasite, but a new study suggests that it was more likely to attack only Mesozoic mammals. See more: *Huang et al. 2013; Zhu et al. 2015*

1996—USA—The oldest tick that may have infected dinosaurs

The tick *Carios jerseyi* was found on a feather preserved in amber in New Jersey. The tick was from the Argasidae family. Currently, these arachnids feed on warm-blooded animals for brief periods. An older and doubtful case is of tiny bodies (70 microns) discovered in a fossil feather (present-day Brazil). It is suspected that they belonged to crustaceans called ostracods, freshwater and saltwater animals that were preserved on the feather by happenstance. See more: *Grimaldi & Case 1995; Martill & Davis 1998, 2001; de la Fuente 2003; Proctor 2003*

1996—Mongolia—The first association between pupae and dinosaurs

Pupae of insects (15–18 mm x 8–9 mm) were found in the vicinity of some dinosaur nests. The pupae were named *Fictovichnus gobiensis* and at first were attributed to beetles but now are considered wasps. See more: *Johnston et al. 1996; Molina 2015*

1996—The first virus disease originated in Mesozoic birds

It has been suggested that hepatitis B was caused by a virus that originated among neoaves about eighty-two million years ago and later infected mammals. This was determined after sequencing the genome of the virus. See more: *Suh et al. 2013*

2000—New Mexico, USA—The oldest bone pathology in theropods

NMMNH P-29168 has fused posterior leg bones (tibia, fibula, talus, and calcaneus). It dates from the Upper Triassic (upper Norian, ca. 218–208.5 Ma). See more: *Heckert et al. 2000, 2003; Molnar 2001*

2000—Alaska, USA—The northernmost theropod with bone pathologies

Several teeth of *Dromaeosaurus albertensis*, *Troodon formosus*, and an unidentified tyrannosaurid, all from the AK collection, had identified anomalies. Its present-day latitude is 70.1° N, but in the Cretaceous, it was 84.1° N. See more: *Fiorillo & Gangloff 2000*

2001—Australia—The southernmost theropod with trauma

QM F34621, a phalanx of *Timimus hermani*, has a fracture. Its present-day latitude is 38.8° S, but its position in the Cretaceous was 76.1° S. See more: *Molnar 2001*

2001—Wyoming, USA—The most damaged theropod part

The majority of theropod fractures are found in thoracic ribs, followed by the caudal vertebrae. See more: *Molnar 2001*

Amazing recovery
Protoceratops andrewsi vs.
Velociraptor mongoliensis
IGM 100/25
Dinosaurs had the ability to recover relatively quickly from bone fractures so severe that they would kill any mammal.
See more: *Anné et al. 2015*

2002—Antarctica—The southernmost theropod bone pathology

Polarornis gregorii TTU P 9265, from Seymour Island (64.3° S), has lesions similar to hypertrophic osteoarthropathy in the tibia. See more: *Chatterjee 2002*

2002—Canada and USA—The largest variety of bone pathologies found in a Cretaceous species

Different individuals of *Tyrannosaurus rex* have reported asymmetry in the humerus, cancer, double-pointed teeth, double-serrated teeth, malocclusion (misalignment of the teeth), bone overgrowths such as osteophytes, and infections caused by fungi, bacteria, and protozoans. See more: *Tanke & Rothschild 2002; Wolff et al. 2009*

2002—Canada and USA—The greatest variety of traumatic bone pathologies found in a Cretaceous species

Cases of amputation, avulsion, cannibalism, fractures, infection, injuries, bites, and damaged serration have been reported or suggested in *Tyrannosaurus rex*. See more: *Tanke & Rothschild 2002*

2002—USA—The largest variety of bone pathologies found in a Jurassic species

Several *Allosaurus fragilis* fossils have tumors, bone overgrowths, osteoarthritis, and infections such as osteomyelitis. See more: *Hanna 2002; Tanke & Rothschild 2002*

2002—USA—The greatest variety of trauma found in a Jurassic species

Cases of abscesses (inflammation), amputation, broken teeth, and long-bone fractures have been found in *Allosaurus fragilis* fossils. See more: *Hanna 2002; Tanke & Rothschild 2002*

2002—The most common trauma in primitive theropods and saurischians

Fractures are the most most recognized damage in theropod species. They have been reported in specimens of *Acrocanthosaurus, Alber-* *tosaurus, Allosaurus, Carcharodontosaurus, Chirostenotes, Daspleto-saurus, Gorgosaurus, Herrerasaurus, Hesperornis, Monolophosaurus, Saurornitholestes, Sinraptor, Tarbosaurus, Tyrannosaurus* and *Velociraptor.* See more: *Tanke & Rothschild 2002*

2002—The most common traumatic bone pathology in theropods

The fusion of different bones occurs for various reasons, such as infections, arthritic diseases, tumors, and repaired fractures. Fusions in caudal vertebrae are more frequent in theropods. Cases have been reported in *Allosaurus, Appalachiosaurus, Beipiaosaurus, Ceratosaurus, Megaraptor, Monolophosaurus, Nomingia* and *Tyrannosaurus.* See more: *Tanke & Rothschild 2002; Porfiri et al. 2007*

2002—The most common taphonomic (fossil damage) marks in theropods and primitive saurischians

Bite marks (perforations or scrapes) have been documented in *Acrocanthosaurus, Albertosaurus, Allosaurus, Carcharodontosaurus, Chirostenotes, Daspletosaurus, Deinocheirus, Deinonychus, Gorgosaurus, Herrerasaurus, Monolophosaurus, Saurornitholestes, Sinraptor, Tarbosaurus, Tyrannosaurus* and *Velociraptor.* See more: *Tanke & Rothschild 2002*

2003—The most common bone pathology in theropods

Neoplasms are the most reported in various theropods, among which are *Acrocanthosaurus, Albertosaurus, Allosaurus, Archaeornithomimus, Carcharodontosaurus, Coelophysis, Deinonychus, Dromiceiomimus, Dryptosaurus, Gorgosaurus, Saurornitholestes, Spinosaurus, Stokesoaurus/Marshosaurus, Struthiomimus, Troodon, Tyrannosaurus* and *Utahraptor.* See more: *Rothschild et al. 2003*

2003—Canada—The specimen of a Cretaceous theropod with the most pathologies

A female specimen of *Gorgosaurus libratus* has numerous pathologies: fractured gastralia, calcified hematomas, a bifurcated toenail, a scapulocoracoid with a healed fracture, a deformity in the greater trochanter of the left femur, tooth loss due to osteomyelitis, and a tumor in the back of the skull that perhaps incapacitated her and caused her death. See more: *Evans & Larson 2003; Glut 2006*

2007—Madagascar—The most extreme amputation in a theropod

FMNH PR 2295, a *Majungasaurus crenatissimus*, was missing ten vertebrae from the tail. The remaining three were merged and infected due to the damage. See more: *Farke & O'Connor 2007*

2009—China—The first evidence of a theropod injured by a sauropod

An individual of *Sinraptor hepingensis* had a fracture at the level of the scapula. Because of its high location, it is speculated that it was produced by the impact of a caudal mace from the tail of a *Mamenchisaurus hochuanensis* or any other contemporary sauropod. In an episode of the documentary *Dinosaur Planet*, it was suggested that the lesion on the skull of *Aucasaurus garridoi* may have been caused by a titanosaur. See more: *Dinosaur Planet, "Alpha's Egg" 2003; Xing et al. 2009*

2014—Canada—Most recent theropod footprints with pathologies

The type specimen of *Bellatoripes fredlundi* had its finger II amputated. It dates from the late Upper Cretaceous (upper Campanian, ca. 83.6–72.1 Ma). See more: *McCrea et al. 2014*

2015—Canada—Theropod footprints with the strangest pathology

PRPRC 2002.01.001 has a very severe curvature in the fingers, which seems to be a serious dislocation that perhaps contributed to the rest of the foot's deformation. See more: *McCrea et al. 2015*

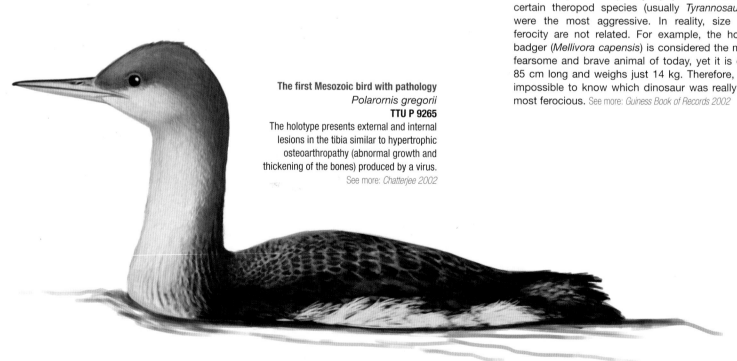

The first Mesozoic bird with pathology
Polarornis gregorii
TTU P 9265
The holotype presents external and internal lesions in the tibia similar to hypertrophic osteoarthropathy (abnormal growth and thickening of the bones) produced by a virus.
See more: *Chatterjee 2002*

MYTH: The most ferocious dinosaur

One idea repeated frequently in popular media is that certain theropod species (usually *Tyrannosaurus*) were the most aggressive. In reality, size and ferocity are not related. For example, the honey badger (*Mellivora capensis*) is considered the most fearsome and brave animal of today, yet it is only 85 cm long and weighs just 14 kg. Therefore, it is impossible to know which dinosaur was really the most ferocious. See more: *Guiness Book of Records 2002*

PATHOLOGICAL CURIOSITIES

It has been proposed that dromaeosaurs were parasitic animals that climbed onto large herbivores, which consequently developed numerous defenses, such as crests, spines, shields, and whips in order to get rid of them. The problem with this interpretation is that the structures existed long before the theropods appeared, and that they are also present in places where the dromaeosaurs were not. The idea is not plausible or true. See more: *Fraser 2014*

The teeth identified as *Paronychodon* were once thought to be pieces with pathology due to their strange shape, but now the teeth are known to belong to a particular type of a troodontid whose skeletal remains are not yet known. See more: *Currie et al. 1990; Sankey 2001; Sankey et al. 2002, 2005; Makovicky & Norell 2004*

Dinosaurs had an amazing ability to recover from bone fractures. This is suggested by the recovery of some from wounds so severe that they would be deadly for mammals. See more: *Anné et al. 2014*

Pathologies in theropods and other primitive saurischians

Non-traumatic osteopathies

Triassic
Coelophysis bauri

Jurassic
Allosaurus fragilis, Ceratosaurus dentisulcatus, Ceratosaurus nasicornis, Compsognathus longipes, Dilophosaurus wetherilli, Marshosaurus or Stokesosaurus?, Megalosaurus bucklandii, Megapnosaurus rhodesiensis, Monolophosaurus jiangi, Poekilopleuron bucklandii, Timimus hermani?, Torvosaurus tanneri, Veterupristisaurus milneri

Cretaceous
Acrocanthosaurus atokensis, Alectrosaurus olseni, Albertosaurus sarcophagus, Appalachiosaurus montgomeriensis, Archaeornithomimus asiaticus, Aucasaurus garridoi, Bambiraptor feinbergi, Bistahieversor sealeyi, Carcharodontosaurus saharicus, Caudipteryx sp., Chirostenotes sp., Confuciusornis sp., Daspletosaurus torosus, Deinonychus antirrhopus, Dromaeosaurus sp., Dromiceiomimus, Dryptosaurus aquilunguis, Falcarius utahensis, Gallimimus bullatus, Citipati osmolskae, Gorgosaurus libratus, Majungasaurus crenatissimus, Megaraptor namunhaiquii, Microraptor gui, Neovenator salerii, Nomingia gobiensis, Pectinodon bakkeri, Polarornis gregorii, Sinornithoides youngi, Saurornitholestes, Spinosaurus aegyptiacus, Struthiomimus altus, Troodon formosus, Tyrannosaurus rex, Velociraptor mongoliensis

Traumatic osteopathies

Triassic
Coelophysis bauri, Herrerasaurus ischigualastensis, Megapnosaurus rhodesiensis, Procompsognathus triassicus

Jurassic
Allosaurus fragilis, Ceratosaurus nasicornis, Compsognathus longipes, Guanlong wucaii, Labrosaurus ferox, Marshosaurus bicentesimus, Monolophosaurus jiangi, Ornitholestes hermanni, Sinraptor dongi, Sinraptor hepingensis

Cretaceous
Acrocanthosaurus atokensis, Albertosaurus sarcophagus, Alectrosaurus olsoni, Becklespinax altispinax, Carcharodontosaurus saharicus, Chirostenotes sp., Citipati osmolskae, Daspletosaurus torosus, Daspletosaurus sp., Deinocheirus mirificus, Deinonychus antirrhopus, Elmisaurus, Gorgosaurus libratus, Hesperonychus elizabethae, Hesperornis, Majungasaurus crenatissimus, Mononykus olecranus, Nanantius eos, Neovenator salerii, Ornithomimus sp., Utahraptor ostrommaysorum, Saurornitholestes, Talos sampsoni, Tarbosaurus bataar, Velociraptor mongoliensis, Troodon formosus, Tyrannosaurus rex, Velocisaurus unicus

Theropod footprints displaying pathologies

Anchisauripus thomasi, cf. Grallator, Eubrontes, Bellatoripes fredlundi

Theropod oolites displaying pathologies

Preprismatoolithus coloradensis, Triprismatoolithus stephensi, Elongatoolithus andrewsi, Elongatoolithus elongatus, Macroolithus yaotunensis

See more: *Tanke & Rothschild 2002, Jackson & Varricchio 2010; Rothschild, Xiaoting & Martin 2012; McCrea et al. 2014, 2015; Ford**

1:1

Pseudopulex magnus
Length: 22.8 mm

RECORD:
The largest Mesozoic flea.
Gao et al. 2012

Saurophthirus longipes
Length: 12 mm

RECORD:
The northernmost Mesozoic parasite. Latitude 53.3° N (paleolatitude 56.0° N).
Ponomarenko 1976

Saurodectes vrsanskyi
Length: 18.5 mm

RECORD:
The largest Mesozoic louse.
Rasnitsyn & Zherikhin 1999

Torirostratus pilosus
Length: 12 mm

RECORD:
The insect with the largest blood traces.
Yao et al. 2014

Tarwinia australis
Length: 7 mm

RECORD:
The southernmost Mesozoic parasite. Latitude 38.6° S (paleolatitude 77.0° S).
Jell & Duncan 1986

The trap with the most theropods
Guanlong wucaii &
Limusaurus inextricabilis
Specimens: IVPP V14531, IVPP V14532, IVPP V 15923, IVPP V 15924 & IVPP V16134
Five theropods were found buried one above the other. It is possible that a young Guanlong was prepared to eat the three trapped *Limusaurus*. Another adult was trapped when it tried to cannibalize the fourth dinosaur and possibly broke his neck. All were victims of the mud trap, which functioned like quicksand.
See more: *Xu et al. 2006, 2009*

Testimony in stone
Theropod footprints

Records: The largest and smallest footprints, the oldest and most recent, the strangest, etc

Fossil impressions are indirect evidence of past organisms' activities. Such impressions can be traces of footprints, bite marks, scratches, evidence of rest, integuments such as skin or feathers, and others.

"When you see a bone, a skeleton, you see its death. But when you see footprints, you see part of its life." —Ignacio Díaz Martinez, Paleontologist

Triassic

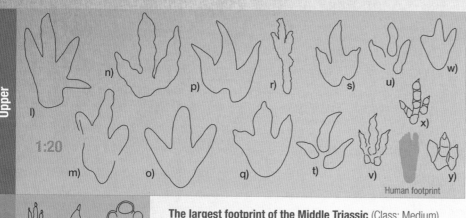

Upper

n) p) r) s) u) w)

l)

x)

1:20

m) o) q) t) v) y)

Human footprint

Middle

d) e) f) g)

h) i) j) k)

Lower

a) b)

c)

The largest footprint of the Middle Triassic (Class: Medium)

Sphingopus ladinicus (d) was longer than *Coelurosaurichnus sassandorfensis* (e) due to its highly developed heel. A 45 cm long footprint (Argentina) was dated from the Middle Triassic, but it is now known that it was from the beginning of the Upper Triassic. See more: *Kuhn 1958; Marsicano et al. 2006, 2015; Avanzini & Wachtler 2012*

The largest footprint of the Lower Triassic (Class: Tiny)

Rotodactylus cursorius (a), from Arizona in the United States, and *R.* sp. (b), from Germany, were dinosauromorph footprints that were 7 cm long, including finger I (hallux). Early avemetatarsalians were very small. See more: *Peabody 1948; Fisher & Kunz 2013*

The largest footprint of the Upper Triassic (Class: Large)

The two longest footprints of the Triassic are up to 50 cm long (l and m), although *Eubrontes* sp. (n), 45 cm long, is remarkable because it lacks the mark of a prominent heel. See more: *Ellenberger 1965; Le Loeuff et al. 2009; Niedzwedtki 2011*

The oldest theropod tail print

The oldest cases are *Gigandipus* and *Hypephus*, both from the Passaic Formation of the Upper Triassic (Norian, ca. 227–208.5 Ma). Some authors consider them to be from the Early Jurassic (New Jersey, USA). See more: *Rainforth 2002; Kim & Lockley 2013*

Godzilla footprints? (right)

A 2 m geological fault (Nova Scotia, Canada) was confused with the imprint of a Triassic predator. Its author would have been gigantic! See more: *Grantham 1989*

1:60

The largest dinosauromorph handprint (left)

1:3

A specimen of *Agialopus wyomingensis* (Wyoming, USA) left an 7 cm impression of its front leg. See more: *Branson & Mehl 1932*

The longest theropod stride of the Triassic

Eubrontes sp., from the Upper Triassic (France), has 3 m long strides. Unfortunately, the specimen was destroyed. See more: *Ellenberger 1965; Ellenberger et al. 1970; Lucas et al. 2006*

The longest primitive saurischian stride of the Triassic

The record is held by large footprints similar to *Gigandipus* from the Nam Phong Formation (Thailand): Its stride was 2.7 m long. It had a very long finger I (hallux), similar to that of primitive saurischians, such as *Herrerasaurus*. See more: *Le Loeuff et al. 2009*

The highest speed of theropod footprints of the Triassic

A speed of 22 km/h was estimated from *Grallator* sp. footprints (Wales, United Kingdom). The prints are 10 cm long, and the strides have an average length of 1.7 m. See more: *Lockley et al. 1996; Kubo & Kubo 2013*

The strangest Triassic footprint! (left)

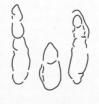

Some dinosauromorph footprints are strange, since their symmetry is strongly ectaxonic, while the majority of dinosaurs had mesaxonic legs. Footprints from the Middle Triassic (Spain) have a much shorter middle finger, something that has never been seen in any skeleton. See more: *Demathieu et al. 1999*

1:2

Notable large Triassic footprints

Species, size (length)	Record, group	Country, reference
Lower Triassic		
a) *Rotodactylus cursorius* 6 cm	Largest North American Lower Triassic (dinosauromorph)	Arizona, USA Peabody 1948
b) *Rotodactylus* sp. 5 cm	Largest European Lower Triassic (dinosauromorph)	Germany Fisher & Kunz 2013
c) *Prorotodactylus mirus* 4.7 cm	Largest European Lower Triassic (avemetatarsalian)	Poland Ptaszynski 2000
Middle Triassic		
d) *Sphingopus ladinicus* 28 cm	Longest European Middle Triassic (dinosauromorph)	Italy Avanzini & Wachtler 2012
e) *Coelurosaurichnus sassandorfensis* 25.6 cm	Largest European Middle Triassic (dinosauromorph)	France Kuhn 1958
f) *Paratrisauropus bronneri* 18 cm	Widest European Middle Triassic (dinosauromorph)	Switzerland Demathieu & Weidmann 1982
g) *Grallator* isp. 8.7 cm	Largest South American Middle Triassic (dinosauromorph)	Argentina Melchor & De Valais 2006
h) *Rotodactylus mckeei* 8 cm	Largest North American Middle Triassic (dinosauromorph)	Arizona, USA Peabody 1948
i) *Grallator* sp. ~6.2 cm	Largest European Middle Triassic (basal saurischian)	France Montenat 1968
j) Unnamed 4.5 cm	Largest European Middle Triassic (avemetatarsalian)	Spain Demathieu et al. 1999
k) *Rotodactylus* sp. 3.4 cm	Largest African Middle Triassic (dinosauromorph)	Morocco Klein et al. 2011
Upper Triassic		
l) Unnamed 50 cm	Largest Asian Upper Triassic (Cimmeria) (primitive saurischian)	Thailand Le Loeuff et al 2009
m) *Eubrontes* sp. 50 cm	Longest European Upper Triassic (theropod)	France Ellenberger 1965
n) *Eubrontes* sp. 45 cm	Largest European Upper Triassic (theropod)	Slovakia Niedzwiedzki 2011
o) *Eubrontes* sp. 43 cm	Largest Oceanian Upper Triassic (theropod)	Australia Staines & Woods 1964
p) Unnamed 43 cm	Largest South American Upper Triassic (theropod)	Brazil Da Silva et al. 2012
q) Unnamed 40 cm	Largest African Upper Triassic (theropod)	Morocco Byron & Dutuit 1981
r) Unnamed 40 cm	Largest European Upper Triassic (primitive saurischian)	Poland Gierlinski et al. 2009
s) *Qemetrisauropus princeps* 35 cm	Largest African Upper Triassic (primitive saurischian)	Lesotho Ellenberger 1972
t) Unnamed 30 cm	Largest South American Upper Triassic (primitive saurischian)	Argentina Marsicano et al. 2007
u) *Atreipus* sp. 27.7 cm	Largest North American Upper Triassic (dinosauromorph)	Utah, USA Lockley & Hunt 1995
v) Unnamed 27 cm	Largest Asian Upper Triassic (Paleoasia) (primitive saurischian)	China Xing et al. 2014
w) *Eubrontes* sp. 27 cm	Largest North American Upper Triassic (theropod)	Virginia, USA Weem 1987
x) Unnamed 24 cm	Largest North American Upper Triassic (primitive saurischian)	Greenland Milan et al. 2004
y) *Dinosauripus* sp. ~20 cm	Largest European Upper Triassic (dinosauromorph)	Germany Rehnelt 1952

The smallest footprint of the Upper Triassic (Class: Tiny)

Among the smallest footprints are those of dinosauromorphs and some theropods that were probably juveniles. *Rotodactylus tumidus* (England) is the smallest of this period. It was only about 2.1 cm long. Another footprint assigned to this species may have been even smaller (ca. 1.94 cm long), but it was not described in detail. See more: *Morton 1897; Haubold 1971; Batty 2008*

The smallest handprint of a dinosauromorph (left)

1:1

The front leg impression of *Atreipus-Grallator* (Morocco) is 6 mm long. It dates from the Middle Triassic and does not rank among the smallest individuals from that area. See more: *Klein et al. 2011*

The most numerous dinosauromorph footprints

Almost a thousand footprints of *Rotodactylus matthesi* are preserved in the Solling Formation (Germany). See more: *Haubold 1998*

The narrowest dinosauromorph footprint

The width of some *Rotodactylus cursorius* footprints (Arizona, USA) corresponds to about 43% of the length of the footprint, including finger I (hallux). See more: *Peabody 1948*

The widest dinosauromorph footprint

Paratrisauropus latus (Switzerland) is approximately 14% wider than it is long. See more: *Demathieu 1982*

The narrowest primitive saurischian footprint

The width of an unnamed footprint from Poland, because it includes the metatarsal mark, is approximately 31% of its total length. See more: *Gierlinski et al. 2009*

The widest primitive saurischian footprint

The "Type 3" footprint (Argentina) is almost as wide as it is long. See more: *Marsicano et al. 2007*

The oldest primitive avemetatarsalian footprints

Prorotodactylus sp. of the Stryczowice Formation (Poland) dates from the Lower Triassic (lower Olenekian or Induan, ca. 251.17–249 Ma). Dinosauromorphs and avemetatarsalians had ectaxonic hind legs. See more: *Niedzwiedzki et al. 2013*

The oldest dinosauromorph footprints

The genus *Rotodactylus* sp. appeared in the Lower Triassic (upper Onelekian or Anisian, ca. 249.2–247.2 Ma) in north-central Pangea (Germany and Poland). See more: *Ficher & Kurz. 2013*

The oldest dinosaur footprints

Grallator sp. (France) might be the oldest primitive saurischian footprint. It dates from the Middle Triassic (Anisian-Ladinian, ca. 242 Ma).
See more: *Montenat 1968*

The oldest theropod footprints

The unnamed footprint "Type 2" (Argentina) is from the Early Triassic (lower Carnian, ca. 237–234 Ma). See more: *Marsicano et al. 2007, 2015*

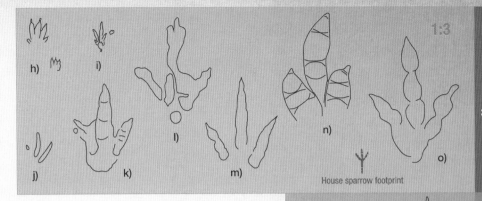

1:3

House sparrow footprint

Upper

Triassic

The smallest footprint of the Middle Triassic (Class: Tiny)

The smallest footprints that have been attributed to dinosauromorphs were found in south-central Pangea (Morocco). They were barely 0.8 and 1.3 cm long, so the smallest could be a juvenile specimen (d). See more: *Klein et al. 2011*

Middle

The smallest footprint of the Lower Triassic (Class: Very small)

The smallest of this time is *Protodactylus* isp. (b), a possible fingerprint of a basal avemetatarsalian with a quadruped gait. It was 2 cm long. Here we compare it with the footprint of a rock pigeon (*Columba livia*), which is 5 cm long. *Plesiothornipos binneyi* (a) is 1.9 cm long, but it is an indeterminate footprint See more: *Harkness 1850; Thulborn 2006; Fisher & Kunz 2013*

Lower

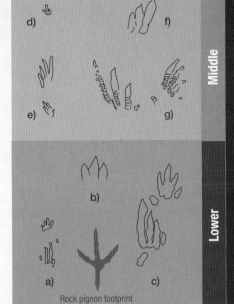

Rock pigeon footprint

Notable small Triassic footprints

Species, measurement (length)	Record, group	Country, reference
Lower Triassic		
a) *Plesiothornipos binneyi* 1.9 cm	Smallest Lower Triassic (doubtful identification)	England Harkness 1850
b) *Protodactylus* isp. 2 cm	Smallest Lower Triassic (avemetatarsalian)	Germany Fisher & Kunz 2013
c) *Rotodactylus* sp. 4 cm	Smallest Lower Triassic (dinosauromorph)	Polonia Niedzwiedzki et al. 2013
Middle Triassic		
d) *Atreipus-Grallator* 0.8 cm	Smallest Middle Triassic and smallest African (dinosauromorph)	Morocco Klein et al. 2011
e) *Prorotodactylus lutevensis* 2.5 cm	Smallest European Middle Triassic (avemetatarsalian)	France Demathieu 1984
f) *Rotodactylus* cf. *cursorius* ~3 cm	Smallest European Middle Triassic (dinosauromorph)	Italy Avanzini & Mietto 2008
g) *Rotodactylus mckeei* 4.1 cm	Smallest North American Middle Triassic (dinosauromorph)	Arizona, USA Peabody 1948
Upper Triassic		
h) *Rotodactylus tumidus* 2 cm	Smallest European Upper Triassic (dinosauromorph)	England Maidwell 1914
i) *Banisterobates boisseaui* 2.4 cm	Smallest North American Upper Triassic (dinosauromorph)	Virginia, USA Frasier & Olsen 1996
j) Indeterminate ~2.4 cm	Smallest North American Upper Triassic (theropod)	Virginia, USA Weems & Kimmel 1993
k) *Grallator pisanus* 7 cm?	Smallest European Upper Triassic (primitive saurischian)	Italy Bianucci & Landini 2005
l) *Saurichnium anserinum* ~8.4 cm	Smallest African Upper Triassic (primitive saurischian)	Namibia Gürich 1926
m) *Grallator* sp. 7.2 cm	Smallest Oceanian Upper Triassic (primitive saurischian)	Australia Da Silva 2012
n) *Grallator* isp. 8.5 cm	Smallest South American Upper Triassic (theropod)	Brazil Silva et al. 2008
o) *Grallator* sp. 10 cm	Smallest North American Upper Triassic (primitive saurischian)	Colorado, USA Gaston et al. 2003

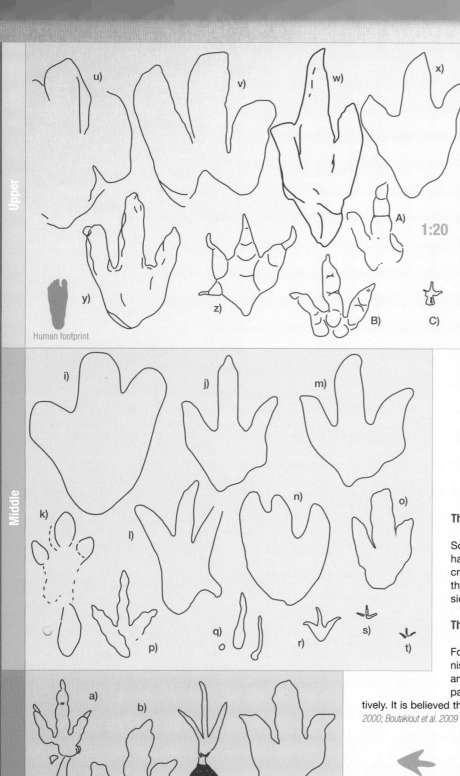

Upper

Middle

Lower

1:20

Human footprint

The largest footprint of the Upper Jurassic (Class: Very large)

There are several tracks of exceptional size in the Iouaridène syncline (v). Some range from 70 to 82 cm long, although the largest reaches 90 cm (u) but includes part of the metatarsal. See more: *Boutakiout et al. 2009*

The largest footprint of the Middle Jurassic (Class: Very large)

Turkmenosaurus kugitangensis (i) is a large footprint (86 cm) of a probable huge megalosauroid. There is a report of some footprints assigned to *Megalosauripus pombali* (Portugal) that are 92 cm long, but they were unfortunately lost. It is possible that they were *Eutynichnium lusitanicum*, since it frequently retains a large part of the metatarsal. See more: *Gomes 1915-1916; Lapparent & Zbyszewski 1957; Amanniazov 1985*

The largest footprint of the Lower Jurassic (Class: Very large)

Megatrisauropus malutensis (a) is 85 cm long, including the full metatarsal marking. In fact, the footprint is 45 cm long from the tip of finger III to finger I. Another footprint from Poland (b) is the largest, 65 cm long. See more: *Ellenberger 1970; Gierlinski 1991; Niedzwedzki 2006*

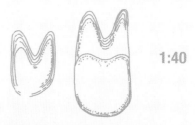

1:40

The strangest footprints of the Jurassic! (above)

Some impressions from the Lower Jurassic (present-day Lesotho) have strange shapes and are large. Lengths vary between 70 and 114 cm. It has been suggested that they might be theropod footprints, but they do not seem to be. They may be marks of excavations or impressions caused by unknown events. See more: *Ellenberger 1974; Thulborn 1990*

The longest theropod footprint tracks of the Jurassic

Forty-one *Megalosaurus uzbekistanicus* footprints were found in Turkmenistan. The total was length of 311 m. Another track has 108 footprints and was 271 m long. Two other separate tracks from Morocco could be part of the same series; each covers 507 and 567 m in length, respectively. It is believed that it could have reached about 2 km in total length. See more: *Lockley et al. 2000; Boutakiout et al. 2009*

The highest speed for theropod footprints of the Jurassic (above)

Since its footprints are 40 cm long and its strides are 5.12 m long, on average, a probable speed of 41 km/h was estimated for an *Eubrontes* (Utah, USA). See more: *Hamblin et al. 2006*

1:10

The oldest theropod footprint with a pathology (right)

A *Eubrontes* track from the Lower Jurassic shows only two fingers on the right track, possibly due to a congenital amputation or defect. *Sauroidichnites* (*Typopus*) *abnormis* has a strange rotation in its legs, but it is not clear that the footprints belong to a dinosaur (Hettangian, ca. 201.3–199.3 Ma) (Massachusetts, USA). See more: *Hitchcock 1844; Abel 1935; Thulborn 1990; Tanke & Rothschild 2002; McCrea et al. 2015*

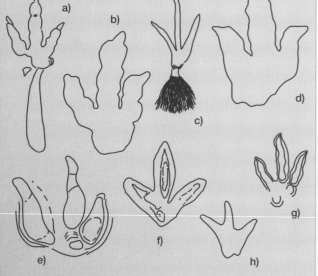

1:20

Restless tail (right)

Tail marks are very rarely preserved. When it does happen, it usually leaves a straight line. *Gigandipus caudatus* (formerly known as *Gigantitherium caudatum*), however, left a sinous tail mark, whose ripples appear regularly (Massachusetts, USA). See more: *Hitchcock 1858; Kim & Lockley 2013*

The longest stride of the Jurassic

A series of 92 tracks found in England includes strides that were from 2.12 to 5.65 m long. See more: *Day et al. 2004*

1:40

The most complete impressions (above)

A specimen of *Eubrontes* (Utah, USA) left impressions of both legs with metatarsals, both hands, belly marks, and the ischiatic callus. It also left impressions of repositioning while resting. See more: *Milner et al. 2009*

1:10

The oldest theropod footprint with membrane (above)

Talmontopus tersi, (France) has marks that seem to belong to interdigital membranes, although some authors suggest that they were actually formed by soft sediment pressure. It dates from the Upper Triassic or the Lower Jurassic (Rhaetian or Hettangian, ca. 208.5–199.3 Ma). See more: *Lapparent & Montenat 1967; Haubold 1986; Falkingham et al. 2008*

The oldest swimming theropod tracks

Eubrontes and *Grallator* (Utah, USA) were reported with evidence that they had advanced in the water. They date from

1:20

the Lower Jurassic (Hettangian ca. 201.3–199.3 Ma). See more: *Milner, Lockley & Kirkland 2006*

The longest swimming theropod footprint of the Jurassic (above)

Characichnos tridactylus is a morphotype footprint that has marks of a theropod or other vertebrate animal, such as a crocodile, which leaves long marks on the substrate while floating in water. The longest specimen from England is from this period and is 35 cm long. See more: *Whyte & Romano 2001; Milán et al. 2010*

Notable large Jurassic footprints

Species, measurement (length)	Record, group	Country, reference
Lower Jurassic		
a) *Megatrisauropus malutensis* 85 cm	Longest Lower Jurassic and longest African (theropod)	Lesoto Ellenberger 1970
b) Unnamed 65 cm	Largest Lower Jurassic and largest European (theropod)	Polonia Niedzwiedzki 2006
c) *Steropoides ingens* 63.5 cm	Longest North American Lower Jurassic	Connecticut, USA Hitchcock 1836
d) cf. *Eubrontes* 60.5 cm	Largest North American Lower Jurassic (theropod)	Arizona, USA Morales & Bulkey 1996
e) *Carmelopodus* sp. 53 cm	Largest African Lower Jurassic (theropod)	Morocco Ishigaki 1988
f) cf. *Eubrontes glenrosensis* 45 cm	Largest Asian Lower Jurassic (theropod)	China Li et al. 2010
g) Unnamed 39 cm	Largest Middle Eastern Lower Jurassic (theropod)	Iran Kellner et al. 2012
h) Unnamed 30 cm	Largest Oceanian Lower Jurassic (theropod)	Australia Scanlon 2006
Middle Jurassic		
i) *Turkmenosaurus kugitangensis* 86 cm	Longest Middle Jurassic and largest Asian (theropod)	Turkmenistan Amanniazov 1985
j) Unnamed 82 cm (72 cm mean length)	Longest European Middle Jurassic (theropod)	England Ellenberger 1970
k) *Megalosauripus* sp. ~76 cm	Longest North American Middle Jurassic (theropod)	Utah, USA Niedzwiedzki 2006
l) Unnamed 71 cm	Largest Oceanian Middle Jurassic (theropod)	Australia Hitchcock 1836
m) Unnamed 70 cm	Largest Middle Jurassic and largest European (theropod)	Portugal Morales & Bulkey 1996
n) *Samandrinda surghani* 60 cm	Largest Indostanian Middle Jurassic (theropod)	Pakistan Malkani 2007
o) *Megalosauripus* sp. 53 cm	Largest North American Middle Jurassic (theropod)	Utah, USA Li et al. 2010
p) Unnamed 46 cm	Largest African (Malagasy) Middle Jurassic (theropod)	Madagascar Kellner et al. 2012
q) *Paravipus didactyloides* 35.6 cm	Largest African Middle Jurassic (theropod)	Niger Ellenberger 1970
r) *Talmontopus* sp. 18 cm	Largest Middle Eastern Middle Jurassic (theropod)	Iran Niedzwiedzki 2006
s) Unnamed ~7 cm	Largest European Middle Jurassic (bird)	England Romano & Whyte 2003
t) Unnamed 5.7 cm	Largest African Middle Jurassic (bird)	Morocco Belvedere et al. 2011
Upper Jurassic		
u) Unnamed 90 cm	Longest Upper Jurassic and longest African (theropod)	Morocco Boutakiout et al. 2009
v) Unnamed 82 cm	Largest Upper Jurassic and largest African (theropod)	Morocco Boutakiout et al. 2009
w) Unnamed 105 cm (82 cm)	Longest European Upper Jurassic (theropod)	Spain García-Ramos et al. 2006
x) *Megalosauripus* sp. 77 cm	Largest European Upper Jurassic (theropod)	Portugal Santos et al. 1995
y) *Megalosauripus uzbekistanicus* 72 cm	Largest Asian Upper Jurassic (theropod)	Tajikistan Gabuniya & Kurbatov 1982
z) Unnamed 54 cm	Largest South American Upper Jurassic (theropod)	Chile Moreno et al. 2004
A) *Allosaurus* 47 cm	Largest North American Upper Jurassic (theropod)	Wyoming, USA Lockley et al. 1998
B) *Allosaurus* 44 cm	Largest North American Upper Jurassic (theropod)	Colorado, USA Farlow 1987
C) Unnamed 12.5 cm	Largest European Upper Jurassic (bird)	France Lange-Badre et al. 1996

The smallest footprint of the Middle Jurassic (Class: Tiny)

Guinness World Records recognized the specimen GLAHM 114913/1 (h) as the smallest dinosaur footprint in the world in June 2004. The 1.78 cm measurement was reported by Neil Clark. Even smaller ones have been reported in different parts of the world. The smallest marks that have been measured, where the big finger is 1.5 cm long (g), are apparently poorly printed. Mesozoic bird footprints, among the oldest known, were found in the same locality. See more: *Clark et al. 2005, Belvedere et al. 2011*

1:1

The smallest footprint of the Lower Jurassic (Class: Tiny)

The smallest specimen of *Grallator* sp. (a) is barely 1.5 cm long. Other footprints reported by Eldon George in Canada are less than 2 cm long. The smallest are approximately 1.67 cm long. They were not formally described, but some sources mention them. See more: *Lessem 1992; Olsen 1995*

The smallest footprints of the Upper Jurassic (Class: Very small)

It is possible that the *Wildeichnus* ichnogenus was left by young theropods or very small species. They all stand out for their diminutive size. *Wildeichnus* sp. (o) is even smaller than *Aquatilavipes* sp. (p), the smallest footprints of Jurassic birds. See more: *Lockley et al. 2006; Gierlinski et al. 2009*

MYTH: The rain that did not exist

Some footprints, especially from the Jurassic, are accompanied by small round marks that are up to 1 cm in diameter. For a long time, the round marks were interpreted as possible raindrops that fell in soft mud and were preserved forever. Nowadays, specialists consider them escaped gas bubbles, since their edges are slightly raised above the surrounding substrate, a very different morphology than would be expected from a few raindrops. See more: *Kirkland & Milner 2006*

Notable small Jurassic footprints

Species, measurement (length)	Record, group	Country, reference
Lower Jurassic		
a) *Grallator* sp. 1.5 cm	Smallest Lower Jurassic and smallest North American (theropod)	New Jersey, USA *Olsen 1995*
b) *Plesiornis pilulatus* 4.6 cm	Smallest European Lower Jurassic (theropod)	Poland *Gierlinski 1996*
c) *Grallator (Masitisisauropus) exiguus* 6.2 cm (var ruber)	Smallest African Lower Jurassic (theropod)	Lesotho *Ellenberger 1970*
d) *Schizograllator otariensis* 23 cm	Shortest length Asian Lower Jurassic (theropod)	Japan *Matsukawa et al. 2005*
e) *Iranosauripus zebarensis* 23 cm	Smallest Middle Eastern Lower Jurassic (theropod)	Iran *Lapparent & Sadat 1975*
f) *Eubrontes (Paracoelurosaurichnus) monax* 25.5 cm	Smallest Asian Lower Jurassic (theropod)	China *Zhen et al.1986*
Middle Jurassic		
g) Unnamed 1.5 cm	Shortest length Middle Jurassic and African (theropod)	Morocco *Belvedere et al. 2011*
h) Unnamed 1.78 cm	Smallest Middle Jurassic and smallest European (theropod)	Scotland *Clark et al. 2005*
i) *Wildeichnus* sp. 2 cm	Smallest African Middle Jurassic (theropod)	Morocco *Gierlinski et al. 2009*
j) *Sarmientichnus scagliai* 3.46 cm	Smallest South American Middle Jurassic (theropod)	Argentina *Casamiquela 1964*
k) Unnamed 6 cm	Smallest North American Middle Jurassic (theropod)	Utah, USA *Lockley et al. 2007*
l) *Wildeichnus isp.* 5.3 cm	Smallest Middle Eastern Middle Jurassic (theropod)	Iran *Abbassi et al. 2015*
m) *Shensipus tungchuanensis* 9.8 cm	Smallest Asian Middle Jurassic (theropod)	China *Young 1966*
n) Unnamed 10.9 cm	Smallest Oceanian Middle Jurassic (theropod)	Australia *Hill et al. 1966*
Upper Jurassic		
o) *Wildeichnus* sp. 3.7 cm	Smallest Upper Jurassic and smallest European (theropod)	Poland *Gierlinski et al. 2009*
p) *Aquatilavipes* sp. 4.4 cm	Smallest Asian Upper Jurassic (bird)	China *Lockley et al. 2006*
q) *Pullornipes aureus* 4.7 cm	Smallest Asian Upper Jurassic (bird)	China *Lockley et al. 2006*
r) *Menglongipus sinensis* 6.7 cm	Smallest Asian Upper Jurassic (theropod)	China *Xing et al. 2009*
s) Unnamed 7.5 cm	Smallest North American Upper Jurassic (theropod)	Wyoming, USA *Foster & Lockley 1995*
t) Unnamed 17 cm	Smallest South American Upper Jurassic (theropod)	Chile *Moreno et al. 2004*

Footprints of an adult with offspring (left)

A *Grallator* (*Kainotrisauropus*) *morijiensis* specimen is accompanied by other similar footprints that are between 18% and 22% of its size. It is believed that the prints were left by an adult accompanied by its offspring (present-day Lesotho). See more: *Ellenberger 1972*

1:10

The narrowest theropod footprint (right)

The width of a specimen of *Sarmientichnus scagliai* (*Casamiquelichnus navesorum*) (j) is approximately 25% of its total length. This is because fingers II and IV are not always preserved. See more: *Casamiquela 1964; Coria & Paulina Carabajal 2004*

1:4

The Jurassic footprint with the greatest finger amplitude (right)

Taupezia landeri (France) has an amplitude of 90 degrees in the fingers II and IV, as happens in footprints of birds and other paravian theropods very closely related to them. See more: *Delair 1963*

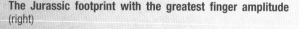

1:10

The oldest raptor footprint? (right)

Some didactyl impressions from Cimmeria (Iran) resemble those of *Velociraptorichnus*. They date from between the Lower and Middle Jurassic (Toarcian-Aalenian, ca. 182.7–170.3 Ma). See more: *Abbassi & Madanipour 2014*

1:10

Uneven steps (right)

Theropod footprints from the Aganane Formation (Morocco) have an uneven gait: a short step followed by a long one, which is interpreted as limping. See more: *Jenny & Josen 1982; Ishigaki 1986*

1:20

Deposits with the most Jurassic footprints

A total of 2,700 were registered in Kugitang Tau, Turkmenistan, and 2,200 in Cabo Epichel, Portugal. See more: *Romanshko 1986, Demathieu et. al 1999*

The largest Jurassic footprint deposit

The area known as Khodja-Pit Ata (Kazakhstan) occupies an area of about 30 km². See more: *Lockley 1997*

Theropods footprints of the Jurassic found at the highest altitude

Various types of theropod footprints were found at 2,700 m altitude in the high mountains of Jebel Waougoulzat (Morocco). See more: *Ishigaki 1988*

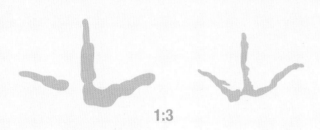

1:3

The oldest bird footprints (above)

Due to the wide spread of the toes, it is possible that some footprints from the Lower Jurassic (England and Morocco) were left by birds. They date from the Bajocian (ca. 170.3–168.3 Ma). Older footprints from Iran (Aalenian, ca. 174.1–170.3 Ma) have great amplitude, but they may have been left by paravian theropods. See more: *Romano & Whyte 2003; Belvedere et al. 2011; Abbassi & Madanipour 2014*

The oldest gregarious theropod footprints

Evidence of theropods that walked in groups of seven or eight individuals was found in the Connecticut River valley (Massachusetts, USA). The footprints date from the Lower Jurassic (Hettangian, ca. 201.3–199.3 Ma). See more: *Ostrom 1972; Lockley & Matsukawa 1999*

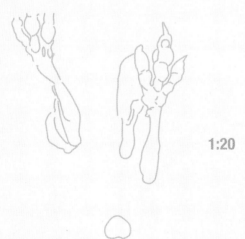

1:20

The oldest theropod footprints with feather impressions (above)

The specimen AC 1/7 of *Grallator* (*Eubrontes*) *minusculus* shows monofilament marks on the belly. Another footprint, AC 51/16 of *Anchisauripus tuberosus* (*Plesiornis mirabilis*), seems to show feathers, but experts question this. Both date from the Lower Jurassic (Hettangian-Sinemurian, ca. 201.3–190.8 Ma) (Massachusetts, USA). See more: *Hitchcock 1865; Lull 1904, 1952; Gierlinski 1994*

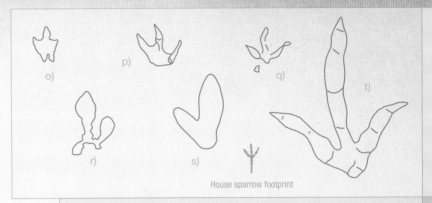

o) p) q) t) r) s)

House sparrow footprint

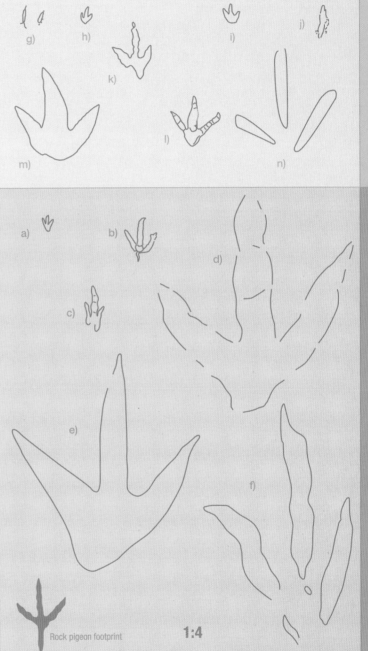

g) h) i) j) k) l) m) n)

a) b) c) d) e) f)

Rock pigeon footprint   1:4

Late · **Early** · Lower Cretaceous

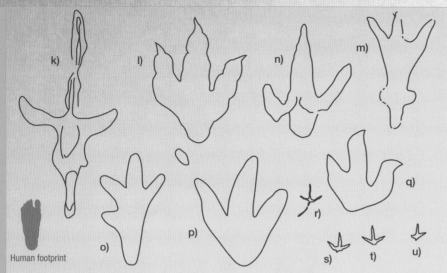

Human footprint

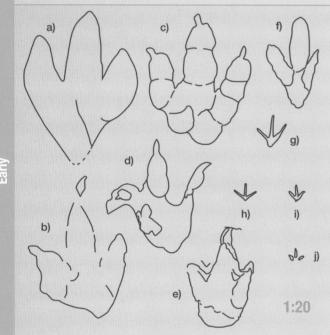

1:20

The largest of the late Lower Cretaceous (Class: Very large)

A footprint similar to *Magnoavipes* (k) has a strangely elongated shape, which increases the size and shape of the real foot. It may have been left in deep, soft mud, as is known to happen when a modern bird treads on a similar substrate. Another footprint of a carcharodontosaurid (l) was shorter but belonged to a larger animal. The longest bird footprint of that age is 17 cm long but was caused by dragging the thumb while moving. The real leg was 9.1 cm long. See more: *Lockley, Matsukawa & Witt 2006; Milán 2006; Martin et al. 2013*

The largest of the early Lower Cretaceous (Class: Very large)

Footprints 81 cm long (a) were considered large ornithopods due to their very robust form, but it is now known that they were from a theropod. *Archaeornithipus meijidei* (g) are long-legged bird footprints that are 16.6 cm long. See more: *Fuentes Vidarte 1996; Gismundi & Chacaltana 2010*

Notable large Cretaceous footprints

Species, measurement (length)	Record, group	Country, reference
Early Lower Cretaceous		
a) Unnamed 81 cm	Largest Early Lower Cretaceous and largest South American (theropod)	Peru Obata et al. 2006
b) Unnamed 70 cm	Longest European Early Lower Cretaceous (theropod)	Spain Casanovas et al. 1995
c) Unnamed 69 cm	Largest European Early Lower Cretaceous (theropod)	Spain Barco et al. 2005
d) *Megalosauropus* sp. ~60 cm	Largest North American Early Lower Cretaceous (theropod)	Canada Long 1988
e) *Megalosauropus broomensis* 53 cm	Largest Oceanian Early Lower Cretaceous (theropod)	Australia McCrea et al. 2011
f) Unnamed 48 cm	Largest Asian Early Lower Cretaceous (theropod)	China Xing et al. 2012
g) *Archaeornithipus meijidei* 16.6 cm	Largest European Early Lower Cretaceous (bird)	Spain Fuentes Vidarte 1996
h) Unnamed 9 cm	Largest African Early Lower Cretaceous (bird)	Niger Ginsburg 1966
i) Unnamed 8.6 cm	Largest North American Early Lower Cretaceous (bird)	Canada McCrea et al. 2014
j) *Koreanaornis dodsoni* 6.3 cm	Largest Asian Early Lower Cretaceous (bird)	China Xing et al. 2011
Late Lower Cretaceous		
k) Unnamed 110 cm	Longest Late Lower Cretaceous and longest North American (theropod)	New Mexico, USA Ellenberger 1970
l) Unnamed 78 cm	Largest Late Lower Cretaceous and largest African (theropod)	Argelia Mahboubi et al. 2007
m) "Ornitholestes" ~61.7 cm	Largest European Late Lower Cretaceous (theropod)	Spain Brancas et al. 1979
n) Unnamed 60 cm	Largest North American Late Lower Cretaceous (theropod)	Mexico Kappus et al. 2011
o) Unnamed 59 cm	Longest Asian Late Lower Cretaceous (theropod)	China Azuma et al. 2006
p) *Chapus lockleyi* 58.2 cm	Largest Asian Late Lower Cretaceous (theropod)	China Li et al. 2006
q) Unnamed 43 cm	Largest Asian Late Lower Cretaceous (Cimmeria) (theropod)	Thailand Le Loeuff et al. 2002
r) Unnamed 17 cm	Largest Oceanian Late Lower Cretaceous (bird)	Australia Martin et al. 2013
s) *Wupus agilis* 10.5 cm	Largest Asian Late Lower Cretaceous (bird)	China Xing et al. 2007
t) *Limiavipes curriei* 10.1 cm	Largest North American Late Lower Cretaceous (bird)	Canada McCrea & Sarjeant 2001
u) Unnamed ~8.65 cm	Largest European Late Lower Cretaceous (bird)	Spain Moratalla & Sanz 1992

The site with the widest variety of dinosaur footprints

Cal Orcko (Bolivia) was found near the quarry of a cement factory in 1985. In 2006, 5,055 footprints were catalogued, divided into 332 different forms. That figure has more than doubled since then. See more: *Meyer 1998*

The site with the most dinosaur footprints

In Shar Tsav, which belongs to the Nemegt Formation (Mongolia), more than 18,000 dinosaur footprints between 6 and 70 cm long have been reported. Most (about 13,500) were identified as theropods, but some ornithischians similar to *Parksosaurus* apparently had very pointed nails and may require revisions. See more: *Ishigaki 2010*

The largest Cretaceous site

The fingerprint mega-formation known as the Glen Rose Formation (Texas, USA) occupies about 100 km². See more: *Lockley 1991*

The longest swimming Cretaceous theropod footprint

A trail of twelve theropod footprints (Spain) shows that the animal was striding while swimming, trying to maintain a straight course while the water diverted it. The largest of the footprints is about 60 cm long. See more: *Ezquerra et al. 2007*

1:20

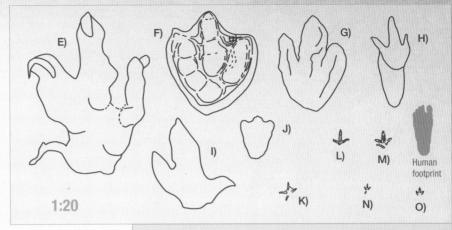

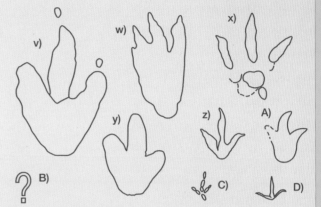

1:20

Late

Early

The largest of late Upper Cretaceous (Class: Very large)

The *Tyrannosauripus pillmorei* footprint (E) may have been left by a *Tyrannosaurus rex*. Other footprints, such as *Tyrannosauropus petersoni*, turned out to be of hadrosaurs. The largest bird footprint is about 20 cm long (K). See more: *Peterson 1924; Haubold 1971; Larson 2003; Lockley & Hunt 1994; Lockley, Janke & Triebold 2011*

The largest of the early Upper Cretaceous (Class: Very large)

The footprint SLPG-D (v) may belong to a large carcharodontosaurid. It is the largest of those reported in rocks from the early Upper Cretaceous. A bird footprint found in Canada, similar to that of *Jindongornipes kimi* but 16 cm long, is the largest of that period. Unfortunately, it has not been formally described. It may have belonged to a large land bird that was 1.85 m long and weighed 30 kg. See more: *Carvalho 2001; Plint en Lockley & Rainforth 2002*

Notable large Cretaceous footprints

Species, measurement (length)	Record, group	Country, reference
Early Upper Cretaceous		
v) Unnamed 78.2 cm	Largest Early Upper Cretaceous and largest South American (theropod)	Brazil *Carvalho 2001*
w) *Macropodosaurus gravis* 56 cm	Longest Asian Early Upper Cretaceous (theropod)	Tajikistan *Gierlinski & Niedzwiedzki 2005*
x) Unnamed ~45.7 cm	Largest North American Early Upper Cretaceous (theropod)	New Jersey, USA *Baird 1989*
y) Unnamed 43 cm	Largest African Early Upper Cretaceous (theropod)	Morocco *Ibrahim et al. 2014*
z) Unnamed 27 cm	Largest Middle Eastern Early Upper Cretaceous (theropod)	Israel *Avnimelech 1966*
A) Unnamed 27 cm	Largest European Early Upper Cretaceous (theropod)	Portugal *Santos 2002*
B) Unnamed 16 cm	Largest North American Early Upper Cretaceous (bird)	Canada *Plint in Lockley & Rainforth 2002*
C) *Koreanaornis* sp. 14.3 cm	Largest African Early Upper Cretaceous (bird)	Tunisia *Contessi & Fanti 2012*
D) Unnamed ~12.2 cm	Largest Middle Eastern Early Upper Cretaceous (bird)	Israel *Avnimelech 1966*
Late Upper Cretaceous		
E) *Tyrannosauripus pillmorei* 86 cm	Largest Late Upper Cretaceous and largest North American (theropod)	South Dakota, USA *Lockley & Hunt 1994*
F) Unnamed 55 cm	Largest Asian Late Upper Cretaceous (theropod)	Mongolia *Ishigaki et al. 2009*
G) *Eubrontes* 51 cm	Largest South American Late Upper Cretaceous (theropod)	Bolivia *Rios 2005*
H) Unnamed ~28 cm	Largest African Late Upper Cretaceous (theropod)	Morocco *Ambroggi & Lapparent 1954*
I) Unnamed 48 cm	Largest European Late Upper Cretaceous (theropod)	Spain *Santisteban & Suñer 2003*
J) Unnamed 22.5 cm	Largest Indian Late Upper Cretaceous (theropod)	India *Mohabey 1986*
K) *Sarjeantopodus semipalmatus* 9.5 cm	Largest North American Late Upper Cretaceous (bird)	Wyoming, USA *Lockley et al. 2003*
L) *Patagonichornis venetiorum* ~9 cm	Longest South American Late Upper Cretaceous (bird)	Argentina *Casamiquela 1987*
M) *Yacoraitichnus avis* 8 cm	Largest South American Late Upper Cretaceous (bird)	Argentina *Alonso & Marquillas 1986*
N) *Koreanaornis* sp. 4.9 cm (3.3 cm)	Longest Asian Late Upper Cretaceous (bird)	South Korea *Huh et al. 2012*
O) *Aquatilavipes* sp. 4 cm	Largest Asian Late Upper Cretaceous (bird)	South Korea *Huh et al. 2012*

The longest Cretaceous theropod footprint track

In Cal Orcko (Bolivia) a track extends for about 581 m. There are more extensive trails in this town than anywhere else in the world. The previous record was 347 m. See more: *Meyer 2011*

MYTH: The larger footprint, the larger the creature that left it

This claim is not completely true, because some dinosaurs had shorter fingers or proportionally smaller or larger feet, and heel or metatarsal impressions sometimes further increase the length of a footprint. In addition, a soft substrate sometimes deforms and increases the length of the footprint.

The strangest footprints of the Cretaceous! (right)

Irenesauripus mclearni footprints (Canada) are associated with marks 81 cm long. Apparently, the theropod was digging with its front or rear claws. See more: *McCrea et al. 2014*

1:20

Cretaceous theropod footprints found at the highest altitude

In Huari Province (Peru), there are footprints at 4,600 m above sea level. See more: *Salas-Gismondi & Chacaltana 2010*

The highest estimated speeds in Cretaceous theropod footprints (left)

1:10

The speed of *Velociraptorichnus* sp. Muz. PIG 1704. II.6 (Poland) has been estimated at 50 km/h from its stride, which corresponds to 12.3 times the size of its feet. Other speeds: 44.1 km/h (China), 42.8 km/h (Texas, USA), 40 km/h (China), 37 km/h (Spain), and *Xiangxipus* sp. 25.7 km/h (South Korea). See more: *Farlow 1981; Viera & Torres 1995; White et al. 2001; Gierlinski 2009; Xu, Liu & Yang 2011; Li, Bai & Wei 2011*

The longest stride of a Cretaceous theropod

The prints Q94/98 (Texas, USA) were 38 cm long and its stride was up to 6.6 m long, so it is estimated that the theropod was advancing at approximately 39.9–43.6 km/h. See more: *Farlow 1981*

The longest stride of a Mesozoic bird

The record is held by *Archaeornithipus meijidei* (Spain). Strides 50–75 cm long were reported. See more: *Fuentes Vidarte 1986*

Lower Cretaceous

Late

Early

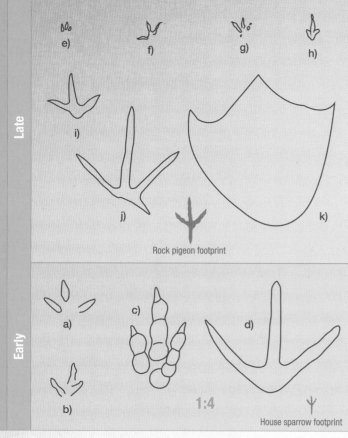

Rock pigeon footprint

1:4

House sparrow footprint

The smallest footprints (right)

Marks left by pecking birds have been associated with the footprints of the specimen K5001 (South Korea). The marks measure a maximum of 4 x 7.6 mm, although there are even smaller ones. See more: *Falk et al. 2010, 2014*

1:1

1:1

The smallest footprint to be identified as a dinosaur

The best-preserved footprints

The footprint CLS-F-7 (Spain) is about 50 cm deep, so you can see the leg as if it were a 3-D sculpture. There is a previous report, but it is still unpublished (South Dakota, USA). See more: *Huerta et al. 1999; Triebold et al. 1999*

Footprints in 4D

In some countermolds made in dense clay (Spain), skin impressions with pentagonally and hexagonally shaped scales can be seen, as can grooves generated by movement. They are known as "four-dimensional footprints" because they preserve movement marks. See more: *Cobos et al. in press*

The places with the most Mesozoic birds footprints

More Mesozoic bird footprints have been found in the Haman and Jindong Formations (South Korea) than in the rest of the world. See more: *Lockley & Rainforth 2002*

The smallest of the late Lower Cretaceous (Class: Tiny)

The smallest footprint attributed to a dinosaur is just 1 cm long and might have belonged to a newborn (Maryland, USA). The smallest *Koreanaornis hamanensis* specimen (South Korea) is only 1.77 cm long, which makes it the smallest bird footprint of the entire Mesozoic. See more: *Kim 1969; Standford et al. 2011*

The smallest of the early Lower Cretaceous (Class: Very small)

A bird footprint (China) that lies between the Jurassic and Cretaceous is 3.4 cm long. Another of a breeding theropod (Spain) measures 3.5 cm but is narrower. See more: *Pascual-Arribas et al. 2014*

Notable small Cretaceous footprints

Species, measurement (length)	Record, group	Country, reference
Early Lower Cretaceous		
a) Unnamed 3.4 cm	Shortest Early Lower Cretaceous and shortest Asian (bird)	China *Lockley*
b) Unnamed 3.5 cm	Smallest Early Lower Cretaceous and smallest European (theropod)	Spain *Pascual-Arribas et al. 2014*
c) *Anchisauripus australis* 10.7 cm	Smallest Early Lower Cretaceous and smallest South American (theropod)	Argentina *Lull 1942*
d) Unnamed 11.8 cm	Smallest Early Lower Cretaceous and smallest Oceanian (theropod)	Australia *Martin et al. 2013*
Late Lower Cretaceous		
e) Unnamed 1 cm	Smallest Late Lower Cretaceous and smallest North American (theropod)	Maryland, USA *Standford et al. 2011*
f) *Koreanaornis hamanensis* 1.55 cm	Smallest Asian Late Lower Cretaceous (bird)	South Korea *Kim 1969*
g) Unnamed 2 cm	Smallest European Late Lower Cretaceous (bird)	Spain *Moratalla et al. 2003*
h) *Neograllator emeienensis* 2.7 cm	Smallest Asian Late Lower Cretaceous (theropod)	China *Zhen et al. 1994*
i) *Paxavipes babcockensis* 3.36 cm	Smallest North American Late Lower Cretaceous (bird)	Canada *McCrea et al. 2015*
j) Unnamed 11.4 cm	Shortest length Oceanian Late Lower Cretaceous (bird)	Australia *Martin et al. 2013*
k) Unnamed 16 cm	Smallest African Late Lower Cretaceous (theropod)	Morocco *Masrour et al. 2013*

The most recent theropod tail prints

Sinuous theropod tail marks were reported in the Jinchuan (China) and Saniri (South Korea) Formations. Both date from the Lower Cretaceous. See more: *Wen et al. 2008; Kim & Chun 2011; Kim & Lockley 2013*

MYTH: The theropod that pursued a sauropod

In some tracks, *Megalosauripus glenrosensis* (Texas, USA) is observed moving in a direction similar to that of the sauropod *Brontopodus birdi*. Since the tracks overlap, it was interpreted as evidence of hunting, but there is no evidence that this is true. It is most likely a coincidence. See more: *Lockley 1991*

The most mysterious footprints

Deogmyeongosauripus are footprints reported in a comparative list of species of the Jindong Formation (South Korea) that date from the late Lower Cretaceous. It is not known what type of dinosaur they belong to, only that they were discovered in 1982.

Other enigmatic names of theropod footprints are *"Carnotaurichnus neuquinus"* (Cretaceous, Argentina), *"Kafirniganosauropus"* (Cretaceous, Tajikistan), and *"Malutitrisauropus"* (Lower Jurassic, Lesotho). In addition, some have been mentioned in popular media, such as *"Cibolosaurus"* from the Lower Cretaceous (Texas, USA) and *"Sucresauripis boliviensis"* (a possible misspelling of *"Sucresauripus boliviense"*), a theropod footprint from the Upper Cretaceous of Bolivia. See more: *Ellenberger 1970; Zhen et al. 1993; Djalilov & Novikov 1993; Leonardi 1994; McLeod 1999; Meyer in Anonymous 2015; Ford**

The smallest of late Upper Cretaceous (Class: Tiny)

Koreanaornis sp. (r) footprints are between 1.9 and 2.5 cm long. The footprint of a smaller non-avian theropod are 2.7 cm long, but some authors interpret it as a bird. See more: *Ambroggi & Lapparent 1954; Huh et al. 2012*

The smallest of the early Upper Cretaceous (Class: Tiny)

Koreanaornis cf. *hamanensis* (l) are bird footprints that are between 2.5 and 2.9 cm long. *Columbosauripus* cf. *amouraensis* (n) are the smallest non-avian theropod footprints of that time. See more: *Haubold 1975; Anfinson et al. 2009*

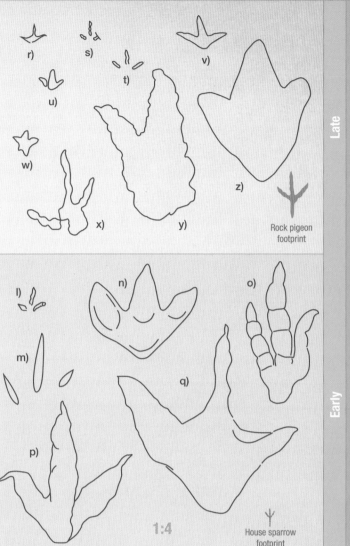

Rock pigeon footprint

House sparrow footprint

1:4

Late

Early

Upper Cretaceous

Notable small Cretaceous footprints

Species, measurement	Record, group	Country, reference
Early Upper Cretaceous		
l) *Koreanaornis* cf. *hamanensis* 2.5 cm	Smallest North American Early Upper Cretaceous (bird)	Utah, USA *Anfinson et al. 2009*
m) Unnamed 7.5 cm	Smallest South American Early Upper Cretaceous (bird)	Argentina *Krapovickas 2010*
n) *Columbosauripus amouraensis* 10 cm	Smallest African Early Upper Cretaceous (theropod)	Algeria *Bellair & Lapparent 1948*
o) Unnamed 15 cm	Smallest South American Early Upper Cretaceous (theropod)	Argentina *Calvo 1991*
p) Unnamed 15 cm	Smallest Asian Early Upper Cretaceous (theropod)	Japan *Matswkawa et al. 2005*
q) Unnamed 20 cm	Smallest North American Early Upper Cretaceous (theropod)	New Mexico, USA *Wolfe 2006*
Late Upper Cretaceous		
r) *Koreanaornis* sp. 1.9 cm	Smallest Asian Late Upper Cretaceous (bird)	South Korea *Lockley & Hunt 1994*
s) Unnamed 2 cm	Smallest North American Late Upper Cretaceous (bird)	Wyoming, USA *Lockley & Rainforth 2002*
t) Unnamed 2 cm	Smallest North American Late Upper Cretaceous (bird)	Wyoming, USA *Lockley & Rainforth 2002*
u) *Barrosopus slobodai* 2.1 cm	Smallest South American Late Upper Cretaceous (bird)	Argentina *Coria et al. 2002*
v) Unnamed ~2.67 cm	Smallest African Late Upper Cretaceous (theropod)	Morocco *Ambroggi & Lapparent 1954*
w) Unnamed 3.5 cm	Smallest African Late Upper Cretaceous (bird)	Morocco *Ambroggi & Lapparent 1954*
x) Unnamed 9 cm	Smallest North American Late Upper Cretaceous (theropod)	Alaska, USA *Fiorillo et al. 2012*
y) *Anchisauripus* sp. 15 cm	Smallest South American Late Upper Cretaceous (theropod)	Bolivia *Ríos 2005*
z) *Velociraptorichnus* sp. 16 cm	Smallest European Late Upper Cretaceous (theropod or bird?)	Poland *Gierlinski 2008*

The footprint with the greatest width of toes (right)

The greatest separation of toes in a footprint occurs in birds. This separation does not usually exceed 90 degrees in theropods, but the maximum reported in a bird, in the *Shandongornipes muxiai* specimen LRH-dz66, from China, is 142 degrees. This ichnogenus is special, because members' toes are arranged in a zygodactyl manner, similar to such birds as the woodpecker, roadrunner, cuckoo, toucan, and some owls. The only other Mesozoic bird that has this characteristic is *Dalingheornis weii*. See more: *Li, Lockley & Liu 2005; Lockley et al. 2007*

1:4

The country with the most Mesozoic bird footprints

There are about 49 locations where bird footprints from the late Lower Cretaceous are known. Forty-three are in South Korea. See more: *Lockley & Rainforth 2002*

The tsunami that caused the most dinosaur footprints to be preserved

A study showed that a giant wave during the late Lower Cretaceous affected what is now the province of Teruel (Spain). It facilitated the conservation of hundreds of dinosaur footprints, forming the largest known mega-occurrence in Europe. See more: *Navarrete et al. 2014*

The widest theropod footprint

A footprint in Praia Santa (Portugal) is approximately 50% wider than it is long, partly because it lacks the heel, which considerably decreases its length. See more: *Anonymous 2003*

The narrowest footprint of a Mesozoic bird

The width of the footprint of *Koreanaornis* sp. (South Korea) is approximately 58% of the total length. It includes finger I (hallux). See more: *Deok 2000*

The widest footprint of a Mesozoic bird

A footprint of *Koreanaornis hamanensis* (South Korea) is approximately 85% wider than it is long. See more: *Kim 1969*

The most recent theropod footprint

There are several footprints from the end of the Mesozoic. Among the most recent is *Tyrannosauripus pillmorei* (New Mexico, USA), which was located 20 m from the sedimentary line that separates the Cretaceous and Cenozoic. See more: *Lockley & Hunt 1994*

HISTORIC RECORDS OF FOOTPRINTS OF DINOSAUROMORPHS, THEROPODS, AND MESOZOIC BIRDS

ca. 1600 BC—Poland—The oldest petroglyphs with dinosaur footprints

Engravings of dancing devils with antennae and horns are integrated on ichnofossils in Kontrewers. See more: *Gierlinski & Kowalski 2006*

900–1000 BC—Spain—The first theropod footprints discovered?

Pre-Romans destroyed some tridactyl footprints that were found on rocks used in the construction of the Necropolis of Revenga de Burgos. See more: *Mayor 2000; Pascual-Marquinez et al. 2010*

ca. 1000–1200—USA—The first representation of a theropod footprint

A pictogram of a tridactyl footprint of "*Eubrontes*" in Utah was considered a footprint of a large bird by the Anazasi people. See more: *Mayor & Sarjeant 2001*

1802—USA—The first interpretation of a dinosaur footprint

The first footprints discovered in the Connecticut River valley (Massachusetts) were identified as belonging to turkeys. In other parts of the world, footprints have been attributed to eagles, crows, pigeons, even gigantic birds. See more: *Brookes 1763; Parkinson 1822*

1835—Massachusetts, USA—The first age estimate of bird footprints

James Deane reported footprints of a 3,000-year-old antediluvian turkey. This track is located near Greenfield, Massachusetts, USA. See more: *Thulborn 1990*

1835—Germany—The first Triassic footprints

Chirotherium was discovered in 1834 by Sickler in Thuringia. They are the first known footprints of the Triassic. It was suggested that they were left by various animals, including dinosaurs. They were left by archosaurs. See more: *Kaup 1835*

1836—USA—The first theropod footprints

Ornithichnites ingens (now *Steropoides*) and *O. tuberosus* (now *Anchisauripus*) left the marks of only three toes in each footprint. See more: *Hitchcock 1836*

1836—USA—The first tetradactyl theropod footprints

Ornithichnites giganteus (later called *Brontozoum* and currently called *Eubrontes*) and *O. diversus* (now *Steropoides*) have big toe marks. See more: *Hitchcock 1836*

1836—USA—The first book of theropod footprints

Edward Hitchcock uses the recent term *ichnology* for the title of his work *Ornithichnology*. See more: *Hitchcock 1836*

1836—USA—The first theropod impressions with feathers

Ornithichnites diversus clarus, *O. diversus platydactylus* and *O. ingens* (now *Steropoides*), from the Lower Jurassic of Massachusetts, show the marks of feather tufts that emerge from their heels. See more: *Hitchcock 1836*

1836—USA—The first Jurassic theropod footprints named in eastern North America

Edward Hitchcock published *Ornithichnites diversus*, *O. giganteus*, *O. ingens*, and *O. tuberosus* (currently distributed in the ichnogenera *Anchisauripus*, *Eubrontes*, and *Steropoides*). These theropod footprints from the Connecticut River valley, Massachusetts, are now known to date from the Lower Jurassic. (They were previously thought to be from the Triassic.) See more: *Hitchcock 1836*

1840—Colombia—The first report of a theropod footprint in northern South America

Carl Degenhardt reported large bird footprints in "Mexico" in 1840. In reality, they were from Colombia. We do not know their age, as their exact location was not reported. They were not illustrated. See more: *Degenhardt 1840; Buffetaut 2000; Guzmán et al. 2014*

1841—USA—The first theropod footprint name change

Edward Hitchcock separates the ichnites of birds (*Ornithoidichnites*), reptiles (*Sauroidichnites* and *Palamopus*), and mammals (*Tetrapodichnites*). See more: *Hitchcock 1841*

1850—England—The first avemetatarsalian footprints in western Europe?

Plesiothornipos binneyi are footprints that were identified as theropods. In fact, it is impossible to identify them, as their age (Middle Triassic) and shape means they could be of any mesaxonic archosaur. See more: *Harkness 1850; Haubold 1971; Thulborn 2006*

1858—USA—The first theropod tail mark

Sinuous *Gigantitherium caudatum* tail prints were reported in the Connecticut Valley, Massachusettss. See more: *Hitchcock 1858; Kim & Lockley 2013*

1867—USA—The first recognition of theropod footprints

Edward Drinker Cope was the first to recognize with certainty that the Connecticut Valley (Massachusetts, USA) footprints belonged to theropods. Edward Hitchcock, their discoverer, never accepted that possibility because he considered dinosaurs to be heavy four-legged animals, while these were biped and graceful. They also showed evidence of feathers. See more: *Cope 1867*

1868—USA—The first theropod footprints from western North America

Footprints of large birds were reported in the Aldmon Formation, Wyoming, USA. Some were actually theropods. See more: *Hayden 1868; Deibert & Breithaupt 2006*

1879—Wales—The first Triassic theropod footprint from western Europe

In Wales, a specimen of "*Brontozoum*" *thomasi* (now *Anchisauripus*) was described. See more: *Sollas 1879; Thulborn 1990*

1881—Algeria—The first Cretaceous theropod footprints in North Africa

The first African dinosaur footprints were left by small theropods. See more: *Le Mesle & Peron 1881; Lockley 1991*

1882—Tajikistan—The first Jurassic theropod footprints in western Asia

Eubrontes tianschanicum was located on the Yagnob River. The name was changed to *Gabirutosaurus* 106 years later, and it was changed again later, to *Gabirutosaurichnus*. See more: *Romanovsky 1882; Gabunia & Kurbatov 1988; Lockley 1991*

1883—England—The first Cretaceous theropod footprint in western Europe

Some footprints are described in Greensand, some of which appear to be of theropods. See more: *Saxby 1846; Thulborn 1990*

1886—Italy—The first dinosauromorph footprints in western Europe?

Footprints of *Ichnites verrucae* (currently *Thecodontichnus*) are very similar to those of *Parachirotherium*, which are considered to be of dinosauromorphs. See more: *Thommasi 1886; Haubold & Klein 2000*

1889—USA—The first large collection of theropod footprints

At the end of the description of the Connecticut Valley footprints, Edward Hitchcock managed to collect some 20,000 fossil tracks, many of which belong to theropods. See more: *Moore 2014*

1895—USA—The first Triassic theropod footprints from eastern North America

The first footprints of the Triassic were located in Virginia. Previous reports from Connecticut, Massachusetts, and New Jersey are now thought to be from the Lower Jurassic. See more: *Mitchell 1895*

1897—England—The first footprints of dinosauromorphs in western Europe

Rhynchosaurus tumidus (now *Rotodactylus*) were considered small lizard footprints. See more: *Morton 1897; Haubold 1971*

1898—USA—The first Jurassic theropod footprints in western North America

Footprints located by Wells in the Morrison Formation in 1898 were described a year later. They were compared with the legs of an *Allosaurus*, which is similar. See more: *Marsh 1899*

1908—USA—The first dinosaur footprint mistaken for a human one

Some footprints from the late Lower Cretaceous found in Texas were impressions of poorly preserved metatarsals. Nonprofessionals interpreted them as belonging to humans. The fact was so relevant that numerous forgeries were manufactured to sell to tourists, although with very obvious anatomical errors. Similar cases occurred in France in 1961 with footprints from the Middle Triassic, and in Turkmenistan in 1984 with prints from the Upper Jurassic. See more: *Amanniazov 1984; Bellini 1984; Lockley 1991*

1916—Portugal—The first Jurassic theropod footprint in western Europe

Disregarding an informal discovery made in the fifteenth century, which can not be identified with certainty, the first official report of theropod footprints was in 1916. See more: *Gomes 1916; Santos & Rodríguez 2008*

1917—USA—The first evidence of Cretaceous theropod footprints in western North America

Several large footprints were discovered in Texas, some of which were related to *Acrocanthosaurus*, a carcharodontosaurid described thirty-three years later. See more: *Shuler 1917; Sternberg 1932; Stovall & Langston 1950; Lockley & Hunt 1994*

1920—Brazil—The first Cretaceous theropod footprints in northern South America

Several theropod footprints of various sizes were discovered in 1920. The findings were published four years later. See more: *Moraes 1924; Leonardi 1994*

1925—Namibia—The first Jurassic theropod footprints in southern Africa

Footsteps were discovered in 1925 and described a year later. The species *Saurichnium anserinum*, *S. damarense*, *S. parallelum*, and *S. tetractis* were named. They were all very different animals, so the ichnogenus is in disuse. See more: *Gürich 1926; Weishampel et al. 2004*

1926—Canada—The first named footprint of a Cretaceous theropod in western North America

Ornithomimipus angustus, from Alberta, are interpreted as footprints left by an ornithomimosaur. See more: *Peterson 1924; Sternberg 1932; Lockley & Hunt 1994*

1929—Argentina—The first Cretaceous theropod footprints in southern South America

Small theropod footprints were published two years after their discovery. See more: *Huene 1931*

1933—Australia—The first Jurassic theropod footprint in Oceania

A brief report precedes two publications on Australia's footprints by thirty-four years. See more: *Ball 1934; Bartholomai 1966; Hill et al.1966*

1935—China—The first incorrect report of a "quadruped" theropod footprint

Changpeipus carbonicus was thought to be the front and back legs of the same animal, but it was actually a combination of footprints of a small specimen with those of a large specimen. See more: *Young 1960*

1939—China—The first Jurassic theropod footprint from East Asia?

Jeholosauripus s-satoi was discovered a year before it was named. It is possible that the footprints were actually left by ornithischian dinosaurs. See more: *Yabe et al. 1940; Park 1926; Zhen et al. 1989*

1940—China—The first Jurassic theropod footprint named in East Asia?

Jeholosauripus s-satoi was discovered a year before it was named. It is possible that the footprints were actually left by ornithischian dinosaurs. See more: *Yabe et al. 1940; Park 1926; Zhen et al. 1989*

1941—Lesotho—The first Jurassic theropod footprint named in southern Africa

The name *Ichnites euskelosauroides* was changed to *Platysauropus robustus* twenty-nine years later. See more: *Huene 1941; Ellenberger 1970*

1942—Argentina—The first Cretaceous theropod footprint named in southern South America

Anchisauripus australis is very similar to the Lower Jurassic footprints described from eastern North America. See more: *Lull 1942*

1948—USA—The first dinosauromorph footprints in western North America

Rotodactylus bradyi, *R. cursorius*, and *R. mckeei* were located in Arizona. See more: *Peabody 1948*

1949—Algeria—The first Cretaceous theropod footprints from North Africa

Footprints of small carnivorous dinosaurs were reported in Algeria. See more: *Bellair & Lapparent 1949*

1951—Georgia—The first Cretaceous theropod footprints named in eastern Europe

Satapliasaurus dsotsenidzei, *S. kandelakii*, and *S. tschabukianii* were named. The country of Georgia is considered within the European continent by some authors, while others include it in Asia. See more: *Gabuniya 1951; Lessertisseur 1955; Weishampel et al. 2004*

1952—Canada—The first dinosauromorph footprints in eastern North America

Gigandipus milfordensis (currently *Atreipus*) was described in Nova Scotia. See more: *Bock 1952*

1952—Australia—The first Cretaceous theropod footprints from Oceania

The presence of footprints in the city of Broome was reported for the first time. See more: *Glauert 1952*

1952—USA—The first Triassic theropod footprint in eastern North America

Anchisauripus gwyneddensis (now *A. milfordensis*) was located in Pennsylvania. The footprints belong to the Upper Triassic. The New Jersey and Connecticut Valley, Massachusetts, footprints precede it, but they are now known to date from the Lower Jurassic. See more: *Bock 1952*

1954—Morocco—The first Mesozoic bird footprint of the Cretaceous in North Africa

Several footprints similar to those of birds (toes spread wide) were found. See more: *Ambroggi & Lapparent 1954*

1954—China—The first footprints of Jurassic theropods in East Asia?

The footprints of *Changpeipus carbonicus* were reported six years before being named. It was preceded by *Jeholosauripus s-satoi*, whose footprints could belong to theropods or ornithischians. See more: *Yabe et al. 1940; Lee 1955; Young 1960*

1955—Germany—The first Cretaceous theropod footprint named in western Europe

Megalosauripus maximus was changed to *Bückebergichnus* three years later and was later corrected to *Bueckeburgichnus*. See more: *Lessertisseur 1955; Kuhn 1958; Thulborn 2001*

1956—Chile—The first Jurassic theropod footprints in southern South America

Footprints were discovered that were formally reported six years later and were not illustrated until 1965. See more: *Galli & Dingman 1962; Dingman & Galli 1965*

1960—China—The first Jurassic theropod footprint named in East Asia

Changpeipus carbonicus (now *Eubrontes*) was described after *Jeholosauripus s-satoi*, but it is possible that the latter are footprints of ornithischian dinosaurs. See more: *Yabe et al. 1940; Young 1960; Lockley et al. 2013*

1962—Israel—The first theropod footprints from the Middle East

Some footprints attributed to ornithomimosaurs could be of any other type of sprinting theropod. See more: *Avnimelech 1962*

1962—Morocco—The first Triassic theropod footprint in North Africa

A medium-size footprint was located in Morocco. It was described nineteen years after its discovery. See more: *Byron & Dutuit 1981*

1964—Argentina—The first dinosauromorph footprints in southern South America

"Coelurosaurs" or small "pre-dinosaurs" from the early Triassic are reported. Discovered by José Bonaparte, the tracks were most likely left by dinosauromorphs. They were not formally reported until about twenty-five years later. See more: *Bonaparte in Leonardi 1989, 1994*

1964—Tajikistan—The earliest Cretaceous theropod footprints in western Asia

The first discovered footprints of therizinosaurs were named *Macropodosaurus gravis*. See more: *Zakharov 1964*

1964—Argentina—The first Triassic theropod footprints from southern South America

There are footprints attributed to "Coelurosaurs" from the Upper Triassic in Santa Cruz. They were reported by Rodolfo Casamiquela in 1964. See more: *Leonardi 1989, 1994*

1964—Argentina—The first named footprints of a Jurassic theropod in southern South America

Sarmientichnus scagliai and *Wildeichnus navesi* were reported in the province of Río Negro. See more: *Casamiquela 1964*

1964—Australia—The first Triassic theropod footprint in Oceania

The presence of *Eubrontes* in Australia was reported. See more: *Staines & Woods 1964; Thulborn 1998*

1964—Argentina—The first monodactyl theropod footprint

On some occasions, *Sarmientichnus scagliai* left imprints of one to three fingers. This is because the middle finger supported most of the weight, as occurs in cursorial theropods such as *Velocisaurus unicus*. See more: *Casamiquela 1964; Bonaparte 1991; Valais 2011*

1966—Israel—The first Mesozoic bird footprints in the Middle East

The footprints found are similar to those of long-legged birds found in other parts of the world. See more: *Avnimelech 1962; Lockley et al. 1992*

1967—France—The first Jurassic theropod footprints in western Europe

The species *Eubrontes veillonensis*, *Grallator maximus*, *G. olonensis*, *G. variabilis*, *Saltopoides igalensis* and *Talmontopus tersi* were named. See more: *Lapparent & Montenat 1967*

1967—Australia—The first Cretaceous theropod footprint in Oceania

Megalosauropus broomensis are medium-size footprints. See more: *Colbert & Merrilees 1967*

1968—France—The first primitive saurischian footprint in western Europe?

Grallator sp. from the Lower Triassic is the oldest dinosaur footprint. See more: *Montenat 1968*

1969—South Korea—The first Mesozoic bird footprint of the Cretaceous in East Asia

Koreanaornis hamanensis are one of the most numerous Mesozoic bird footprints. See more: *Kim 1969*

1970—Lesotho—The first Triassic theropod footprints in South Africa

In a monumental book that extended over an additional two volumes, a large number of theropod footprints were described: *Grallator bibractensis*, *G. damanei*, *G. matsiengensis*, *Kainotrisauropous morijiensis*, *K. moshoeshoei*, *Megatrisauropus malutiensis*, *Neotrisauropus deambulator*, *N. lacunensis*, *N. leribeensis*, *Neotripodiscus makoetlani*, *Otouphepus declivis*, *O. magnificus*, *O. minor*, *O. palustris*, *Platysauropus robustus*, "Platisauropus" *(Platysauropus) ingens*, *Prototrisauropodiscus minimus*, *Prototrisauropus angustidigitus*, *P. crassidigitus* and *P. rectilineus*. See more: *Ellenberger 1970, 1972, 1974*

1970—Lesotho—The first primitive saurischian footprints in southern Africa

The footprints *Qemetrisauropus minor* and *Q. princeps* have characteristics similar to those of primitive saurischians. See more: *Ellenberger 1970, 1972, 1974*

1971—Algeria—The first Jurassic theropod footprint in North Africa?

Jurassic carnivorous dinosaur footprints were reported. Another precedes them, but it is possible that they are from the Cretaceous. See more: *Ginsburg et al. 1966; Bassoullet 1971*

1971—Algeria—The first Cretaceous theropod footprint named in northern Africa

Columbosauripus amouraensis is a species of an ichnogenus that was originally created in North America thirty-nine years earlier. See more: *Haubold 1971*

1971—Australia—The first Jurassic theropod footprints in Oceania

Changpeipus bartholomaii is a species of a genus described in China eleven years earlier, but it is doubtful whether it belongs to the same ichnogenus. See more: *Haubold 1971; Lockley et al. 2013*

1975—Iran—The first Jurassic theropod footprint named in the Middle East

Iranosauripus zerabensis dates from the Lower Jurassic. See more: *Lapparent & Sadat 1975*

1975—Croatia—The first Cretaceous theropod footprints in eastern Europe

Various footprints were found in Croatia. Additional discoveries dating from the late Lower Cretaceous and the late Upper Cretaceous came later. See more: *Gogala 1975*

1975—Guyana—The first Jurassic theropod footprint in northern South America?

Footprints attributed to a bird or a small theropod were reported. The discovery precedes another discovery in Colombia, but as it was not illustrated or described, it is impossible to know its true nature. See more: *Degenhardt 1840; Montalvao et al. 1975*

1982—China—The first Cretaceous theropod footprints in eastern Asia

Xiangxipus chenxiensis and *X.youngi* are footprints of probable oviraptorosaurs. See more: *Zheng 1982; Gierlinski & Lockley 2013*

1983—Hungary—The first Jurassic theropod footprint in eastern Europe

Komlosaurus carbonis were interpreted as the footprints of an ornithischian, but twenty-eight years later, it became synonymous with *Kayentapus*, a set of theropod tracks. See more: *Kordos 1983; Osi et al. 2011*

1985—Thailand—The first Cretaceous theropod footprints from Cimmeria

Carnivorous dinosaur footprints of medium size were described. See more: *Buffetaut et al. 1985*

1986—China—The first named footprint of a Cretaceous theropod in East Asia

The unofficial name of "Deinonychosaurichnus" was given. Nine years later, it was changed to *Velociraptorichnus*. See more: *Zhen et al. 1987; Zhen et al. 1994*

1986—China—The first didactyl theropod footprints of the Cretaceous

Footprints that only have two fingers were informally named "Deinonychosaurichnus." The name was later changed to *Velociraptorichnus*. See more: *Zhen et al. 1987; Zhen et al. 1994*

1986—India—The first Cretaceous theropod footprints in South Asia (Hindustan)

A footprint that was attributed to a sauropod has the typical form of some theropod footprints. See more: *Mohabey 1986*

1986—Argentina—The first Mesozoic bird footprint of the Cretaceous in southern South America

In the collection of the Department of Paleontology of the Faculty of Natural Sciences of the University of Salta, the name of *Yacoraitichnus avis* appeared as "Iacoraitichnus avis." See more: *Alonso & Marquillas 1986; Leonardi 1994*

1987—China—The first Triassic theropod footprint in eastern Asia?

It was suggested that the footprints of *Pengxianpus cifengensis*, described twenty-six years earlier and originally identified as primitive sauropodomorphs, were theropod. Both dinosaurs left very similar footprints, so it is difficult to determine. See more: *Young & Young 1987; Xing et al. 2013*

1987—USA—The first primitive saurischian footprint in eastern North America?

Grallator sp., from Virginia, could belong to a primitive saurischian. See more: *Weems 1987*

1988—Morocco—The first didactyl footprint of a Jurassic theropod

Some *Argoides* footprints had two to three fingers. They were produced by cursorial theropods. See more: *Ishigaki 1988; Ishigaki & Lockley 2010*

1990—China—The first Mesozoic bird footprint of the Jurassic in East Asia

Bird tracks, reported in the Tuchengzi Formation, were named *Pullornipes aureus* and cf. *Aquatilavipes* in 2006. See more: *Wang et al. 1990; Lockley et al. 2006*

1994—China—The first Cretaceous theropod footprints named in eastern Asia

Grallator emeiensis and *Velociraptorichnus sichuanensis* are impressions of small theropods. See more: *Zhen et al. 1994*

1994—Bolivia—The largest Cretaceous site of dinosaur footprints is discovered

Klaus Pedro Schütt discovered an area that currently houses more than 10,000 footprints, including hundreds of tracks several meters long. It is known as the "Farallón de Cal Orcko." Unfortunately, landslides have destroyed many of them. See more: *Meyer 2003*

1996—France—The first Mesozoic bird footprint of the Jurassic in western Europe?

An unnamed footprint that seems to be of a bird has marks of what might be a feather plume behind the heel. Another footprint discovered thirty-three years earlier, *Taupezia landeri*, has a very open toe width but could have belonged to a paravian theropod. See more: *Delair 1963; Lange-Badre et al 1996*

1996—Spain—The first Mesozoic bird footprints of the Cretaceous in western Europe

Archaeornithipus meijidei are footprints of large, long-legged birds. See more: *Fuentes-Vidarte 1996*

1996—Peru—The first Cretaceous theropod footprint in northern South America

Ornithomimipus jaillardi may have belonged to ceratosaur theropods, as there may have been no ornithomimosaurs in South America. See more: *Jaillard et al. 1993*

1997—USA—The first Cretaceous theropod footprint in eastern North America

Megalosauropus sp. is a footprint from the Lower Cretaceous of Virginia. See more: *Weems & Bachman 1997*

1998—Germany—The largest deposit of dinosauromorph footprints is reported

Between 600 and 1,000 *Rotodactylus* footprints are preserved in each square meter of the Solling Formation, dating from the Middle Triassic. They were collected in the 1960s See more: *Haubold 1998*

1998—Australia—The first primitive saurischian footprint in Oceania?

cf. *Grallator* sp. could belong to a primitive saurischian. See more: *Thulborn 1998*

2000—Poland—The last named dinosauromorph footprint

Prorotodactylus mirus are the oldest known footprints of primitive avemetatarsalians. As of 2016, none has been named more recently. See more: *Ptaszynski 2000*

2001—England—The first ichnogenus that shows dinosaurs and other animals floating

The *Characichnos* footprints show theropods, crocodiles, turtles, or amphibians that were floating at the moment the impression was made. See more: *Whyte & Romano 2001; Kukihara et al. 2010*

2003—USA—The first primitive saurischian footprint in western North America?

Grallator sp. from the Triassic of Colorado, could belong to a primitive saurischian. See more: *Gaston et al. 2003*

2004—Algeria—The first dinosauromorph footprint in northern Africa

Rotodactylus cf. *bessieri* is reported in the Middle Triassic. See more: *Kotanski et al. 2004*

2006—China—The first Mesozoic bird footprint from the Jurassic named in eastern Asia

Pullornipes aureus and cf. *Aquatilavipes* were named in China. They are between the Upper Jurassic and Lower Cretaceous, so their ages may vary. See more: *Wang et al. 1990; Lockley et al. 2006*

2006—Thailand—The first named Cretaceous theropod footprint from Cimmeria?

Siamopodus khaoyaiensis may be a theropod or an ornithopod footprint. See more: *Lockley et al. 2006*

2006—Spain—The largest theropod footprint

A mold of a footprint 105 cm long includes part of the metatarsal, making the foot only 82 cm long. See more: *García-Ramos et al. 2006; Boutakiout et al. 2009; García-Ortiz de Landaluce et al. 2009*

2007—Pakistan—The first Jurassic theropod footprint named in southern Asia (Hindustan)

Samandrinda surghari lacks an appropriate description, so it is a dubious name. See more: *Malkani 2007*

2007—Argentina—The first primitive saurischian footprint in southern South America?

A publication describes dinosaur footprints, some of which were probably primitive saurischians and theropods. See more: *Marsicano et al. 2007*

2007—Spain—The theropod that swam against the current

Footprints suggest movement against the current at an angle of 30 degrees. See more: *Esquerra 2007*

2008—Brazil—The first Triassic theropod footprint in southern South America

Grallator isp., 8.5 cm long, was located in southern Brazil. It is the smallest footprint of a Triassic dinosaur from South America. See more: *Silva et al. 2008*

2010—USA—The first evidence of excavation activity of a theropod

Some fossils have the mark of a false fingernail typical of dromaeosaurids and troodontids. They were found associated with the burrow of a Mesozoic mammal. See more: *Simpson et al. 2010*

2010—Portugal—The first formal article about world-record dinosaur footprints

García-Ortiz de Landaluce, Esperanza, José Manuel Ortega-Girela, Alberto Hurtado-Reyes, and Ignacio Díaz-Martínez, 2009, "Revision of the Biggest and Smallest Theropod, Ornithopod and Sauropod Footprints of La Rioja (Spain) and Its Comparison with the World Record: The Guinness World Record," *Paleolusitana: Journal of Paleontology and Paleoecology*, no. 1, 201–9 (in Spanish).

2011—Niger—The first Jurassic theropod footprint named in northern Africa

Paravipus didactyloides is a didactyl footprint. Although it has been suggested that it may have been an ancient dromaeosaurid, it is similar to *Argoides*, a cursorial theropod footprint that also presents impressions of two, three, and even four fingers. See more: *Ishigaki 1988; Ishigaki & Lockley 2010; Mudroch et al. 2011*

2011—Morocco—The first Jurassic bird footprints in northern Africa?

Footprints similar to birds of the Middle Jurassic could be of paravians. See more: *Belvedere et al 2011*

2011—Spain—The first three-dimensional theropod footprint

A footprint was preserved in a deep mold, keeping the shape of the podotheca as if it were a sculpture. See more: *Adorna-Fernandes et al. 2004*

2013—Australia—The first Mesozoic bird footprint in Oceania

Some medium-size bird footprints were reported, including one that dragged its leg. See more: *Martin et al. 2013*

2013—Spain—The first footprints with movement marks

A few countermolds of footprints with sliding grooves from walking were left in dense mud. They are known as "four-dimensional footprints" because they reveal the movement of the theropod that produced them.

2013—China—The longest trail of a floating theropod

Footprints discovered in 1991 reveal the presence of a theropod floating in shallow water. It had to advance 15 m by means of its hind legs. These marks are known as *Characichnos* sp. See more: *Xing et al. 2013*

2014—China—The first primitive saurischian footprint of eastern Asia?

A footprint similar to *Grallator* is the oldest in Asia. It could have been a primitive saurischian. See more: *Xing et al. 2014*

2015—Canada—The last named Mesozoic bird footprint

Paxavipes babcockensis resembles footprints of *Barrosopus slobodai* from Argentina. See more: *McCrea et al. 2015*

2015—China—The last named theropod footprint

Velociraptorichnus zhangi are the only dromaeosaurid footprints that apparently left some tridactyl impressions. Due to their highly developed finger II claws, all of the others were didactyl. See more: *Xing et al. 2015*

     1:1

How is a dinosaur like a horseshoe crab?

The footprints of lumilids are so similar to those of theropods that one is often mistaken for the other. One Middle Triassic specimen from England was baptized "Coelurosaurichnus sp." and was believed to be one of the smallest footprints ever discovered.

Symmetrical and asymmetrical footprints
Eustreptospondylus oxoniensis &
Sciurumimus albersdoerferi
BMMS BK 11 y OUM J13558
Among all megalosauroids, these are the only two species that preserved almost complete legs. Thanks to this, it is known that they were similar to the *Megalosauripus* footprints, which had asymmetrical proportions between the fingers. In contrast, the finger lengths of other large contemporary theropods, such as the allosauroids, were more symmetrical. See more: *Walker 1964; Lockley, Meyer & Santos 1998; Rauhut et al. 2012; Molina-Pérez et al. manuscript in preparation*

Avemetatarsalian footprints

N.º	Ichnospecies	Country	Era	Bibliography
1	*Agialopus wyomingensis*	Wyoming, USA	UT	Branson & Mehl 1932
2	*Anchisauripus bibractensis*	France	MT	Demathieu 1971
3	*Atreipus acadianus*	Canada	UT	Olsen & Baird 1986
4	*Atreipus metzneri*	Germany, France	UT	Heller 1952; Olsen & Baird 1986
5	*Atreipus milfordensis*	Pennsylvania, USA	UT	Bock 1952
6	*Atreipus sulcatus*	New Jersey, USA	UT	Baird 1957
7	*Banisterobates boisseaui*	Virginia, USA	UT	Frasier & Olsen 1996
8	*Coelurosaurichnus arntzeniusi*	Germany	UT	Rehnelt 1952
9	*Coelurosaurichnus kehli*	Germany	UT	Beurlen 1950
10	*Coelurosaurichnus kronbergeri*	Germany	UT	Rehnelt 1952, 1959
11	*Coelurosaurichnus largentierensis*	France	MT	Courel & Demathieu 1976
12	*Coelurosaurichnus moeni*	Germany	UT	Beurlen 1950
13	*Coelurosaurichnus palyssii*	France	MT	Gand 1976
14	*Coelurosaurichnus perriauxi*	France	MT	Demathieu & Gand 1972
15	*Coelurosaurichnus ratumensis*	Netherlands	MT	Demathieu & Oosterink 1988
16	*Coelurosaurichnus sabinensis*	France	MT	Gand 1976
17	*Coelurosaurichnus sassendorfensis*	France	UT	Kuhn 1958
18	*Coelurosaurichnus schlauersbachense*	France	UT	Weiss 1934
19	*Coelurosaurichnus schlehenbergense*	Germany	UT	Rehnelt 1950
20	*Coelurosaurichnus toscanus*	Italy	MT	Huene 1941
21	*Coelurosaurichnus ziegelangernensis*	Germany	UT	Sarjeant 1996; Kuhn 1958
22	*Parachirotherium postchirotheroides*	France	MT	Kuhn 1958
23	*Paratrisauropus bronneri*	Switzerland	MT	Demathieu & Weidmann 1982
24	*Paratrisauropus latus*	Switzerland	MT	Demathieu & Weidmann 1982
25	*Paratrisauropus mirus*	Switzerland	MT	Demathieu & Weidmann 1982
26	*Plesiothornipos binneyi*	England	MT	Harkness 1850
27	*Prorotodactylus lutevensis*	France	MT	Demathieu 1984; Ptaszynski 2000
28	*Prorotodactylus mirus*	Poland	LT	Ptaszynski 2000; Brusatte et al. 2010
29	*Rotodactylus bessieri*	France	MT	Demathieu 1984
30	*Rotodactylus bradyi*	Arizona, USA	MT	Peabody 1948
31	*Rotodactylus cursorius*	Arizona, USA	MT	Peabody 1948
32	*Rotodactylus kronachensis*	Germany	MT	Demathieu & Leitz 1982
33	*Rotodactylus lucasi*	Italy	MT	Demathieu et al. 1999
34	*Rotodactylus matthesi*	Germany	MT	Haubold 1967, 1971
35	*Rotodactylus mckeei*	Arizona, USA	MT	Peabody 1948
36	*Rotodactylus tumidus*	England	UT	Morton 1897; Maidwell 1914; Haubold 1971
37	*Rotodactylus velox*	France	MT	Demathieu & Gand 1974
38	*Sphingopus ferox*	France	MT	Demathieu 1966
39	*Sphingopus ladinicus*	Italy	MT	Avanzini & Wachtler 2012
40	*Thecodontichnus verrucae*	Italy	UT	Tommasi 1886

Impressions of the oldest avemetatarsalian hand

Prorotodactylus isp. (a) of the Stryczowice Formation (Poland) dates from the Lower Triassic (lower Olenekian, ca. 251.17–249 Ma). See more: *Niedzwiedzki et al. 2013*

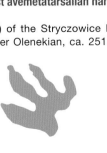

1:1

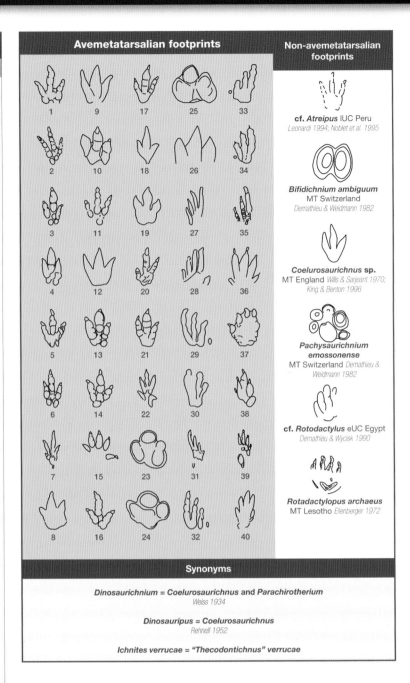

Avemetatarsalian footprints

Non-avemetatarsalian footprints

cf. *Atreipus* IUC Peru
Leonardi 1994; Noblet et al. 1995

Bifidichnium ambiguum
MT Switzerland
Demathieu & Weidmann 1982

Coelurosaurichnus sp.
MT England *Wills & Sarjeant 1970; King & Benton 1996*

Pachysaurichnium emossonense
MT Switzerland *Demathieu & Weidmann 1982*

cf. *Rotodactylus* eUC Egypt
Demathieu & Wycisk 1990

Rotadactylopus archaeus
MT Lesotho *Ellenberger 1972*

Synonyms

Dinosaurichnium = Coelurosaurichnus and Parachirotherium
Weiss 1934

Dinosauripus = Coelurosaurichnus
Rehnell 1952

Ichnites verrucae = "Thecodontichnus" verrucae

The only ichnospecies that is known just by handprints

Coelurosaurichnus ratumensis (Netherlands). See more:*Demathieu & Oosterink 1988*

1:1

The most robust avemetatarsalian handprint

Rotodactylus velox (France), unlike most of its relatives, had very wide front and back legs. See more: *Demathieu & Gand 1974*

1:2

Footprints of primitive saurischians

N.º	Ichnospecies	Country	Era	Bibliography
1	Coelurosaurichnus grancieri	France	UT	Courel & Demathieu 2000
2	Cridotrisauropus ginsburgi	France	UT	Ellenberger 1965
3	Cridotrisauropus minor	France	UT	Ellenberger 1965
4	Grallator pisanus	Italy	UT	Bianucci & Landini 2005
5	Prototrisauropus rectilineus	Lesotho	UT	Ellenberger 1972
6	Qemetrisauropus minor	Lesotho	UT	Ellenberger 1972
7	Qemetrisauropus princeps	Lesotho	UT	Ellenberger 1972
8	Saurichnium anserinum	Namibia	UT	Gurich 1926

Theropod footprints

N.º	Ichnospecies	Country	Era	Bibliography
1	Abelichnus astigarrae	Argentina	eUC	Calvo 1991
2	Anchisauripus australis	Argentina	ILC	Lull 1942
3	Anchisauripus danaus	Massachusetts, USA	LJ	Hitchcock 1845
4	Anchisauripus exsertus	Connecticut & Massachusetts, USA	LJ	Hitchcock 1858
5	Anchisauripus gwyneddensis	Pennsylvania, USA	UT	Bock 1952
6	Anchisauripus hitchcocki	Connecticut, USA	LJ	Lull 1904
7	Anchisauripus minusculus	USA, France	LJ	Hitchcock 1858; Demathieu & Sciau 1995
8	Anchisauripus parallelus	Connecticut & Massachusetts, USA	LJ	Hitchcock 1865
9	Anchisauripus sillimani	Connecticut, Massachusetts & New Jersey, USA	LJ	Hitchcock 1845
10	Anchisauripus thomasi	England	LJ	Sollas 1879
11	Anchisauripus tuberatus	Massachusetts, USA	LJ	Hitchcock 1836; Reichenbach 1852
12	Anchisauripus tuberosus	USA, Hungary	LJ	Hitchcock 1843; Gierliski 1996
13	Anticheiropus hamatus	Massachusetts, USA	LJ	Hitchcock 1865
14	Anticheiropus pilulatus	Massachusetts, USA	LJ	Hitchcock 1865
15	Argoides macrodactylus	Massachusetts & New Jersey, USA	LJ	Hitchcock 1841
16	Argoides minimus	Massachusetts, USA	LJ	Hitchcock 1836; Lull 1904
17	Argoides redfieldii	Massachusetts, USA	LJ	Hitchcock 1844; Reichenbach 1852
18	Bellatoripes fredlundi	Canada	lUC	McCrea et al 2014
19	Bressanichnus patagonicus	Argentina	eUC	Calvo 1991
20	Buckeburgichnus maximus	Germany	eLC	Lessertisseur 1955
21	Carmelopodus untermannorum	Utah, USA	MJ	Lockley et al. 1988, 1998
22	Changpeipus bartholomaii	Australia	MJ	Haubold 1971
23	Changpeipus carbonicus	China	MJ	Young 1960; Lu et al. 2007
24	Changpeipus pareschequier	China	LJ	Xing et al. 2009
25	Chapus lockleyi	China	ILC	Li et al. 2006
26	Chodjapilesaurus krimholzi	Turkmenistan	UJ	Amanniazov 1985
27	Chonglongpus hei	China	MJ	Young & Young 1987
28	Chongqingpus microiscus	China	MJ	Young & Young 1987
29	Chongqingpus nananensis	China	MJ	Young & Young 1987
30	Chongqingpus yemiaoxiensis	China	MJ	Young & Young 1987
31	Coelurosaurichnus romanovskyi	Tajikistan	UJ	Gabuniya & Kurbatov 1988
32	Coelurosaurichnus tatricus	Slovakia	UT	Michalik & Sykora 1976
33	Columbosauripus amouraensis	Algeria	eUC	Haubold 1971
34	Columbosauripus ungulatus	Canada	ILC	Sternberg 1932
35	Deferrariischnium mapuchensis	Argentina	eUC	Calvo 1991

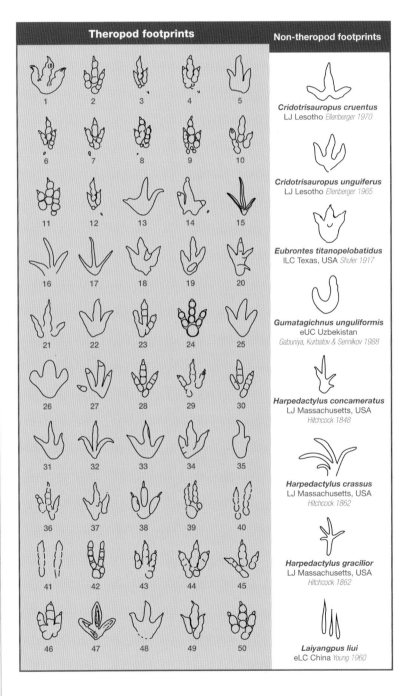

Primitive saurischian footprints

(1, 2, 3, 4, 5, 6, 7, 8)

Non-primitive saurischian footprints

Typopus (Sauroidichnites) abnormis
LJ Massachusetts, USA
Hitchcock 1844

Theropod footprints

(1–50)

Non-theropod footprints

Cridotrisauropus cruentus
LJ Lesotho *Ellenberger 1970*

Cridotrisauropus unguiferus
LJ Lesotho *Ellenberger 1965*

Eubrontes titanopelobatidus
ILC Texas, USA *Shuler 1917*

Gumatagichnus unguliformis
eUC Uzbekistan
Gabuniya, Kurbatov & Sennikov 1988

Harpedactylus concameratus
LJ Massachusetts, USA
Hitchcock 1848

Harpedactylus crassus
LJ Massachusetts, USA
Hitchcock 1862

Harpedactylus gracilior
LJ Massachusetts, USA
Hitchcock 1862

Laiyangpus liui
eLC China *Young 1960*

36	*Deuterotrisauropus deambulator*	France	UT	Ellenberger 1965
37	*Deuterotrisauropus socialis*	Lesotho	LJ	Ellenberger 1972
38	*Dilophosauripus williamsi*	USA, France	LJ	Welles 1971; Demathieu & Sciau 1995
39	*Dromaeopodus shandongensis*	China	lLC	Li et al. 2008
40	*Dromaeosauripus hamanensis*	South Korea	lLC	Kim et al. 2008
41	*Dromaeosauripus jinjuensis*	South Korea	lLC	Kim et al. 2012
42	*Dromaeosauripus yongjingensis*	China	lLC	Xing et al. 2013
43	*Eubrontes approximatus*	Connecticut, USA	LJ	Hitchcock 1865
44	*Eubrontes divaricatus*	USA, France	LJ	Hitchcock 1841; Demathieu & Sciau 1999
45	*Eubrontes expansum*	Massachusetts, USA	LJ	Hitchcock 1841
46	*Eubrontes giganteus*	Connecticut and Massachusetts, USA	LJ	Hitchcock 1836; 1845
47	*Eubrontes glenrosensis*	China	LJ	Li et al. 2010
48	*Eubrontes glenrosensis*	Texas, USA	lLC	Shuler 1935
49	*Eubrontes loxonyx*	Massachusetts, USA	LJ	Hitchcock 1848
50	*Eubrontes platypus*	China, USA	LJ	Lull 1904; Zhen et al. 1989
51	*Eubrontes soltykovensis*	Hungary, Poland	LJ	Gierlinski 1991, 1996
52	*Eubrontes veillonensis*	France	LJ	Lapparent & Montenat 1967
53	*Fuscinapedis woodbinensis*	Texas, USA	eUC	Lee 1997
54	*Gabirutosaurichnus tianschanicum*	Tayikistan	UJ	Romanovsky 1882; Gabunia & Kurbatov 1988
55	*Gigandipus caudatus*	Connecticut, USA	LJ	Hitchcock 1855
56	*Grallator andeolensis*	France	LJ	Gand et al. 2000
57	*Grallator cuneatus*	Connecticut, USA	LJ	Hitchcock 1858
58	*Grallator cursorius*	Connecticut, USA	LJ	Hitchcock 1858
59	*Grallator damanei*	Lesotho	LJ	Ellenberger 1970
60	*Grallator formosus*	Massachusetts, USA	LJ	Hitchcock 1858
61	*Grallator gracilis*	Massachusetts, USA	LJ	Hitchcock 1865
62	*Grallator limnosus*	China	LJ	Zhen, Li & Rao 1985
63	*Grallator madseni*	Arizona, USA	UT	Irby 1995
64	*Grallator maximus*	France	UT	Lapparent & Montenat 1967
65	*Grallator molapoi*	Lesotho	LJ	Ellenberger 1970
66	*Grallator oloensis*	France	UT	Lapparent & Montenat 1967
67	*Grallator sauclierensis*	France	LJ	Demathieu & Sciau 1992
68	*Grallator tenuis*	USA, India, Poland	LJ	Hitchcock 1841; Gierlinski 1996; Pienkowski et al. 2015
69	*Grallator variabilis*	France	UT	Lapparent & Montenat 1967
70	*Grallator zviersi*	Poland	LJ	Gierlinski 1996
71	*Hadrosaurichnus australis*	Argentina	lUC	Alonso 1980
72	*Hispanosauropus hauboldi*	Spain	UJ	Mensink & Mertmann 1984
73	*Hunanpus jiuquwanensis*	China	lLC	Zeng 1982
74	*Hyphepus fieldi*	Massachusetts, USA	LJ	Hitchcock 1858
75	*Iberosauripus grandis*	Spain	UJ	Cobos et al. 2014
76	*Iranosauripus zebarensis*	Iran	LJ	Lapparent & Sadat 1975
77	*Irenesauripus acutus*	Canada	lLC	Sternberg 1932
78	*Irenesauripus occidentalis*	Canada	lLC	Sternberg 1932
79	*Irenichnites gracilis*	Canada	lLC	Sternberg 1932; McCrea et al. 2015
80	*Jinlijingpus nianpanshanensis*	China	LJ	Young & Young 1987
81	*Kainotrisauropus morijiensis*	Lesotho	LJ	Ellenberger 1970
82	*Kainotrisauropus moshoshoei*	Lesotho	LJ	Ellenberger 1970
83	*Karkushosauropus karkushensis*	Tajikistan	UJ	Novikov & Dzhalilov 1993
84	*Kayentapus hailiutuensis*	China	LJ	Li et al. 2010
85	*Kayentapus hopii*	Arizona, USA	LJ	Welles 1971
86	*Kayentapus minor*	USA, Poland	LJ	Lull 1915; Niedzwiedzki 2006

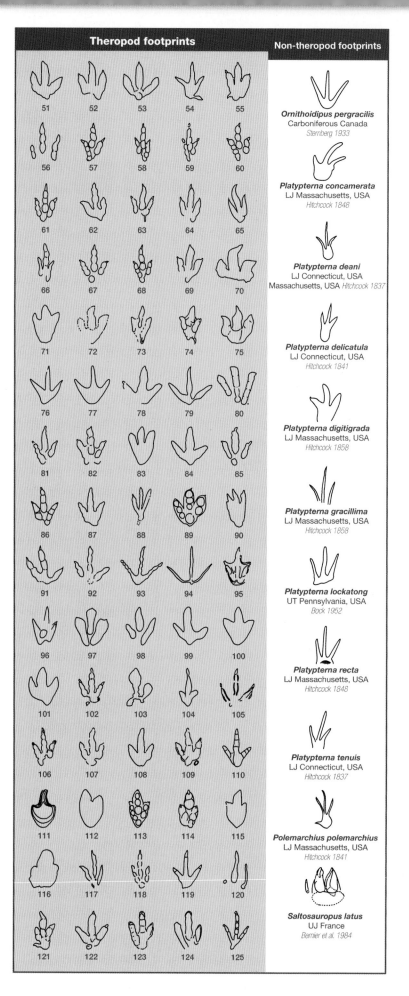

Theropod footprints

Non-theropod footprints

Ornithoidipus pergracilis
Carboniferous Canada
Sternberg 1933

Platypterna concamerata
LJ Massachusetts, USA
Hitchcock 1848

Platypterna deani
LJ Connecticut, USA
Massachusetts, USA *Hitchcock 1837*

Platypterna delicatula
LJ Connecticut, USA
Hitchcock 1841

Platypterna digitigrada
LJ Massachusetts, USA
Hitchcock 1858

Platypterna gracillima
LJ Massachusetts, USA
Hitchcock 1858

Platypterna lockatong
UT Pennsylvania, USA
Bock 1952

Platypterna recta
LJ Massachusetts, USA
Hitchcock 1848

Platypterna tenuis
LJ Connecticut, USA
Hitchcock 1837

Polemarchius polemarchius
LJ Massachusetts, USA
Hitchcock 1841

Saltosauropus latus
UJ France
Bernier et al. 1984

87	*Kayentapus minor*	Virginia, USA	UT	Weems 2006
88	*Lapparentichnus oleronensis*	France	UJ	Haubold 1971
89	*Lufengopus dongi*	China	LJ	Lu et al. 2006
90	*Macropodosaurus gravis*	Tajikistan	eUC	Zakharov 1964
91	*Magnoavipes asiaticus*	China	ILC	Matsukawa et al. 2014
92	*Magnoavipes caneeri*	Colorado, USA	eUC	Lockley et al. 2001
93	*Magnoavipes denaliensis*	Alaska, USA	lUC	Fiorillo et al. 2011
94	*Magnoavipes lowei*	Texas, USA	eUC	Lee 1997
95	*Masitisisauropus angustus*	Lesotho	LJ	Ellenberger 1974
96	*Masitisisauropus exiguus*	Lesotho	LJ	Ellenberger 1970
97	*Megaichnites jizhaishiensis*	China	LJ	Young & Young 1987
98	*Megalosauripus brionensis*	Croatia	eLC	Haubold 1971
99	*Megalosauropus broomensis*	Australia	eLC	Glauert 1953; Colbert & Merrilees 1967
100	*Megalosauropus teutonicus*	Germany	UJ	Smith 1959; Kaever & Lapparent 1974
101	*Megalosauropus uzbekistanicus*	Tajikistan, Uzbekistan	UJ	Gabunia & Kurbatov 1982
102	*Megatrisauropus malutiensis*	Lesotho	LJ	Ellenberger 1970
103	*Menglongipus sinensis*	China	UJ	Xing et al. 2009
104	*Neograllator emeienesis*	China	ILC	Zhen et al. 1994; Lockley et al. 2008
105	*Neotripodiscus makoetlani*	Lesotho	LJ	Ellenberger 1970
106	*Neotrisauropus deambulator*	Lesotho	LJ	Ellenberger 1970; Wilson et al. 2009
107	*Neotrisauropus lambereshei*	Lesotho	LJ	Ellenberger 1970
108	*Neotrisauropus leribeensis*	Lesotho	LJ	Ellenberger 1970
109	*Neotrisauropus mokanametsongensis*	Lesotho	LJ	Ellenberger 1974
110	*Ornithomimipus angustus*	Canada	lUC	Sternberg 1926
111	*Ornithomimipus jaillardi*	Peru	lUC	Jaillard et al. 1993
112	*Otouphepus declivis*	Lesotho	LJ	Ellenberger 1970
113	*Otouphepus magnificus*	Connecticut, USA	LJ	Cushamn 1904
114	*Otouphepus minor*	Connecticut, USA	LJ	Lull 1915
115	*Otouphepus palustris*	Lesotho	LJ	Ellenberger 1970
116	*Otouphepus poolei*	Pennsylvania, USA	UT	Bock 1952
117	*Paracoelurosaurichnus monax*	China	LJ	Zhen Li & Rao 1986
118	*Paragrallator matsiengensis*	Lesotho	LJ	Ellenberger 1970
119	*Paragrallator yangi*	China	ILC	Li & Zhang 2000
120	*Paravipus didactyloides*	Niger	MJ	Mudroch et al. 2011
121	*Picunichnus donnotoi*	Argentina	eUC	Calvo 1991
122	*Platysauropus ingens*	Lesotho	LJ	Ellenberger 1974
123	*Platysauropus robustus*	Lesotho	LJ	Huene 1941; Ellenberger 1970
124	*Platytrisauropus lacunensis*	Lesotho	LJ	Ellenberger 1970
125	*Plesiornis pilulatus*	USA, Poland	LJ	Hitchcock 1858; Gierlinski & Pienkowski 1999
126	*Prototrisauropodiscus minimus*	Lesotho	LJ	Ellenberger 1972
127	*Prototrisauropus angustidigitus*	Lesotho	LJ	Ellenberger 1972
128	*Prototrisauropus crassidigitus*	Lesotho	LJ	Ellenberger 1972
129	*Prototrisauropus graciosus*	Lesotho	LJ	Ellenberger 1972
130	*Regarosauropus manovi*	Tajikistan	UJ	Dzhalilov & Novikov 1995
131	*Salfitichnus mentoor*	Argentina	lUC	Alonso & Marquillas 1986
132	*Saltopoides igalensis*	France	LJ	Lapparent & Montenat 1967
133	*Samandrinda surghani*	Pakistan	MJ	Malkani 2007
134	*Sarmientichnus scagliai*	Argentina	MJ	Casamiquela 1964; Coria & Paulina Carabajal 2004
135	*Satapliasaurus dsocenidzei*	Georgia	eLC	Gabunia 1951; Dzhalilov & Novikov 1993
136	*Satapliasaurus kandelakii*	Georgia	eLC	Gabunia 1951; Dzhalilov & Novikov 1993
137	*Satapliasaurus tschabukianii*	Georgia	eLC	Gabunia 1951; Dzhalilov & Novikov 1993

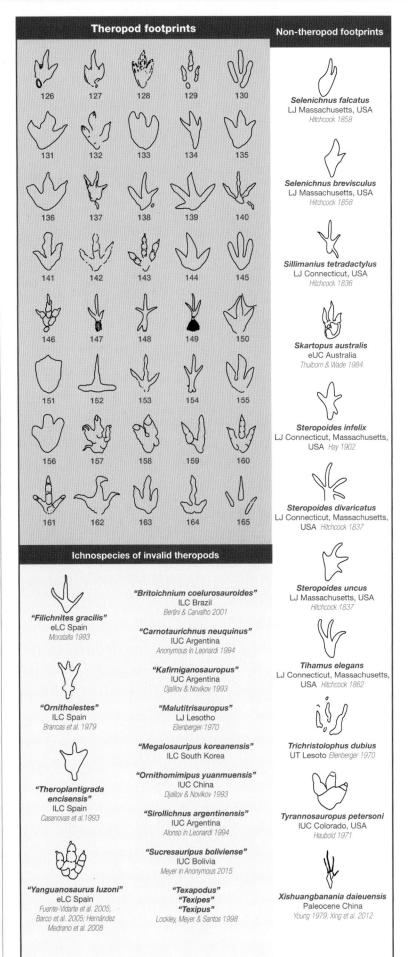

Theropod footprints

126 127 128 129 130
131 132 133 134 135
136 137 138 139 140
141 142 143 144 145
146 147 148 149 150
151 152 153 154 155
156 157 158 159 160
161 162 163 164 165

Ichnospecies of invalid theropods

"*Filichnites gracilis*" eLC Spain *Moratalla 1993*

"*Ornitholestes*" ILC Spain *Brancas et al. 1979*

"*Theroplantigrada encisensis*" ILC Spain *Casanovas et al. 1993*

"*Yanguanosaurus luzoni*" eLC Spain *Fuente-Vidarte et al. 2005; Barco et al. 2005; Hernández Medrano et al. 2008*

"*Britoichnium coelurosauroides*" ILC Brazil *Bertini & Carvalho 2001*

"*Carnotaurichnus neuquinus*" IUC Argentina *Anonymous in Leonardi 1994*

"*Kafirniganosauropus*" IUC Argentina *Djallov & Novikov 1993*

"*Malutitrisauropus*" LJ Lesotho *Ellenberger 1970*

"*Megalosauripus koreanensis*" ILC South Korea

"*Ornithomimipus yuanmuensis*" IUC China *Djallov & Novikov 1993*

"*Sirollichnus argentinensis*" IUC Argentina *Alonso in Leonardi 1994*

"*Sucresauripus boliviense*" IUC Bolivia *Meyer in Anonymous 2015*

"*Texapodus*" "*Texipes*" "*Texipus*" *Lockley, Meyer & Santos 1998*

Non-theropod footprints

Selenichnus falcatus LJ Massachusetts, USA *Hitchcock 1858*

Selenichnus brevisculus LJ Massachusetts, USA *Hitchcock 1858*

Sillimanius tetradactylus LJ Connecticut, USA *Hitchcock 1836*

Skartopus australis eUC Australia *Thulborn & Wade 1984*

Steropoides infelix LJ Connecticut, Massachusetts, USA *Hay 1902*

Steropoides divaricatus LJ Connecticut, Massachusetts, USA *Hitchcock 1837*

Steropoides uncus LJ Massachusetts, USA *Hitchcock 1837*

Tihamus elegans LJ Connecticut, Massachusetts, USA *Hitchcock 1862*

Trichristolophus dubius UT Lesoto *Ellenberger 1970*

Tyrannosauropus petersoni IUC Colorado, USA *Haubold 1971*

Xishuangbanania daieuensis Paleocene China *Young 1979; Xing et al. 2012*

138	*Saurexallopus cordata*	Canada	eUC	McCrea et al. 2014
139	*Saurexallopus lovei*	Wyoming, USA	lUC	Harris et al. 1996, 1997
140	*Saurexallopus zerbstorum*	Wyoming, USA	lUC	Lockley et al. 2003
141	*Saurichnium damarense*	Namibia	UT	Gurich 1926
142	*Schizograllator otariensis*	Japan	LJ	Matsukawa et al. 2005
143	*Schizograllator xiaohebaensis*	China	LJ	Zhen et al. 1985
144	*Shensipus tungchuanensis*	China	MJ	Young 1966
145	*Shirkentosauropus shirkentensis*	Tajikistan	UJ	Djalilov & Novikov 1993
146	*Stenonyx lateralis*	Connecticut, USA	LJ	Hitchcock 1858; Lull 1904
147	*Steropezum elegantus*	Massachusetts, USA	LJ	Hitchcock 1836, 1841
148	*Steropoides diversus clarus*	Massachusetts, USA	LJ	Hitchcock 1836, 1845
149	*Steropoides ingens*	Massachusetts and New Jersey, USA	LJ	Hitchcock 1836, 1889
150	*Talmontopus tersi*	France	LJ	Lapparent & Montenat 1967
151	*Taponichnus donottoi*	Argentina	lUC	Alonso & Marquillas 1986
152	*Taupezia landeri*	England	UJ	Delair 1963
153	*Toyamasauripus masuiae*	Japan	ILC	Matsukawa et al. 1997
154	*Triaenopus lulli*	Connecticut, USA	LJ	Thorpe 1929
155	*Tuojiangpus shuinanensis*	China	MJ	Young & Young 1987
156	*Turkmenosaurus kugitangensis*	Turkmenistan	MJ	Amanniazov 1985
157	*Tyrannosauripus pillmorei*	New Mexico, USA	lUC	Lockley & Hunt 1994
158	*Velociraptorichnus sichuanensis*	China	ILC	Zhen et al. 1994
159	*Velociraptorichnus zhangi*	China	ILC	Xing et al. 2015
160	*Weiyuanpus zigongensis*	China	LJ	Gao 2007
161	*Wildeichnus navesi*	Argentina	MJ	Casamiquela 1964
162	*Xiangxipus chenxiensis*	China	ILC	Zeng 1982; Gierlinski & Lockley 2013
163	*Youngichnus xiyangensis*	China	LJ	Zhen et al. 1985
164	*Zhengichnus jinningensis*	China	LJ	Zhen et al. 1985
165	*Zizhongpus wumaensis*	China	MJ	Yang & Yang 1987

Trinominal theropod and primitive saurischian names

Anchisauripus exsertus branfordi
Connecticut, USA *Thorpe 1929*

Prototrisauropus rectilineus var gravis
UT Lesoto *Ellenberger 1972*

Prototrisauropus rectilineus var lentus
UT Lesoto *Ellenberger 1972*

Kainotrisauropus moshoeshoei var decoratus
LJ Lesoto *Ellenberger 1970*

Kainotrisauropus moshoeshoei var ingens
LJ Lesoto *Ellenberger 1970*

Kainotrisauropus moshoeshoei var profundus
LJ Lesoto *Ellenberger 1970*

Masitisisauropus angustus var cursor
LJ Lesoto *Ellenberger 1974*

Masitisisauropus exiguus var ruber
LJ Lesoto *Ellenberger 1970*

Ornithoidichnites tuberosos dubius
LJ Connecticut, USA *Hitchcock 1836*

Footprints identified as theropods that may be ornithopods

Asianopodus pulvinicalx
eLC Japan *Matsukawa et al. 2005*

Eutynichnium lusitanicum
UJ Portugal *Nopcsa 1923*

Paracorpulentapus zhangsanfengi
lUC China *Xing et al. 2014*

Eutynichnium lusitanicum
ILC China *Li et al. 2011*

Gypsichnites pacensis
ILC Canada *Sternberg 1932*

Paraornithopus fabrei
LJ France *Demathieu 2002; Demathieu et al. 2003*

Boutakioutichnium atlasicus
UJ Morocco *Nouri et al. 2011*

Jialingpus yuechienensis
UJ China *Zhen et al. 1983*

Siamopodus khaoyasensis
eLC Thailand *Lockley et al. 2006*

Changpeipus xuiana
MJ China *Lu et al. 2007*

Komlosaurus carbonis
LJ, Hungary *Kordos 1983*

Siamopodus xui
ILC China *Xing et al. 2014*

Corpulentapus lilasia
ILC China *Li et al. 2011*

Minisauripus chuanzhuensis
ILC China *Zhen et al. 1994*

Therangospodus oncalensis
eLC Spain *Lockley et al. 2000*

Eutynichnium gomesi
UJ Portugal *Antunes 1976*

Minisauripus zhenshuonani
ILC South Korea *Lockley et al. 2008*

Therangospodus pandemicus
eLC Argentina, Turkmenistan, Utah, USA *Lockley 1989; Pazos et al. 2012*

Synonyms

Bellona = Eubrontes
Reichenbach 1852

Berecynthia = Argoides
Reichenbach 1852

Brontozoum = Eubrontes
Hitchcock 1847

Casamiquelichnus navesorum = Sarmientichnus
Coria & Paulina Carabajal 2004

Changpeipus luanpingeris = C. carbonicus
Young 1979

Changpeipus xuiana = C. carbonicus
Lu et al. 2007

Cybele = Anchisauripus
Reichenbach 1852

"Deinonychosaurichnus" = Velociraptorichnus
Zhen et al. 1987

"Exallopus" = Saurexallopus
Harris, Johnson, Hicks & Tauxe 1996

Gigantitherium = Gigandipus
Hitchcock 1858

Ichnites euskelosauroides = Platysauropus
Huene 1941

Leptonyx = Stenonyx
Hitchcock 1865

Macropodosauropus = Macropodosaurus
Dzhaliov & Novikov 1993

Pelargides = Anchisaurus
Reichenbach 1852

Plastisauropus = Platysauropus
Ellenberger 1970

Satapliasauropus = Satapliasaurus
Dzhaliov & Novikov 1993

"Shulerpodus" = Eubrontes titanopelobatidus
Farlow, Pittman & Hawthorne 1989

Names from academic theses

"Eubrontes jossephbarratti"
Rainforth 2005

"Eubrontes neoparallelus"
Rainforth 2005

"Grallator curvensis"
Peng 2004

"Graliator laohukouensis"
Peng 2004

"Grallator leei"
Peng 2004

"Plesiornichnus"
Rainforth 2005

"Riojapodus amei"
Diaz-Martinez 2013

"Velociraptorichnus huanghensis"
Peng 2004

Indeterminate dinosaur ichnospecies

"Deogmyeongosauripus"
ILC South Korea
Anonymous in Zhen et al. 1993

Mesozoic bird footprints

N°	Ichnospecies	Country	Era	Bibliography
1	*Aquatilavipes izumiensis*	Japan	eLC	Azuma & Takeyama 1991
2	*Aquatilavipes swiboldae*	Canada	lLC	Currie 1981
3	*Archaeornithipus meijidei*	Spain	eLC	Fuentes Vidarte 1996
4	*Barrosopus slobodai*	Argentina	lUC	Coria et al. 2002
5	*Dongyangornipes sinensis*	China	lUC	Azuma et al. 2012
6	*Goseongornipes markjonesi*	South Korea	lLC	Lockley et al. 2006
7	*Gruipeda vegrandiunus*	Alaska, USA	lUC	Fiorillo et al. 2011
8	*Gyeongsangornipes lockleyi*	South Korea	lLC	Kim et al. 2013
9	*Hwangsanipes choughi*	South Korea	lUC	Yang et al. 1995
10	*Ignotornis gajinensis*	South Korea	lLC	Kim et al. 2012
11	*Ignotornis mcconnelli*	Colorado, USA	lLC	Meehl 1931
12	*Ignotornis yangi*	South Korea	lLC	Kim et al. 2006
13	*Jindongornipes kimi*	South Korea	lLC	Lockley et al. 1992
14	*Koreanaornis anhuiensis*	China	lLC	Jin & Yan 1994
15	*Koreanaornis dodsoni*	China	lLC	Xing et al. 2011
16	*Koreanaornis hamanensis*	South Korea	lLC	Kim 1969
17	*Koreanaornis sinensis*	China	lLC	Zhen et al. 1994
18	*Limiavipes curriei*	Canada	lLC	McCrea & Sarjeant 2001; McCrea et al. 2014
19	*Muguiornipes robusta*	China	lLC	Xing et al. 2011
20	*Patagonichnornis venetiorum*	Argentina	lUC	Casamiquela 1987
21	*Paxavipes babcockensis*	Canada	lLC	McCrea et al. 2014
22	*Pullornipes aureus*	China	UJ	Lockley et al. 2006
23	*Sarjeantopodus semipalmatus*	Wyoming, USA	lUC	Lockley, Nadon & Currie 2003
24	*Shandongornipes muxiai*	China	lLC	Li et al. 2005
25	*Tatarornipes chabuensis*	China	lLC	Lockley et al. 2012
26	*Uhangrichnus chuni*	South Korea	lUC	Yang et al. 1995
27	*Yacoraitichnus avis*	Argentina	lUC	Alonso & Marquillas 1986
28	*Wupus agilis*	China	lLC	Xing et al. 2007, 2015

The longest web-footed Mesozoic bird footprint

Sarjeantopodus semipalmatus (Wyoming, USA) is 9.5 cm long, including finger I (halux). See more: *Lockley et al. 2003*

The largest web-footed Mesozoic bird footprint

Uhangrichnus chuni (Alaska, USA) is about 7.4 cm long. See more: *Fiorillo et al. 2011*

The smallest web-footed Mesozoic bird footprint

The footprints of *Gyeongsangornipes lockleyi* (South Korea) measure 2.8–3.4 cm long. See more: *Kim et al. 2013*

The oldest web-footed Mesozoic bird footprint

Ignotornis yangi (China) dates from the Aptian (ca. 125–113 Ma). See more: *Kim et al. 2006*

1:4

Mesozoic bird footprints

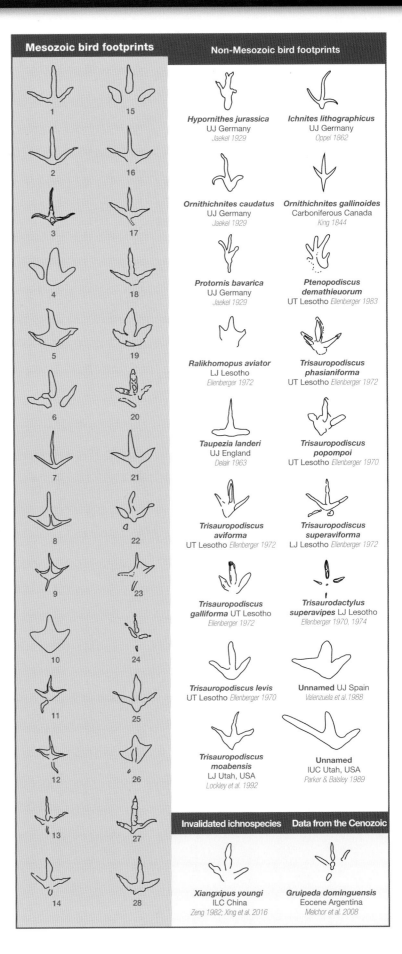

Non-Mesozoic bird footprints

Hypornithes jurassica
UJ Germany
Jaekel 1929

Ichnites lithographicus
UJ Germany
Oppel 1862

Ornithichnites caudatus
UJ Germany
Jaekel 1929

Ornithichnites gallinoides
Carboniferous Canada
King 1844

Protornis bavarica
UJ Germany
Jaekel 1929

Ptenopodiscus demathieuorum
UT Lesotho *Ellenberger 1983*

Ralikhomopus aviator
LJ Lesotho
Ellenberger 1972

Trisauropodiscus phasianiforma
UT Lesotho *Ellenberger 1972*

Taupezia landeri
UJ England
Delair 1963

Trisauropodiscus popompoi
UT Lesotho *Ellenberger 1970*

Trisauropodiscus aviforma
UT Lesotho *Ellenberger 1972*

Trisauropodiscus superaviforma
LJ Lesotho *Ellenberger 1972*

Trisauropodiscus galliforma UT Lesotho
Ellenberger 1972

Trisaurodactylus superavipes LJ Lesotho
Ellenberger 1970, 1974

Trisauropodiscus levis
UT Lesotho *Ellenberger 1970*

Unnamed UJ Spain
Valenzuela et al. 1988

Trisauropodiscus moabensis
LJ Utah, USA
Lockley et al. 1992

Unnamed
IUC Utah, USA
Parker & Balsley 1989

Invalidated ichnospecies

Xiangxipus youngi
ILC China
Zeng 1982; Xing et al. 2016

Data from the Cenozoic

Gruipeda dominguensis
Eocene Argentina
Melchor et al. 2008

Chronicle and dinomania
The history and culture of dinosaurs

Records: Historical and cultural.

Experts have made compilations of historical facts related to dinosaur paleontology, pointing out outstanding fossils, revolutionary discoveries, scientific advances, new lines of research, or events that have influenced the use and study of the discipline. Several authors have suggested an organization based on different chronological divisions to represent the progress and trends that have occurred over the course of history. It is not our intent to duplicate the content of these compilations, but to complement them and apply them to the only topic that interests us ... records!

CHRONOLOGY OF THEROPOD RECORDS

The Heroic, Classic, Modern, and Renaissance periods were created by Michael K. Brett-Surman in 1997. Here we add Archaic, created by José Luis Sanz (1999), and propose one more, Mythological. In addition, to organize the exciting data better, each period is divided into two subgroups. See more: *Brett-Surman 1997; Sanz 1999; Pereda-Suberbiola et al. 2010; Spalding & Sarjeant 2012*

Mythological period	Early Mythological (ca. 130,000–3,300 BC)
	Late Mythological (ca. 3,300 BC–ca. AD 1670)
Archaic period	Early Archaic (1671–1762)
	Late Archaic (1763–1823)
Heroic period	Early Heroic (1824–57)*
	Late Heroic (1858–96)*
Classical period**	Early Classical (1897–1928)
	Late Classical (1929–38)
Modern period**	Early Modern (1938–46)
	Late Modern (1947–68)
Renaissance period	Early Renaissance (1969–95)
	Late Renaissance (1996–present)

* Ancient period according to *Sanz 2007*
** First Modern period according to *Sanz 2007*
Second Modern period according to *Sanz 2007*

Early Mythological subperiod (ca. 130,000–3,300 BC), approximately 126,700 years

THE FIRST DINOSAUR REMAINS DISCOVERED IN PREHISTORY!

In antiquity, many fantastic explanations were offered for dinosaur fossil remains and footprints. Considering the ignorance of their true nature or age, and that knowledge then was transmitted mainly by the spoken word and in brief writings, the explanations were quite practical.

ca. 10,000 BC—Mongolia—Object with oldest dinosaur oolites
Late Paleolithic or Early Neolithic people pierced dinosaur eggshells for use as necklace ornaments. See more: *Andrews 1943; Pauc & Buffetaut 1994*

ca. 6200–5400 BC—China—The representation of an older dragon
In 1994, a huge red-brown dragon almost 20 m long was found at Xinglongwa Culture sites. The first forms showed a pig-dragon with the body of a snake. It has not been proven that the myth of the dragon comes from discoveries of dinosaur remains, as is popularly believed, but this association likely exists. See more: *Yang 2007; Dong in Dong & Milner 1988; Mayor 2005*

Late Mythological subperiod (ca. 3,300 BC–AD 1670), ca. 4,970 years

DRAGONS DID EXIST!

Writing begins, and some early documents mention the existence of fossils, but legends and beliefs in fantastic beings remain ways to explain fossilized remains.

ca. 300 BC—China—The first fossil bones assigned to dragons
Chinese historian Cheng Qu reported on "Hua Yang Zhi Goo" dragon bones, but there are doubts about whether they belonged to dinosaurs or large mammals. See more: *Dong in Dong & Milner 1988; Needham 1959*

ca. 1133—Belgium—The first fossil skull attributed to a dragon in Europe
In a church in Mons, Belgium, an extinct crocodile skull was exhibited as the remains of the dragon "Le Lumeçon" defeated by Gilles de Chin near the Haine River. See more: *Liétard-Rouzé 2010*

1614—England—The first European dragon based on the remains of dinosaurs
It is suspected that a "strange and monstrous three-meter dragon" may have been derived from dinosaur bones, due to where they were found. See more: *Buffetaut 1991; Spalding & Sarjeant 2012*

1669—Spain—The first dinosaur bones?
In 1981, it was reported that the "bones of giants" had been found in 1669 and 1671. Unfortunately, they were not described or conserved, so their true nature can't be confirmed. See more: *Mares 1681; Pereda-Suberbiola et al. 2010*

Early Archaic subperiod (1671–1762), 91 years

THE FIRST DINOSAUR COLLECTORS!

Fossils begin to be considered giant animals or petrified humans. The first remains are now recognizable as dinosaurs. The binominal classification system is created.

1699—England—The first listed theropod fossil
A tooth illustrated by Edward Lhuyd is assigned the number 1328, referring to the collection to which it belonged. See more: *Lhuyd 1699*

THE EXISTENCE OF GIANT REPTILES IS PROVEN!

The study of dinosaurs is formalized, although they are still considered huge reptiles, similar to modern lizards.

1821—England—The first mention of *Megalosaurus*
In the first publication on this theropod, it is referred to as a great lizard. See more: *Conybeare 1821;*
Spalding & Sarjeant 2012

DINOSAUR PALEONTOLOGY BEGINS!

Dinosaurs are thought to belong to a group different from modern reptiles. The first formal names of dinosaurs are generated. The first dinosaur tracks are interpreted as giant birds.

1824—England—The first scientific description of a theropod
Even though *Megalosaurus* did not receive a species name, the official nomenclature of dinosaurs began with its publication by William Buckland. See more: *Buckland 1824*

1839—Switzerland—The first fossil bird assigned to the Mesozoic
Protornis glarniensis was a coraciform bird that was named in an article five years after it was discovered. It was considered to be from the Upper Cretaceous but is now known to come from the Oligocene. See more: *Meyer 1839, 1844; Peyer 1957*

1855—France—The oldest theropod discovered in marine deposits?
"*Megalosaurus*" *terquemi* is based on only one tooth and one Lower Jurassic claw. Its identification is doubtful. It could have been a theropod, a sauropodomorph, or another type of archosaur. See more: *Terquem 1855; Gervais 1859; Lapparent 1967; Huene 1926; Huene & Maubeuge 1954; Buffetaut, Cuny & Le Loeuff 1991*

DINOSAURS STAND ON TWO LEGS, AND DINOMANIA RISES!

The bipedal nature of some dinosaurs is recognized. A conflict known as the Bone Wars sparks an increase in the collection and revision of fossil specimens. The first expeditions in search of dinosaurs are launched. Dinosaur fever re-emerges at the end of the period. Discoveries are made on different continents.

1859—England—The first publication on the origin of birds
The first scientific studies on the origin of birds emerged almost immediately after publication of the theory of natural selection and the discovery of the first skeleton of *Archaeopteryx lithographica*. See more: *Darwin 1859; Owen 1863; Gegenbaur 1863; Cope 1867; Huxley 1868*

1861—Germany—The first theropod considered a "missing link"
The first skeleton recognized as *Archeopteryx lithographica* was published shortly after Charles Darwin's book. Historically, it was a lucky coincidence. Since the skeleton presented both avian and "reptilian" characters, as Darwin predicted, its impact was great. See more: *Darwin 1859; Buffetaut 1991*

1862—Germany—The first auctioned Mesozoic bird fossil
Karl Häberlein demanded £700 for the fossil of *Archeopteryx lithographica*, an exorbitant sum at the time. The English bought it and delivered it to Richard Owen for study. A second, more complete specimen was sold in 1877 for £1,000, although the asking price was £1,800. See more: *Desmond 1990*

1863—Germany—The first theory about the origin of birds
A relationship between birds and dinosaurs was suggested for the first time by the embryologist and anatomist Karl Gegenbaur. Five years later, Thomas Huxley reached the same conclusion. See more: *Gegenbaur 1863*

1863—The scientific debates about dinosaurs that have lasted the longest
The theory that the digits of bird wings correspond to fingers I, II, and III was proposed in 1825. Thirty-eight years later, it was suggested that they actually correspond to fingers II, III, and IV. The discussion continued for 123 years (1863–1986), until it was settled by numerous embryological, anatomical, and phylogenetic tests. A debate about the origin of birds lasted 118 years (1868–1986), and was unnecessarily extended to 146 years. A group of dissident ornithologists remain doubtful that birds rose from dinosaurs, although much evidence is against them. See more: *Merkel 1825; Owen 1863; Huxley 1868; Gauthier 1986; Vargas & Fallon 2005; Young 2009; Díaz-Aros 2010; Feduccia & Czerkas 2014; Smith et al. 2015*

1863—Germany—The first fossil bird that was considered a "missing link"
Archaeopteryx lithographica was described by Richard Owen as an intermediate between birds and reptiles. In addition, Charles Darwin pointed to it as irrefutable proof of the existence of evolution. See more: *Owen 1863; Darwin 1866; Huxley 1868*

1868—Germany—The first dinosaur to be considered a "missing link"
Detailed anatomical comparisons were made between several *Archeopteryx* fossils and modern animals. The result was that *Archeopteryx* was more similar to *Compsognathus longipes*, although the latter was not considered a dinosaur until 1896. See more: *Huxley 1868; Marsh 1896*

1870—USA—The first expedition in search of fossils
The first expedition carried out exclusively for paleontological collection was directed by Othniel Charles Marsh. He was accompanied by students from Yale University. See more: *Moore 2014*

1871—Canada—The first mention of a mythological being based on dinosaur fossils
Jean-Baptiste mentioned a place where local tribes regarded fossils as the remains of the "grandparents of the American bison." See more: *Currie & Koppelhus 2005*

1877—USA—The first large-scale dinosaur conflict
The Bone Wars, a feud between Cope and Marsh, began in 1877. Their dispute had disdvantages and benefits for the study of dinosaurs. It turned Cope and Marsh into such bitter enemies that they did not hesitate to use dishonest methods—bribes, robberies, the destruction of bones, insults, slander, scandals, and refusals to give scientific credit to colleagues. The benefits were the description of 89 new species of dinosaurs judged between valid and doubtful (153 in total), the reconstruction of complete skeletons, and an increase in the popularity of dinosaurs outside the scientific community. The Bone Wars ended in 1892, although the personal dispute lasted until Cope died in 1897. See more: *Wallace 1999; Preston 1993*

PALEOBIOLOGY BEGINS!

Paleobiological studies are applied to dinosaurs. The second dinosaur fever arises with Barnum Brown, and media interest increases. The first skeletons of real bones are built.

1898—USA—The first cabin made of dinosaur bones
There were so many dinosaur bones and other extinct species in Colorado that a shepherd built the famous cabin near Bone Cabin Quarry. See more: *Levy 2010*

1902—USA—The first censored fossilized Mesozoic bird
A group of people urged Professor Othniel Charles Marsh to hide the specimen of *Ichthyornis*, a bird with teeth, because its existence provided evidence that greatly supported the theory of evolution proposed by Charles Darwin in his book *On the Origin of Species by Means of Natural Selection, or the Preservation of Favoured Races in the Struggle for Life.* See more: *Darwin 1859; Clarke 2004*

1905—Canada—The first theropods found in a group
Ten specimens of *Albertosaurus sarcophagus* were found in Alberta. See more: *Osborn 1915*

1905—USA—The first mounted theropod skeleton
Mounted and displayed publicly, the specimen AMNH 5753 is an *Allosaurus fragilis* that appears inclined over the tail of a *Brontosaurus*. The specimen was never formally described. See more: *Breithaupt et al. 1999*

1913—USA—The first non-avian dinosaur report from the Tertiary
At the moment, all non-avian dinosaur fossils are exclusively from the Mesozoic, except for some pieces that were shifted to more recent sediments due to natural causes. See more: *Lee 1913; Farlow et al. 1987*

1915—USA—The first theory about theropods with "four wings"
A proposal on the origin of flight in birds suggested the presence of rectrices in the four members of a hypothetical dinosaur called "Tetrapteryx." This was corroborated for the first time in a specimen of *Microraptor* in 2003. See more: *Beebe 1915*

1915—USA—The first protected dinosaur site
President Woodrow Wilson inaugurated the Dinosaur National Monument, located on the border of Colorado and Utah. It now covers an area of 853.26 km^2 and includes a wall containing more than 1,500 bones.

1917—USA—The described theropod that was destroyed the fastest
The holotype of *Podokesaurus holyokensis* was consumed by fire only six years after it had been discovered and five years after it had been donated to Mount Holyoke College. Fortunately, plaster molds were stored in the Museum of Natural History in New York. See more: *Talbot 1911; Colbert 1964*

THE ORIGIN OF BIRDS FROM DINOSAURS IS REAFFIRMED!

Anatomical evidence is discovered linking dinosaurs and birds. The first pathologies are interpreted in dinosaur footprints.

1933—USA—The first theropod to present a clavicle
Since a theropod fossil with the clavicle had not been found, birds were thought to have descended from crocodilelike animals or other archosaurs. Dinosaurs were thought not to have clavicles. *Segisaurus halli* was the first theropod to have one. A clavicle was later found in dinosaurs. Birds' furcula, or wishbone, arose from the fusion of the clavicles in dinosaurs. See more: *Camp 1936*

THE WORST TIME FOR STUDYING DINOSAURS!

World War II causes scientific study in many areas to stagnate, and dinosaur paleontology was no exception.

1944—Germany—Fossil remains of theropods destroyed in World War II
Many fossils that were mounted or on display were pulverized under bombardment, including holotype specimens of *Bahariasaurus*, *Carcharodontosaurus*, and *Spinosaurus* in the Bavarian State Museum in Munich. See more: *Weishampel 1990*

1948—The year the most dinosauromoph ichnospecies were named
Three footprint species were named in the years 1948, 1950, 1976, and 1982.

DINOSAURS ARE FASHIONABLE AGAIN!

Scientific interest in the study of dinosaurs is reborn. The number of scientific articles on dinosaur paleontology increase. New expeditions to collect fossils are launched.

1947—USA—The first discovery of hundreds of theropods together
Hundreds of *Coelophysis* were found at Ghost Ranch, New Mexico. You can even see two different morphotypes, male and female. Coelophysis is the Triassic dinosaur with the most preserved specimens. See more: *Colbert 1989; Rinehart 2009*

1966—USA—The first town called Dinosaur
Artesia is a Colorado town that changed its name to take advantage of its proximity to Dinosaur National Monument. It even has Brontosaurus Boulevard, Stegosaurus Highway, Triceratops Terrace, and Tyrannosaurus Track. See more: *Moore 2014*

1968—USA—The first suggestion of ectothermia in non-avian dinosaurs
Bob Bakker presents tests to prove that dinosaurs were "warm-blooded" animals, causing a controversy that still continues. See more: *Bakker 1968; Grady et al. 2014; Démic 2015*

DINOSAUR BLOOD GETS WARM!

New evidence supports the phylogenetic relationship between birds and dinosaurs. *Deinonychus* becomes one of the most important discoveries of dinosaur paleontology. It is evident that dinosaurs had metabolisms similar to those of birds and mammals. Cladistics are applied for the first time to dinosaur phylogeny. Most new species rely on materials better than fragments.

1969—USA—The first study on the encephalization quotient in dinosaurs
Dinosaur intelligence is estimated allometrically from the relationship of the mass of the real brain to the mass of the brain predicted for an animal of a given size. See more: *Jerison 1969; Buchholtz 2012*

1970—The year the most theropod species were named
In 1970, twenty-five footprints were named. Some are no longer considered valid. On the other hand, eight genera were created in 1985 and 1987.

1971—Mongolia—The first theropod preserved in battle
On the Polish-Mongol expedition into the Gobi Desert, a *Velociraptor mongoliensis* was found locked on to a *Protoceratops andrewsi*. They may have been buried by a dune or a sandstorm. See more: *Kielan-Jaworowska & Barsbold 1972; Carpenter 1998*

1972—The year the most ichnospecies of primitive saurischians were named
Five species and three footprint genera were named.

1973—Lebanon—The first theropod remains in amber
A feather conserved in amber dates from the early Lower Cretaceous. See more: *Schlee 1973*

1973—Australia—The first theropod preserved in opal
Kakuaru kujani ("serpent of rainbow light") is an opalized tibia that was held in a private collection for thirty years. See more: *Huene 1932; Molnar & Pledge 1980*

1976—England—The first biomechanical study of dinosaur locomotion
A published article proposed formulas to estimate the sprinting speed of dinosaurs. See more: *Alexander 1976*

1985—China—The theropod fossil discovered at the greatest altitude
"Megalosaurus tibetensis" and "Ngexisaurus dapukaensis" were located at an altitude of 4,900 m, but they were not formally published. *Cryolophosaurus ellioti* and an indeterminate coelophysoid were found in Antarctica at an altitude of 4,000 m. See more: *Zhao 1985; Hammer & Hickerson 1994; Xing, Harris & Currie 2011*

1986—The first cladistic phylogeny applied to dinosaurs
Cladistics were applied to the phylogeny of dinosaurs by Jacques Gauthier, showing that the birds were a group derived from theropod dinosaurs. See more: *Gauthier 1986*

1986—Canada—The first symposium on dinosaur systematics
The Dinosaur Systematics Symposium was held June 3–5, 1986. The topics were sexual dimorphism, ontogeny, variation, and other systematic and anatomical aspects. See more: *Carpenter & Currie 1990*

1987—USA—The first mistaken report of a Mesozoic bird of the Triassic
Protoavis texensis came to be considered a primitive bird. After further analysis, it was shown to be a mixture of bones from different animals, including juvenile theropods. See more: *Chatterjee 1987, 1991*

1990—USA—The most abundant theropod of the Triassic
More than one hundred specimens of *Coelophysis bauri* are known, from juveniles to adults. See more: *Colbert 1990; Rinehart et al. 2009*

1992—USA—The first theropod to be scanned
A skull of *Tyrannosaurus rex* was scanned by computed tomography, making *Tyrannosaurus* the first dinosaur to be imaged. See more: *Bakker 1992*

1992—USA—The longest legal conflict over a theropod fossil
The *Tyrannosaurus rex* specimen "Sue" (FMNH PR 2081) was the subject of a legal battle that began in 1992 and ended in 1995 in favor of Maurice Williams, who auctioned the fossil in 1997. The nickname was in honor of the discoverer, Susan Hendrickson. See more: *Fiffer 2001*

1993—Germany—The first heraldry with a theropod
The coat of arms of Bedheim contains a red *Liliensternus* under five organ pipes that symbolize the local church.

1993—USA—The first attempt to identify sex in a theropod
It was proposed that chevron size would help identify the males of *Tyrannosaurus rex*. Although this evidence was questioned, the presence of spinal-cord tissue in the MOR 1125 corroborated that females were more robust. See more: *Fischman 1993; Carpenter 1992; Larson 1994; Erickson et al. 2005; Schweitzer, Wittmeyer & Horner 2005*

1993—USA—The first biomolecular sample of a theropod
Proteins were revealed in a *Tyrannosaurus rex*. Part of the sample was degraded when extracted. See more: *Schweitzer 1993; Schweitzer, Starkey & Horner 1994; Tanke & Rothschild 2002*

1993—USA—The first evidence of collagen in a theropod
Evidence of collagen fibers was found in a well-preserved specimen of *Tyrannosaurus rex*. See more: *Schweitzer 1993*

1994—USA—The first studies of isotopes in a theropod
Oxygen isotopes were studied in the fossil remains of a *Tyrannosaurus rex* to determine its body temperature in life. See more: *Anonymous 1992; Barrick & Showers 1994*

THE FEATHER REVOLUTION BEGINS!

Non-avian dinosaurs possessing filaments and feathers are discovered. Scientific publications consider electronic formats as valid material. New technology is applied to the study of dinosaurs.

1996—China—The first feathered theropod
The body of *Sinosauropteryx prima* was completely covered with short filaments. Fossil materials had not proven this feature, but it had been suspected. See more: *Bakker 1986; Paul 1988; Ji & Ji 1996*

1996—Canada—The highest encephalization quotient in a dinosaur is estimated
According to a study, *Dromiceiomimus brevitertius* had an encephalization ratio between 7.1 and 8.6, surpassing Troodon (*Stenonychosaurus inequalis*), which had a maximum level of 7.1. An even higher estimate, 12.68, was made for *Bambiraptor feinbergi*, but it is a juvenile. See more: *Hurlburt 1996; Hurlburt et al. 1996*

1996—The year the most Mesozoic bird species were named
Six species and five genera of fossil eggs were named.

1997—USA—The first fraud of a Mesozoic bird fossil
A Chinese farmer found an incomplete fossil of *Yanornis martini* in 1997 and sold it to dealers. To increase its value, the skeleton was mixed with the remains of a *Microraptor zhaoianus*. Called "Archaeoraptor liaoningensis," the specimen was published in *National Geographic* magazine, causing a scandal over the improper handling of fossils, early publication in informal media, and illegal trafficking. See more: *Sloan 1999; Simons 2000*

1998—USA—The most valuable Mesozoic bird fossil
The fossil of *Archeopteryx lithographica* BMNH 37001 (London specimen) was valued at ten million marks. The first sale price was £700 (about $1,000), paid to Karl Häberlein by the Museum of Natural History in London, where it remains to this day. See more: *Naish 2011; Rolfe, Milner & Hay 1988*

1998—USA—The first space theropod
A fossil skull of *Coelophysis* was sent to the space shuttle *Endeavour* to be transferred to the Russian space station *Mir*, traveling more than six million kilometers. See more: *Chure 2009*

2000—China—The highest concentration of Mesozoic birds
An unusual number of specimens of *Confuciusornis sanctus* was found: 40 individuals per 100 m². Due to volcanic ash deposits, it is believed that many of the birds died simultaneously. In 2010, the Shandong Tianyu Natural Museum owned 536 specimens. It is known that many copies never reached the museum and ended up being sold illegally or informally. If all the fossils were reunited, the number would surely exceed 1,000.. See more: *Wang et al. 2000; Zheng et al. 2010*

2001—England—The deepest theropod fossil discovery
It is probable that remains of the first *Megalosaurus bucklandii* were recovered at a depth of 12 m. See more: *Cadbury 2001*

2003—Poland—The most abundant dinosauromorph of the Triassic
There are more than twenty specimens of *Silesaurus opolensis*. See more: *Dzik 2003; Piechowski, Talandia & Dzik 2014*

2004—Australia—The highest payment for an opalized theropod fossil
The South Australian Museum paid $22,000 for the opalized tibia of *Kakuru kujani* SAM P17926. See more: *Naish 2011; Rolfe, Milner & Hay 1988*

2004—China—The largest museum in the world dedicated to dinosaurs
The Shandong Tianyu Natural History Museum, which is located in Linyi, Pingyi County, occupies an area of 28,000 m² and houses 1,106 dinosaurs. It has its own research center, 4-D cinema, and twenty-eight exhibition halls. See more: *Guiness Record 2010*

2005—USA—The first organic material in a theropod
Structures reminiscent of cells, blood vessels, and fibers were found inside the femur of a *Tyrannosaurus rex*. See more: *Schweitzer et al. 2005*

2005—Utah, USA—The most abundant theropods of the Cretaceous
Hundreds of *Falcarius utahensis* may be in an 8 km² site in the town of Cedar Mountain. See more: *Kirkland et al. 2005*

2008—Argentina—First formal online work on an invalidated theropod
Aerosteon was described in an electronic publication in 2008. At that time, however, in order for a formal document to be considered valid, several paper copies had to be printed according to the rules of the International Code of Zoological Nomenclature. This made the name invalid until a year later. The rules have since been changed to favor formal work online.

2009—The year the most theropod species were named
Twenty-one species and eighteen genera were named in 2009. Twenty genera were named in 2010, but three are invalid today.

2010—China—The most abundant Mesozoic bird
From 1993 to 2000, nearly 1,000 specimens of *Confuciusornis sanctus* were found. Many were sold illegally. The Tianyu Natural History Museum in Shandong reported a collection of 536 specimens. See more: *Zheng et al. 2010*

2010—The year the most Mesozoic bird species were named
Thirteen species and twelve genera were named in 2010. An additional twelve genera and two invalid species were named in 2015.

2010—Utah, USA—The most abundant theropod of the Jurassic
The Tianyu Natural History Museum in Shandong reported that it had 255 specimens of *Anchiornis huxleyi* in its collections in 2010. See more: *Zheng et al. 2014*

2010—Romania—The last country where a Mesozoic bird was found
The specimen Nven 1 was an enantiornithean bird consisting of a humerus. Other reports of previous Mesozoic birds were made, but they turned out to be non-avian theropods, among which are *Elopteryx*, *Heptasteornis*, and *Bradycneme*. See more: *Andrews 1913; Harrison & Walker 1975; Wang et al. 2011; Brusatte et al. 2013*

2013—Zambia—The last country where a basal dinosauromorph was found and named
Lutungutali sitwensis is one of the last dinosauromorphs reported in Africa. See more: *Martínez et al. 2013; Peecook et al. 2013; Martínez et al. 2013*

2013—Chile—The last country where a basal dinosauromorph was found
The specimen SGO.PV.22250 was reported. Its publication is pending. See more: *Rubilar-Rugers et al. 2013*

2013—Saudi Arabia—The last country where a non-avian theropod was found
The teeth SGS 0061 and SGS 0090 belong to a small abelisaurid. A spinosaurid tooth (UM10575) was reported in Malaysia in 2014, but there is still no formal publication. See more: *Kear et al. 2013*

2014—France—The first Mesozoic bird fossil projected by a 3-D laser
The skeleton and feathers of *Archeopteryx lithographica* were projected in 3-D laser using a technique known as camera obscura, which could open new avenues for fossil observation.

2015—USA—The highest body temperature in a theropod
Solid, extensive analysis revealed that *Tyrannosaurus rex* had a body temperature of 11°C, higher than tuna (6.9°C) and lower than the leatherback sea turtle (*Dermochelys coriacea*, 13°C). *Tyrannosaurus rex* may have needed less food in proportion to its size, but its growth was fast. It had the ability to perform dynamic activities, and heat loss was slow. See more: *Grady et al. 2014*

2015—Chile—The last country where a non-avian theropod was named
Chilesaurus diegosuarezi was the first non-avian theropod to receive a scientific name in Chile, although in 1929 it had been named *Neogaeornis wetzeli*, a neornithic bird. The first remains of a theropod of the Upper Jurassic were reported in 1961. See more: *Lambrecht 1929; Biese 1961; Novas et al. 2015*

2015—Brazil—The last country where a Mesozoic bird was named
Cratoavis cearensis was named ten years after the first Mesozoic bird found in Brazil, UFRJ-DG 06-Av, was reported. See more: *Alvarenga & Nava 2005; Carvalho et al 2015a, 2015b*

2015—Spain—The largest online search engine for dinosaur articles
It is the most powerful search tool on the Internet for scientific publications on dinosaurs. It offers about 13,560 records. See more: http://www.aragosaurus.com/?seccion=basearag (Accessed December 31, 2015)

2015—The largest virtual dinosaur database in the network
The Paleobiology Database is an online source of information about the distribution and classification of fossil animals, plants, and microorganisms. Created in 2000 by the National Science Foundation and the Australian Research Council, it includes information on diverse dinosaur species as well as their footprints.

2015—The theropod with the greatest number of results on the internet
According to Google, the term "Tyrannosaurus" gets 1.62 million results. (Accessed December 31, 2015)

POPULAR AND FORMAL

Dinomania is a sociocultural phenomenon sparked by a fascination with dinosaurs, which can become fashionable and, therefore, create commercial opportunities. The phenomenon can end up distorting the natural characteristics of dinosaurs, turning them into popular monsters whose behavior or appearance strays far from reality. On the other hand, it promotes interest among the general public in the results of formal investigations and the dissemination of truthful information.

SCULPTURES OF THEROPODS

ca. 1100 BC—China—First bronze engraving of a dragon
The distinctive horns of Chinese dragons appear in artwork of the Shang Dynasty. See more: *Andersson 1934; Spalding & Sarjeant 2012*

1852—England—The first theropod sculptor
Benjamin Waterhouse Hawkins was a passionate natural-history sculptor who made several life-size sculptures of *Megalosaurus*. His work was commissioned in 1852 by the Crystal Palace Company. See more: *Mayor 2005*

1854—England—First sculpture of a theropod to be exhibited
The first theropod sculpture was created by the artist Benjamin Waterhouse Hawkins and exhibited in the Crystal Palace of London (England). It shows *Megalosaurus* as its appearance and size were thought to be at that time. See more: *McCarthy & Mick 1994*

1871—USA—First vandalism against dinosaur sculptures
William Magear "Boss" Tweed was a corrupt politician who ordered the destruction of paleo-sculptor Benjamin Waterhouse Hawkins's sculptures. They were housed in a museum named Paleozoic Park in New York. See more: *Sanz 2007*

1907—Germany—The first dinosaur sculpture in battle
A statue of a *Stegosaurus* fighting a *Ceratosaurus* stands in the Tierpark Hagenbeck, in Hamburg. It depicts one of the most classic fight scenes in popular culture. See more: *Lambert 1990*

1945—Mexico—First fake dinosaur sculptures
A fake find was reported in the state of Guanajuato. Controversial representations of dinosaurs, the so-called Acámbaro Figures were supposedly sculpted in the pre-classical Chupícuaro Culture (800 BC–AD 200), but it was determined in 1972 that they were recent pieces. Possessing anatomical errors typical of the time, they were made for sale by peasants inspired by films. See more: *Pezatti 2005*

1966—Peru—First fake dinosaur recordings
To sell engraved stones, scenes were carved and offered as evidence of humans from another galaxy who had cohabited with dinosaurs. In general, they are copies of illustrations made in other countries showing dinosaurs hunting humans. Some have pencil traces and include corrugated cardboard. See more: *Paris 1998*

1986—USA—The largest concrete sculpture of a theropod
On the road between Los Angeles and Palm Springs, California, is Mr. Rex, a sculpture that stands over 20 m high and weighs more than 100 t. Its structure is hollow, and the surrounding countryside can be admired from inside its jaws. See more: *Knight 2014*

2000—Canada—The largest theropod statue
In the town of Drumheller, a fiberglass and steel sculpture of a *Tyrannosaurus rex* was constructed that is 25 m tall and 46 m long and weighs 72 t. See more: *Blake 1999*

2007—USA—The most expensive theropod fossil replica
Several copies of the replicated skeleton of *Tyrannosaurus rex* BHI 3033 have been sold for $100,000 each. See more: *Worthington & Beech 2002*

2013—USA—The largest theropod built from balloons
Larry Most created a sculpture of an *Acrocanthosaurus* for display at the Virginia Museum of Natural History. The sculpture is 20 m long and 6 m high and consists of a series of balloons tied together. See more: *Campbell 2015*

The first reconstruction of a theropod (*Megalosaurus*). It had a crocodile's head and a hump, and it was quadruped. The reconstruction error was eventually recognized.

THEROPODS AND MESOZOIC BIRDS IN LITERATURE

ca. 1100 BC—Iraq—The oldest poem about a dragon
Enûma Elish, a Babylonian poem, mentions dragons in the Mesopotamian creation myth. See more: *Batto 1992*

1820—England—First poem dedicated to a paleontologist
Dedicated to Professor William Buckland:

> *Where shall we our great Professor inter*
> *That in peace may rest his bones?*
> *If we hew him in a rocky sepulchre*
> *He'll rise and break the stones*
> *And examine each stratum that lies around*
> *For he's quite in his element underground.*

See more: *Whatley 1820*

1853—England—The first theropod in a work of fiction
Megalosaurus is mentioned in the novel *Bleak House*. It is described as an animal 12 m long. See more: *Dickens 1853*

1864—France—The first novel about a paleontological expedition
In the novel *Voyage au center de la terre* (*Journey to the Center of the Earth*), by Jules Verne, various extinct animals appear but no dinosaurs. Dinosaurs, however, are always present in film adaptations. See more: *Verne 1863*

1869—USA—The first reconstruction of a bipedal theropod
The first image of a bipedal theropod was of *Dryptosaurus aquilunguis* (formerly known as *Laelaps aquilunguis*). It was interpreted as a jumping animal similar to a kangaroo. See more: *Cope 1869; Marsh 1877*

1877—Colorado, USA—The first named dinosaur that was never illustrated
Apatodon mirus was thought to be a mammalian tooth, but it was actually an indeterminate dinosaur bone. It is tentatively synonymous with *Allosaurus*, but since the piece was lost and never drawn, it is impossible to confirm. See more: *Marsh 1877; Baur 1890; Mortimer*

1886—Holland—The first ichnological bibliographic compilation
The first bibliographic compilation of fossil footprints was collected between 1828 and 1886. See more: *Winkler 1886; Vintaned 2008*

1893—England—The first dinosaur cartoons
The cartoonist and caricaturist Edward T. Reed created *Prehistoric Peeps*, a series of illustrations showing humans with dinosaurs, pterosaurs, and giant snake species in comical situations. See more: *Bissette 2010*

1905-1906—USA—The least suitable nickname for a theropod
On many occasions, a fossil gained a nickname at the same time it received a formal name. These nicknames usually refer to some aspect of the individual. For example, the *New York Times* called *Tyrannosaurus rex* the "boxer of antiquity," a nickname that's easily forgiven, considering that arms were not found in the first specimens. A challenger for the title is the "real devourer of men of the jungle," coined by the same newspaper. See more: *Anonymous 1905, 1906*

1912—Scotland—The first dinosaur novel
The Lost World is a novel by Sir Arthur Conan Doyle about an expedition to a South American plateau inhabited by dinosaurs, animals, and prehistoric men. Among theropods that appear in the novel are *Allosaurus* or *Megalosaurus*. It was the first book that managed to place dinosaurs in the popular imagination for the first time. See more: *Conan Doyle 1912*

1915—Russia—The first novel with underground dinosaurs
The novel *Plutonia*, by Vladimir Obruchev, is about an underground world with dinosaur fauna. In its predecessor, *Voyage au center de la terre* (*Journey to the Center of the Earth*), by Jules Verne, various extinct animals appear, but no dinosaurs. See more: *Obruchev 1915*

1932—USA—The first comic strip with dinosaurs
Alley Oop was a comic strip created by the artist Vincent T. Hamlin, who wrote and drew for over four decades. It was published in about 800 newspapers. See more: *Szymanczyk 2014*

1934—England—The first professional dinosaur book for the general public
The paleontologist William Elgin Swinton wrote a book that included an alphabetical list of English dinosaurs. See more: *Swinton 1934*

1935—USA—The first dinosaur book for children
The *Book of Prehistoric Animals*, printed in color, was sixty-four pages long and included a map showing dinosaurs from around the world. See more: *Raymond & Carter 1935*

1954—Japan—The first dinosaur cartoons
Godzilla's films have been accompanied by comic strips and black-and-white adaptations of them. The first were made in Japan. They did not appear in the United States until 1976. In both cases, some appear without credits. See more: *Kalat 2010*

1959—Guatemala—The shortest dinosaur story
In the micro- book *The Dinosaur*, by Augusto Monterroso, is a statement that says: "When he woke up, the dinosaur was still there." See more: *Monterroso 1959*

1963—USA—The name of a bird that is based on an onomatopoeia
The genus name *Torotix* comes from an imitation of the cry of a flamingo, which was believed to be a relative. The word was created by the Greek playwright Aristophanes in his work *The Birds*, where it is even extended to Torotorotorotorotix and Torotorotorotorolililix. See more: *Brodkob 1963*; http://www.curioustaxonomy.net

In science fiction, dinosaurs usually appear as monsters.

1969—USA—Tarzan's first comic with dinosaurs
Tarzan at the Earth's Core, by Edgar Rice Burroughs, depicts a world of dinosaurs and reptilian men who have no feelings of friendship, sympathy, or love. It was based on Burrough's novel *Pellucidar*, published fifty-five years earlier. See more: *Burroughs 1914; Bleiler 1948*

1972—USA—The first dinosaur dictionary
The Dinosaur Dictionary is a 218-page-long book in which dinosaurs are described from A to Z. See more: *Glut 1972*

1977—Japan—The first "dinosauroid" work of fiction
Aritsune Toyota is a Japanese science fiction writer who wrote about humanoid reptiles that descended from theropods similar to *Dromaeosaurus*. Carl Sagan speculated on the same subject with *Saurornitholestes* the same year. A year later, Henry Jerison made a speech on the subject before the American Psychological Association, using *Dromiceiomimus* as a model. A publication by Russell and Séguin using *Troodon* (*Stenonychosaurus*) finally had the most media impact. See more: *Toyota 1977; Sagan 1977; Jerison 1978; Russell & Séguin 1982; Kaneko 1997*

1984—USA—The first theropod "Transformer"
In "The Autobot's Last Stand," number 4 of the comic *Transformers*, dinobots appear for the first time. The leader was a *Tyrannosaurus rex* named Grimlock. See more: *Furman 2008*

1988—USA—The first book on dinosaurs of the future
The New Dinosaurs: An Alternative Evolution is a work of fiction that speculates about the possible evolution of dinosaurs if they had not become extinct. The genus Megalosaurus is an example. It survives with two species in Madagascar: *M. modernus* and *M. nanus*. See more: *Dixon 1988*

1990—USA—The best-selling dinosaur novel
The novel *Jurassic Park*, by Michael Crichton, became one of the most-read books of the 1990s. It is considered a "jewel of scientific fantasy," touching on such diverse topics as chaos theory, genetics, and paleontology. See more: *Crichton 1990*

1991—China—The first dinosaur atlas
Dinosaur Atlas, by Deng (Ying) Pa Ke, is a pocket book.

1992—UK & USA—The first hardcover dinosaur atlas
The Great Dinosaur Atlas by William Lindsay, is a book with over 350 illustrations in full color. See more: *Lindsay 1992*

1993—England—The dinosaur magazine with the most volumes
Dinosaurs! Discover the Giants of the Prehistoric World is a collection of 104 installments divided into twelve volumes with commentary by the paleontologist David Norman.

ILLUSTRATIONS AND PHOTOGRAPHS OF DINOSAUROMORPHS, THEROPODS, AND MESOZOIC BIRDS

1677—England—The first illustration of a theropod bone
A piece found in 1671 was given to Robert Plott in 1676, who illustrated it and, because of its large size, considered it an elephant piece brought by the Romans. See more: *Plott 1677*

1699—England—The first illustrated theropod tooth
The naturalist Edward Lhuyd illustrated a theropod tooth, interpreting it as a curious stone that imitated organic forms. See more: *Lhuyd 1699; Delair & Sarjeant 2002*

1763—England—The first reillustration of a theropod
Brookes refigured Plott's illustration eighty-six years later. See more: *Brookes 1763*

1832—USA—The first professional illustration of a theropod footprint
On May 20, 1832, James Deane sent a letter to Edward Hitchcock in which he had illustrated a fossil print in India ink. See more: *Hebert 2014*

1833—England—The first reconstruction of a theropod
The Penny Magazine illustrated *Megalosaurus* for the first time, depicting it as a large aquatic monitor lizard. See more: *Debus 2009*

1854—England—The first reconstruction of a theropod sculpture
In May 1854, Waterhouse Hawkins and Richard Owen illustrated *Megalosaurus* as it would look in life for the first time. See more: *Hawkins 1854; Owen 1854*

1861—Germany—The first illustration of a Mesozoic bird
Karl Häberlein, the owner of the first skeletal fossil of *Archeopteryx lithographica*, did not allow anyone to take data or make drawings before it was sold, but Albert Oppel examined and memorized it, so he could sketch it later with amazing precision. See more: *Buffetaut 1991*

1863—France—The first illustrated theropod fight
The artist Louis Figuier depicted a *Megalosaurus* attacking an *Iguanodon* that is biting it. See more: *Figuier 1863*

1892—Germany—The first cards with theropods
The Liebig Continental Trading Cards cigar company created Prehistoric Animals in Different Ages, a series in which *Megalosaurus* is represented. See more: *Moore 2014*

1897—USA—The first illustration of a jumping theropod
Charles Robert Knight reconstructed an illustration of two *Dryptosaurus aquilungus* (formerly known as *Laelaps*). The first jumps over the second, as the second prepares to defend itself. See more: *Paul 1966*

The first international scientific dinosaur-illustration competition was "International Dinosaur Illustration Contest (CIID)," which was held in 2000–01. Ten countries participated.

1899—USA—The first photograph of a theropod footprint in a formal publication
Two footprints from the Upper Jurassic appeared on a sheet. The smaller one seems to have belonged to a theropod. See more: *Marsh 1899*

1899—USA—The first photograph of a theropod in a formal publication
Specimens of *Allosaurus* (some referred to as probable *Megalosaurus*) were shown in several photographs. Graphic descriptions used to be very detailed illustrations. See more: *Osborn 1899*

1905—USA—The first illustration of a theropod with a human for scale
In the description of *Tyrannosaurus rex*, a human skeleton is used to show its size visually. See more: *Osborn 1905*

1926—Denmark—The first illustration of a theropod running without dragging its tail
In the book *The Origin of Birds*, the Danish artist Gerhard Heilmann illustrated a running *Compsognathus* for the first time. In all previous illustrations, dinosaurs were shown dragging their tails. See more: *Heilmann 1927*

1932—Democratic Republic of the Congo—The first false photograph of a theropod
The *Rhodesia Herald* newspaper published a photo of a supposed carnivorous dinosaur. Named "Kasai rex," it appeared to be devouring a rhinoceros. It was described as reddish with blackish stripes and about 13 m long. The fraud was verified when it was noticed that the photograph was a montage of a lizard; white cut lines could be seen on the rhinoceros corpse. The fraud was orchestrated by John Johnson. See more: *Coghlan 2004*

1941—USA—The highest price paid for a dinosaur drawing
The Carnegie Museum of Natural History paid $100,000 for the paleontologist Henry Fairfeld Osborn's original illustration, made between 1905 and 1906, of a *Tyrannosaurus rex* skeleton drawn to scale with a human. See more: *Davidson 2008*

1947—USA—The largest mural with theropods
The artist Rudolph Franz Zallinger painted the mural *The Age of Reptiles* between 1943 and 1947. It is almost 34 m long and 4.9 m tall. It shows about 300 million years of evolution, in which theropods such as *Allosaurus*, *Compsognathus*, *Podokesaurus*, *Struthiomimus*, and *Tyrannosaurus* appear. See more: *Volpe 2001*

1947—USA—The largest mural with a Mesozoic bird
Two *Archeopteryx* appear in the mural *The Age of Reptiles*, by Rudolph Franz Zallinger. The work is about 34 m long and 4.9 m tall. See more: *Volpe 2001*

1949—USA—The first Pulitzer Prize for a literary work on dinosaurs
Rudolph Franz Zallinger received a Pulitzer Prize for his mural *The Age of Reptiles*. Some paleontologists, including Robert Bakker and Peter Dodson, have commented that this work inspired them to study dinosaur paleontology. See more: *Volpe 2001; Wallace 2004*

1954—USA—The first dinosaur skeleton illustrated with a black silhouette
The illustration of a *Struthiomimus altus* skeleton appears surrounded by a silhouette. See more: *Scheele 1954; Hartmann**

1956—Czech Republic—The most influential modern dinosaur paleoartist
The paleontological reconstructions of Czech painter and illustrator Zdenek Michael František Burian were influential for six decades. His first great book was *Prehistoric Animals*, printed in then Czechoslovakia and translated into many languages. Stephen Gould considers it one of the most influential visual books of the twentieth century. Burian's work inspired and continues to inspire numerous artists of extinct animals. See more: *Prokop 1990*

1975—USA—The first theropod drawn with feathers
An illustration by Sarah Landry depicts a *Syntarsus rhodesiensis* with a crest of feathers on the back of the skull, predicting the presence of feathers. See more: *Bakker 1975; Xu et al. 2009*

1985—England—The first color illustration of a feathered Mesozoic bird
Shortly after the presence of feathers in *Avimimus portentosus* was revealed, it was illustrated in full color by the paleoartist John Sibbick. See more: *Norman 1985*

1995—Brazil—The first theropod to appear on postage stamps
Before *Irritator challengeri* was officially described, it appeared printed on postage stamps as "Angaturama limai," its synonym. See more: *Mortimer**

2010—USA—The book with the most skeletal dinosaur reconstructions
The Princeton Field Guide to Dinosaurs, by Gregory S. Paul, has more than 200 skeletal reconstructions of various dinosaur species, some from different perspectives, juvenile specimens, and cases of sexual dimorphism. See more: *Paul 2010*

2010—USA—The book with the most skeletal theropod reconstructions
The Princeton Field Guide to Dinosaurs, by Gregory S. Paul, includes more than eighty reconstructions of various species of theropods and other basal saurischians. See more: *Paul 2010*

THEROPODS IN MOVIES

1905—England—The first movie with dinosaurs
The silent short film *Prehistoric Peeps*, directed by Lewin Fitzhamon, is about a scientist who dreams of prehistoric adventures. Dinosaurs and cavemen appear represented by people in disguise. See more: *Debus 2013*

1908—England—The first animation with dinosaurs?
Film-effects pioneer Walter R. Booth made the animated short *The Prehistoric Man* using his hand's shadows. Presumably, a dinosaur appears in it, but the film is lost, so it is impossible to verify. See more: *Klossner 2006*

1914—USA—The first feature film with a theropod
Brute Force is a silent film directed by D.W. Griffith in which a theropod similar to *Ceratosaurus* appears but has very limited movements—opening and closing its mouth and leaning forward. See more: *Berry 2005*

1914—USA—The dinosaur movie with the longest name
The Dinosaur and the Missing Link: A Prehistoric Tragedy (forty-eight characters) was directed by Willis H. O'Brien. It is also the first film to integrate the word dinosaur into the title. Some sources date it to 1914 and 1915, others to 1917. See more: *Archer 1993*

1914—USA—The first stop-motion dinosaur animator
Willis H. O'Brien was the first filmmaker to recreate the movement of dinosaurs, giving life to numerous characters. His work was recognized at the Oscars in 1950, when he won the Special Effects award. See more: *Archer 1993*

1914—USA—The first stop-motion dinosaur movie
The Dinosaur and the Missing Link: A Prehistoric Tragedy, directed by Willis H. O'Brien, animates dinosaurs for the first time, using photographs that change frame by frame. See more: *Archer 1993*

1918—USA—The first film in which there is a battle between dinosaurs
There is a fight between a *Tyrannosaurus* and a *Triceratops* in *The Ghost of Slumber Mountain*, directed by Willis H. O'Brien. The theropod wins the battle. See more: *Berry 2005*

1925—USA—The first dinosaur movie adapted from a novel
The Lost World directed by Harry Hoyt, animated by Willis H. O'Brien, and modeled by Mexican student Marcel Delgado, caused a sensation because people thought its animated models of dinosaurs were real. The movie was based on the homonymous novel by Sir Arthur Conan Doyle, written in 1912. See more: *Berry 2005*

1925—England/France—The first film projected on a commercial flight
The first in-flight movie was about dinosaurs. *The Lost World*, directed by Harry Hoyt, was shown on an Imperial Airways flight from London to the American continent. See more: *García Tsao 2012*

1929—USA—The first heroic theropod
In the movie *Death of the Moon*, written by Alexander M. Phillips, a *Tyrannosaurus* saves the world from aliens. See more: *Debus 2006*

1933—USA—The first talking dinosaur movie
Originally called "Kong" or *The Eighth Wonder*, the movie *King Kong*, directed by Merian C. Cooper and Ernest B. Schoedsack, included synchronized audio, a novelty that had begun to appear in cinema in 1927. The premiere took place on March 2 at Radio City Music Hall. See more: *Cooper & Schoedsack 1933; Pettigrew 1999*

1933—USA—The first film where a theropod is defeated in a fight
King Kong, the famous giant gorilla of the fictional species "Gigantopithecus giganteus," wins a battle against a *Tyrannosaurus rex*. He succeeds in unhooking the jaw in a scene that was repeated by Peter Jackson in his 2005 adaptation, in which Kong fights a "Vastanosaurus rex." See more: *Cooper & Schoedsack 1933; Pettigrew 1999; Blenkin et. al 2005*

1933—USA—Dinosaur movie with the most critical acclaim
In April 2004, *Empire* magazine named *King Kong*, directed by Merian C. Cooper and Ernest B. Schoedsack, the best monster movie of all time. It was also considered one of the one hundred best films of all time by *Time* magazine. It has a score of 7.3 on the website *FilmAffinity*, 8 on IMDb, and 98% on *Rotten Tomatoes* (86% audience score). See more: *Cooper & Schoedsack 1933*

1940—USA—The first film with dinosaurs represented by live animals
One Million BC/ The Cave Dwellers directed by Hal Roach Jr. and Hal Roach, was the highest-grossing movie of the year. The *New York Times* called it a "masterpiece of imaginative fiction." The crocodiles, iguanas, and lizards that represented the dinosaurs were mistreated during filming, provoking criticism. See more: *Klossner 2006*

1940—USA—The first animated color theropod movie
Fantasia by Walt Disney Productions, included several themes. In the fourth—*The Rite of Spring*, by Igor Stravinsky—there is a scene in which a *Tyrannosaurus* fights a *Stegosaurus*. It was the second animated feature film produced by Walt Disney. See more: *Culhane 1999*

1941—USA—The first animated dinosaur film to win an Oscar
Fantasia received two honorary awards in 1941 for its music. See more: *Culhane 1999*

1951—USA—The first movie of an island with dinosaurs
Lost Continent directed by Sam Newfield, is about an expedition to an island in the South Pacific that turns out to be populated by dinosaurs. See more: *Berry 2005*

1955—USA—The first space movie with dinosaurs
King Dinosaur, directed by Bert Gordon, places dinosaur fauna on Planet Nova. See more: *Berry 2005*

1954—Japan—The most popular fictional theropod
Godzilla or Gojira is the most famous fictional dinosaur in the world. He has appeared in twenty-eight movies and two remakes. He also appears in animations, songs, publications, and other popular media. He is defined in the films as "Godzillasaurus," a supposed theropod dinosaur possessing *Iguanodon* arms, *Stegosaurus* plates, and a sauropod tail. Along with its immense size, it has the ability to breathe fire. See more: *Christiansen 2000; Carpenter 2010*

1956—USA—The first western with dinosaurs
The Beast of Hollow Mountain was directed and produced by Edward Nassoure and Ismael Rodriguez. The script, by Willis H. O'Brien and Robert Hill, deals with an *Allosaurus* that attacked cattle and settlers. It was filmed in Mexico by Churubusco Studios. By mistake, some sources mention that this record belongs to *The Valley of Gwangi* or *The Valley Where Time Stood Still*, which was directed by Jim O'Connolly thirteen years later. See more: *Berry 2005*

1960—USA—Dinosaur movies with the most similar name
Only two letters and an exclamation mark differentiate *Dinosaurus!*, by Irvin S. Yeaworth, and *Dinosaur*, an animated movie produced by Walt Disney. See more: *Berry 2005*

1960—USA—The first dinosaur-movie remake
The Lost World was adapted for the cinema again, thirty-five years after the first version. This time, it was directed by Irwin Allen. See more: *Berry 2005*

1961—England—The movie of a fictional theropod with the shortest name
Gorgo by director Eugène Lourié, is a science fiction movie. Gorgo is an giant ancient monster that is captured and then rescued by its mother. See more: *Berry 2005*

1964—Spain—The movie with an invisible dinosaur
El Sonido de la Muerte (The sound of death) or *El Sonido Prehistórico* (The prehistoric sound), directed by J.A. Nieves Conde, was about archaeologists who find a dinosaur egg in Greek ruins. The carnivore was invisible, so it was extremely dangerous. See more: *Moore 2014*

1966—Mexico—Tarzan's first film with dinosaurs
Track of the Dinosaur, directed by Lawrence Dobkin, starred Ron Ely. It was filmed in Mexico City by Churubusco Studios. See more: *Schneider 2012*

1984—USA—The first dinosaur movie in go motion
In *Prehistoric Beast*, directed by Phil Tippett, a variant of the stop-motion technique was applied. It consisted of moving the models while taking pictures consecutively. This added realism to the animation of a *Tyrannosaurus* that fought against a *Monoclonius*. See more: *Berry 2005*

1992—USA—The first film with anthropomorphic dinosaurs
Adventure in Dinosaur City, directed by Brett Thompson, is about three teenagers who are absorbed by television in a world populated by civilized dinosaurs and pterosaurs. See more: *Berry 2005*

1993—USA—The most successful dinosaur movie
Jurassic Park, directed by Steven Spielberg, cost about $63 million and became the highest-grossing

dinosaur movie. It was shown in 2,566 cinemas worldwide and grossed more than $1.029 billion. See more: http://www.boxofficemojo.com/movies/?id=jurassicpark.htm (Accessed on December 31, 2015)

1993—USA—The dinosaur film with the most awards
Jurassic Park, directed by Steven Spielberg, earned twenty-three awards, including three Oscars (and fourteen nominations). It also won awards from BAFTA, BMI, Blue Ribbon, Bram Stoker, Cinema Audio Society, Golden Reel, Grammy, Hugo, Golden Screen, Czech Lion, Mainichi Film Concours, MTV Movie, Saturn, People's Choice, Sierra, Young Artist, the Japanese Academy of Cinematography, the Mainichi Eiga Concours festivals, and others. See more:http://www.imdb.com/title/tt0107290/awards (Accessed on December 31, 2015)

1993—USA—The most critically acclaimed dinosaur movie
Review sites rated *Jurassic Park*, directed by Steven Spielberg, and *King Kong*, directed by Merian C. Cooper and Ernest B. Schoedsack, highest among movies with dinosaurs. *Jurassic Park* has a score of 7 on *FilmAffnity*, 8.1 on IMDb, and 93% on *Rotten Tomatoes* (91% audience score), while *King Kong* (1933) obtained scores of 7.3, 8, and 98% (86%), respectively. See more: (Accessed on December 31, 2015)

1998—France—The dinosaur horror movie with the longest name
Terror of Prehistoric Bloody Monsters from Space is a science fiction comedy directed by Richard J. Thompson. It is also known as *Jurassic Trash*. See more: *Thompson 1998*

2001—USA—The theropod movie with the shortest name
Director Jim Wynorski's *Raptor* is an action, fantasy, and horror movie. See more: *Berry 2005*

2013—USA—The dinosaur film with the least critical acclaim
Opinions differ on movie-review websites. Anthony Fankhauser's *Jurassic Attack* has a score of 1.4 on *FilmAffinity*, and it was least favored on IMDb, too, earning a score of 2.2. The film that obtained the lowest percentage on *Rotten Tomatoes*, however, is *A Sound of Thunder*, directed by Peter Hyams (6%, 18% audience score), beating Dan Bishop and Shlomo May-Zur's *Raptor Ranch* (13%). One must consider that *Jurassic Attack* was not rated on this site. See more: (Accessed on December 31, 2015)

2014—Japan—The largest fictional dinosaur
In the beginning, Godzilla was 50 m tall. Then, to make him stand out among buildings, his size was increased. His height grew to 55 m in 1999, 80 m in 1984, and 100 m in 1991, and in the movie *Godzilla 2000*, directed by Gareth Edwards, he stood 108 m tall and weighed 90,000 t. See more: *Christiansen 2000*

2015—USA—The most expensive dinosaur film series
The first four *Jurassic Park* films have cost about $379 million combined.

2015—USA—The most successful dinosaur film series
The four Jurassic Park films have grossed more than $3.6 billion collectively.

2015—USA—The most expensive dinosaur movie
Jurassic World had a budget of $150 million, which makes it the dinosaur movie with the largest budget.

2015—USA—The highest-grossing dinosaur movie
Jurassic World grossed over $1.606 billion

2015—Argentina—The first classification of films with paleontological background
It is proposed that paleontological films be classified in four subgenres based on their theme: Caveman, Dinosauria, Godzillians, and Mamiferoid. See more: *Echecury 2015*

Godzilla is usually described as a huge mutant dinosaur.

THEROPODS AND MESOZOIC BIRDS IN THE THEATER, ON TV, AND IN OTHER MEDIA

1897—France—The first play with a Mesozoic bird
The parody *Ubu cocu* (Ubu the cuckold or *Archeopteryx*), by Alfred Jarry, in which the extinct bird turns out to be the fruit of adultery, is one of the precursors of surrealism. See more: *Buffetaut 1992*

1956—USA—The first documentary with theropods
The Animal World presents past and present species, including *Allosaurus, Ceratosaurus,* and *Tyrannosaurus.* See more: *Webber 2004*

1960—USA—First animated series with dinosaurs for television
The Flintstones, the most important series produced by William Hanna and Joseph Barbera, premiered on September 30, 1960. Among the items popularized by the series were bronto burgers. See more: *Klossner 2006*

1960—USA—The longest-lasting animated series with dinosaurs
Running for six seasons, 166 episodes, and spinning off specials, movies, and other subseries, *The Flintstones* was the most-watched animated show. It was overtaken by *The Simpsons* in 1997. See more: *Klossner 2006*

1961—USA—The first real character to appear in an animated series with dinosaurs
In episode 6, season 1 of *The Flintstones*, the British actor Cary Grant appears as Gary Granite, parodying himself. See more: *Booker 2006*

1977—USA—The first animated series featuring a theropod
In a series by Hanna-Barbera, Godzilla was accompanied by his son, called Godzuki or Godzooky, who was small and could fly. See more: *Kalat 2010*

1983—USA—The first video game starring a dinosaur
The objective of *Godzilla*, a game for the Commodore 64, was to kill the gigantic green monster that attacks Japan. See more: *Presley & Deckel 1984*

1988—USA—The animated dinosaur series with the most sequels
The animated film *The Land before Time*, directed by Don Bluth, relates the adventures of juvenile dinosaurs. It spawned thirteen sequels, and others that are in progress. See more: *Simpson 2009*

1991—China—The first dinosaur CDs in Asia
Some of the first titles that appeared were *Adventure Dinosaur Island*, by Cao Mei; *Dinosaurs*, by Yue Han Ma La Mu; *Dinosaur Park*, by Hao Hai Zi Tu Shu Bian Ji Bu; *Dinosaur World Encyclopedia*, by Ji Jiang Hong; *Encyclopedia of Dinosaurs*, by Xing Tao; *Prehistoric Dinosaur Tracks*, by Deng Xing Lida; and *Super Dinosaur*, by Sheng Shi Hua Nian. See more: www.amazon.com/ (Accessed on December 31, 2015)

1992—USA—The first dinosaur CD in North America
The multimedia CD *Dinosaur Art*, produced by Doodle Art, went to market on June 18, 1992. See more: http://www.amazon.com/ (Accessed on December 31, 2015)

1994—USA—The first dinosaur series in which all the characters die
The TV series *Dinosaurs*, created by Michael Jacobs and Jim Henson, premiered in 1991. It related the adventures of the Sinclair family, very particular dinosaurs. At the end of the series, we learn how they became extinct. See more: Klossner 2006

1995—USA—The cartoon dinosaur movie with the longest name
The Land Before Time III: The Time of the Great Giving (42 characters), was directed by Roy Allen Smith. See more: *Allen Smith & Dowlatabadi 1995*

1997—Spain—The first dinosaur CD in Europe
The first may have been *Historia natural de los dinosaurios* (Natural history of the dinosaurs), by paleontologists José Luis Sanz, Joaquín Moratalla, and Bernardino P. Pérez-Moreno. See more: www.amazon.com/ (Accessed December 31, 2015)

1998—Mexico—The first CD of dinosaurs in Latin America
Deinos Saurus by the paleontologist René Hernández Rivera, was produced by the Coordination of Academic Services of the National Autonomous University of Mexico.

1999—USA—The most successful dinosaur documentary
The BBC documentary series *Walking with Dinosaurs* consists of six episodes, plus specials. It was successful with audiences in most of the countries where it was televised and is considered one of the hundred best documentaries in history. It spent the most money per minute of any documentary. See more: *Guiness Records 2000*

2001—USA—The first documentary about the pathologies of a theropod
"The Ballad of Big Al," a special episode of *Walking with Dinosaurs*, deals with the numerous wounds suffered by *Allosaurus fragilis* MOR 693 during his life. The episode won two Emmys for music. See more: *Bryant 2004*

2013—Alaska—The first Mesozoic narrator bird
The enantiornithean bird *Alexornis antecedens* appears in the animated film *Walking with Dinosaurs* as a main character and narrator. Named Alex, he was voiced by the Colombian actor John Leguizamo. See more: *Glass 2013*

2013—Japan—The most-watched dinosaur joke on the Internet
According to YouTube, the most-viewed joke video is "Dinosaur T-REX," with more than 14 million views. See more: www.youtube.com/watch?v=R3MKomHLYF4 (Accessed December 31, 2015)

2013—USA—The realistic dinosaur video with the most online views
The Best of Discovery's Dinosaurs is a series of animated videos. One is "Dinosaur Revolution, which has been viewed more than 100 million times since it was uploaded on July 3, 2013. See more: www.youtube.com (Accessed December 31, 2015)

2014—USA—The most-watched dinosaur video on the Internet
According to YouTube, the most-watched dinosaur video is "Finger Family Crazy Dinosaur Family Nursery Rhyme | Funny Finger Family Songs for Children in 3D," with over 130 million views. See more: www.youtube.com/watch?v=ZTfFvnQIeqQ (Accessed December 31, 2015)

2015—Dinosaurs with the highest number of search engine results
According to Google, searching for the word dinosaur produces 106 million results. Searching for *Dinosaurs* and *Dinosauria* produce 54.8 million and 518,000 hits, respectively. See more: www.google.com (Accessed December 31, 2015)

To animate dinosaurs accurately and realistically, it is important to know their anatomy. Among the most common mistakes are flexible tails, downward-facing hands, and unrealistic sizes.

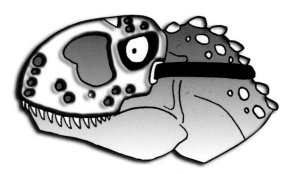

THEROPODS AND MESOZOIC BIRDS IN MUSIC

1967—England—The first rock band with a theropod's name
Tyrannosaurus Rex was a psychedelic and folk rock band started by Marc Bolan. It was the first glam rock group in history. The name of the group was abbreviated to T. rex in 1970. See more: *Melton 2014*

1977—USA—The first song dedicated to a fictional dinosaur
The psychedelic and heavy-metal rock group Blue Öyster Cult performed the song "Godzilla" on the album *Spectres*. See more: *Tsutsui 2006*

1981—USA—The first song that mentions the word dinosaur
According to the song "She Said," from the album *Psychedelic Jungle* by the punk rock band The Cramps: "She looked like a dinosaur 'bout to jump outta that seat." See more: *Britannica Educational Publishing 2013*

1983—Argentina—The first song with the title "The Dinosaurs"
The song "Los Dinosaurios" (The dinosaurs), composed and performed by Charly García, appeared on the album *Modern Clicks*. It protests the civic-military dictatorship that ravaged Argentina between 1976 and 1983. It is considered one of the best songs in Latin music history. See more: *Dylan & Knobloch 2013*

1985—Spain—The song that repeats the word dinosaur the most
On the album *Young Blood* by the heavy metal group Sobredosis, the word dinosaur is repeated twelve times in the 4:49-minute song "Dinosaur."

The name *Masiakasaurus knopfleri* is dedicated to the singer Mark Knopfler.

1985—USA—The first album titled *Dinosaur*
American alternative rock band Dinosaur Jr. was formed by J. Mascis and Lou Barlow. It was one of the groups that most influenced grunge music. The band's first album is called *Dinosaur*. See more: *Buckley 2003*

1987—Mexico—The first musical group with the name Dinosaurs (and the shortest)
Led by Juan Ramírez, the musical group Los Dinnos changed its name to Los Dinnos Aurios in 1995. They later went back to Los Dinnos. See more: http://losdinnos.com/

2005—USA—Songs with theropod genera
The group Fourth Grade Security Risk created songs that refer to dinosaurs, including "The Struthiomimus Strut" or "Gasosaurus," for the album *Dinosaur Tracks*.

2005—Songs with Mesozoic bird genera
The band Fourth Grade Security Risk composed the songs "Archeopteryx" and "Ichthyornis."

2008—USA—The first song with the full name of a theropod
The musician Rod Zombie (Robert Bartleh Cummings) performed the instrumental song "Tyrannosaurus Rex." See more: *McIver 2015*

2010—Mongolia—The first Mesozoic bird dedicated to a musical group
The species name *Hollanda luceria*, a bird, honors Lucero, a punk band from Memphis, Tennessee, United States. See more: *Bell et al. 2010*

2010—China—The first Mesozoic bird dedicated to a singer
The name *Qiliania graffini* honors Greg Graffin, the lead vocalist of the punk group Bad Religion. He is also a professor of paleontology at the University of California. See more: *Ji et al. 2011*

2011—Madagascar—The first theropod dedicated to a singer
The name *Masiakasaurus knopfleri* honors Mark Knopfler, front man of the English rock band Dire Straits. Its discoverers were listening to the song "Sultans of Swing" when it was found. See more: *Sampson et al. 2001*

2011—Spain—The first rock band with the name of a Mesozoic bird
A progressive rock and kraut-rock band is called Archeopteryx Ultraavantgarda. See more: https://archaeopteryx.bandcamp.com/

OTHER CULTURAL RECORDS OF THEROPODS AND MESOZOIC BIRDS

1905—USA—The name of the most famous theropod arises
The dinosaur was named *Tyrannosaurus rex* and *Dyamosaurus imperiosus* in the same publication. The name that had priority is the one mentioned first. See more: *Osborn 1905*

1905—USA—The first dinosaur tea
A tea called "Dinosaur Tea" was offered at the presentation of the first mounted skeleton of an *Allosaurus fragilis*, discovered in 1898. See more: *Breithaupt et al. 1999*

1986—USA—The most famous antagonism between dinosaurs
The rivalry between *Triceratops* and *Tyrannosaurus* was depicted in the mural painted by Charles R. Knight in 1942 for the Field Museum of Natural History. In 1986, the paleontologist Robert Bakker declared, "There has never been a more spectacular meeting between a predator and its prey." See more: *Bakker 1986*

1989—Canada—The first clothing brand named after a Mesozoic bird
Arc'teryx is a clothing and sporting goods company. Its name comes from *Archaeopteryx*. See more: http://arcteryx.com

1990—Chile—The first asteroid named after a theropod
Asteroid 9951, discovered by E.W. Elst from La Silla Observatory, was named Tyrannosaurus on November 15. See more: Minor planet center

1991—Chile—The first asteroid named after a Mesozoic bird
Asteroid 9860, discovered by E.W. Elst from La Silla Observatory, was named Archeopteryx on August 6. See more: Minor planet center

1995—Canada—The first professional sports team with a theropod
The Toronto Raptors basketball team was founded in Ontario. The team's mascot is a red dromaeosaur. See more: *Gilbert 2006*

2000—USA—The most popular theropod
The paleontologist Robert Bakker has described *Tyrannosaurus rex* as the "most popular of dinosaurs, among people of all ages, all cultures and all nationalities." A Google search yields 725,000 results. The theropod that most closely matches its popularity is *Velociraptor* (450,000 hits on Google, accessed December 31, 2015). *Tyrannosaurus rex* is the only dinosaur commonly known to the public by its full scientific (binomial) name, and even its scientific abbreviation, *T. rex*, is well known. See more: *Bakker 2000; Brochu & Ketcham 2003;* https://www.google.com (Accessed December 31, 2015)

The shortest way to write *dinosaur*
According to Google Translate, the shortest way to write *dinosaur* with the Latin or Roman alphabet is *dinozò* in the Haitian Creole language. In other script systems, it occupies a smaller space (恐龙 in simplified Chinese, 恐龍 in traditional Chinese, 공룡 in Korean and 恐竜 in Japanese). See more: https://translate.google.com/

The longest word for *dinosaur*
Using Google Translate, it can be verified that the longest way to write *dinosaur* is *Dinoszaurusz* (in Hungarian). See more: https://translate.google.com/

AUTHORS OF FORMAL PUBLICATIONS ABOUT DINOSAURS

AUTHORS UNDER THE MAGNIFYING GLASS

The achievements of dinosaur paleontologists are not measured by the number of dinosaurs they name throughout their careers, but in the contribution to knowledge made by their research, reflections, experiments, and comparisons with modern animals. Little by little over time, the tendency to name incomplete or doubtful remains has been reduced, avoiding the creation of problematic nominations or unnecessary synonyms. There are more scientists now than at all other times combined. See more: *Asimov 1979*

1736—England—The first invalid name for a theropod
A distal fragment of a theropod femur was called "Scrotum humanum" for descriptive purposes. It has sometimes been used as synonymous with *Megalosaurus bucklandii*, but it is impossible to determine if that piece belonged to this megalosaurid or another contemporary theropod. See more: *Brookes 1763; Parkinson 1822*

1866—England—The first name for a theropod by an anonymous author
An unknown author mentioned the name *Calamospondylus oweni*, attributing the authorship to the Reverend William Fox. See more: *Anonymous 1866*

1893—USA—The paleontologist who named the most invalid dinosaurs
Edward Drinker Cope named sixty-four species, of which only nine (14%) are currently considered valid. See more: *Benton 2008*

1897—England—The paleontologist who named the most dinosaurs for the longest period of time
Richard Owen named dinosaurs from 1841 to 1897, a duration of fifty-six years. See more: *Benton 2008*

1899—USA—The paleontologist who named the most dinosaurs
Othniel Charles Marsh named eighty species, of which twenty-nine (36%) are currently considered valid. See more: *Benton 2008*

1899—USA—The paleontologist who named the most invalid theropods
Othniel Marsh named *Hallopus victor, Laopteryx priscus, Macelognathus vagans, Meniscoessus caperatus*, and *Tripriodon caperatus*. None is a theropod. See more: *Benton 2008*

1932—Germany—The paleontologist who has renamed the most theropod species
Friedrich von Huene made six name changes: *Altispinax dunkeri* (Dames 1884) Huene 1923, *Betasuchus bredai* (Seeley 1883) Huene 1932, *Caudocoelus sauvagei* (Nopcsa 1928) Huene 1932, *Erectopus superbus* (*Megalosaurus superbus* Sauvage 1882) Huene 1923, *Proceratosaurus bradleyi* (Woodward 1910) Huene 1926, and "Megalosauridorum" *terquemi* (Huene 1926) Huene 1932. See more: *Benton 2008*

1934—Germany—The paleontologist who has named the most dubious theropods
Friedrich von Huene described seventeen species whose validity is in doubt (*Avipes dillstedtianus, Coeluroides largus, Compsosuchus solus, Dolichosuchus cristatus, Dryptosauroides grandis, Halticosaurus longotarsus, Jubbulpuria tenuis, Laevisuchus indicus,* "Megalosaurus" *nethercombensis,* "Megalosaurus" *lydekkeri,* "Megalosaurus" *terquemi, Ornithomimoides barasimlensis, O. mobilis, Rapator ornitholestoides, Thecocoelurus daviesi, Velocipes guerichi,* and *Walgettosuchus woodwardi*). He also created *Altispinax* and *Betasuchus* to name other dubious remains. See more: *Benton 2008*

1972—Lesotho—The paleontologist who has named the most basal saurischian footprints
Paul Ellenberger named seven species from footprints similar to those produced by the basal saurischian *Guaibasaurus*. See more: *Ellenberger 1970, 1972*

1986—China—The paleontologist who has received the most dedications among the names of theropod footprints
Xiangxipus youngi and *Youngichnus xiyangensis* were dedicated to the paleontologist Yang Zhongjian, who is also known as Chung Chien Young. See more: *Zhen 1982; Zhen et al. 1986*

1993—Lesotho—The paleontologist who has named the most theropod footprints
Paul Ellenberger named twenty-three species from theropod tracks. See more: *Ellenberger 1970, 1972, 1974; Jaillard 1993*

1993—Lesotho—The paleontologist who has named the most dinosaur footprints
Paul Ellenberger created ninety-five name combinations to identify a large number of dinosaur footprints. It is important to note that some are considered doubtful because they lack a sufficient description. See more: *Ellenberger 1970, 1972, 1974; Jaillard 1993*

2006—Argentina—The paleontologist with a highest percentage of valid dinosaurs
José Fernando Bonaparte named twenty-seven species, of which only one is not currently valid (96% success). See more: *Benton 2008*

2006—China—The paleontologist who has received the most dedications among theropod names
The paleontologist Pinying Dong Zhiming has been honored in the names of three theropods (*Caudipteryx dongi, Sinraptor dongi,* and *Sinornithomimus dongi*) and a fingerprint (*Lufengopus dongi*), and in one invalid name ("Monolophosaurus dongi"). See more: *Grady 1993; Currie & Zhao 1994; Zhou & Wang 2000; Kobayashi & Lu 2003; Lu et al. 2006*

2009—China—The paleontologist who named the most Mesozoic bird eggs
Xiaosi Fang has named eight species of avian oolites, either alone or as part of a team.

2013—USA—The paleontologist who has named the most dinosauromorphs
Sterling J. Nesbitt has participated in various teams of researchers, coauthoring the descriptions of *Asilisaurus kongwe, Diodorus scytobrachion, Dromomeron romeri, D. gregorii,* and *Lutungutali sitwensis.* See more: *Benton 2008*

2013—China—The paleontologist who has named the most dinosaur eggs
Zikui Zhao has named thirty-one species of oolites, either alone or as part of a team.

2013—China—The paleontologist who has named the most theropod eggs
Zikui Zhao has named seventeen species of theropod oolites.

2014—China—The paleontologist who named the most valid dinosaurs
Dong Zhiming named fifty-five species, of which forty are currently valid (72% success). See more: *Benton 2008*

2015—China—The paleontologist who has named the most Mesozoic birds
Xiaolin Wang has named twenty-seven valid species, plus another two invalid species and a dubious non-avian theropod. If we add Mesozoic birds, we have a total of thirty-two species of Mesozoic theropods. He is still working, so this number may change. See more: *Benton 2008*

2015—China—The paleontologist who has named the most Mesozoic birds
Zhou Zhonghe has named more than thirty-two species of birds. See more: *Benton 2008*

2015—China, USA—Paleontologists who have named the most Mesozoic bird footprints
Martin G. Lockley and Lida Xing have participated as coauthors in the descriptions of eleven new species of Mesozoic bird footprints, in the same work teams.

2015—China—The most common surname among Mesozoic bird paleontologists
Nine paleontologists who have named Mesozoic bird species share the surname Wang: Jinqi Wang, Li Wang, Lixia Wang, Min Wang, Ren-Feil Wang, Xia Wang, Xiaolin Wang, Xuri Wang, and Yan Wang.

2015—China—The paleontologist who has received the most Mesozoic bird dedications
Professor and fossil curator Lianhai Hou has five species dedicated to him: *Changmaornis houi, Cuspirostrisornis houi, Houornis caudatus, Longicrusavis houi* and *Pengornis houi.* See more: *Hou 1997; Zhou, Clarke & Zhang 2008; O'Connor et al. 2010; Wang et al. 2013; Wang & Liu 2015*

In 2008, before the manuscript of *Epidexipteryx* was reviewed by scientists, it appeared by mistake on the *Nature* magazine website. This caused confusion.

Dinosauromorphs

First dinosauromorph named by an author
Saltopus elginensis Huene 1910

First dinosauromorph named by two authors
Eucoelophysis baldwini Sullivan & Lucas 1989

First dinosauromorph named by three authors
Diodorus scytobrachion Kammerer, Nesbitt & Shubin 2012

First dinosauromorph named by four authors
Agnosphitys cromhallensis Fraser, Padian, Walkden & Davis 2002

First dinosauromorph named by six authors
Dromomeron gregorii Nesbitt, Irmis, Parker, Smith, Turner & Rowe 2009

First dinosauromorph named by seven authors
Dromomeron romeri Irmis, Nesbitt, Padian, Smith, Turner, Woody & Downs 2007

First dinosauromorph named by nine authors
Ignotosaurus fragilis Martinez, Apaldetti, Alcober, Columbi, Sereno, Fernandez, Santi-Malnis, Correa & Abelin 2013

Dinosauromorph footprints

First primitive saurischian footprints named by an author
Rhynchosauroides tumidus (now *Rotodactylus*) Morton 1897

First primitive saurischian footprints named by two authors
Agialopus wyomingensis Branson & Mehl 1932

Dinosauromorph footprints named by the most authors
Agialopus wyomingensis Branson & Mehl 1932, *Banisterobates boisseaui* Frasier & Olsen 1996, *Coelurosaurichnus largentierensis* Courel & Demathieu 1976, *Coelurosaurichnus metzneri* Heller 1952, *Coelurosaurichnus perriauxi* Demathieu & Grand 1972, *Coelurosaurichnus ratumensis* Demathieu & Oosterink 1988 and *Paratrisauropus bronneri*, *P. latus*, and *P. mirus* Demathieu & Weidmann 1982

Primitive saurischians

First primitive saurischian named by an author
"*Megalosaurus*" *obtusus* Henry 1876, "*Thecodontosaurus*" *primus* Huene 1905, "*Zanclodon*" *silesiacus* Jaekel 1910 (identification in doubt) "*Thecodontosaurus*" *alophos* Haughton 1932

First primitive saurischian named by two authors
Sanjuansaurus gordilloi Alcober & Martínez 2010

First primitive saurischian named by three authors
Guaibasaurus candelariensis Bonaparte, Ferigolo & Ribeiro 1999

First primitive saurischian named by four authors
Eoraptor lunensis Sereno, Forster, Rogers & Monetta 1993

First primitive saurischian named by five authors
Caseosaurus crosbyensis Hunt, Lucas, Heckert, Sullivan & Lockley 1998

Primitive saurischian footprints

First primitive saurischian footprints named by an author
Saurichnium anserinum Gurich 1926

Theropods

First theropod named by an author
Megalosaurus Buckland 1824

First theropod named by two authors

Megalosaurus insignis Eudes-Deslongchamps & Lennier in Lennier 1870
Dromaeosaurus albertensis Matthew & Brown 1922

First theropod named by three authors
Compsognathus corallestris Bidar, Demay & Thomel 1972
Gallimimus bullatus Osmólska, Roniewicz & Barsbold 1972

First theropod named by four authors
Yangchuanosaurus shangyouensis Dong, Chang, Li & Zhou 1978

First theropod named by five authors
Camposaurus arizonensis Hunt, Lucas, Heckert, Sullivan & Lockley 1998

First theropod named by six authors
Genusaurus sisteronis Accaire, Beaudoin, Dejax, Fries, Michard & Taquet 1995

First theropod named by seven authors
Erliansaurus bellamanus Xu, Zhang, Sereno, Zhao, Kuang, Han & Tan 2002

First theropod named by eight authors
Guanlong wucaii Xu, Clark, Forster, Norell, Erickson, Eberth, Jia & Zhao 2006

First theropod named by nine authors
Deltadromeus agilis Sereno, Dutheil, Iarochene, Larsson, Lyon, Magwene, Sidor, Varricchio & Wilson 1996
Anchiornis huxleyi Xu, Zhao, Norell, Sullivan, Hone, Erickson, Wang, Han & Guo 2008

First theropod named by ten authors
Linhenykus monodactylus Xu, Sullivan, Pittman, Choiniere, Hone, Upchurch, Tan Xiao, Tan & Han 2011

First theropod named by eleven authors
Linheraptor exquisitus Xu, Choiniere, Pittman, Tain, Xiao, Li, Tan, Clarke, Norell, Hone & Sullivan 2010

First theropod named by twelve authors
Suchomimus tenerensis Sereno, Beck, Dutheil, Gado, Larsson, Lyon, Marcot, Rauhut, Sadleir, Sidor Varricchio, Wilson & Wilson 1998

First theropod named by fifteen authors
Limusaurus inextricabilis Xu, Clark, Mo, Choiniere, Forster, Erickson, Hone, Sullivan, Eberth, Nesbitt, Zhao, Hernández, Jia, Han & Guo 2009

Theropod footprints

First theropod footprints named by an author
Ornithichnites diversus clarus, *O. diversus platydactylus*, *O. giganteus*, *O. ingens*, *O. minimus*, *O. tetradactylus*, and *O. tuberosus* Hitchcock 1836

First theropod footprints named by two authors
Eubrontes veillonensis, *Grallator olonensis*, *Grallator maximus*, *Grallator variabilis*, *Saltopoides igalensis*, and *Talmontopus tersi* Lapparent & Montenat 1967
Megalosauropus broomensis Colbert & Merrilees 1967

First theropod footprints named by three authors
Grallator limnosus, *Paracoelurosaurichnus monax*, *Schizograllator xiaohebaensis*, *Youngichnus xiyangensis*, and *Zhengichnus jinningensis* Zhen, Li & Rao 1985

First theropod footprints named by four authors
"*Deinonychosaurichnus*" Zhen, Zhen, Chen & Zhu 1987 (invalid name)
Exallopus lovei Harris, Johnson, Hicks & Tauxe 1996

First theropod footprints named by five authors
Velociraptorichnus sichuanensis Zhen, Li, Zhang, Chen & Zhu 1994

First theropod footprints named by six authors
"*Theroplantigrada encisensis*" Casanovas, Ezquerra, Fernández, Pérez-Lorente, Santafé & Torcida 1993 (invalid name)
Paravipus didactyloides Mudroch, Richter, Joger, Kosma, Idle & Maga 2011

First theropod footprints named by seven authors
Ornithomimipus jaillardi Jaillard, Cappetta, Ellenberger, Feist, Grambast-Fessard, Lefranc & Sige 1993

First theropod footprints named by eight authors
Saltosauropus latus Bernier, Barale, Bourseau, Buffetaut, Demathieu, Gaillard, Gall & Wenz 1984 (not a theropod)
Dromaeosauripus youngjingensis Xing, Li, Harris, Bell, Azuma, Fujita, Lee & Currie 2012

Theropod eggs

First theropod egg named by an author
Oolites elongatus Young 1954

First theropod egg named by two authors
Oolithes chiangchiungtigensis, *Oolithes irenensis*, and *Oolithes laminadermus* Chao & Chiang 1974

First theropod egg named by three authors
Continuoolithus canadensis, *Dispersituberoolithus exilis*, *Porituberoolithus warnerensis*, and *Tristraguloolithus cracioides* Zelenitsky, Hills & Currie 1996

First theropod egg named by five authors
Shixingoolithus erbeni and *Stromatoolithus pinglingensis* Zhao, Ye, Li, Zhao & Yan 1991

First theropod egg named by six authors
Dendroolithus fengguangcunensis and *Paraspheroolithus sanwangbacunensis* Fang, Zhang, Zhang, Lu, Han & Li 2005

First theropod egg named by seven authors
Tipoolithus achloujensis García, Tabuce, Cappeta, Marandat, Bentaleb, Benabdallah & Vianey-Liaud 2003

First theropod egg named by nine authors
Macroelongatoolithus xixiaensis Li, An, Zhu, Zhang, Liu, Qu, You, Liang, Li 1995

First theropod egg named by eleven authors
Lanceolithus huangtangensis, *L. xiapingensis*, *Macroolithus lashuyuanensis*, *Pinnatoolithus nanxiongensis*, *P. sangequanensis*, and *Pinnatoolithus shitangensis* Fang, Li, Zhang, Zhang, Zhang, Lin, Guo, Cheng, Li, Zhang & Cheng 2009

First theropod egg named by twelve authors
Dendroolithus furcatus, *D. dendriticus*, *D. sanlimiaoensis*, and *D. zhaoyingensis* Xiaosi, Liwu, Cheng, Zou, Pang, Qang, Chen, Zhen, Wang, Liu, Xie & Jin 1998

Mesozoic birds

First Mesozoic bird named by an author
Archaeopteryx lithographica Meyer 1861

First Mesozoic bird named by two authors
Wyleyia valdensis Harrison & Walker 1973 (identification in doubt)
Horezmavis eocretacea and *Judinornis nogontsovensis* Nesov & Borkin 1983

First Mesozoic bird named by three authors
Pasquiaornis hardiei and *P. tankei* Tokaryk, Cumbaa & Storer 1997

First Mesozoic bird named by four authors
Confuciusornis sanctus Hou, Zhou, Gu & Zhang 1995

First Mesozoic bird named by five authors
Confuciusornis dui Hou, Martin, Zhou, Feduccia & Zhang 1999

First Mesozoic bird named by six authors
Dalingheornis liweii Zhang, Hou, Hasegawa, O'Connor, Martin & Chiappe 2006
Dapingfangornis sentisorhinus Li, Duan, Hu, Wang, Cheng & Hou 2006

First Mesozoic bird named by seven authors
Eoalulavis hoyasi Sanz, Chiappe, Perez-Moreno, Buscalioni, Moratalla, Ortega & Poyato-Ariza 1996

First Mesozoic bird named by eight authors
Shenzhouraptor sinensis Ji, Ji, You, Zhang, Yuan, Ji, Li & Li 2002

First Mesozoic bird named by nine authors
Archaeornithura meemannae Wang, Zheng, O'Connor, Lloyd, Wang, Wang, Zhang & Zhou 2015
Qiliania graffini Ji, Atterholt, O'Connor, Lamanna, Harris, Li, You & Dodson 2011

Mesozoic bird footprints

First Mesozoic bird footprints named by an author
Ignotornis mcconelli Mehl 1931

First Mesozoic bird footprints named by two authors
Yacoraitichnus avis Alonso & Marquillas 1986

First Mesozoic bird footprints named by three authors
Sarjeantopodus semipalmatus Lockley, Nadon & Currie 2003

First Mesozoic bird footprints named by four authors
Aquatilavipes izumiensis Azuma, Akakawa, Tomida & Currie 2002
Barrosopus slobodai Coria, Currie, Eberth & Garrido 2002

First Mesozoic bird footprints named by five authors
"*Aquatilavipes sinensis*" Zhen, Li, Zhang, Chen & Zhu 1987
Jindongornipes kimi Lockley, Yang, Matsukawa, Fleming & Lim 1992

First Mesozoic bird footprints named by six authors
Koreanaornis dodsoni and *Muguiornipes robusta* Xing, Harris, Jia, Luo, Wang & An 2011

First Mesozoic bird footprints named by seven authors
Pullornipes aureus Lockley, Matsukawa, Ohira, Li, Wright, White & Chen 2006

First Mesozoic bird footprints named by eight authors
Paxavipes babcockensis McCrea, Buckley, Plint, Lockley, Matthews, Noble, Xing & Krawetz 2015

First Mesozoic bird footprints named by ten authors
Koreanaornis gansuensis Xing, Buckley, Li, Lockley, Zhang, Marty, Wang, Li, McCrea & Peng 2015

Mesozoic bird eggs

First Mesozoic bird eggs named by an author
Laevisoolithus sochavai and *Subtiliolithus microtuberculatus* Mikhailov 1991

First Mesozoic bird eggs named by two authors
Ageroolithus fontllongensis Vianey-Liaud & López-Martínez 1997

First Mesozoic bird eggs named by three authors
Dispersituberoolithus exilis and *Tristraguloolithus cracioides* Zelenitsky, Hills & Currie 1996

First Mesozoic bird eggs named by five authors
Parvoblongoolithus jinguoensis Zhang, Jin, O'Connor, Wang & Xie 2014

First Mesozoic bird eggs named by seven authors
Tipoolithus achloujensis García, Tabuce, Cappeta, Marandat, Bentaleb, Benabdallah & Vianey-Liaud 2003

SCIENTIFIC ARTICLES ABOUT DINOSAUROMORPHS, THEROPODS, AND MESOZOIC BIRDS

1665—France—The first scientific journal
Journal des sçavans did not contain entirely scientific material. It published from January 5, 1665, until 1792 and reappeared in 1797 under a new name, *Journal des savants*, and became a literary magazine. See more: *Hallam 1842*

1665—England—The first exclusively scientific magazine
Philosophical Transactions of the Royal Society is the scientific bulletin that has been in circulation the longest. It continues to be published even today. It began on March 6, 1665. Its first publication about a theropod was Platt, 1759. See more: *Platt 1758*

1699—England—The first publication with an illustration of a theropod tooth
A theropod tooth was illustrated in Lhuyd, *Lithophylacii Britannici Ichnographia, sive lapidium aliorumque fossilium Britannicorum singulari figure insignium* (London: Gleditsch and Weidmann, 1699).

1759—England—The first publication of post-Linnaean dinosaurs
Platt, 1759, "An Account of the Fossil Thigh Bone of a Huge Animal, Dug Up at Stonesfield near Woodstock in Oxfordshire," *Philosophical Transactions of the Royal Society of London* 50, no. 2, 524-27.

1822—England—The first publication of a Jurassic theropod
Parkinson, 1822, *Outlines of Oryctology: An Introduction to the Study of Fossil Organic Remains, Especially of Those Found in the British Strata; Intended to Aid the Student in His Inquiries Respecting the Nature of Fossils, and Their Connection with the Formation of the Earth* (London: The Author).

1829—England—The first publication of a Cretaceous theropod
Murchison, 1829, "Geological Sketch of the North-western Extremity of Sussex, and the Adjoining Parts of Hants and Surrey," *Transactions of the Geological Society of London*, Second Series 2, 97–105.

1836—USA—The first publication of theropod tracks
Hitchcock, 1836, "Ornithichnology—Description of the Foot Marks of Birds (Ornithichites) on New Red Sandstone in Massachusetts," *American Journal of Science* 29, no. 2, 307–40.

1837—England—The first publication on gastroliths
Riley and Stutchbury, 1837, *On an Additional Species of the Newly Discovered Saurian Animals in the Magnesian Conglomerate of Durham Down, near Bristol,* Report of the Sixth Meeting of the British Association for the Advancement of Science, v. F., 1836 (1837), 94.

1844—France—The first publication of theropod coprolites
Hitchcock, 1844, "Report on Ichnolithology, or Fossil Footmarks, with a Description of Several New Species, and the Coprolites of Birds from the Connecticut River Valley, and a Supposed Footmark for the Valley of the Hudson River," *American Journal of Science*, Second Series 47, 292–322.

1857—Germany—The first publication of a Mesozoic bird
von Meyer, 1857, "Beiträge zur näheren Kenntniss fossiler Reptilien," *Neues Jahrb. Min. Geol. Pal.*, 532–43.

1858—Germany—The first publication of a Triassic theropod
Quenstedt, 1858, *Der Jura* H. (Tübingen: Laupp'schen), 1–842.

1858—USA—The first great work on dinosaur footprints
Ichnology of New England, a 220-page monograph, includes the description and illustration of numerous tracks, including those of many dinosaurs. See more: *Hitchcock 1858*

1861—Germany—The first publication on stomach contents in a theropod
Wagner, 1861, "Neue Beiträge zur Kenntnis der urweltlichen Fauna des lithographischen Schiefers: V. *Compsognathus longipes* Wagner." Abh. Bayer. Akad. Wiss. 9, 30–38.

1901—Germany—The first publication of skin preserved in a theropod
Huene, 1901. "Der vermutliche Hautpanzer des *Compsognathus longipes* WAGN," Neues Jahrbuch für Mineralogie, *Geologie und Paläontologie*, no. 1 (1901), 157–60.

1901—Germany—The first publication on dinosaur eggs
Huene, 1909, "Skizz zu einer Systematik und Stammesgeschichte der Dinosaurier," *Centralblatt für Mineralogy, Geologie und Palaontologie* (Stuttgart: Abhandlungen), 12–22.

1913—USA—The first publication on the possible presence of non-avian dinosaurs in the Cenozoic
Read, 1913, "Recent Discovery of Dinosaurs in the Tertiary," *American Journal of Science,* Fourth Series 25, 531–34.

1913—USA—The first scientific journal dedicated to a paleontologist
Copeia, a quarterly magazine that covered ichthyology and herpetology, was born on December 27. The magazine was published by the American Society of Ichthyologists and Herpetologists. The name was a tribute to Edward Drinker Cope.

1920—USA—The first article about blood in dinosaurs
Moodie, 1920, "Concerning the Fossilization of Blood Corpuscles," *The American Naturalist* 54, 460–64.

1962—USA—The first weight estimate in theropods
Colbert, 1962, "The Weights of Dinosaurs," *American Museum* Novitates, no. 2076.

1965—USA—The first publication on the possible endothermy of dinosaurs
The paleontologist Dale Russell speculated that dinosaurs were "hot-blooded" in 1965, long before the discovery of *Deinonychus* by John Ostrom, which generated many studies. See more: *Russell 1965*

1975—USA—The first book on homeothermy in dinosaurs
Adrian J. Desmond wrote about dinosaur metabolism in *The Hot-Blooded Dinosaurs: A Revolution in Palaeontology*, which suggests that they were very active, just like mammals. See more: *Desmond 1975*

1986—USA—The most-discussed dinosaur science book
The contents of Robert Bakker's book *The Dinosaur Heresies* sparked controversy, but with the passage of time, it was shown that Bakker was right on some questions, helping to modernize the study of dinosaurs. Dinosaurs are currently considered fast and active animals. See more: *Bakker 1968, 1986*

1986—USA—The first book of non-avian theropods covered with feathers
The Dinosaur Heresies, the revolutionary book by Robert Thomas Bakker, was the first to illustrate a *Deinonychus* with feathers. Contrary to prevailing wisdom at that time, it presented observations demonstrating that it was impossible that dinosaurs were ectothermic (cold-blooded). See more: *Bakker 1986*

1988—USA—The longest book on theropods
Predatory Dinosaurs of the World: A Complete Illustrated Guide, 464 pages long, describes non-avian theropods. See more: *Paul 1988*

1989—USA—The dinosaur bibliography with the most citations
A Bibliography of the Dinosauria (Exclusive of the Aves) 1677–1986, by Daniel J. Chure and John S. McIntosh, gathers more than 6,000 citations in seventy-six pages. See more: *Chure & McIntosh 1989*

1990—The longest book on dinosaur footprints
Dinosaur Tracks by Tony Thulborn, is 410 pages long. See more: *Thulborn 1990*

1996—USA—The longest book on dinosaur eggs
Dinosaur Eggs and Babies edited by Kenneth Carpenter, has 392 pages. See more: Carpenter 1996. See more: *Carpenter 1996*

1996—Brazil—First theropod name dedicated to a fictitious person
The name of *Irritator challengeri* is dedicated to Challenger, the irritating professor in Sir Arthur Conan Doyle's novel *The Lost World*. See more: *Conan Doyle 1912; Martill et al. 1996*

1997—USA—The largest dinosaur encyclopedia
Dinosaurs: The Encyclopedia, by Donald F. Glut, began as a 1,088-page book but continued with supplements in 1999 (456 pages), 2001 (686 pages), 2003 (726 pages), 2006 (761 pages), 2007 (798 pages), 2009 (715 pages), and 2012 (876 pages). In sum, the work totals 6,106 pages. See more: *Glut 1997, 1999, 2001, 2003, 2006, 2007, 2009, 2012*

2002—USA—The most extensive collection of pathologies in dinosaurs
In 96 pages in *Dinosores: An Annotated Bibliography of Dinosaur Paleopathology and Related Topics*, 1838–2001, Darren H. Tanke and Bruce M. Rothschild compiled, classified, and analyzed more than 940 citations related to diseases, trauma, mass deaths, extinction, and similar subjects. See more: *Tanke & Rothschild 2002*

2002—USA—The longest book on Mesozoic birds
Mesozoic Birds: Above the Heads of Dinosaurs, by Luis M. Chiappe and Lawrence M. Witmer, is 536 pages long. See more: *Chiappe & Witmer 2002*

2003—China—The first publication of gastroliths in Mesozoic birds
Zhou and Zhang, 2003, "Anatomy of the Primitive Bird *Sapeornis chaoyangensis* from the Early Cretaceous of Liaoning, China." *Canadian Journal of Earth Sciences* 40, 731–47.

2004—USA—The longest article on Mesozoic birds
A review of the genera *Ichthyornis* and *Apatornis* conducted by the paleontologist Julia A. Clarke contains 179 pages and sixty-six figures. *Guildavis* and *Austiornis*, two new genera, and *Laceornis marshi*, a new species, were named. See more: *Clarke 2004*

2004—USA—The dinosaur book with the most extensive biogeographical map
The 89-page section "Dinosaurian Distribution," by David B. Weishampel, in the second edition of the book *The Dinosauria*, edited by Weishampel, Peter Dodson, and Halszka Osmólska, compiles the geographical distribution of dinosaur bones and tracks around the world. See more: *Weishampel et al 2004*

2005—USA—The longest thesis on dinosaur footprints
The 1,316-page-long doctoral thesis by Emma Clare Rainforth reviews the taxonomy of dinosaur footprints from the Lower Jurassic of Connecticut and Massachusetts. See more: *Rainforth 2005*

2005—USA—The longest thesis on dinosaur eggs
The doctoral thesis of Gerald Grellet-Tinner was 234 pages long. He compared the evolution of theropod eggs and birds. See more: *Grellet-Tinner 2005*

2005—USA—The longest theropod thesis
The doctoral thesis of Thomas David Carr, 1,270 pages long, deals with phylogeny in advanced tyrannosauroids. See more: *Carr 2005; Carr & Williamson 2010*

2006—China—The first publication on stomach contents in a Mesozoic bird
Dalsätt, Zhou, Zhang, and Ericson, 2006, "Food Remains in *Confuciusornis sanctus* Suggest a Fish Diet," *Naturwissenschaften* 93 (9), 444–46.

2008—USA—The longest book on a theropod species
Tyrannosaurus rex, the Tyrant King, 456 pages long, is about this species and some of its relatives. The book was edited by Peter Larson and Kenneth Carpenter. See more: *Larson & Carpenter 2008*

2009—USA—The longest thesis on Mesozoic birds
Jingmai Kathleen O'Connor's Ph.D. dissertation is 234 pages long. It is a review of enantiornithean birds. See more: *O'Connor 2009*

2009—Germany—The longest thesis on dinosauromorphs
Regina Fechner's doctoral thesis in natural sciences is 211 pages long. It deals with the biomechanics of the pelvis and legs of dinosauromorphs that evolved into the sauropods. See more: *Fechner 2009*

2011—Italy—The longest formal work on a theropod
The exhaustive analysis of *Scipionyx samniticus* carried out by Cristiano Dal Sasso and Simone Maganuco is 282 pages long and has two supplements of 39 pages each, for a total of 360 pages. See more: *Dal Sasso & Maganuco 2011*

2012—USA—The first biomechanical study on theropod jumping
The jumping capacity of dromaeosaurs was reviewed, along with the role played by their rigid tails in jumping. See more: *Alexander 2012*

2012—USA—The longest dinosaur book
The second edition of *The Complete Dinosaur*, edited by M.K. Brett-Surman, Thomas R. Holtz Jr., and James O. Farlow, is 1,128 pages long. See more: *Brett-Surman, Holtz Jr. & Farlow 2012*

2013—England—The longest dinosauromorph article
Max C. Langer and Jorge Feruglio's review of *Sacisaurus agudoensis* was 39 pages long. See more: *Langer & Ferigolo 2013*

BIBLIOGRAPHY OF DINOSAURS FROM A TO Z

Paleontologist names from A to Z

The alphabetical list of dinosaur paleontology citations would begin and end with:

Abbassi, Nasrollah. Iranian paleontologist (1970)
Zittel, Karl Alfred. German paleontologist (1839-1904)

Authors of publications on dinosaurs from A to Z

If an alphabetical list of authors who have written formal articles on dinosaurs were made, it would begin and end with these two citations:

Abbey 1977. "Petrified dinosaur blood." *Lapidary Jour.* 31 (8):1858.

Zug, Vitt & Caldwell 2001. *Herpetology: An Introductory Biology of Amphibians and Reptiles.* 2nd ed. (San Diego: Academic Press), 630 pages.

Authors of publications on dinosaur footprints from A to Z

Abbassi 2006. "New Early Jurassic dinosaur footprints from Shemshak Formation, Harzavil Village, western Alborz, north Iran." *J. Geosci.* 62, 31-40.

Zhen, Li, Zhang, Chen & Zhu 1987. "Bird and Dinosaur Footprints from the Lower Cretaceous of Emei County, Sichuan:

Abstracts of the First International Symposium on Nonmarine Cretaceous Correlation," *IGCVP Project* 245, 37–38.

Authors of publications on dinosaur eggs from A to Z

Amo-Sanjuan, Canudo & Cuenca-Bescos 2000. "First Record of Elongatoolithid Eggshells from the Lower Barremian (Lower Cretaceous) of Europe (Cuesta Corrales 2, Galve Basin, Teruel, Spain)" *in First International Symposium on Dinosaur Eggs and Babies, Extended Abstracts*, edited by A. M. Bravo and T. Reyes, 7–14.

Zou, Wang & Wang 2013. "A New Oospecies of Parafaveoloolithids from the Pingxiang Basin, Jiangxi Province, China," *Vertebrata PalAsiatica* 51, no. 2, 102–6.

Authors of publications on dinosauromorphs from A to Z

Arcucci 1986. "New Materials and Reinterpretation of *Lagerpeton chanarensis* Romer (Thecodontia, Lagerpetonidae nov.) from the Middle Triassic of La Rioja, Argentina," *Ameghiniana* 23, 233–42.

Small 2009. "A Late Triassic Dinosauromorph Assemblage from the Eagle Basin (Chinle Formation), Colorado, U.S.A.," *Journal of Vertebrate Paleontology* 29, 182A.

Authors of publications on theropods from A to Z

Abel 1911. "Die Vorfahren der Vögel und ihre Lebensweise" *Verhandlungen der Zoologischen-botanischen Gesellschaft* (Vienna) 61, 144–91.

Zinke & Rauhut 1994. "Small Theropods (Dinosauria, Saurischia) from the Upper Jurassic and Lower Cretaceous of the Iberian Peninsula," *Berl. Geowiss. Abhandl.* 13, 163–77.

Authors of publications on primitive saurischians from A to Z

Adams 1875. "On a Fossil Saurian Vertebra (*Arctosaurus osborni*) from the Arctic Regions," *Proceedings of the Royal Irish Academy* 2, no. 2, 177–79.

Yates 2007. "Solving a Dinosaurian Puzzle: The Identity of *Aliwalia rex* Galton," *Historical Biology* 19, no. 1, 93–123.

Authors of publications on Mesozoic birds from A to Z

Abbott 1992. "Archeopteryx Fossil Disappears from Private Collection," *Nature* 357, 6.

Zhou, Zhou & O'Connor 2014. "A New Piscivorous Ornithuromorph from the Jehol Biota," *Historical Biology* 26, no. 5, 608–18.

Author of the publication where a primitive avemetatarsalian was named

Woodward 1907. "On a New Dinosaurian Reptile (*Scleromochlus taylori*, gen. et sp. nov.) from the Trias of Lossiemouth, Elgin," *Quarterly Journal of the Geological Society* 63, 140–n.p.

Authors of publications where a dinosauromorph has been named from A to Z

Arcucci 1987. "A New Lagosuchidae (Thecodontia-Pseudosuchia) of Los Chañares Fauna (Reptilian Chañarense, Middle Triassic), La Rioja, Argentina" *Ameghiniana* 24, 89–94.

Woodward 1907. "On a New Dinosaurian Reptile (*Scleromochlus taylori*, gen. et sp., nov.) from the Trias of Lossiemouth, Elgin" *Quarterly Journal of the Geological Society* 63, 140–n.p.

Authors of publications where a primitive saurischian has been named from A to Z

Adams 1875. "On a Fossil Saurian Vertebra (*Arctosaurus osborni*) from the Arctic Regions," *Proceedings of the Royal Irish Academy* 2, no. 2, 177–79.

Reig 1963. "The Presence of Saurischian Dinosaurs in the 'Estratos de Ischigualasto' [Upper Mesotriassic] of the Provinces of San Juan and La Rioja [Argentine Republic]" *Ameghiniana* 3, 3–20.

Authors of publications where a theropod has been named from A to Z

Accarie, Beaudoin, Dejax, Fries, Michard & Taquet 1995. "De couverte d'un Dinosaure Theropode nouveau (*Genusaurus sisteronis* ng, sp.) dans l'Albien marin de Sisteron (Alpes de Haute-Provence, France) et extension au Cretace inferieur de la lignee ceratosaurienne," *Compte rendu hebdomadaire des seances de l'Académie des Sciences Paris, Second Series* 320, 327–34.

Zhou & Wang 2000. "A New Species of *Caudipteryx* from the Yixian Formation of Liaoning, Northeast China," *Vertebrata PalAsiatica* 38, no. 2, 111–27.

Authors of publications where a Mesozoic bird has been named from A to Z

Agnolin & Martinelli 2009. "Fossil Birds from the Late Cretaceous Los Alamitos Formation, Río Negro Province, Argentina," *Journal of South American Earth Sciences* 27, 42–49.

Zhou, Zhou & O'Connor 2014. "A New Piscivorous Ornithuromorph from the Jehol Biota," *Historical Biology* 26, no. 5, 608–18.

According to the International Code of Zoological Nomenclature, the alphabetical list of authors who have created scientific names of any organism begins and ends with:

Abe—Tokiharu Abe (1911–96)
Zur Strassen—Otto Zur Strassen (1869–1961)
See more: http://en.wikipedia.org/wiki/List_of_authors_of_names_published_under_the_ICZN

It is almost impossible to prepare a complete list of dinosaur species. Although these have been created in formal works, academic theses, conferences, low-circulation magazines, and other publications, they are difficult to obtain.

List of Theropods

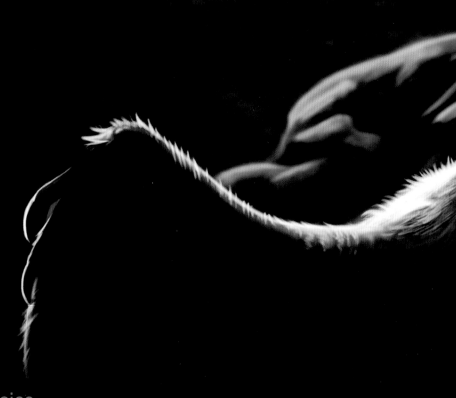

Records: Largest specimens of each species

Today there are more than 10,000 species of birds, but there may have been many fewer species of theropods in the Mesozoic Era. There are currently around 700 accepted species of non-avian theropods and Mesozoic birds combined, as well as some doubtful and some invalid species. While their number is growing each year, some species now recognized as distinct could be identified as the same in future.

Complete list of valid dinosauromorphs and theropods

The graphic tables that follow present the different species of dinosauromorphs, primitive saurischians, theropods and Mesozoic birds named as of December 31, 2015. These are divided into different phylogenetic groups or assigned to the closest one possible.

Estimations of body mass and length are calculated using one or more volumetric models, extrapolating from very similar specimens to the most incomplete. As many of the bone remains are far from complete (e.g., single teeth), in order to avoid overestimating the lowest estimate is used in all cases.

Bibliographic citations do not always refer to the author who created the genus; they sometimes refer to the author who described the largest (or record) specimen of a species.

Names in quotation marks and italics are those incorrectly assigned to valid names.

Names in quotation marks and without italics are invalid names.

The following abbreviations are used in this book

LT = Lower Triassic
MT = Middle Triassic
UT = Upper Triassic
LJ = Lower Jurassic
MJ = Middle Jurassic
UJ = Upper Jurassic
eLC = Early Lower Cretaceous
lLC = Late Lower Cretaceous
eUC = Early Upper Cretaceous
lUC = Late Upper Cretaceous

WNA = western North America
ENA = eastern North America
WE = western Europe
EE = eastern Europe
WA = western Asia
EA = eastern Asia
CIM = Cimmeria
ME = Middle East
IND = Hindustan
NSA = northern South America
SSA = Southern South America
NAf = North Africa
SAf = Southern Africa
OC = Oceania
AN = Antarctica

(n.d.) = Doubtful name
(n.n.) = Invalid name

Note: Lengths reported here do not include feathers and represent the total length of an animal measured along the vertebral centra.

Sample graphics used in the tables:

All dinosauromorphs and theropods are represented by silhouettes scaled down as indicated in the upper right of the table. Each bar at the top is equal to a meter in length. Human beings measure 1.8 meters in height.

AVEMETATARSALIA

Avemetatarsalians similar to *Scleromochlus* - Jumpers, short bodies, small game hunters.

N.º	Genus and species	Largest specimen	Material and data	Size	Mass
1	*Scleromochlus taylori* UT, WE (Scotland)	BMNH R3556 Undetermined age	Skull and partial skeleton - *Woodward 1907* There are seven specimens of similar size. *Benton 1999*	Length: **18.5 cm** Hip height: **7 cm**	18 g

1:10

Dinosauromorphs similar to *Lagerpeton* - Jumpers, small game hunters.

N.º	Genus and species	Largest specimen	Material and data	Size	Mass
1	*Lagerpeton chanarensis* UT, SSA (Argentina)	UPLR 06 Undetermined age	Partial skeleton - *Sereno & Arcucci 1993* Previously known as MLP 64-XI-14-10	Length: **52 cm** Hip height: **20 cm**	290 g
2	"Ischigualasto form" UT, SSA (Argentina)	PVSJ 883 Undetermined age	Incomplete femur - *Martínez et al. 2013* The complete femur could measure 109 to 139 mm long.	Length: **73 cm** Hip height: **28 cm**	800 g
3	*Dromomeron gregorii* UT, WNA (Arizona & Texas - USA)	GR 239 Adult	Tibia - *Nesbitt et al. 2009* Some specimens vary in size.	Length: **77 cm** Hip height: **30 cm**	950 g
4	*Dromomeron romeri* UT, WNA (New Mexico - USA)	GR 238 Adult	Femur - *Irmis et al. 2007* First North American lagerpetid	Length: **85 cm** Hip height: **33 cm**	1.2 kg

Indeterminate dinosauromorphs - Bipeds, ectaxonic feet, small game hunters or omnivores.

N.º	Genus and species	Largest specimen	Material and data	Size	Mass
5	*Agnosphitys cromhallensis* UT, WE (England)	VMNH 1745 Juvenile?	Ilium - *Fraser et al 2002* Other assigned pieces may have belonged to basal dinosaurs. *Langer et al. 2013*	Length: **34 cm** Hip height: **11 cm**	50 g
6	*Lagosuchus talampayensis* (n.d.) MT, SSA (Argentina)	UPLR 09 Juvenile	Partial skeleton - *Romer 1971* Valid or chimera? *Sereno & Arcucci 1994*	Length: **40 cm** Hip height: **13 cm**	80 g
7	*Marasuchus lilloensis* MT, SSA (Argentina)	PVL 4671 Undetermined age	Partial skeleton - *Romer 1972* The Chañares Formation dates from the Upper Triassic. *Marsicano et al. 2015*	Length: **65 cm** Hip height: **21 cm**	225 g
8	*Lewisuchus admixtus* MT, SSA (Argentina)	PULR 01 (UNLR 01) Undetermined age	Partial skeleton - *Romer 1972, Bittencourt et al. 2014* Portions of the original material belong to *Chanaerosuchus*. Primitive silesaurid?	Length: **1.1 m** Hip height: **34 cm**	1.4 kg
9	*Pseudolagosuchus major* (n.d.) MT, SSA (Argentina)	PVL 13454 Undetermined age	Partial skeleton - *Arcucci 1987* *Lewisuchus admixtus*? *Nesbitt et al. 2010*	Length: **1.4 m** Hip height: **42 cm**	2.7 kg

Dinosauromorphs similar to *Saltopus* - Mesaxonic feet, small game hunters or omnivores.

N.º	Genus and species	Largest specimen	Material and data	Size	Mass
10	*Saltopus elginensis* UT, WE (Scotland)	NHMUK R3915 Undetermined age	Partial skeleton - *Huene 1910* Basal silesaurid? *Benton & Walker 2000*	Length: **50 cm** Hip height: **15 cm**	110 g
11	*Avipes dillstedtianus* (n.d.) MT, WE (Germany)	Uncatalogued Undetermined age	Incomplete metatarsus- *Huene 1932* Dinosauriform or dinosaur? *Rauhut & Hungerbuhler 2000*	Length: **90 cm** Hip height: **27 cm**	630 g

1:50

Dinosauromorphs similar to *Silesaurus* - Bipeds, acultative quadrupeds, large stomach, phytophagous.

N.º	Genus and species	Largest specimen	Material and data	Size	Mass
1	*Ignotosaurus fragilis* UT, SSA (Argentina)	PVSJ 884 Undetermined age	Ilium - *Martínez et al. 2013* First known South American silesaurid.	Length: **93 cm** Hip height: **30 cm**	2.5 kg
2	*Sacisaurus agudoensis* UT, SSA (southern Brazil)	MCN PV10021 Undetermined age	Tibia - *Langer & Ferigolo 2013* Considered ornithischia by Ferigolo & Langer 2006	Length: **1.15 m** Hip height: **37 cm**	3.15 kg
3	*Lutungutali sitwensis* MT, SAf (Zambia)	NHCC LB32 Undetermined age	Partial pelvis and caudal verterbra - *Peecook et al. 2013* Was known as "N'tawere form."	Length: **1.4 m** Hip height: **43 cm**	8.7 kg
4	*Eucoelophysis* sp. UT, WNA (New Mexico - USA)	TMP84-63-33 Undetermined age	Partial skeleton - *Rinehart et al. 2009* More recent than *Eucoelophysis baldwini*.	Length: **1.55 cm** Hip height: **55 cm**	11.5 kg
5	*Technosaurus smalli* UT, WNA (Texas - USA)	TTU-P 11127 Undetermined age	Tibia - *Nesbitt & Chatterjee 2008* This piece may belong to *Technosaurus*.	Length: **1.65 cm** Hip height: **53 cm**	13 kg
6	*Silesaurus opolensis* UT, EE (Poland)	ZPAL Ab III 361/27 Adult	Incomplete femur - *Piechowski et al. 2015* Females were larger.	Length: **1.7 m** Hip height: **54 cm**	15 kg
7	*Diodorus scytobrachion* UT, NAf (Morocco)	MHNM-ARG 30 Undetermined age	Partial jaw and teeth - *Kammerer et al. 2012* Distinguished by its forward-leaning teeth.	Length: **1.75 cm** Hip height: **55 cm**	15.5 kg
8	*Eucoelophysis baldwini* UT, WNA (New Mexico - USA)	NMMNH P-31293 Adult?	Tibia - *Heckert et al. 2003* Was considered a dinosaur. *Ezcurra 2006*	Length: **1.8 m** Hip height: **57 cm**	17.7 kg
9	*Asilisaurus kongwe* MT, SAf (Tanzania, USA)	NMT RB Adult?	Humerus, femur, and tibia - *Nesbitt et al. 2010* Several mixed specimens.	Length: **1.9 m** Hip height: **60 cm**	19 kg

Indeterminate dinosauriforms or dinosaurs - Bipeds, robust bodies, probably phytophagous.

N.º	Genus and species	Largest specimen	Material and data	Size	Mass
10	*Teyuwasu barberenai* UT, SSA (southern Brazil)	JVP 16:728 Undetermined age	Partial skeleton - *Huene 1938, Kischlat 1999* It is similar to silesaurids and was very massive. *Ezcurra 2012*	Length: **2.3 m** Hip height: **70 cm**	75 kg

1:100

DINOSAURIA (primitive saurischians)

Indeterminate primitive saurischians - Light bodies, small or medium game hunters.

N.º	Genus and species	Largest specimen	Material and data	Size	Mass
1	*"Thecodontosaurus" alophos* MT, SAf (Tanzania)	SAM-PKK10654 Undetermined age	Cervical and thoracic vertebrae - *Haughton 1932* *Nyasasaurus parrigtoni? Nesbitt et al. 2013*	Length: 2.45 m Hip height: 80 cm	19 kg
2	*Nyasasaurus parrington* MT, SAf (Tanzania)	NHMUK R6856 Undetermined age	Incomplete humerus and vertebrae - *Nesbitt, et al 2013* Described in a thesis forty-six years prior. *Charig 1967*	Length: 2.6 m Hip height: 85 cm	23 kg
3	*"Megalosaurus" obtusus* (n.d.) UT, WE (France)	Uncatalogued Undetermined age	Tooth - *Huene 1942* Dinosaur and rauisuchid mix? *Weishampel et al. 2004*	Length: 2.65 m Hip height: 87 cm	25 kg
4	cf. *Spondylosoma absconditum* (n.d.) MT, SSA (southern Brazil)	GPIT 479/30/11 Undetermined age	Incomplete humerus - *Huene 1942* Dinosaur and rauisuchid mix? *Langer 2004*	Length: 2.9 m Hip height: 95 cm	33 kg
5	*"Thecodontosaurus" primus* (n.d.) MT, EE (Poland)	Uncatalogued Undetermined age	Vertebra - *Huene 1905* Dinosaur or *Tanystropheus antiquus?*	Length: 3.2 m Hip height: 1.05 m	45 kg
6	*"Zanclodon" silesiacus* (n.d.) MT, EE (Poland)	Uncatalogued Undetermined age	Tooth - *Jaekel 1910* Rauisuchid or theropod? *Carrano et al. 2012*	Length: 3.9 m Hip height: 1.3 m	80 kg

Primitive saurischians similar to *Staurikosaurus* - Light bodies, small or medium game hunters.

N.º	Genus and species	Largest specimen	Material and data	Size	Mass
7	*Staurikosaurus pricei* UT, SSA (southern Brazil)	MCZ 1669 Undetermined age	Partial jaw and skeleton - *Colbert 1970* First dinosaur named in Brazil.	Length: 2.1 m Hip height: 60 cm	13 kg
8	*Sanjuansaurus gordilloi* UT, SSA (Argentina)	Uncatalogued Undetermined age	Partial skeleton - *Alcober & Martínez 2010* Contemporary of *Herrerasaurus ischigualastensis*	Length: 3.6 m Hip height: 1 m	63 kg

Primitive saurischians similar to *Herrerasaurus* - Robust bodies, small to big game hunters.

N.º	Genus and species	Largest specimen	Material and data	Size	Mass
9	*Ischisaurus cattoi* (n.d.) MT, SSA (Argentina)	MLP 61-VIII-2-3 Juvenile?	Incomplete femur - *Reig 1963* *Herrerasaurus* juvenile? *Romer 1966, Cooper 1981*	Length: 2.8 m Hip height: 80 cm	50 kg
10	*Herrerasaurus ischigualastensis* UT, SSA (Argentina)	MCZ 7064 Adult	Partial skeleton - *Brinkman & Sues 1987* *Frenguellisaurus* is likely an adult *Herrerasaurus ischigualastensis*.	Length: 3.8 m Hip height: 1.25 m	190 kg
11	*Frenguellisaurus ischigualastensis* UT, SSA (Argentina)	PVSJ 53 Adult	Skull, incomplete bones, and vertebrae - *Novas 1986* *Frenguellisaurus* might be an *Herrerasaurus*, but it is more recent. *Sereno & Novas 1992*	Length: 5.3 m Hip height: 1.55 m	360 kg

Primitive saurischians similar to *Chindesaurus* - Light, small to medium game hunters.

N.º	Genus and species	Largest specimen	Material and data	Size	Mass
12	*Caseosaurus crosbyensis* (n.d.) UT, WNA (Texas - USA)	UMMP 8870 Undetermined age	Ilium - *Case 1927* *Chindesaurus bryansmalli? Langer 2004*	Length: 2.4 m Hip height: 70 cm	19 kg
13	*Chindesaurus bryansmalli* UT, WNA (Arizona, New Mexico - USA)	PEFO 10395 Undetermined age	Partial skeleton - *Long & Murry 1995* More derived than herrerasaurids. *Yates 2006*	Length: 2.4 m Hip height: 70 cm	19 kg

Primitive saurischians similar to *Guaibasaurus* - Light bodies, small game hunters or possible omnivores.

N.º	Genus and species	Largest specimen	Material and data	Size	Mass
14	*Guaibasaurus candelariensis* UT, SSA (southern Brazil)	MCN-PV Undetermined age	Metacarpus - *Bonaparte et al. 2007* Its ungual phalanges were not pointed.	Length: 3 m Hip height: 90 cm	35 kg

Primitive saurischians similar to *Eoraptor* - Robust bodies, small game hunters or omnivores.

N.º	Genus and species	Largest specimen	Material and data	Size	Mass
15	*Alwalkeria maleriensis* UT, IND (India)	ISI R 306 Undetermined age	Femur and vertebrae - *Chatterjee 1987* Other remains did not belong to it. *Remes & Rauhut 2005*	Length: 1.1 m Hip height: 38 cm	2 kg
16	*Eoraptor lunensis* UT, SSA (Argentina)	PVSJ 559 Adult	Partial skull and skeleton - *Sereno et al. 2013* Sauropodomorph or basal saurischian? *Martinez et al. 2011; Otero et al. 2015*	Length: 1.5 m Hip height: 50 cm	5 kg

1:150

THEROPODA

Theropods similar to *Eodromaeus* - Light bodies, small game hunters.

N.º	Genus and species	Largest specimen	Material and data	Size	Mass
1	cf. *Agrosaurus macgillivrayi* (n.d.) UT, WE (England)	BMNH 49984e Undetermined age	Tooth - *Seeley 1891* Tooth. *Galton 2007*	Length: 1.25 m Hip height: 37 cm	2.8 kg
2	*Daemonosaurus chauliodus* UT, WNA (New Mexico - USA)	CM 76821 Subadult	Skull and jaw - *Sues et al. 2011* Very large teeth for its size.	Length: 1.6 m Hip height: 47 cm	4.1 kg
3	*Eodromaeus murphi* UT, SSA (Argentina)	PVSJ 561 Undetermined age	Partial skeleton - *Martinez et al. 2011* More robust head than *Daemonosaurus* and *Tawa*.	Length: 1.6 m Hip height: 48 cm	4.2 kg
4	*Tawa hallae* UT, WNA (New Mexico - USA)	GR 155 Undetermined age	Incomplete pelvis - *Nesbitt et al. 2009* Coelophysoid skull and basal saurischian pelvis.	Length: 1.8 m Hip height: 54 cm	6.2 kg
5	*Arctosaurus osborni* (n.d.) UT, WNA (western Canada)	NMI 62 1971 Undetermined age	Incomplete cervical vertebrae - *Adam 1875* Theropod or arcosauromorph? *Nesbitt et. al 2007*	Length: 1.95 m Hip height: 58 cm	7.7 kg
6	*Halticosaurus longotarsus* UT, WE (Germany)	SMNS 12353 Undetermined age	Incomplete skull and skeleton - *Huene 1908* The vertebrae are short and primitive. *Rauhut & Hungerbuhler 2000*	Length: 2.7 m Hip height: 80 cm	21 kg

Theropods similar to *Procompsognathus* - Light bodies, fused pelvis, small game hunters.

#	Species / Locality	Specimen / Age	Material / Notes	Size	Mass
7	Protoavis texensis (n.n.) UT, WNA (Texas - USA)	TTU P 9200 Juvenile	Incomplete femur - *Chatterjee, 1991* It was a mix of several animals, including a coelophysoid. *Hutchinson 2001, Nesbitt et al. 2005*	Length: 58 cm Hip height: 17 cm	280 g
8	cf. Megapnosaurus LJ, WNA (Mexico)	IGM 6624 adult	Partial skeleton - *Munter 1999; Hernández 2002* Assigned to cf. *Megapnosaurus*. *Gudiño-Maussán & Guzmán 2014*	Length: 91 m Hip height: 26 cm	1.1 kg
9	Procompsognathus triassicus UT, WE (Germany)	SMNS 12591 Undetermined age	Partial skull and skeleton - *Fraas 1912, Fraas 1913* Its head was shorter than that of other coelophysoids. *Sereno & Wild 1992*	Length: 1 m Hip height: 28 cm	1.3 kg
10	Panguraptor lufengensis LJ, EA (China)	LFGT-0103 Adult	Cervical vertebrae I - *You, Azuma, Wang, Wang & Dong 2014* The skull might not have belonged to it.	Length: 1.65 m Hip height: 47 cm	6.4 kg
11	Pterospondylus trielbae (n.d.) UT, WE (Germany)	Uncatalogued Undetermined age	Partial skeleton - *Jaekel 1913* Valid genus? *Rauhut & Hungerbuhler 2000*	Length: 1.8 m Hip height: 50 cm	8.7 kg
12	Megapnosaurus rhodesiensis LJ, SAf (South Africa, Zimbabwe)	QG 1 Adult	Partial skeleton - *Raath, 1969* Its original name was changed in an entomological magazine. *Ivie et al. 2001*	Length: 2.1 m Hip height: 58 cm	13 kg
13	Podokesaurus holyokensis LJ, ENA (Massachusetts - USA)	BSNH 13656 Undetermined age	Ribs, pubis, and tibia - *Colbert & Baird 1958* The holotype specimen was destroyed in a fire. A tooth exists.	Length: 2.25 m Hip height: 63 cm	16 kg
14	Lepidus praecisio UT, WNA (Texas - USA)	TMM 41936-1.1 Adult	Maxilla with teeth - *Sterling et al. 2015* The name means "fascinating fragments."	Length: 3.15 m Hip height: 90 cm	44 kg

Theropods similar to *Coelophysis* - Larger and heavier bodies than primitive coelophysinae, fused pelvis, small game hunters.

#	Species / Locality	Specimen / Age	Material / Notes	Size	Mass
15	"Tanystropheus" willistoni (n.d.) UT, WNA (New Mexico - USA)	AMNH 2726 Undetermined age	Incomplete ilium - *Padian 1986* "Longosaurus" longicollis synonym?	Length: 2.6 m Hip height: 62 cm	23 kg
16	Coelophysis bauri UT, WNA (New Mexico - USA)	UCMP 129618 Adult	Partial skeleton - *Padian,1986* NMMNH P-54620 is another of similar size.	Length: 2.75 m Hip height: 70 cm	32 kg
17	Longosaurus longicollis (n.d.) UT, WNA (New Mexico - USA)	AMNH 2717 Undetermined age	Partial skeleton - *Welles 1984* More recent than *Coelophysis bauri*.	Length: 2.9 m Hip height: 75 cm	38 kg

Theropods similar to *Segisaurus* - Relatively strong teeth, light bodies, fused pelvis, small or large game hunters.

#	Species / Locality	Specimen / Age	Material / Notes	Size	Mass
18	Segisaurus halli LJ, WNA (New Mexico - USA)	UCMP 32101 Subadult	Partial skeleton - *Camp 1936* First theropod with a conserved clavicle.	Length: 1.55 m Hip height: 40 cm	4.4 kg
19	Kayentavenator elysiae (n.d.) LJ, WNA (Arizona - USA)	UCMP V128659 Juvenile	Partial skeleton - *Gay 2010* Is it "Syntarsus" kayentakatae? *Ezcurra 2012*	Length: 1.9 m Hip height: 48 cm	8.4 kg
20	Camposaurus arizonensis UT, WNA (Arizona - USA)	UCMP 34498 Undetermined age	Incomplete tibia and fibula, talocalcaneal - *Hunt et al 1998* Different to *Coelophysis*. *Nesbitt et al. 2007*	Length: 2.1 m Hip height: 52 cm	11 kg
21	"Syntarsus" kayentakatae LJ, WNA (Arizona - USA)	MNA V2623 Adult	Partial skeleton - *Rowe 1989* Has tougher teeth than its relatives.	Length: 3 m Hip height: 73 cm	30 kg

Theropods similar to *Liliensternus* - Light bodies, small game hunters.

#	Species / Locality	Specimen / Age	Material / Notes	Size	Mass
22	"Megalosaurus" cloacinus (n.d.) UT, WE (Germany)	SMNS 52457 Undetermined age	Teeth - *Quenstedt 1858* Probably an indeterminate theropod. *Carrano et al. 2012*	Length: 2.4 m Hip height: 63 cm	13.5 kg
23	"Comanchesaurus kuesi" (n.n.) UT, WNA (New Mexico - USA)	NMMNH P4569 Undetermined age	Thoracic vertebrae - *Nesbitt, Irmis & William 2007* Thesis name, probably a coelophysoid. *Mortimer**	Length: 3.55 m Hip height: 90 cm	44 kg
24	Velocipes guerichi (n.n.) UT, EE (Poland)	BMNH R38058 Undetermined age	Fibula - *Huene 1932; Rauhut & Hungerbuhler 2000* Indeterminate theropod *Czepinski et al. 2014*	Length: 3.6 m Hip height: 95 cm	53 kg
25	Dolichosuchus cristatus (n.d.) UT, WE (Germany)	BMNH R38058 Undetermined age	Tibia and metatarsus - *Huene 1932* Synonym of *Liliensternus*? *Welles 1984*	Length: 4 m Hip height: 1 m	60 kg
26	Zupaysaurus rougieri UT, SSA (Argentina)	PULR-076 Undetermined age	Partial skull and skeleton - *Arcucci & Coria 1997* Lacked crests on the skull. *Ezcurra 2007*	Length: 4.2 m Hip height: 1.05 m	70 kg
27	Liliensternus liliensterni UT, WE (Germany)	MB.R.2175 Undetermined age	Partial skull and skeleton - *Huene 1934* The holotype is composed of two specimen. *Rauhut & Hungerbuhler 2000*	Length: 4.8 m Hip height: 1.25 m	110 kg
28	Lophostropheus airelensis UT, WE (France)	Caen University coll. Undetermined age	Teeth and partial skeleton - *Cuny & Galton 1993* May have been from the Lower Jurassic.	Length: 5.2 m Hip height: 1.3 m	136 kg
29	Gojirasaurus quayi UT, WNA (Texas - USA)	UCM 47221 Undetermined age	Pubis and tibia - *Parrish & Carpenter 1986* May have been a basal coelophysoid. *Ezcurra, 2012*	Length: 5.6 m Hip height: 1.45 m	177 kg

1:150

253

Theropods similar to *Dilophosaurus* - Light bodies, small to medium game hunters or piscivores.

N.º	Genus and species	Largest specimen	Material and data	Size	Mass
1	Dracovenator regenti LJ, SAf (South Africa)	BMNH R4840/1 Undetermined age	Incomplete skull - *Yates 2005* There is a juvenile specimen. *Munyikwa & Raath 1999*	Length: 5 m Hip height: 1.25 m	185 kg
2	Dilophosaurus wetherilli LJ, WNA (Arizona - USA)	UCMP 77270 Adult	Partial skull and skeleton - *Welles 1995; vide Welles and Pickering 1995* It had long and narrow teeth.	Length: 6.3 m Hip height: 1.55 m	350 kg

Theropods similar to *Sinosaurus* - Relatively robust bodies, small or medium game hunters.

N.º	Genus and species	Largest specimen	Material and data	Size	Mass
3	"Megalosaurus" woodwardi (n.d.) UT, WE (England)	BMNH 41352 Undetermined age	Teeth - *Lydekker 1909* Tetanuran? *Carrano et al. 2012*	Length: 1.75 m Hip height: 50 cm	11 kg
4	Sinosaurus triassicus LJ, EA (China)	IVPP V48 Undetermined age	Tooth - *Ösi & Mártón 2006* D. sinensis synonym? *Dong 2003*	Length: 5 m Hip height: 1.4 m	255 kg
5	"Dilophosaurus" sinensis (n.d.) LJ, EA (China)	LDM-LCA 10 Undetermined age	Partial skull and skeleton - *Dong 2003* More recent than *Sinosaurus triassicus*.	Length: 5.5 m Hip height: 1.55	350 kg
6	Cryolophosaurus ellioti LJ, AN (Antarctica)	FMNH PR1821 Undetermined age	Partial skull and skeleton - *Hammer & Hickerson 1994* Basal tetanuran? *Carrano et al. 2012*	Length: 7.7 m Hip height: 2 m	780 kg
7	"Sinosaurus shawanensis" (n.n.) LJ, EA (China)	IVPP V31 Undetermined age	Throacic vertebra - *Young 1948* More recent than *Sinosaurus triassicus*. Synonym of *Dilophosaurus sinensis*?	Length: 9.2 m Hip height: 2.55 m	1.7 t

Theropods similar to *Monolophosaurus* - Light bodies, small to medium game hunters.

N.º	Genus and species	Largest specimen	Material and data	Size	Mass
8	Chuandongocoelurus primitivus MJ, EA (China)	CCG 20010 Juvenile	Partial skeleton - *He 1984* Considered an adult. *Norman 1990; Benson et al. 2014*	Length: 2 m Hip height: 50 cm	14 kg
9	Monolophosaurus jiangi MJ, EA (China)	IVPP 84019 Undetermined age	Partial skull and skeleton - *Zhao & Currie 1994* Its likeness to the tyrannosauroid *Guanlong* is convergent. *Carrano et al. 2012*	Length: 7.5 m Hip height: 1.85 m	710 kg

Theropods similar to *Chilesaurus* - Light bodies, herbivores.

N.º	Genus and species	Largest specimen	Material and data	Size	Mass
10	Chilesaurus diegosuarezi UJ, SSA (Chile)	SNGM Undetermined age	Astragalus - *Salgado et al. 2008; Novas et al. 2015* The phylogenetic position is in doubt. *Mortimer**	Length: 3.5 m Hip height: 82 cm	100 kg

1:250

AVEROSTRA

Averostra similar to *Tachiraptor* - Light bodies, small game hunters.

N.º	Genus and species	Largest specimen	Material and data	Size	Mass
1	Tachiraptor admirabilis LJ, NSA (Venezuela)	IVIC-P-2867 Undetermined age	Incomplete ischium - *Langer et al. 2014* Related to the basal trunk of Ceratosauria.	Length: 3.5 m Hip height: 90 cm	42 kg
2	"Newtonsaurus" cambrensis UT, WE (England)	SMNH 52457 Undetermined age	Jaw with teeth - *Newton 1899* Primitive Ceratosauria? *Carrano et al. 2012*	Length: 5.1 m Hip height: 1.3 m	130 kg

Averostra similar to *Elaphrosaurus* - Light bodies, small teeth, small game hunters.

N.º	Genus and species	Largest specimen	Material and data	Size	Mass
6	Walgettosuchus woodwardi (n.d.) ILC, OC (Australia)	BMNH R3717 Undetermined age	Incomplete caudal verterbra - *Powell 1979* Indeterminate? *Agnolin et al. 2010*	Length: 3.9 m Hip height: 1.15 m	95 kg
7	Camarillasaurus cirugedae ILC, WE (Spain)	MPG-KPC1-46 Adult	Partial skeleton and teeth - *Sanchez-Hernandez & Benton 2014* The most derived.	Length: 5.5 m Hip height: 1.6 m	270 kg
3	"Elaphrosaurus philtippettensis" (n.n.) UJ, WNA (Colorado - USA)	USNM 8415 Undetermined age	Humerus - *Galton 1982* Informally named. *Pickering 1995*	Length: 4.6 Hip height: 1.3 m	100 kg
4	Elaphrosaurus bambergi UJ, SAf (Tanzania)	HMN Gr.S. 38-44 Undetermined age	Partial skeleton - *Janensch 1920* Specimen HMN Gr.S. 38-44 is a partial skeleton.	Length: 7.5 m Hip height: 2.1 m	210 kg
5	Spinostropheus gauthieri MJ, NAf (Niger)	MNHN Undetermined age	Ulna - *Lapparent 1960* Was not illustrated and may have been destroyed.	Length: 8.5 m Hip height: 2.4 m	600 kg

Averostra similar to *Limusaurus* - Light bodies, ample bellies, toothless, small game hunters or omnivores.

N.º	Genus and species	Largest specimen	Material and data	Size	Mass
8	Limusaurus inextricabilis UJ, WA (China)	IVPP V15924 Subadult	Partial skeleton - *Xu et al. 2009* Its appearance was similar to that of ornithomimosaurs.	Length: 1.8 m Hip height: 70 cm	15 kg
9	Unnamed MJ, WA (China)	CCG 20011 Subadult	Vertebrae and incomplete scapula - *He 1984* It is considered a large specimen of *Chuandongocoelurus primitivus*. *Mortimer**	Length: 4 m Hip height: 1.5 m	155 kg

1:250

CERATOSAURIDAE

Probable ceratosaurids similar to *Sarcosaurus* - Light bodies, flat and wide (anteposterior) teeth, small or medium game hunters.

Nº	Genus and species	Largest specimen	Material and data	Size	Mass
1	*Sarcosaurus woodi* LJ, WE (England)	BMNH R4840/1 Adult	Incomplete pelvis and fragments - *Andrews 1921* Basal ceratosaurid? *Paul 1998; Ezcurra 2012*	Length: 3.35 m Hip height: 1 m	71 kg
2	*Lukousaurus yini* LJ, EA (China)	V263 Undetermined age	Incomplete humerus - *Young 1948* Primitive ceratosaurid?	Length: 4 m Hip height: 1.2 m	120 kg
3	*Sarcosaurus andrewsi* (n.d.) LJ, WE (England)	BMNH R3542 Undetermined age	Tibia - *Huene 1932* *Sarcosaurus woodi*? *Carrano & Sampson 2004*	Length: 5 m Hip height: 1.45 m	210 kg
4	"*Merosaurus newmani*" (n.n.) LJ, WE (England)	BMNH 39496 Undetermined age	Partial femur and tibia - *Owen 1861* It is considered *Scelidosaurus* remains. *Pickering 1995*	Length: 6.5 m Hip height: 1.9 m	500 kg

Ceratosaurids similar to *Ceratosaurus* - Robust bodies, wide and very flat teeth, small or medium game hunters or piscivores.

Nº	Genus and species	Largest specimen	Material and data	Size	Mass
5	*Fosterovenator churei* UJ, WNA (Wyoming - USA)	YPM VP 058267 D Juvenile	Tibia - *Dalman 2014* Tetanuran? *Mortimer**	Length: 2.95 m Hip height: 90 cm	85 kg
6	"*Ceratosaurus sulcatus*" (n.d.) UJ, WNA (Wyoming - USA)	YPM 1936 Undetermined age	Maxillar tooth - *Marsh 1896* Could be a valid species of *Ceratosaurus*	Length: 4.3 m Hip height: 1.3 m	200 kg
7	*Ceratosaurus meriani* UJ, WE (Switzerland)	MH 350 Undetermined age	Premaxillar tooth - *Greppin 1870* Valid species?	Length: 4.9 m Hip height: 1.5 m	390 kg
8	"*Labrosaurus*" *stechowi* UJ, SAf (Tanzania)	MB R 1083 Undetermined age	Premaxillar tooth - *Janensch 1920* Similar to *Ceratosaurus*	Length: 5.2 m Hip height: 1.6 m	465 kg
9	*Ceratosaurus nasicornis* UJ, WNA (Colorado - USA)	USNM 4735 Adult	Semicomplete skull and skleleton - *Marsh 1884* It is the type specimen of the genus.	Length: 5.5 m Hip height: 1.65 m	550 kg
10	*Ceratosaurus magnicornis* UJ, WNA (Colorado - USA)	MWC 1 Undetermined age	Partial skull and skeleton - *Madsen & Welles 2000* It had a larger crest, a more robust skull, and shorter legs than *Ceratosaurus nasicornis*.	Length: 5.6 m Hip height: 1.7 m	560 kg
11	*Ceratosaurus dentisulcatus* UJ, WE (Portugal)	SHN(JJS)-65 Undetermined age	Femur, tibia, and fibula - *Malafaia et al. 2014* Has been found in Europe and North America.	Length: 5.8 m Hip height: 1.75 m	630 kg
12	*Ceratosaurus roechlingi* UJ, SAf (Tanzania)	MB R 2162 Undetermined age	Incomplete caudal vertebra - *Janensch 1925* Other remains belong to abelisaurids. *Rauhut 2011*	Length: 6.25 m Hip height: 1.9 m	790 kg
13	*Genyodectes serus* ILC, SSA (Argentina)	MLP 26-39 Undetermined age	Incomplete skull - *Woodward 1901* It had more premaxillar teeth than *Ceratosaurus*.	Length: 6.25 Hip height: 1.9 m	790 kg
14	*Ceratosaurus dentisulcatus* UJ, WNA (Utah - USA)	UMNH 5278 Undetermined age	Partial skull and skeleton - *Madsen & Welles 2000* Very different from *Ceratosaurus nasicornis*	Length: 6.8 m Hip height: 2.05 m	1 t

1:200

ABELISAUROIDEA

Probable abelisauroids - Relatively light bodies, small game hunters.

N.º	Genus and species	Largest specimen	Material and data	Size	Mass
1	*Ozraptor subotaii* MJ, OC (Australia)	UWA 82469 Undetermined age	Incomplete tibia - *Long & Molnar 1998* May not have been a basal abelisauroid. *Rauhut 2012*	Length: 2.3 m Hip height: 65 cm	13 kg
2	*Berberosaurus liassicus* LJ, NAf (Morocco)	MHNM-Pt9 Subadult	Partial skeleton - *Allain et al. 2007* Ceratosaurid or abelisauroid? *Carrano & Sampson 2008; Hendrickx & Mateus 2014*	Length: 5.1 m Hip height: 1.65 m	220 kg
3	*Eoabelisaurus mefi* MJ, SSA (Argentina)	MPEF PV 3990 Adult	Partial skeleton and skull - *Pol & Rauhut 2012* Abelisaurid or abelisauroid? *Novas et al. 2013*	Length: 7.5 m Hip height: 2.1 m	445 kg

Abelisauroids similar to *Noasaurus* - Relatively light bodies, small game hunters or piscivores.

	Genus and species	Largest specimen	Material and data	Size	Mass
4	*Ligabueino andesi* ILC, SSA (Argentina)	MACN-N 42 Juvenile	Partial skeleton - *Bonaparte 1996* Noasaurid?	Length: 70 cm Hip height: 18 cm	440 g
5	"*Sidormimus*" (n.n.) ILC, NAf (Niger)	Uncatalogued Undetermined age	Partial skeleton - *Sereno 2010* Unpublished but mentioned in Sereno et al. 2004	Length: 1 m Hip height: 27 cm	1.45 kg
6	*Velocisaurus unicus* IUC, SSA (Argentina)	MUCPv 41 Undetermined age	Incomplete posterior extermity - *Bonaparte 1991* Its central finger is much larger than the others.	Length: 1.5 m Hip height: 41 cm	4.7 kg
7	*Vitakrisaurus saraiki* IUC, IND (Pakistan)	MSM-303-2 Undetermined age	Metatarsus and phalanges - *Malkani 2010* "Vitakrisaurids" are synonyms of noasaurids.	Length: 1.55 m Hip height: 42 cm	5.3 kg
8	*Compsosuchus solus* eUC, IND (India)	GSI K27/578 Undetermined age	Atlas and axis - *Huene & Matley 1933* Intederminate noasaurid. *Carrano et al. 2011*	Length: 2 m Hip height: 56 cm	12 kg
9	*Jubbulpuria tenuis* IUC, IND (India)	GSI K27/614 Undetermined age	Distal caudal vertebra -- *Huene & Matley 1933* *Laevisuchus indicus*? *Wilson 2012*	Length: 2.6 m Hip height: 73 cm	28 kg
10	*Noasaurus leali* eUC, NSA (Argentina)	MACN-PV 622 Adult	Cervical vertebra - *Frankfurt & Chiappe 1999* Initially considered an oviraptorid. *Agnolin & Martinelli 2007*	Length: 3 m Hip height: 80 cm	38 kg

N.º	Genus and species	Largest specimen	Material and data	Size	Mass
11	Laevisuchus indicus IUC, IND (India)	GSI K27/696 Undetermined age	Cervical vertebra - *Huene & Matley 1933* Specimen GSI K27/588 was of similar size.	Length: 3.1 m Hip height: 82 cm	39 kg
12	Ornithomimoides barasimlensis (n.d.) IUC, IND (India)	GSI K27/541 Undetermined age	Caudal vertebra - *Huene & Matley 1933* *Coeluroides largus?* Wilson 2012	Length: 3.3 m Hip height: 85 cm	44 kg
13	Betasuchus bredai IUC, WE (Netherlands)	BMNH 42997 Undetermined age	Incomplete femur - *Huene 1926, 1932* "Ornithomimidorum" is not a genus; it means that it was similar to ornithomimosaurs.	Length: 4 m Hip height: 1.05 m	83 kg
14	Genusaurus sisterornis ILC, WE (France)	MNHN, Bev.1 Undetermined age	Partial skeleton - *Accarie et al. 1995* Abelisaurid? *Juárez-Valieri et al. 2004*	Length: 4.4 m Hip height: 1.2 m	110 kg
15	Masiakasaurus knopfleri IUC, SAf (Madagascar)	FMNH PR 2457 Subadult	Skull - *Sampson et al. 2001* Both are light and robust specimens.	Length: 4.6 m Hip height: 1.25 m	128 kg
16	Dahalokely tokana eUC, SAf (Madagascar)	UA 8678 Subadult	Tooth - *Farke & Sertich 2013* Abelisaurid? *Farke & Sertich 2009*	Length: 4.8 m Hip height: 1.3 m	150 kg
17	Ornithomimoides mobilis IUC, IND (India)	GSI K20/610 Undetermined age	Caudal vertebra - *Huene & Matley 1933* *Coeluroides largus?* Wilson 2012	Length: 5 m Hip height: 1.4 m	160 kg
18	Coeluroides largus IUC, IND (India)	GSI K27/562 Undetermined age	Caudal vertebra - *Huene & Matley 1933* Different from *Laevisuchus indicus.* Wilson 2012	Length: 5.9 m Hip height: 1.65 m	265 kg
19	Austrocheirus isasii IUC, SSA (Argentina)	MPM-PV 10003 Young adult	Partial skeleton - *Ezcurra et al. 2010* Giant noasaurid?	Length: 9.3 m Hip height: 2.5 m	1 t
20	Dryptosauroides grandis IUC, IND (India)	GSI Undetermined age	Caudal vertebra - *Huene & Matley 1933* Very similar to *Ornithomimoides.* Novas et al. 2004	Length: 10 m Hip height: 2.8 m	1.5 t

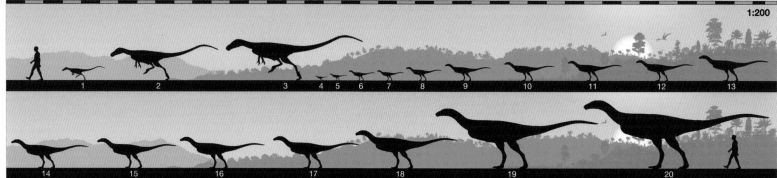

1:200

1 2 3 4 5 6 7 8 9 10 11 12 13

14 15 16 17 18 19 20

ABELISAURIDAE

Abelisaurids similar to *Rugops* - Relatively robust bodies, medium game hunters.

N.º	Genus and species	Largest specimen	Material and data	Size	Mass
1	Tarascosaurus salluvicus IUC, WE (France)	FSL 330202 Adult	Thoracic vertebra - *Le Loeuff & Buffetaut 1991* The measurements of specimen FSL 330203 are unknown.	Length: 3.1 m Hip height: 98 cm	90 kg
2	Rugops primus eUC, NAf (Niger)	MNN IGU1 Undetermined age	Incomplete skull and teeth - *Sereno et al. 2004* It had a very wrinkled face.	Length: 5.3 m Hip height: 1.7 m	410 kg
3	Xenotarsosaurus bonapartei IUC, SSA (Argentina)	PVL 612 Undetermined age	Partial posterior extremity and vertebrae - *Martinez et al 1987* It is primitive. *Tortosa et al. 2013*	Length: 5.4 m Hip height: 1.73 m	430 kg
4	Kryptops palaios ILC, NAf (Niger)	MNN GAD1 Adult	Partial maxilla - *Sereno & Brusatte 2008* Other assigned teeth belonged to carcharodontosaurids. *Carrano et al 2012*	Length: 5.8 m Hip height: 1.85 m	550 kg

Abelisaurids similar to *Majungasaurus* - Robust bodies, medium to large game hunters.

N.º	Genus and species	Largest specimen	Material and data	Size	Mass
5	"Massospondylus" rawesi (n.d.) IUC, IND (India)	BMNH R4190 Undetermined age	Tooth - *Lydekker 1890* Abelisaurid? *Carrano et al. 2012*	Length: 2.8 m Hip height: 70 cm	60 kg
6	Orthogoniosaurus matleyi (n.d.) IUC, NAf (India)	GI Undetermined age	Tooth - *Das-Gupta 1931* Abelisaurid? *Carrano et al. 2012*	Length: 3.9 m Hip height: 95 cm	145 kg
7	Indosaurus matleyi IUC, IND (India)	GSI K27/565 Undetermined age	Partial skull - *Huene 1932* Badly preserved material. *Novas et al. 2004*	Length: 5.6 m Hip height: 1.4 m	420 kg
8	cf. "Megalosaurus" crenatissimus IUC, NAf (Egypt)	MGUP MEGA002 Undetermined age	Tooth - *Gemmellaro 1921* There were other pieces. *Smith & Lamanna 2006*	Length: 5.7 m Hip height: 1.45 m	470 kg
9	Arcovenator escotae IUC, WE (France)	MHNA-PV-2011.12.1-5 & 15 Undetermined age	Partial skull and tibia - *Tortosa et al. 2014* Majungasaurine?	Length: 7.2 m Hip height: 1.85 m	950 kg
10	Majungasaurus crenatissimus IUC, SAf (Madagascar)	MNHN.MAJ 4 Adult	Incomplete skull - *Sues & Taquet 1979* It was considered a pachycephalosaurid.	Length: 8.1 m Hip height: 2 m	1.3 t
11	Rahiolisaurus gujaratensis IUC, IND (India)	ISIR 436 Adult	Ilium - *Novas, Chaterjee, Rudra & Datta 2010* Different from those described by Huene & Matley 1933.	Length: 9.2 m Hip height: 2.3 m	2 t
12	Indosuchus raptorius IUC, IND (India)	GSI K20/350 Subadult?	Partial skull - *Huene & Matley 1933* Might include *Indosaurus matleyi*	Length: 9.7 m Hip height: 2.4 m	2.3 t
13	Rajasaurus narmadensis IUC, IND (India)	Uncatalogued Undetermined age	Ilium, sacrum, and partial tibia - *Matley 1923* The short tibia may belong to another specimen.	Length: 10.5 m Hip height: 2.6 m	3 t

Abelisaurids similar to *Ekrixinatosaurus* - Long and robust bodies, short legs, medium to big game hunters.

N.º	Genus and species	Largest specimen	Material and data	Size	Mass
14	Ekrixinatosaurus novasi eUC, SSA (Argentina)	MUCPv-294 Undetermined age	Partial skeleton - *Calvo et al. 2004* Similar to *Skorpiovenator* but proportioned like *Majungasaurus*. *Juarez-Valieri et al. 2011*	Length: 8.2 m Hip height: 2 m	1.4 t

Abelisaurids similar to *Skorpiovenator* - Bipedal, relatively robust bodies, medium to large game hunters.

N.º	Genus and species	Largest specimen	Material and data	Size	Mass
15	cf. Loncosaurus argentinus IUC, SSA (Argentina)	Uncatalogued Undetermined age	Tooth - *Huene 1929* The piece was attributed to the ornithopod *Loncosaurus*.	Length: 3.3 m Hip height: 1.15 m	150 kg
16	"Carnosaurus" (n.n.) eUC, SSA (Argentina)	MACN Undetermined age	Tooth - *Huene 1929* Invalid name; it refers to a "carnosaur."	Length: 3.4 m Hip height: 1.2 m	170 kg

N.º		Largest specimen	Material and data	Size	Mass
17	"Bayosaurus pubica" IUC, SSA (Argentina)	MCF-PVPH-237 Undetermined age	Incomplete pelvis and vertebrae - *Coria et al. 2006* Unofficial name, published by mistake.	Length: **4.1 m** Hip height: **1.4 m**	295 kg
18	*Quilmesaurus curriei* IUC NSA (Argentina)	MPCA-PV-100 Undetermined age	Incomplete femur and tibia - *Coria 2001* Carnotaurine? *Juárez Valieri et al. 2004*	Length: **4.6 m** Hip height: **1.6 m**	430 kg
19	*Ilokelesia aguadagrandensis* eUC, SSA (Argentina)	Uncatalogued Undetermined age	Caudal vertebra - *Coria & Salgado 2000* It has been suggested as a noasaurid.	Length: **5.8 m** Hip height: **2 m**	840 kg
20	*Skorpiovenator bustingorryi* eUC, SSA (Argentina)	MMCH-PV 48 Undetermined age	Skull and partial skeleton - *Canale et al 2009* Very differently proportioned to *Carnotaurus*	Length: **6 m** Hip height: **2.1 m**	950 kg
21	*Abelisaurus comahuensis* IUC, SSA (Argentina)	MC 11098 Undetermined age	Incomplete skull - *Bonaparte & Novas 1985* Its head was more elongated than that of other abelisaurids.	Length: **7.2 m** Hip height: **2.5 m**	1.65 t
22	*Pycnonemosaurus nevesi* eUC, NSA (northern Brazil)	DGM 859-R Undetermined age	Partial skeleton and teeth - *Kellner & Campos 2002* Carnotaurine? Its tail was very long.	Length: **9.3 m** Hip height: **3.2 m**	3.6 t

Abelisaurids similar to *Carnotaurus* - Relatively light bodies, long legs, small game hunters.

N.º		Largest specimen	Material and data	Size	Mass
23	*Aucasaurus garridoi* IUC, SSA (Argentina)	MCF-PVPH-236 Adult	Skull and partial skeleton - *Coria et al. 2000* It lacked horns, unlike *Carnotaurus*	Length: **5.3 m** Hip height: **1.8 m**	600 kg
24	*Carnotaurus sastrei* IUC, SSA (Argentina)	MACN-CH 894 Undetermined age	Pelvis, sacrum, and tibia - *Bonaparte 1985* It had the shortest face of all abelisaurids.	Length: **7.7 m** Hip height: **2.4 m**	1.85 t

1:250

MEGALOSAUROIDEA

Probable megalosauroids - Robust bodies, medium to big game hunters.

N.º	Genus and species	Largest specimen	Material and data	Size	Mass
1	*Sciurumimus albersdoerferi* UJ, WE (Germany)	BMMS BK 11 Undetermined age	Skull, partial skeleton, and filaments - *Rauhut et al. 2012* Coelurosaur? *Godefroit et al. 2013*	Length: **72 cm** Hip height: **13.5 cm**	250 g
2	"*Morosaurus*" *marchei* UJ, WE (Portugal)	BMNH R283 Undetermined age	Caudal vertebra - *Sauvage 1898* Considered a Cretaceous sauropod.	Length: **4.2 m** Hip height: **1.2 m**	175 kg
3	"*Metriacanthosaurus reynoldsi*" MJ, WE (England)	Uncatalogued Undetermined age	Tooth - *Reynolds 1939* There are other attributed bones.	Length: **5.6 m** Hip height: **1.6 m**	410 kg
4	*Chienkosaurus ceratosauroides* UJ, EA (China)	IVPP V193 Undetermined age	Ulna - *Young 1942* Metriacantosaurid?	Length: **5.7 m** Hip height: **1.65**	420 kg
5	*Cruxicheiros newmanorum* MJ, WE (England)	WARMS G15770 Undetermined age	Incomplete femur - *Benson & Radley 2010* Another more complete specimen of WARMS G15771.	Length: **6 m** Hip height: **1.7 m**	505 kg
6	*Kaijiangosaurus lini* MJ, EA (China)	CCG Undetermined age	Femur - *He 1984* The holotype could be of another genus. *Carrano et al. 2012*	Length: **6 m** Hip height: **1.7 m**	510 kg
7	"*Saltriosaurus*" LJ, WE (Italy)	MSNM V3664 Undetermined age	Incomplete tibia - *Dalla Vecchia 2001* Not a piatnizkysaurid. *Carrano et al. 2012*	Length: **6.7 m** Hip height: **1.9 m**	680 kg
8	"*Saurocephalus*" *monasterii* UJ, WE (Germany)	Uncatalogued Undetermined age	Tooth - *Muenster 1836* Identified as a theropod by Winfolf 1997	Length: **8.3 m** Hip height: **2.3 m**	1.4 t
9	"*Allosaurus*" *tendangurensis* UJ, SAf (Tanzania)	MB R 3620 Undetermined age	Incomplete tibia - *Janensch 1925* Uncertain classification. *Carrano et al. 2012*	Length: **9.9 m** Hip height: **2.8**	2.2 t
10	*Dandakosaurus indicus* LJ, IND (India)	GSI coll. Undetermined age	Vertebrae and incomplete ischium - *Yadagiri 1982* Uncertain classification. *Carrano et al. 2012*	Length: **10 m** Hip height: **2.8 m**	2.3 t

Megalosauroids similar to *Piatnitzkysaurus* - Relatively robust bodies, medium to big game hunters.

N.º	Genus and species	Largest specimen	Material and data	Size	Mass
11	*Marshosaurus bicentesimus* UJ, WNA (Colorado, Utah - USA)	UUVP 3454 Undetermined age	Incomplete dentary - *Madsen 1976* Includes *Marshosaurus* sp.? *Turner & Peterson 1999*	Length: **4.4 m** Hip height: **1.3 m**	225 kg
12	*Condorraptor currumili* MJ, SSA (Argentina)	MPEF-PV 1672 Juvenile?	Incomplete skull - *Rauhut 2005* Was MPEF 1717 an adult? *Rauhut 2007*	Length: **4.5 m** Hip height: **1.35 m**	280 kg
13	*Piatnitzkysaurus floresi* MJ, SSA (Argentina)	MACN CH 895 Adult?	Partial skeleton - *Bonaparte 1986* The holotype is a subadult. *Bonaparte 1979*	Length: **4.7 m** Hip height: **1.4 m**	320 kg

Megalosaurids similar to *Eustreptospondylus* - Relatively robust bodies, small to medium game hunters.

N.º	Genus and species	Largest specimen	Material and data	Size	Mass
14	*Eustreptospondylus oxoniensis* MJ, WE (England)	OUM J13558 Subadult	Partial skeleton - *Walker 1964* The only tetanurine with an almost complete foot.	Length: **4.5 m** Hip height: **1.3 m**	230 kg
15	*Streptospondylus altdorfensis* MJ, WE (France)	MNHN 8605 Subadult	Incomplete pubis - *Meyer 1832* Indeterminate megalosaur? *Carrano et al. 2012*	Length: **5.3 m** Hip height: **1.6 m**	410 kg

16	"Streptospondylus" cuvieri MJ, WE (England)	BMMS BK 11 Undetermined age	Teeth and fragments - *Owen 1842* *Streptospondylus altdorfensis*?	Length: **5.7 m** Hip height: **1.7 m**	**500 kg**

Megalosaurids similar to *Afrovenator* - Long skulls, relatively robust bodies, medium to big game hunters.

17	Dubreuillosaurus valesdunensiss MJ, WE (France)	MNHN 1998-13 Juvenile	Skull and partial skeleton - *Allain 2002* Not the same genus as *Poekilopleuron*. *Allain 2005*	Length: **4 m** Hip height: **1.1 m**	**165 kg**
18	Magnosaurus nethercombensis MJ, WE (England)	OUM J12143 Adult	Dentary and partial skeleton - *Huene 1923* The smallest basal tetanurean.	Length: **4.5 m** Hip height: **1.25 m**	**220 kg**
19	Leshansaurus qianweiensis UJ, EA (China)	QW200701 Undetermined age	Skull and partial skeleton - *Li, et al. 2009* Assigned to this family by Carrano et al. 2012.	Length: **5.4 m** Hip height: **1.45 m**	**390 kg**
20	Piveteausaurus divesensis MJ, WE (France)	MNHN 1920-7 Adult	Incomplete skull - *Taquet & Welles 1977* Adult *Dubreuillosaurus*?	Length: **5.5 m** Hip height: **1.5 m**	**420 kg**
21	Afrovenator abakensis MJ, NAf (Niger)	UC UBA 1 Undetermined age	Skull and partial skeleton - *Sereno et al. 1994* Was thought to be much more recent.	Length: **6.8 m** Hip height: **1.9 m**	**790 kg**
22	Poekilopleuron bucklandii MJ, WE (France)	MNHN 1897-2 Undetermined age	Incomplete tibia - *Eudes-Deslongchamps 1837* The material was destroyed during World War II.	Length: **6.8 m** Hip height: **1.9 m**	**800 kg**

1:200

Megalosaurids similar to *Torvosaurus* - Long skulls, robust bodies, medium to big game hunters.

1	cf. *Megalosaurus insignis* MJ, WE (Portugal)	Uncatalogued Undetermined age	Incomplete tibia - *Lapparent and Zbyszewski, 1957* It is doubtful that it belongs to a "*Megalosaurus*" insignis	Length: **4.6 m** Hip height: **1.2 m**	**250 kg**
2	Duriavenator hesperis MJ, WE (England)	BMNH R 332 Undetermined age	Partial skeleton - *Waldman 1974* The material was checked by Benson 2008.	Length: **5.4 m** Hip height: **1.4 m**	**380 kg**
3	"Megalosaurus phillipsi" (n.n.) UJ, WE (England)	OUMJ29886 Undetermined age	Tibia - *Huene 1926* Other remains were described by Phillips in 1871.	Length: **7.7 m** Hip height: **2 m**	**1.15 t**
4	"Brontoraptor" (n.n.) UJ, ONA (Wyoming - USA)	TATE 0012 Adult	Partial skeleton - *Siegwarth et al. 1994* The study was not published.	Length: **8.3 m** Hip height: **2.15 m**	**1.5 t**
5	cf. *Megalosaurus bucklandii* UJ, WE (England)	BMNH R1027 Undetermined age	Sacrum - *Owen 1857, Lydekker 1888* The material was considered to be from the Lower Jurassic.	Length: **8.8 m** Hip height: **2.3 m**	**1.65 t**
6	Megalosaurus bucklandii MJ, WE (England)	SDM 44.10 Undetermined age	Thoracic vertebra - *Reynolds 1939* It is the largest of several fragmented specimens.	Length: **9.8 m** Hip height: **2.55 m**	**2.4 t**
7	Megalosaurus insignis UJ, WE (France)	Museum du Havre Undetermined age	Tooth - *Valenciennes 1863* Uncertain classification.	Length: **10 m** Hip height: **2.6 m**	**2.5 t**
8	cf. "*Megalosaurus*" pombali UJ, WE (Portugal)	Uncatalogued Undetermined age	Caudal vertebrae - *Lapparent & Zbyszewski 1957* *Torvosaurus gurneyi*?	Length: **10.7 m** Hip height: **2.75 m**	**3 t**
9	cf. *Megalosaurus bucklandii* MJ, WE (England)	Uncatalogued Undetermined age	Incomplete femur - *Phillips 1871* More recent than *Megalosaurus bucklandii*	Length: **10.7 m** Hip height: **2.75 m**	**3 t**
10	Teinurosaurus sauvagei (n.d.) UJ, WE (France)	MGB 500 Undetermined age	Caudal vertebra - *Sauvage 1897, Nopcsa 1928* Basal ceratosaur, tetanuran or celurosaurid?	Length: **11.4 m** Hip height: **2.9 m**	**3.6 t**
11	Torvosaurus gurneyi UJ, WE (Portugal)	ML 1100 Undetermined age	Jaw, teeth, and caudal vertebra - *Hendrickx & Mateus 2014* ML 632 is a similar specimen. *Mateus et al. 2006*	Length: **11.7 m** Hip height: **3 m**	**4 t**
12	Torvosaurus tanneri UJ, WNA (Colorado, Utah, Wyoming - USA)	BYUVP 2003 Undetermined age	Incomplete dentary - *Siegwarth et al. 1996* BYUVP 4882 is another specimen of similar size. *Jensen 1985*	Length: **11.9 m** Hip height: **3.1 m**	**4.1 t**
13	Edmarka rex (n.d.) UJ, WNA (Wyoming - USA)	CPS 1010 Undetermined age	Incomplete pubis - *Bakker et al. 1992* *Torvosaurus tanneri*? *Holtz et al. 2004; Carrano et al. 2012*	Length: **12 m** Hip height: **3.1 m**	**4.2 t**

1:250

SPINOSAURIDAE

Spinosaurids similar to *Baryonyx* - Serrated teeth, light bodies, small game hunters and/or piscivores.

Nº	Genus and species	Largest specimen	Material and data	Size	Mass
1	"Weenyonyx" (n.n.) ILC, WE (England)	Uncatalogued Juvenile	Tooth - *Munt and Hutt in Naish 1997* *Baryonyx walkeri?*	Length: 1.8 m Hip height: 45 cm	12 kg
2	*Suchosaurus cultridens* (n.d.) eLC, WE (England)	BMNH R4415 Undetermined age	Toth - *Mantell 1827* An older species of *Baryonyx*?	Length: 5.5 m Hip height: 1.6	555 kg
3	*Ostafrikasaurus crassiserratus* UJ, SAf (Tanzania)	MB R 1084 Undetermined age	Premaxillary tooth - *Janensch 1925* May have been a spinosaurid or ceratosaurid. *Buffetaut 2011; Chure 2000 in Mortimer**	Length: 8.4 m Hip height: 2.1 m	1.15 t
4	cf. *Suchosaurus cultridens* ILC, WE (Spain)	CMP-3-758 Undetermined age	Tooth - *Canudo et al. 2008* Was considered a crocodile. *Bataller 1960*	Length: 8.6 m Hip height: 2.15 m	1.4 t
5	cf. *Suchosaurus cultridens* eLC, WE (England)	BMNH R4415 Undetermined age	Tooth - *Mantell 1827* Indeterminate megalosaurid. *Azuma 1991*	Length: 9.3 m Hip height: 2.3 m	1.7 t
6	*Suchosaurus girardi* (n.d.) ILC, WE (Portugal)	MML 1190 Undetermined age	Dentary and partial skeleton - *Mateus et al. 2011* *Suchosaurus* and *Baryonyx walkeri* may be synonymous.	Length: 9.3 m Hip height: 2.3 m	1.7 t
7	*Baryonyx walkeri* ILC, WE (England)	MIWG.6527 Adult?	Hand phalanx - *Naish et al. 2001* The holotype specimen BMNH R9951 is a subadult. *Charig & Milner 1986*	Length: 9.7 m Hip height: 2.4 m	2 t
8	*Cristatusaurus lapparenti* ILC, NAf (Niger)	MNHN GDF 365 Adult	Premaxilla - *Taquet & Russell 1998* May belong to the genus *Baryonyx*	Length: 12 m Hip height: 3.2 m	4 t

Spinosaurids similar to *Ichthyovenator* - Smooth cylindrical teeth, short torsi, light bodies, small game hunters or piscivores.

Nº	Genus and species	Largest specimen	Material and data	Size	Mass
9	*Siamosaurus fusuiensisis* ILC, WE (China)	MDS BK 10 Undetermined age	Teeth - *Hou et al. 1975* Was considered a pliosaur.	Length: 5.1 m Hip height: 1.45 m	250 kg
10	*Siamosaurus suteethorni* ILC, CIM (Thailand)	DMR TF 2043a Undetermined age	Tooth - *Buffetaut & Ingavat 1986* More ancient than *Siamosaurus* sp. from Thailand.	Length: 5.1 m Hip height: 1.45 m	255 kg
11	*Irritator challengerii* ILC, NSA (northern Brazil)	SMNS 58022 Undetermined age	Incomplete skull - *Martill et al. 1996* It has been suggested that *Angaturama* is another piece of the same specimen.	Length: 8.7 m Hip height: 2.45 m	1.3 t
12	*Sigilmassasaurus brevicollis* eUC, NSA (northern Brazil)	Uncatalogued Undetermined age	Caudal vertebra - *Medeiros & Schultz 2002; Medeiros et al. 2014* *Oxalaia quilombensis?*	Length: 9.1 m Hip height: 2.6 m	1.6 t
13	*Ichthyovenator laosensis* ILC, CIM (Laos)	MDS BK 10 Undetermined age	Partial skeleton - *Allain et al 2012* Teeth similar to Baryonychinae and cervical vertebrae of spinosaurines. *Allain et al. 2014*	Length: 10.5 m Hip height: 2.95 m	2.4 t
14	*Sigilmassasaurus brevicollis* ILC, NAf (Morocco)	NMC 41852 Undetermined age	Incomplete humerus - *Russell, 1996* Older than *Spinosaurus aegyptiacus*. *Evers et al. 2015*	Length: 13 m Hip height: 3.7 m	4.7 t
15	*Oxalaia quilombensis* eUC, NSA (northern Brazil)	MN 6117-V Undetermined age	Incomplete premaxilla - *Kellner et al. 2011* Approaches *Spinosaurus* in size.	Length: 13.3 m Hip height: 3.75 m	5 t
16	"*Spinosaurus*" *maroccanus* eUC, NAf (Egypt, Morocco, Niger)	BSPG 2011 I 118 Undetermined age	Cervical vertebrae - *Evers et al. 2015* More recent than *Sigilmassasaurus brevicollis*.	Length: 14.4 m Hip height: 4.1 m	6.5 t

Spinosaurids similar to *Spinosaurus* - Aquatic, quadrapeds on land, long torsi, relatively light bodies, small game hunters or piscivores.

	Genus and species	Largest specimen	Material and data	Size	Mass
1	cf. *Spinosaurus aegyptiacus?* eUC, NAf (Niger)	Uncatalogued Undetermined age	? - *Brusatte & Sereno 2007* Juvenile of *Spinosaurus aegyptiacus?*	Length: 6.4 m Hip height: -	450 kg
2	*Spinosaurus aegyptiacus* eUC, NAf (Argelia, Egypt, Morocco)	MSNM V4047 Adult	Incomplete skull - *Dal Sasso et al. 2005* A smaller uncataloged specimen has a large crest.	Length: 16 m Hip height: -	7.5 t

1:300

ALLOSAUROIDEA

Allosauroids similar to *Yangchuanosaurus* - Relatively robust bodies, medium to large game hunters.

N.º	Genus and species	Largest specimen	Material and data	Size	Mass
1	*Xuanhanosaurus qilixiaensis* MJ, EA (China)	IVPP V6729 Undetermined age	Partial skeleton - *Dong 1984* Primitive megalosauroid? *Benson 2008*	Length: 4.8 m Hip height: 1.3 m	265 kg
2	*Yangchuanosaurus zigongensis* MJ, EA (China)	ZDM 9011 Undetermined age	Partial skeleton - *Gao 1993* Was known as *Szechuanosaurus zigongensis*	Length: 6.5 m Hip height: 1.75 m	490 kg
3	*Szechuanosaurus campi* (n.d.) UJ, EA (China)	IVPP V235 Undetermined age	Teeth - *Young 1942* Metriacantosaurid?	Length: 7.3 m Hip height: 2 m	1 t
4	*Yangchuanosaurus shangyounensis* UJ, EA (China)	CV 00216 Adult	Skull and partial skeleton - *Dong, Zhou & Zhang 1983* The skull is shorter than the femur in juveniles; this is reversed in adults.	Length: 10.5 Hip height: 2.9 m	2.9 t

Allosauroids similar to *Sinraptor* - Relatively robust bodies, medium to large game hunters.

N.º	Genus and species	Largest specimen	Material and data	Size	Mass
5	cf. "*Cryptodraco*" sp. MJ, WE (England)	R.1617 Undetermined age	Thoracic vertebra - *Lydekker 1890* Same location as *Metriacanthosaurus parkeri*. *Delair 1959; Weishampel et al. 2004*	Length: 4 m Hip height: 1.1 m	155 kg
6	*Shidaisaurus jinae* MJ, EA (China)	LDM-LCA 9701-IV Undetermined age	Partial skeleton - *Wu et al. 2009* Its name means "golden age."	Length: 7.1 m Hip height: 1.9 m	950 kg
7	*Metriacanthosaurus parkeri* UJ, WE (England)	OUM J.12144 Undetermined age	Partial skeleton - *Huene 1923* Was thought to be a spinosaurid due to its tall vertebrae.	Length: 7.5 m Hip height: 2.05 m	1.1 t
8	*Sinraptor hepingensis* UJ, EA (China)	ZDM 0024 Undetermined age	Partial skeleton - *Currie & Zhao 1994* Another specimen has damage thought to be caused by a mamenchisaurid. *Xing et al 2009*	Length: 9.2 m Hip height: 2.5 m	2 t
9	*Sinraptor dongi* UJ, EA (China)	IVPP 15310 Adult?	Teeth - *Xu & Clark 2008* The holotype IVPP 10600 was a subadult. *Currie & Zhao 1991*	Length: 11.5 m Hip height: 3.1 m	3.9 t

Indeterminate allosauroids - Relatively robust bodies, medium to big game hunters.

N.º	Genus and species	Largest specimen	Material and data	Size	Mass
10	*Erectopus superbus* ILC, WE (France)	MNHN 2001-4 Undetermined age	Jaw and partial skeleton - *Sauvage 1882* Different; may form a family of its own.	Length: 5 m Hip height: 1.35 m	315 kg
11	"*Mifunesaurus*" (n.n.) eUC, EA (Japan)	YNUGI 10003 Undetermined age	Tooth - *Hisa 1985* Basal tetanurae or allosauroid? *Mortimer**	Length: 6 m Hip height: 1.6 m	550 kg
12	cf. *Erectopus superbus* ILC, WE (Portugal)	Uncatalogued Undetermined age	Incomplete tooth - *Lapparent & Zbyszewski 1957* Similar to *Erectopus*?	Length: 7.4 m Hip height: 2 m	1 t
13	aff. *Erectopus superbus* eUC, NAf (Egypt)	IPHG 1912 VIII 85 Undetermined age	Incomplete tibia - *Stromer 1934* Doubtful identification.	Length: 7.5 m Hip height: 2 m	1.1 t
14	cf. "*Megalosaurus*" *insignis* ILC, WE (France)	SV3 Undetermined age	Foot phalanx - *Parent 1893* Indeterminate theropod.	Length: 8.2 m Hip height: 2.2 m	1.4 t
15	aff. *Morosaurus marchei* (n.d.) ILC, WE (Portugal)	IPHG 1912 VIII 85 Undetermined age	Incomplete tibia - *Sauvage 1897/1898* Was considered a sauropod.	Length: 9.3 m Hip height: 2.5 m	2.1 t

1:200

Allosauroids similar to *Allosaurus* - Relatively robust bodies, medium to big game hunters.

N.º	Genus and species	Largest specimen	Material and data	Size	Mass
1	*Creosaurus atrox* UJ, WNA (Wyoming - USA)	YPM 1890 Subadult	Skull and partial skeleton - *Marsh 1878* Unlike *Allosaurus fragilis*. *Mortimer**	Length: 6.1 m Hip height: 1.6 m	590 kg
2	*Labrosaurus lucaris* (n.d.) UJ, WNA (Wyoming - USA)	YPM 1931 Undetermined age	Partial skeleton - *Marsh, 1879* *Allosaurus fragilis*?	Length: 7.2 m Hip height: 1.95 m	1 t
3	*Allosaurus lucasi* UJ, WNA (Colorado - USA)	YPM VP 57589 Adult	Incomplete skull - *Dalman 2014* *Allosaurus fragilis*?	Length: 7.6 m Hip height: 2.05 m	1.15 t
4	"*Camptonotus*" *amplus* (n.d.) UJ, WNA (Wyoming - USA)	YPM 1879 Undetermined age	Partial skeleton - *Marsh, 1879* *Allosaurus fragilis* or *Saurophaganax*?	Length: 7.6 m Hip height: 2.05 m	1.2 t

N.º	Genus and species	Largest specimen	Material and data	Size	Mass
5	Allosaurus europaeus UJ, WE (Portugal)	ML 415 Undetermined age	Skull and partial skeleton - *Mateus et al 2006* *Allosaurus fragilis?*	Length: 7.8 m Hip height: 2.1 m	1.3 t
6	Antrodemus valens (n.d.) UJ, WNA (Colorado - USA)	USNM 218 Undetermined age	Incomplete caudal vertebra - *Leidy 1870* *Allosaurus fragilis?* Impossible to verify	Length: 7.9 m Hip height: 2.15 m	1.4 t
7	cf. Allosaurus fragilis UJ, WNA (USA)	SMA 0005 Adult	Partial skeleton - *Osborn 1899* More recent than *Allosaurus fragilis*	Length: 9.6 m Hip height: 2.6 m	2.3 t
8	Epanterias amplexus (n.d.) UJ, WNA (Colorado - USA)	MNH 5767 Undetermined age	Partial skeleton - *Cope 1878* *Allosaurus* or *Saurophaganax?*	Length: 10.4 m Hip height: 2.8 m	2.9 t
9	Allosaurus fragilis UJ, WNA (Colorado, Montana, New Mexico, Oklahoma, South Dakota, Utah, Wyoming - USA)	NMMNH 26083 Adult	Femur - *Williamson & Chure 1996* Similar in size to *Epantherias amplexus*; they may be synonymous.	Length: 10.4 m Hip height: 2.8 m	2.9 t
10	Saurophaganax maximus UJ, WNA (New Mexico, Oklahoma - USA)	OMNH 1935 Undetermined age	Humerus - *Chure 1995* May have been of the *Allosaurus maximus* genus.	Length: 12 m Hip height: 3.25 m	4.5 t

1:250

CARCHARODONTOSARIDAE

Carcharodontosaurids similar to *Neovenator* - Relatively robust bodies, medium to large game hunters.

N.º	Genus and species	Largest specimen	Material and data	Size	Mass
1	Neovenator sp. eLC, WE (France)	ANG 10-51 Undetermined age	Tooth - *Neraudeau et al 2012* Possible neovenatorid.	Length: 6 m Hip height: 1.6 m	550 kg
2	"Ornithocheirus" hilsensis (n.d.) eLC, WE (Germany)	Uncatalogued Undetermined age	Foot phalanx - *Koken 1883* Doubtful identification.	Length: 7 m Hip height: 1.9 m	950 kg
3	Neovenator salerii ILC, WE (England)	MIWG 6352 Subadult	Partial skeleton - *Hutt, Martill & Barker 1996* Similar appearance to *Allosaurus*	Length: 9.2 m Hip height: 2.5 m	2 t
4	cf. Neovenator salerii ILC, WE (England)	MIWG 4199 Adult?	Foot phalanx - *Hutt 2001* Adult specimen of *Neovenator salerii?*	Length: 10 m Hip height: 2.65 m	2.4 t

Carcharodontosaurids similar to *Acrocanthosaurus* - Robust bodies, medium to big game hunters.

N.º	Genus and species	Largest specimen	Material and data	Size	Mass
5	Concavenator corcovatus ILC, WE (Spain)	MCCM-LH 6666 Adult	Skull and partial skeleton - *Ortega et al. 2010* It was an adult. *Mortimer com. pers.*	Length: 5.2 m Hip height: 1.9 m	400 kg
6	Shaochilong maortuensis eUC, EA (China)	IVPP V2885.1 Adult	Incomplete skull - *Hu 1964* Has a frontal bone similar to *Labocania*.	Length: 5.4 m Hip height: 1.5 m	525 kg
7	Becklespinax altispinax eLC, WE (England)	BMNH R1828 Undetermined age	Tooth - *Owen 1855* First dinosaur found with a hump.	Length: 6 m Hip height: 2 m	640 kg
8	"Prodeinodon" kwangshiensis (n.d.) ILC, EA (South Korea)	IVPP V4795 Undetermined age	Tooth - *Hou, Yeh & Zhao 1975* Probable carcharodontosaurid? *Mortimer**	Length: 6.4 m Hip height: 1.8 m	790 kg
9	Erectopus cf. superbus eLC, EE (Romania)	UAIC (SCM1) 615 Undetermined age	Tooth - *Simionescu 1913* Carcharodontosaurid. *Csiki-Sava et al. 2016*	Length: 7.3 m Hip height: 2.05 m	1.3 t
10	Kelmayisaurus petrolicus UJ, EA (China)	IVPP V4022 Undetermined age	Jaw and incomplete dentary - *Dong 1973* Carcharodontosaurid? *Brusatte et al. 2012*	Length: 7.9 m Hip height: 2.2 m	1.55 t
11	"Osteoporosia gigantea" (n.n.) eUC, NAf (Morocco)	JP Cr340 Undetermined age	Tooth - *Singer** Possible synonym of *Sauroniops pachytholus?* *Mortimer**	Length: 7.9 m Hip height: 2.2 m	1.55 t
12	Veterupristisaurus milneri ILC, EA (China)	MB R 1938 o ST 270 Undetermined age	Caudal vertebra - *Rauhut 2011* "*Megalosaurus*" *ingens* subadult?	Length: 8 m Hip height: 2.25 m	1.65 t
13	Wakinosaurus satoi (n.d.) eLC, EA (Japan)	KMNH VP 000,016 Undetermined age	Tooth - *Okazaki 1992* Similar to "*Prodeinodon*" *kwangshiensis*	Length: 8.1 Hip height: 2.25 m	1.7 t
14	Eocarcharia dinops ILC NAf (Niger)	MNN GAD2 Undetermined age	Postorbital - *Sereno & Brusatte 2008* Material assigned to *Kryptops* is from *Eocarcharia*.	Length: 8.4 m Hip height: 2.35 m	1.9 t
15	Datanglong guangxiensis ILC, WE (South Korea)	GMG 00001 Adult	Thoracic vertebra - *Mo et al. 2014* Carcharodontosaurid?	Length: 8.6 m Hip height: 2.4 m	2.1 t
16	Tyrannosaurus lanpingensis (n.d.) eLC, EA (China)	IVPP Undetermined age	Tooth - *Ye 1975* Carcharodontosaurid?	Length: 9 m Hip height: 2.5 m	2.3 t
17	Allosaurus medius UJ, ENA (Maryland - USA)	USNM 4972 Undetermined age	Tooth - *Marsh 1888* Similar to *Acrocanthosaurus*	Length: 9.2 m Hip height: 2.5 m	2.5 t
18	Sauroniops pachytholus eUC, NAf (Morocco)	MPM 2594 Undetermined age	Frontal bone - *Cau, Dalla Vecchia & Fabbri 2013* Coexisted with *Carcharodontosaurus* but is more basal.	Length: 10.2 m Hip height: 2.95 m	3.8 t
19	Acrocanthosaurus atokensis ILC, WNA (Oklahoma, Utah, Texas, Wyoming - USA)	OMNH 10168 Undetermined age	Skull and partial skeleton - *Currie & Carpenter 2000* It had very tall vertebrae compared with its relatives.	Length: 11.5 m Hip height: 3.3 m	4.9 t
20	"Megalosaurus" ingens UJ, SAf (Tanzania)	MNHUK R6758 Undetermined age	Tooth - *Charig 1979, Chapman vide Galton & Molnar 2011* Reclassified by *Rauhut 1995*	Length: 12.6 m Hip height: 3.6 m	6.4 t

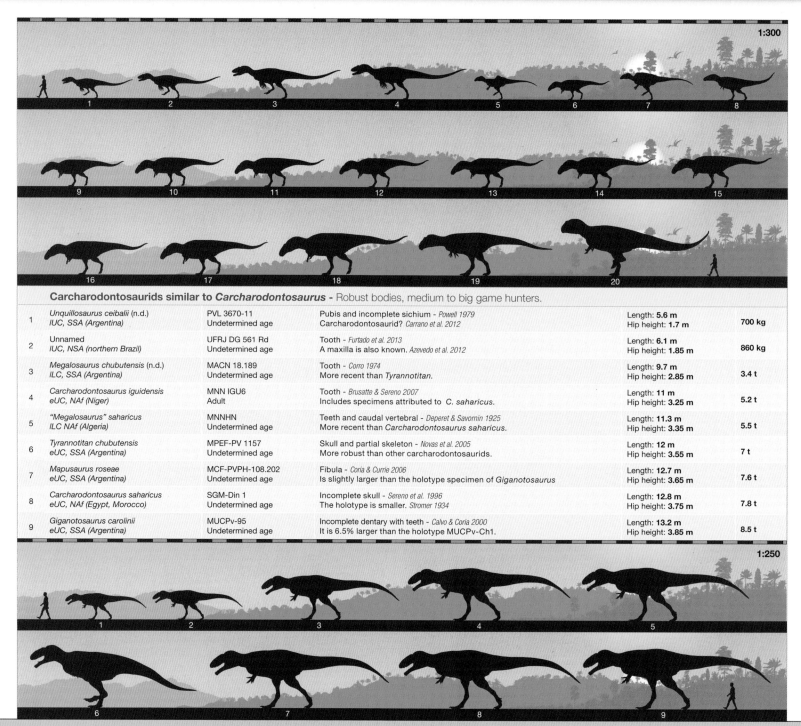

Carcharodontosaurids similar to *Carcharodontosaurus* - Robust bodies, medium to big game hunters.

1	*Unquillosaurus ceibalii* (n.d.) IUC, SSA (Argentina)	PVL 3670-11 Undetermined age	Pubis and incomplete sichium - *Powell 1979* Carcharodontosaurid? *Carrano et al. 2012*	Length: **5.6 m** Hip height: **1.7 m**	**700 kg**
2	Unnamed IUC, NSA (northern Brazil)	UFRJ DG 561 Rd Undetermined age	Tooth - *Furtado et al. 2013* A maxilla is also known. *Azevedo et al. 2012*	Length: **6.1 m** Hip height: **1.85 m**	**860 kg**
3	*Megalosaurus chubutensis* (n.d.) ILC, SSA (Argentina)	MACN 18.189 Undetermined age	Tooth - *Corro 1974* More recent than *Tyrannotitan*.	Length: **9.7 m** Hip height: **2.85 m**	**3.4 t**
4	*Carcharodontosaurus iguidensis* eUC, NAf (Niger)	MNN IGU6 Adult	Tooth - *Brusatte & Sereno 2007* Includes specimens attributed to *C. saharicus*.	Length: **11 m** Hip height: **3.25 m**	**5.2 t**
5	"*Megalosaurus*" *saharicus* ILC NAf (Algeria)	MNNHN Undetermined age	Teeth and caudal vertebral - *Deperet & Savornin 1925* More recent than *Carcharodontosaurus saharicus*.	Length: **11.3 m** Hip height: **3.35 m**	**5.5 t**
6	*Tyrannotitan chubutensis* eUC, SSA (Argentina)	MPEF-PV 1157 Undetermined age	Skull and partial skeleton - *Novas et al. 2005* More robust than other carcharodontosaurids.	Length: **12 m** Hip height: **3.55 m**	**7 t**
7	*Mapusaurus roseae* eUC, SSA (Argentina)	MCF-PVPH-108.202 Undetermined age	Fibula - *Coria & Currie 2006* Is slightly larger than the holotype specimen of *Giganotosaurus*	Length: **12.7 m** Hip height: **3.65 m**	**7.6 t**
8	*Carcharodontosaurus saharicus* eUC, NAf (Egypt, Morocco)	SGM-Din 1 Undetermined age	Incomplete skull - *Sereno et al. 1996* The holotype is smaller. *Stromer 1934*	Length: **12.8 m** Hip height: **3.75 m**	**7.8 t**
9	*Giganotosaurus carolinii* eUC, SSA (Argentina)	MUCPv-95 Undetermined age	Incomplete dentary with teeth - *Calvo & Coria 2000* It is 6.5% larger than the holotype MUCPv-Ch1.	Length: **13.2 m** Hip height: **3.85 m**	**8.5 t**

COELUROSAURIA

Coelurosaurs similar to *Gasosaurus* - Relatively light bodies, small to medium game hunters.

N.º	Genus and species	Largest specimen	Material and data	Size	Mass
1	*Gasosaurus constructus* MJ, EA (China)	IVPP V7264 Undetermined age	Partial skeleton - *Dong & Tang 1985* Probable coelurosaur. *Holtz 2000*	Length: **4.5 m** Hip height: **1.05 m**	**150 kg**
2	*Lourinhanosaurus antunesi* UJ, WE (Portugal)	ML 370 Subadult	Partial skeleton - *Hutt, Martill & Barker 1996* Coelurosaur? *Carrano, et al. 2012*	Length: **5.7 m** Hip height: **1.4 m**	**310 kg**
3	*Siamotyrannus isanensis* ILC, CIM (Thailand)	PW9-1 Undetermined age	Pelvis and vertebrae - *Buffetaut et al. 1996* Basal coelurosaur? *Samathi 2015*	Length: **10 m** Hip height: **2.5 m**	**1.75 t**

Indeterminate coelurosaurs - Light bodies, small game hunters.

1	cf. *Nedcolbertia* eLC, WE (England)	BMNH R36539 Undetermined age	Incomplete femur- *Naish 2000* Undetermined	Length: **1 m** Hip height: **32 cm**	**1.1 kg**

N.º	Genus and species	Largest specimen	Material and data	Size	Mass
2	"Tonouchisaurus mongoliensis" (n.n.) eLC, EA (Mongolia)	Uncatalogued Juvenile?	Partial skeleton - *Anonymous 1994* It is not a tyrannosauroid.	Length: **1.1 m** Hip height: **35 cm**	1.5 kg
3	Aorun zhaoi MJ EA (China)	IVPP V15709 Juvenile	Skull and partial skeleton - *Choiniere et al. 2013* Received the name "Farragochela" in a thesis. *Choiniere 2010*	Length: **1.1 m** Hip height: **35 cm**	1.5 kg
4	Mirischia asymetrica ILC, NSA (northern Brazil)	SMNK 2349 PAL Subadult	Teeth - *Naish et al. 2004* Compsognathid or tyrannosauroid?	Length: **1.65 m** Hip height: **53 cm**	5 kg
5	Iliosuchus incognitus MJ, WE (England)	OUM J29780 Juvenile?	Ilium - *Huene 1932* The specimen OUM J28971 is not *Iliosuchus*. *Benson 2009*	Length: **1.8 m** Hip height: **57 cm**	6.3 kg
6	Aristosuchus pusillus (n.d.) ILC, WE (England)	BMNH R178 Undetermined age	Sacral vertebrae and incomplete pubis - *Seeley 1887* May have been a tyrannosauroid. *Naish 2011*	Length: **1.9 m** Hip height: **60 cm**	8 kg
7	Tugulusaurus faciles ILC, WE (China)	IVPP V4025 Undetermined age	Partial skeleton - *Dong 1973* Had a very long tibia.	Length: **2.3 m** Hip height: **75 cm**	14 kg
8	"Beelemodon" (n.n.) UJ, WNA (Wyoming - USA)	TATE Undetermined age	Teeth fragments- *Bakker 1997* Doubtful identification. *Mortimer**	Length: **2.3 m** Hip height: **75 cm**	14 kg
9	Coelurus fragilis UJ, WNA (Utah, Wyoming - USA)	YPM 1991 Undetermined age	Caudal vertebrae - *Marsh 1879* Represented the invalid "Coeluridae" family.	Length: **2.5 m** Hip height: **80 cm**	17 kg
10	Xinjiangovenator parvus ILC, EA (China)	IVPP V 4024-2 Undetermined age	Fragments and teeth - *Dong 1973* Originally considered *Phaedrolosaurus* remains. *Rauhut & Xu 2005*	Length: **2.7 m** Hip height: **85 cm**	22 kg
11	Nedcolbertia justinhofmanni ILC, WNA (Utah - USA)	CEUM 5073 Undetermined age	Caudal vertebrae and incomplete bones - *Kirkland et al. 1998* CEUM 5072 specimen is of a similar size.	Length: **3 m** Hip height: **95 cm**	30 kg
12	Bicentenaria argentina eUC, SSA (Argentina)	MPCA 865 Adult	Incomplete skull - *Novas et al. 2012* Was first published in print that same year.	Length: **3.1 m** Hip height: **1 m**	33 kg
13	Zuolong salleei UJ, EA (China)	IVPP V15912 Subadult	Skull and partial skeleton - *Choiniere et al. 2010* Hilmar Sallee donated funds for this research.	Length: **3.35 m** Hip height: **1.05 m**	43 kg
14	Sinocalliopteryx gigas ILC, EA (China)	CAGS-IG-T1 Undetermined age	Skull, partial skeleton, and filaments - *Ji et al. 2007* Compsognathid or tyrannosauroid? *Mortimer**	Length: **3.5 m** Hip height: **1.1 m**	48 kg
15	"Kagasaurus" (n.n.) eLC EA (Japan)	FPM 85050-1 Undetermined age	Tooth - *Manabe, Hasegawa & Azuma 1989* Informally named by Hisa in 1998.	Length: **3.9 m** Hip height: **1.25 m**	67 kg
16	"Dryosaurus" grandis ILC, ENA (Maryland - USA)	USNM 5701 Undetermined age	Caudal vertebrae - *Gilmore 1920* It was considered an ornithischian. *Lull 1911*	Length: **4.3 m** Hip height: **1.4 m**	90 kg
17	cf. Sinocalliopteryx sp. ILC, EA (China)	LDRCv2 Undetermined age	Tooth - *Xing et al. 2012* Encrusted in a *Dongbeititan* rib.	Length: **4.4 m** Hip height: **1.4 m**	97 kg
18	"Arkansaurus fridayi" (n.n.) ILC, ENA (Arkansas - USA)	UAM 74-16-2 Undetermined age	Partial posterior member - *Sattler 1983* The species "fridayi" was created 15 years later. *Braden 1998*	Length: **5.3 m** Hip height: **1.7 m**	170 kg

Coelurosaurs similar to *Compsognathus* - Light bodies, small game hunters.

N.º	Genus and species	Largest specimen	Material and data	Size	Mass
19	Juravenator starki UJ, WE (Germany)	JME Sch 200 Juvenile	Skull and partial skeleton - *Gohlich & Chiappe 2006* Has evidence of feathers.	Length: **74 cm** Hip height: **19 cm**	250 g
20	Sinosauropteryx prima ILC, EA (China)	NIGP 127587 Adult	Skull, partial skeleton, and filaments - *Chen, Dong & Zhen 1998* The first theropod in which filaments were found.	Length: **90 cm** Hip height: **33 cm**	1.1 kg
21	Compsognathus longipes UJ, WE (Germany, France)	MNHN CNJ 79 Adult?	Skull and partial skeleton - *Bidar et al. 1972* C. corallestris is an adult of C. longipes.	Length: **1.5 m** Hip height: **40 cm**	2.3 kg

Coelurosaurs similar to *Huaxiagnathus* - Light bodies, small game hunters.

N.º	Genus and species	Largest specimen	Material and data	Size	Mass
22	Huaxiagnathus orientalis ILC, EA (China)	NGMC 98-5-003 Adult?	Caudal vertebra - *Hwang et al. 2004* Is very different physically from *Compsognathus*	Length: **2.1** Hip height: **55 cm**	10 kg

Coelurosaurs similar to *Ornitholestes* - Light bodies, small game hunters.

N.º	Genus and species	Largest specimen	Material and data	Size	Mass
23	Scipionyx samniticus ILC, WE (Italy)	SBA-SA 163760 Juvenile	Skull and partial skeleton - *Dal Sasso & Signore 1998* Was three years old.	Length: **53 cm** Hip height: **10 cm**	100 g
24	Ornitholestes hermanni UJ, WNA (Wyoming - USA)	AMNH 619 Undetermined age	Skull and partial skeleton - *Osborn 1903* More derived than other basal coelurosaurs.	Length: **2.1 m** Hip height: **55 cm**	12 kg

1:150

MEGARAPTORA

Megaraptors similar to *Chilantaisaurus* - Relatively robust bodies, medium to big game hunters.

N.º	Genus and species	Largest specimen	Material and data	Size	Mass
1	Fukuiraptor kitadaniensis ILC, EA (Japan)	FPDM-V98081540 Adult	Tooth - *Currie & Azuma 2006* FPDM 9712201-9712228 is a partial skeleton. *Azuma & Currie 2000*	Length: **4.3 m** Hip height: **1.45 m**	590 kg
2	"Chilantaisaurus" sibiricus eLC, WA (eastern Russia)	FMNH PR2716 Subadult	Incomplete metatarsus - *Riabinin 1915* It is the most ancient known megaraptor.	Length: **8.4 m** Hip height: **2.3 m**	1.4 t
3	"Capitalsaurus" potens ILC, ENA (Maryland - USA)	USNM 3049 Undetermined age	Incomplete caudal vertebra - *Lull 1911* Doubtful identification.	Length: **11.6 m** Hip height: **3 m**	3.9 t

N.º	Genus and species	Largest specimen	Material and data	Size	Mass
4	*Siats meekerorum* eUC, WNA (Utah - USA)	FMNH PR2716 Subadult	Partial skeleton - *Zanno & Makovicky 2013* Was considered a neovenatorid.	Length: **11.7 m** Hip height: **3.2 m**	**3.9 t**
5	*Chilantaisaurus tashuikouensis* eUC, EA (China)	IVPP V2884 Undetermined age	Tooth and partial skeleton - *Hu 1964* Primitive megaraptor? *Benson et al. 2010*	Length: **11.9 m** Hip height: **3.3 m**	**4.1 t**

Megaraptors similar to *Australovenator* - Relatively light bodies, medium to big game hunters.

6	*Rapator ornitholestoides* ILC, OC (Australia)	BMNH R3718 Undetermined age	Metacarpus - *Huene 1932* More recent and unlike *Australovenator wintoniensis*	Length: **4.8 m** Hip height: **1.45 m**	**270 kg**
7	*Australovenator wintonensis* eUC, OC (Australia)	AODF 604 Undetermined age	Dentary, teeth, and partial skeleton - *Hocknull et al. 2009* More recent than *Australovenator* sp.	Length: **5.7 m** Hip height: **1.7 m**	**450 kg**

Megaraptora similar to *Megaraptor* - Relatively light bodies, medium to big game hunters.

8	"Mangahouanga" (n.n.) IUC, OC (Nueva Zelanda)	NZGS CD1 Undetermined age	Caudal vertebra - *Molnar 1980* Indeterminate; its name was created by accident. *Lambert 1983*	Length: **3.5 m** Hip height: **1.05 m**	**130 kg**
9	*Aerosteon riocoloradensis* IUC, SSA (Argentina)	MCNA-PV-3137 Subadult	Partial skeleton - *Sereno et al. 2008* It had very developed air sacs.	Length: **7.5 m** Hip height: **1.7 m**	**1 t**
10	*Megaraptor namunhuaiquii* eUC, SSA (Argentina)	MCF-PVPH 79 Adult	Ulna, phalanges, and metatarsus - *Novas 1998* It had very developed claws.	Length: **8.3 m** Hip height: **2.35 m**	**1.3 t**
11	*Orkoraptor burkei* eUC, SSA (Argentina)	MPM-Pv 3457 Undetermined age	Skull and partial skeleton - *Novas et al. 2008* Its name refers to La Leona River in Aoniken.	Length: **8.4 m** Hip height: **2.4 m**	**1.4 t**

Megaraptors similar to *Deltadromeus* - Light bodies, small game hunters.

12	cf. *Bahariasaurus ingens* eUC, NAf (Niger)	MNHN Undetermined age	Caudal vertebrae - *Lapparent 1960* *Bahariasaurus ingens*?	Length: **4.1 m** Hip height: **1.05 m**	**175 kg**
13	*Deltadromeus agilis* eUC, NAf (Morocco)	SGM-Din 2 Undetermined age	Partial skeleton - *Sereno et al. 1996* Other material of larger but different animals have been attributed to it. *Mortimer**	Length: **7.2 m** Hip height: **2.1 m**	**1 t**
14	"Nototyrannus violantei" (n.n.) eUC, SSA (Argentina)	Uncatalogued Undetermined age	Partial skeleton - *Anonymous 2011* Undescribed material.	Length: **7.6 m** Hip height: **2.15 m**	**1.1 t**
15	*Bahariasaurus ingens* eUC, NAf (Egypt)	IPHG 1922 X 47 Undetermined age	Partial skeleton - *Stromer 1934* A theropod unlike *Deltadromeus*. *Mortimer 2015*	Length: **12.2 m** Hip height: **3.5 m**	**4.6 t**

1:250

TYRANNOSAUROIDEA

Tyrannosauroids similar to *Guanlong* - Light bodies, small to medium game hunters.

N.º	Genus and species	Largest specimen	Material and data	Size	Mass
1	*Proceratosaurus bradleyi* MJ, WE (England)	BMNH R4860 Undetermined age	Incomplete skull and fragments - *Woodward 1910* Was considered a ceratosaurid. *Huene 1926*	Length: **2.7 m** Hip height: **95 cm**	**95 kg**
2	cf. *Iliosuchus incognitus* MJ, WE (England)	OUM J28971 Undetermined age	Tooth - *Larser & Heimhofer 2013* Similar to *Proceratosaurus*	Length: **3.4 m** Hip height: **1.2 m**	**115 kg**
3	*Guanlong wucaii* UJ, EA (China)	IVPP V14531 Adult	Skull and partial skeleton - *Xu et al. 2006* Convergent similarities with *Monolophosaurus*	Length: **4 m** Hip height: **1.4 m**	**180 kg**
4	*Kileskus aristocutus* MJ, WA (western Russia)	ZIN PH 5/117 Undetermined age	Incomplete jaw - *Averianov et al. 2010* As ancient as *Proceratosaurus bradleyi*	Length: **5.2 m** Hip height: **1.85 m**	**700 kg**

Tyrannosauroids similar to *Yutyrannus* - Relatively robust bodies, medium to big game hunters.

5	"Megalosaurus" hungaricus (n.d.) IUC, EE (Romania)	MAFI Ob 3106 Undetermined age	Tooth - *Nopcsa 1902* Tyrannosauroid or dromaeosaurid? *Carrano et al. 2012*	Length: **2.7 m** Hip height: **70 cm**	**47 kg**
6	"Poekilopleuron" schmidti (n.d.) eUC, WA (western Russia)	FPDM-V98081540 Undetermined age	Cervical rib and and gastralia? - *Kiprijanow 1883* An attributed metacarpus was of a sauropod. *Mortimer**	Length: **4.6 m** Hip height: **1.2 m**	**220 kg**
7	*Prodeinodon mongoliensis* (n.d.) eLC, EA (Mongolia)	Uncatalogued Undetermined age	Tooth - *Huene 1923* Was also referred to as *P. mongoliense*	Length: **4.8 m** Hip height: **1.25 kg**	**265 kg**
8	cf. "Megalosaurus" pannoniensis IUC, WE (Portugal)	Uncatalogued Undetermined age	Fragments - *Lapparent & Zbyszewski 1957* Doubtful assignment.	Length: **6 m** Hip height: **1.55 m**	**525 kg**
9	*Altispinax dunkeri* (n.d.) ILC, WE (Germany)	UM 84 Undetermined age	Tooth - *Huene 1923* Is *Prodeinodon* a synonym of *Altispinax*? *Ruiz-Omeñaca & Canudo 2003; Ford**	Length: **6.3 m** Hip height: **1.6 m**	**600 kg**
10	*Phaedrolosaurus ilikensis* (n.d.) ILC, EA (China)	IVPP V 4024-1 Undetermined age	Tooth - *Dong 1973* Was considered a giant dromeosaurid.	Length: **6.3 m** Hip height: **1.6 m**	**600 kg**
11	cf. "Megalosaurus" pannoniensis IUC, WE (France)	Uncatalogued Undetermined age	Premaxillary tooth- *Lapparent 1947* Doubtful identification.	Length: **6.8 m** Hip height: **1.75 m**	**750 kg**
12	cf. "Megalosaurus" pannoniensis IUC, EE (Hungary)	V.01 Undetermined age	Tooth - *Osi & Apesteguía 2008* More ancient than "*Megalosaurus*" *pannoniensis*	Length: **6.8 m** Hip height: **1.75 m**	**750 kg**
13	"Megalosaurus" pannoniensis IUC, WE (Austria)	PIUW Undetermined age	Incomplete tooth - *Osi et al .2010* Similar to *Altispinax*	Length: **6.85 m** Hip height: **1.75 m**	**780 kg**
14	*Sinotyrannus kazouensis* ILC, EA (China)	KZV-001 Undetermined age	Tooth - *Ji, Ji & Zhang 2009* It is much bigger and more recent than its relatives.	Length: **7.5 m** Hip height: **2 m**	**1.2 t**

#	Name / Location	Specimen / Age	Remains / Notes	Size	Weight
15	Yutyrannus huali / ILC, EA (China)	ZCDMV5000 / Adult	Skull, skeleton, and filaments - *Xu et al. 2012* / The theropod with the largest filaments.	Length: 8.5 m / Hip height: 2.2 m	1.5 t
16	cf. Prodeinodon mongoliensis (n.d.) / ILC, EA (Mongolia)	Uncatalogued / Undetermined age	Tooth, tibia, and incomplete fibula - *Bohlin 1953* / Not known with certainty if it is a *Prodeinodon*	Length: 9.8 m / Hip height: 2.5 m	2.3 t

1:250

Tyrannosauroids similar to *Dilong* - Light bodies, small to medium game hunters.

#	Name / Location	Specimen / Age	Remains / Notes	Size	Weight
1	Aviatyrannis jurassica / UJ, WE (Portugal)	IPFUB Gui Th 1 / Undetermined age	Ilium - *Rauhut 2003* / Probably a juvenile.	Length: 1.3 m / Hip height: 40 cm	5.5 kg
2	Santanaraptor placidus / ILC, NSA (northern Brazil)	MN 4802-V / Juvenile	Tooth - *Kellner 1996* / Tyrannosauroid? *Porfiri et al. 2014*	Length: 1.7 m / Hip height: 50 cm	12 kg
3	Dilong paradoxus / ILC, EA (China)	IVPP V14243 / Subadult	Skull, partial skeleton, and filaments - *Xu et al. 2004* / The specimen IVPP V11579 is less ancient.	Length: 1.8 m / Hip height: 55 cm	15 kg
4	"Zunityrannus" (n.n.) / eUC, WNA (New Mexico - USA)	AZMNH / Undetermined age	Skull and partial skeleton - *Wolfe & Kirkland 1998* / The name appears in a 2011 BBC documentary.	Length: 1.9 m / Hip height: 57 cm	18 kg
5	Calamospondylus oweni / eLC, WE (England)	IPFUB Gui Th 1 / Undetermined age	Sacrum and incomplete ilium - *Fox in Anonymous 1866* / *Aristosuchus* or *Calamosaurus*?	Length: 2.4 m / Hip height: 72 cm	24 kg
6	Calamosaurus foxii / ILC, WE (England)	BMNH R901 / Subadult	Cervical vertebrae - *Lydekker 1889* / Similar to *Dilong paradoxus*. *Naish 2011*	Length: 2.45 m / Hip height: 75 cm	29 kg
7	Kakuru kujani / ILC, OC (Australia)	SAM P17926 / Undetermined age	Tibia and incomplete fibula - *Molnar & Pledge 1980* / Abelisauroid or tyrannosauroid? *Mortimer**	Length: 3.25 m / Hip height: 95 cm	85 kg
8	Nuthetes destructor / eLC, WE (England)	DORCM G 913 / Undetermined age	Partial skull and teeth - *Owen 1854* / Was considered a dromeosaurid.	Length: 3.4 m / Hip height: 1 m	100 kg
9	Stokesosaurus clevelandi / WNA (South Dakota, Utah - USA)	UUVP 2320 / Undetermined age	Ilium - *Madsen 1974* / Proceratosaurid?	Length: 3.4 m / Hip height: 1 m	125 kg
10	Eotyrannus lengi / ILC, WE (England)	MIWG 1997.550 / Subadult	Tooth - *Hutt et al. 2001* / Tyrannosauroid or megaraptor? *Porfiri et al. 2014*	Length: 4.3 m / Hip height: 1.3 m	200 kg
11	Timimus hermani / ILC, OC (Australia)	NMV P186303 / Undetermined age	Femur - *Rich & Vickers-Rich 1994* / Probable tyrannosauroid? *Mortimer**	Length: 4.3 m / Hip height: 1.3 m	200 kg
12	Tanycolagreus topwilsoni / UJ, WNA (Wyoming - USA)	UUVP 2999 / Adult?	Premaxilla - *Madsen 1974* / The specimen TPII 2000-09-29 is a subadult.	Length: 4.8 m / Hip height: 1.45 m	280 kg
13	cf. Stokesosaurus clevelandi / UJ, WNA (Colorado - USA)	BYUVP 4862 / Undetermined age	Ischium - *Britt 1991* / May have been a *Stokesosaurus clevelandi*	Length: 5.8 m / Hip height: 1.7 m	500 kg
14	Juratyrant langhami / UJ, WE (England)	OUMNH J.3311 / Undetermined age	Tooth - *Benson 2008* / Unlike *Stokesosaurus*. *Brusatte & Benson 2013*	Length: 6.7 m / Hip height: 1.9 m	760 kg

Tyrannosauroids similar to *Bagaraatan* - Relatively light bodies, small to medium game hunters.

#	Name / Location	Specimen / Age	Remains / Notes	Size	Weight
15	Bagaraatan ostromi / IUC, EA (Mongolia)	ZPAL MgD-I/108 / Adult	Jaw and partial skeleton - *Osmólska 1996* / Quite unlike other theropods.	Length: 3.2 m / Hip height: 80 cm	80 kg

Tyrannosauroids similar to *Dryptosaurus* - Relatively robust bodies, medium to big game hunters.

#	Name / Location	Specimen / Age	Remains / Notes	Size	Weight
16	Xiongguanlong baimoensis / ILC, EA (China)	FRDC-GS JB16-2-1 / Adult	Skull and partial skeleton - *Li et al. 2010* / Among the most primitive.	Length: 4.6 m / Hip height: 1.35 m	275 kg
17	Alectrosaurus olseni / IUC, EA (China)	AMNH 6554 / Undetermined age	Partial skeleton - *Gilmore 1933* / Was a very fast hunter. *Carr 2005*	Length: 5.8 m / Hip height: 1.75 m	575 kg
18	Dryptosaurus macropus / IUC, ENA (New Jersey - USA)	AMNH 2550 / Undetermined age	Tooth - *Cope 1868* / More ancient than *Dryptosaurus aquilunguis*	Length: 6.7 m / Hip height: 2 m	870 kg
19	Dryptosaurus aquilunguis / IUC, ENA (New Jersey - USA)	ANSP 9995 / Adult	Skull and partial skeleton - *Cope 1866* / The first theropod reconstructed as a biped.	Length: 7.1 m / Hip height: 2.1 m	1.1 t
20	Diplotomodon horrificus (n.d.) / IUC, ENA (New Jersey - USA)	ANSP 9680 / Undetermined age	Tooth - *Leidy 1865* / *Dryptosaurus aquilunguis*?	Length: 7.4 m / Hip height: 2.3 m	1.2 t
21	Embasaurus minax (n.d.) / eLC, EA (Kazakhstan)	Uncatalogued / Subadult	Thoracic and cervical vertebra - *Riabinin 1931* / Is very large, despite being very ancient.	Length: 8.4 m / Hip height: 2.55 m	1.7 t

Tyrannosauroids similar to *Appalachiosaurus* - Relatively robust bodies, medium to big game hunters.

#	Name / Location	Specimen / Age	Remains / Notes	Size	Weight
22	"Futabasaurus" (n.n.) / IUC, EA (Japan)	Uncatalogued / Undetermined age	Incomplete tibia - *Hasegawa et al. 1987* / Tyrannosaurid? *Lambert 1990*	Length: 2.4 m / Hip height: 78 cm	53 kg
23	Chingkankousaurus fragilis (n.d.) / IUC, EA (China)	IVPP V636 / Undetermined age	Incomplete scapula - *Young 1958* / Tyrannosauroid or tyrannosaurid? *Brusatte et al. 2013*	Length: 6.1 m / Hip height: 2 m	890 kg
24	Appalachiosaurus montgomeriensis / IUC, ENA (Alabama - USA)	RMM 6670 / Subadult	Skull and semicomplete skeleton - *Carr et al. 2005* / Is not a tyranosaurid, despite its appearance.	Length: 6.4 m / Hip height: 2.1 m	1.1 t
25	Labocania anomala / IUC, WNA (Mexico)	LACM 20877 / Undetermined age	Skull and partial skeleton - *Molnar 1974* / Tyrannosauroid or tyrannosaurid? *Ramírez-Velasco et al. 2015*	Length: 8.2 m / Hip height: 2.7 m	2.6 t
26	Bistahieversor sealeyi / IUC, WNA (New Mexico - USA)	NMMNH P-27469 / Adult	Skull and partial skeleton - *Carr 2005; Carr & Williamson 2010* / Tyrannosaurid or tyrannosauroid? *Loewen et al. 2013; Brusatte et al. 2015*	Length: 9 m / Hip height: 3 m	3.3 t

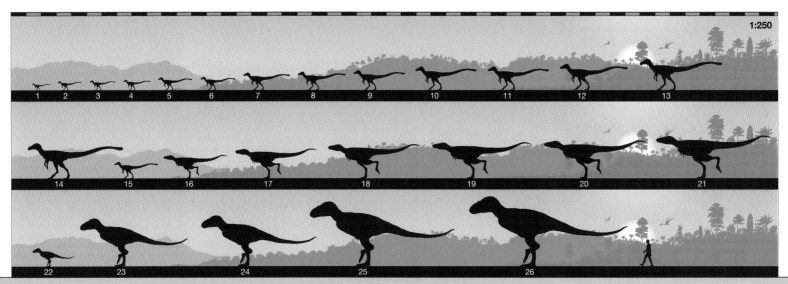

1:250

TYRANNOSAURIDAE

Tyrannosaurids similar to *Alioramus* - Light bodies, small to medium game hunters.

N.º	Genus and species	Largest specimen	Material and data	Size	Mass
1	*Alioramus altai* IUC, EA (Mongolia)	IGM 100/1844 Juvenile	Skull and partial skeleton - *Brusatte et al. 2009* More recent than *Alioramus remotus*	Length: 5 m Hip height: 1.6 m	385 kg
2	*Alioramus remotus* IUC, EA (Mongolia)	GI 3141/1 Juvenile	Skull and partial skeleton - *Kurzanov 1976* Tyrannosauroid or tyrannosaurid? *Loewen et al. 2013; Brusatte et al. 2015*	Length: 5.5 m Hip height: 1.75 m	500 kg
3	*Qianzhousaurus sinensis* IUC, EA (China)	GM F10004-1/8 Undetermined age	Skull and partial skeleton - *Lu et al. 2014* *Aliramus sinensis? Mortimer**	Length: 6.3 m Hip height: 2 m	750 kg

Indeterminate tyrannosaurids - Robust bodies, medium to big game hunters.

N.º	Genus and species	Largest specimen	Material and data	Size	Mass
4	*"Ornithomimus" tenuis* (n.d.) IUC, WNA (Montana - USA)	USNM 5814 Juvenile	Incomplete metatarsus - *Marsh 1890* Probably a juvenile of a known genus.	Length: 3.3 m Hip height: 1.2 m	110 kg
5	*Aublysodon lateralis* IUC, WNA (Montana - USA)	AMNH 3956 Juvenile	Tooth - *Cope 1876* May be a juvenile of a known tyrannosaurid.	Length: 4.3 m Hip height: 1.55 m	245 kg
6	*"Ornithomimus" grandis* (n.d.) IUC, WNA (Montana - USA)	Uncatalogued Undetermined age	Metatarsus - *Leidy 1856* Was thought to be an ornithomimosaur because of its arctometatarsal.	Length: 7.15 m Hip height: 2.4 m	1.6 t
7	*Deinodon horridus* (n.d.) IUC, WNA (Montana - USA)	Uncatalogued Undetermined age	Tooth - *Leidy 1856* The first tyrannosaurid, unidentifiable.	Length: 8 m Hip height: 2.65 m	2.2 t
8	*"Dryptosaurus" kenabekides* (n.d.) IUC, WNA (Montana - USA)	ANSP 9530 Undetermined age	Tooth - *Leidy 1856* *Deinodon horridus?*	Length: 9 m Hip height: 3 m	3.2 t

Tyrannosaurids similar to *Albertosaurus* - Robust bodies, medium to big game hunters.

N.º	Genus and species	Largest specimen	Material and data	Size	Mass
9	*"Albertosaurus" periculosus* (n.d.) IUC, EA (China)	PIN Subadult	Tooth - *Riabinin 1930* More recent and primitive than *Tarbosaurus bataar*	Length: 6.7 m Hip height: 2.2 m	1.3 t
10	*"Albertosaurus" incrassatus* IUC, WNA (Montana - USA)	AMNH 3962 Subadult	Skull and partial skeleton - *Cope 1876* *Albertosaurus* or *Gorgosaurus?*	Length: 6.8 m Hip height: 2.25 m	1.4 t
11	*Gorgosaurus libratus* IUC, WNA (western Canada, Montana - USA)	AMNH 5458 Adult	Skull and partial skeleton - *Russell 1970* More recent and unlike *Albertosaurus*	Length: 8.8 m Hip height: 2.9 m	2.9 t
12	*Albertosaurus sarcophagus* IUC, WNA (western Canada)	CMN 5600 Undetermined age	Incomplete skull - *Osborn 1905* Some specimens attributed to this species are actually of *Gorgosaurus*	Length: 9.7 m Hip height: 3.2 m	4 t

1:250

Tyrannosaurids similar to *Daspletosaurus* - Light bodies, small to medium game hunters.

N.º	Genus and species	Largest specimen	Material and data	Size	Mass
1	*Deinodon falculus* (n.d.) IUC, WNA (Montana - USA)	AMNH 3959 Juvenile	Premaxillary teeth - *Cope 1876* Has never been illustrated.	Length: 1.3 m Hip height: 48 cm	7 kg
2	*Aublysodon mirandus* (n.d.) IUC, WNA (Montana - USA)	ANSP 9533 Juvenile	Premaxillary tooth - *Leidy 1868* Juvenile *Daspletosaurus?*	Length: 1.55 m Hip height: 57 cm	12 kg
3	*Deinodon hazenianus* (n.d.) IUC, WNA (Montana - USA)	UW 34823 Juvenile	Premaxillary tooth - *Cope 1876* There are several questionable pieces in Montana.	Length: 2.8 m Hip height: 1 m	70 kg
4	*Nanuqsaurus hoplundi* IUC, WNA (Alaska - USA)	DMNH 21461 Adult	Incomplete skull - *Fiorillo & Tykoski 2014* Lived in colder climates than its relatives.	Length: 6 m Hip height: 2 m	900 kg

5	*Teratophoneus curriei* IUC, WNA (Utah - USA)	BYU 13719 Undetermined age	Femur - *Carr et al. 2011* Named in a thesis. *Carr 2005*	Length: **6.4 m** Hip height: **2.15 m**	1.15 t
6	*Lythronax argestes* IUC, WNA (Utah - USA)	UMNH VP 20200 Undetermined age	Skull and partial skeleton - *Loewen et al. 2013* Close to *Tarbosaurus* and *Tyrannosaurus*	Length: **6.8 m** Hip height: **2.3 m**	1.4 t
7	*Daspletosaurus torosus* IUC, WNA (western Canada)	CMN 8506 Adult	Skull and partial skeleton - *Russell 1970* AMNH 5438 y UA 11 are of the same size.	Length: **8.8 m** Hip height: **2.95 m**	2.8 t
8	"*Alamotyrannus brinkmani*" (n.n.) IUC, WNA (New Mexico - USA)	USNM Undetermined age	Vertebrae - *Gilmore 1916* Similar in size to *Tyrannosaurus*	Length: **10.7** Hip height: **3.6 m**	5 t

1:250

Tyrannosaurids similar to *Tarbosaurus* - Robust bodies, medium to big game hunters.

1	*Shanshanosaurus houyanshanensis* IUC, EA (China)	IVPP V4878 Juvenile	Partial skeleton - *Dong 1977; Zhai et al. 1978* There is a large specimen contemporary of *Shanshanosaurus*. It may have been *Tarbosaurus bataar* or any other tyrannosaurid of the time. *Zhai et al. 1978*	Length: **2.1 m** Hip height: **70 cm**	23 kg
2	*Raptorex kriegsteini* (n.d.) IUC, EA (Mongolia or China?)	LH PV18 Juvenile	Tibia and metatarsus - *Sereno et al. 2009* *Tarbosaurus bataar* hatchling? *Fowler et al. 2011*	Length: **2.6 m** Hip height: **85 cm**	40 kg
3	cf. *Tarbosaurus bataar* IUC, EA (Mongolia)	ZPAL MgD-I/30 Undetermined age	Tibia and metatarsus - *Holtz 1994* More recent than *Tarbosaurus bataar*?	Length: **7.7 m** Hip height: **2.3 m**	2 t
4	cf. *Szechuanosaurus campi* (n.d.) IUC, EA (China)	IVPP V756 Undetermined age	Mandible and teeth - *Camp 1935* The reference with *S. campi* is wrong. *Young 1942*	Length: **8.1 m** Hip height: **2.45 m**	2.4 t
5	*Tarbosaurus luanchuanensis* (n.d.) IUC, EA (China)	IVPP V4733 Undetermined age	Teeth - *Dong 1979* *Tarbosaurus bataar*?	Length: **9.1 m** Hip height: **2.75 m**	3.5 t
6	*Tarbosaurus zhuchengensis* (n.d.) IUC, EA (China)	NGMC V1777 Undetermined age	Teeth - *Hu 1973* *Zhuchengtyrannus magnus*?	Length: **9.4 m** Hip height: **2.8 m**	3.7 t
7	"*Tarbosaurus*" aff. *bataar* IUC, WA (Kazakhstan)	IZK 33/MP-61 Undetermined age	Incomplete dentary - *Khozatsky 1957; Nesov 1995* The most ancient tyrannosaurid. *Averianov et al. 2012*	Length: **9.5 m** Hip height: **2.9 cm**	3.8 t
8	*Zhuchengtyrannus magnus* IUC, EA (China)	ZCDM V0031 Undetermined age	Jaw and dentary - *Hone et al. 2011* Contemporary of *Tarbosaurus zhuchengensis*.	Length: **9.6 m** Hip height: **2.9 m**	4 t
9	*Tarbosaurus bataar* IUC, EA (Mongolia)	PIN 551-1 Adult	Skull and vertebrae - *Maleev 1955* Several specimens were identified as new genera or species.	Length: **10 m** Hip height: **3 m**	4.5 t

Tyrannosaurids similar to *Tyrannosaurus* - Robust bodies, medium to big game hunters.

10	*Nanotyrannus lancensis* (n.d.) IUC, WNA (Montana - USA)	BMRP 2002.4.1 Juvenile	Skull and partial skeleton - *Henderson 2005* *Tyrannosaurus rex* juvenile or another species?	Length: **5.5 m** Hip height: **2 m**	520 kg
11	cf. *Tyrannosaurus rex* IUC, WNA (Mexico)	ERNO 8549 Undetermined age	Teeth - *Serrano-Brañas et al. 2014* Probably a subadult *Tyrannosaurus rex*.	Length: **9.1 m** Hip height: **2.75 m**	2.5 t
12	*Tyrannosaurus rex* IUC, WNA (Montana, New Mexico, South Dakota, North Dakota, Texas, Wyoming - USA, western Canada)	UCMP 137538 Adult	Foot phalanx - *Longrich et al. 2010* The specimen UCMP 137538 has caused confusion. It has been said that its phalanx is 17% longer than the homologous one from "Sue" FMNH PR2081. This is incorrect, as the pieces were measured in different ways. After a morphometric and photographic analysis, it was discovered that the phalanx of UCMP 137538 is slightly wider than that of "Sue" (by ~1.5%). So it is possible that the animal was just barely larger than "Sue."	Length: **12.3 m** Hip height: **3.75 m**	8.5 t

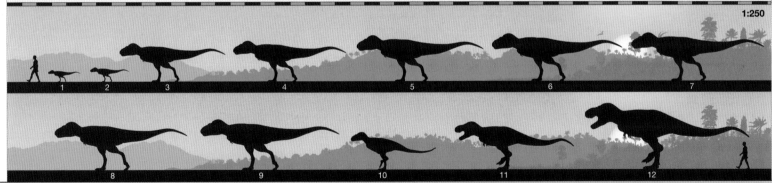

1:250

ORNITHOMIMOSAURIA

Ornithomimosaurs similar to *Nqwebasaurus* - Light bodies, small game hunters.

N.º	Genus and species	Largest specimen	Material and data	Size	Mass
1	*Lepidocheirosaurus natatilis* (n.d) MJ, WA (eastern Russia)	PIN 5435/1 Undetermined age	Metacarpals and phalanges - *Alifanov & Saveliev 2015* Ornithischian *Kulindadromeus zabaikalicus*? *Mortimer**	Length: **1.1 m** Hip height: **32 cm**	1.9 kg
2	*Nqwebasaurus thwazi* eLC, SAf (South Africa)	AM 6040 Subadult	Skull and partial skeleton - *de Klerk et al. 1997* Basal alvarezsauroid? *Dal Sasso & Maganuco 2011; Choiniere et al. 2012; Lee et al. 2014*	Length: **1.2 m** Hip height: **35 cm**	2.5 kg
3	*Aniksosaurus darwini* eUC, SSA (Argentina)	MTD-PV 1/17 Undetermined age	Incomple ulna - *Martínez & Novas 2006* Similar to *Nqwebasaurus*; alvarezsauroid? *Cau 2009*	Length: **2.6 m** Hip height: **80 cm**	25 kg

Ornithomimosaurs similar to *Harpymimus* - Light bodies, small conical teeth, small game hunters, omnivores or herbivores.

| 4 | *Hexing qingyi*
ILC, EA (China) | JLUM-JZ07b1
Adult | Skull and partial skeleton - *Jin et al. 2012*
Its head was very large for its size. | Length: **1.2 m**
Hip height: **45 cm** | 4.5 kg |

5	Shenzhousaurus orientalis ILC, EA (China)	NGMC 97-4-002 Undetermined age	Skull and partial skeleton - *Jin et al. 2003* It had fewer teeth than *Pelecanimimus*	Length: **1.7 m** Hip height: **60 cm**	12 kg
6	Kinnareemimus khonkaenensis ILC, CIM (Thailand)	PW5A-110 Undetermined age	Incomplete metatarsus - *Buffetaut et al. 2009* The most ancient arctometatarsal dinosaur; may have been an alvarezsaurid.	Length: **1.8 m** Hip height: **64 cm**	13 kg
7	Pelecanimimus polyodon ILC, WE (Spain)	LH 7777 Undetermined age	Skull and partial skeleton - *Perez-Moreno et al. 1994* The theropod with the most dentary pieces.	Length: **1.9 m** Hip height: **70 cm**	17 kg
8	Harpymimus okladnikovi ILC, EA (China)	IGM 100/29 Adult	Caudal vertebrae and incomplete bones - *Barsbold & Perle 1984* The specimen CEUM 5072 is of a similar size. *Kobayashi & Barsbold 2005*	Length: **4.45 m** Hip height: **1.6 m**	215 kg

Ornithomimosaurs similar to *Garudimimus* - Relatively graceful, toothless, small game hunters, omnivores or herbivores.

9	Garudimimus brevipes eUC, EA (Mongolia)	GIN 100/13 Undetermined age	Skull and partial skeleton - *Barsbold 1981* The "bone crest" was a broken bone. *Kobayashi 2004*	Length: **2.8 m** Hip height: **1.05 m**	90 kg
10	Valdoraptor oweni eLC, WE (England)	BMNH R2559 Undetermined age	Incomplete metatarsals - *Lydekker 1889* It was an ornithomimosaur. *Allain et al. 2014*	Length: **4.8 m** Hip height: **1.8 m**	435 kg
11	Beishanlong grandis ILC, EA (China)	FRDC-GS GJ (06) 01-18 Subadult	Partial skeleton - *Makovicky et al. 2010* Almost as large as *Gallimimus*.	Length: **5 m** Hip height: **1.85 m**	500 kg
12	"Sanchusaurus" (n.n.) ILC, EA (Japan)	GMNH-PV-028 Undetermined age	Sacral vertebra - *Hasegawa 1984* May have been closer to *Harpymimus* or *Garudimimus*	Length: **5.9 m** Hip height: **2.9 m**	815 kg

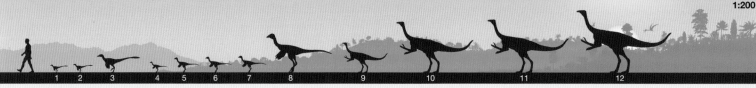

1:200

Ornithomimosaurs similar to *Deinocheirus* - Robust bodies, toothless, omnivores or herbivores.

| 1 | Deinocheirus mirificus
IUC, EA (Mongolia) | IGM 100/127
Adult | Skull and partial skeleton - *Lee et al. 2014*
The most derived hunch-backed theropod. | Length: **12 m**
Hip height: **4.4 m** | 7 t |
| 2 | Deinocheirus sp.
eUC, EA (Mongolia) | Uncatalogued
Undetermined age | Ulna? - *Suzuki et al. 2010*
The piece was not described. It is too large to be a *Deinocheirus* ulna. It may belong to another dinosaur group, or it may even be another type of long bone. More recent than *D. mirificus*. | Length: **16 m?**
Hip height: **5.7 m?** | 16 t? |

1:250

Ornithomimosaurs similar to *Ornithomimus* - Light bodies, short torso, toothless, omnivores or herbivores.

1	"Grusimimus tsuru" (n.n.) ILC, EA (Mongolia)	GIN 960910KD Subadult	Partial skeleton - *Barsbold 1997 (unpublished)* In a species list. *Kobayashi 2004*	Length: **2.15 m** Hip height: **75 cm**	27 kg
2	"Coelosaurus" antiquus IUC, ENA (New Jersey - USA)	ANSP 9222 Adult	Tibia - *Leidy 1865* The name was already used by Owen 1854.	Length: **2.8 m** Hip height: **1 m**	60 kg
3	cf. "Coelosaurus" antiquus IUC, ENA (Maryland - USA)	USNM 256614 Undetermined age	Caudal vertebra - *Baird 1986* Impossible to compare with *Coelosaurus*	Length: **2.85 m** Hip height: **1 m**	63 kg
4	Qiupalong henanensis IUC, EA (China)	HGM 41HIII-0106 Undetermined age	Metacarpus - *Xu et al. 2011* It is similar to North American genera.	Length: **2.85 m** Hip height: **1 m**	63 kg
5	"Saltillomimus rapidus" (n.n.) IUC, WNA (Mexico)	SEPCP 16/237 Adult	Partial skeleton - *Martinez 2014* Described in a thesis. *Aguillon Martinez 2010*	Length: **2.9 m** Hip height: **1.05 m**	68 kg
6	Dromiceiomimus brevitertius IUC, WNA (western Canada)	CMN 12228 Undetermined age	Partial skeleton - *Russell 1972* The name *Dromiceiomimus brevitertius* predates *Ornithomimus edmontonicus*, but the latter is used more frequently.	Length: **3.7 m** Hip height: **1.3 m**	140 kg
7	Archaeornithomimus cf. asiaticus IUC, WA (Tajikistan)	PIN 3041/2 Undetermined age	Incomplete humerus - *Alifanov et al. 2006; Ryan 1997* It is not *Archaeornithomimus bissektensis*.	Length: **3.7 m** Hip height: **1.3 m**	140 kg
8	Sinornithomimus dongi IUC, EA (China)	IVPP-V11797-19 Adult	Ulna and femur - *Kobayashi & Lu 2003* Several were preserved in natural traps.	Length: **3.75 m** Hip height: **1.35 m**	150 kg
9	"Archaeornithomimus" bissektensis eUC, WA (Uzbekistan)	ZIN PH 190/16 Adult?	Nail phalanx - *Sues & Averianov 2015* The species was created from the juvenile specimen CCMGE 479/12457. *Nesov 1995*	Length: **3.75 m** Hip height: **1.35 m**	150 kg
10	Dromiceiomimus samueli IUC, WNA (western Canada)	CMN 12441 Adult	Partial skeleton - *Makovicky 1995* More recent than *D. brevitertius*.	Length: **3.95 m** Hip height: **1.4 m**	165 kg
11	Anserimimus planinychus IUC, EA (China)	ZPAL MgD-I/65 Undetermined age	Partial skeleton - *Bronowicz 2005* The largest specimen is described in a thesis.	Length: **4.4 m** Hip height: **1.55 m**	245 kg

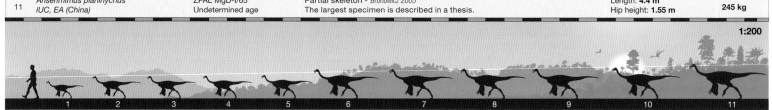

1:200

Ornithomimosaurs similar to *Gallimimus* - Light bodies, short torso, toothless, omnivores or herbivores.

1	Ornithomimus velox IUC, WNA (Colorado, Montana - USA)	YPM 542 Undetermined age	Incomplete frontal limb - *Marsh 1890* The specimen AMNH 5884 (4.9 m and 200 kg) is attributed to it. *Osborn 1916*	Length: **2.45 m** Hip height: **80 cm**	22 kg
2	cf. Struthiomimus altus IUC, WNA (Mexico)	BENC 1/2-0081 Undetermined age	Incomplete femur - *Torres-Rodriguez 2006* Indeterminate genus. *Ramírez-Velasco et al. 2015*	Length: **3.25 m** Hip height: **1.1 m**	77 kg
3	Tototlmimus packardensis IUC, WNA (Mexico)	ERNO 8553 Undetermined age	Partial frontal limb - *Serrano-Brañas et al. 2016* Some sources mention it in 2015.	Length: **3.8 m** Hip height: **1.25 m**	125 kg

N.º	Genus and species	Largest specimen	Material and data	Size	Mass
4	*Archaeornithomimus asiaticus* *IUC, EA (China)*	AMNH 21787 Undetermined age	Cervical vertebra - *Smith & Galton 1990* There are several more complete specimens.	Length: **3.9 m** Hip height: **1.3 m**	140 kg
5	*"Gallimimus" mongoliensis* *IUC, EA (Mongolia)*	IGM 950818 Undetermined age	Dentary and partial skeleton - *Kobayashi & Barsbold 2006* Does not belong to the genus *Gallimimus*	Length: **4.1 m** Hip height: **1.35 m**	150 kg
6	*Struthiomimus altus* *IUC, WNA (western Canada)*	UCMZ 1980.1 Undetermined age	Partial skeleton - *Nicholls & Russell 1981* The first genus to be discovered complete.	Length: **4.5 m** Hip height: **1.5 m**	210 kg
7	*Struthiomimus sedens* *IUC, WNA (Montana, Wyoming - USA,* *western Canada)*	BHI 1266 Undetermined age	Skull and partial skeleton - *Farlow 2001* Largest North American ornithomimosaur.	Length: **5.8 m** Hip height: **1.95 m**	420 kg
8	*Gallimimus bullatus* *IUC, EA (Mongolia)*	IGM 100/11 Adult	Skull and partial skeleton - *Osmólska et al. 1972* It has the maximum size for optimal sprinting.	Length: **6 m** Hip height: **2 m**	500 kg

1:200

1 2 3 4 5 6 7 8

ALVAREZSAUROIDEA

Alvarezsauroids similar to *Haplocheirus* - Light bodies, small game hunters.

N.º	Genus and species	Largest specimen	Material and data	Size	Mass
1	*Haplocheirus sollers* *UJ, EA (China)*	IVPP V15988 Subadult	Skull and partial skeleton - *Choiniere et al. 2010* The most ancient alvarezsauroids.	Length: **2.1 m** Hip height: **60 cm**	15 kg

Alvarezsauroids similar to *Alvarezsaurus* - Light bodies, propubic pelvis, omnivores, insectivores, or herbivores.

2	*Alnashetri cerropoliciensis* *eUC, SSA (Argentina)*	MPCA-477 Undetermined age	Partial frontal limb - *Makovicky et al. 2012* More derived than *Alvarezsaurus*	Length: **70 cm** Hip height: **25 cm**	500 g
3	*Alvarezsaurus calvoi* *IUC, SSA (Argentina)*	MUCPv 54 Subadult	Femur - *Bonaparte 1991* The first described austral alvarezsaur.	Length: **1.05 m** Hip height: **32 cm**	1.7 kg
4	*Achillesaurus manazzonei* *IUC, SSA (Argentina)*	MACN-PV-RN 1116 Adult	Partial skeleton - *Martinelli & Vera 2007* Very similar in size to *Patagonykus puertai*.	Length: **2.8 m** Hip height: **90 cm**	30 kg
5	*Patagonykus puertai* *eUC, SSA (Argentina)*	PVPH 37 Undetermined age	Partial skeleton - *Novas 1993* The first discovered large alvarezsaurid.	Length: **2.8 m** Hip height: **90 cm**	30 kg
6	*Bonapartenykus ultimus* *IUC, SSA (Argentina)*	MPCA 1290 Adult	Partial skeleton - *Agnolin et al. 2012* Was a female, as she had two eggs inside her.	Length: **2.9 m** Hip height: **95 cm**	34 kg

Alvarezsauroids similar to *Shuvuuia* - Light bodies, opistopubic pelvis, omnivores, insectivores, or herbivores.

7	*Parvicursor remotus* *IUC, EA (Mongolia)*	PIN 4487/25 Adult	Partial skeleton - *Karhu & Rautian 1996* Is one of the smallest non-avian terrestial theropods.	Length: **50 cm** Hip height: **19 cm**	185 g
8	*Albinykus baatar* *IUC, EA (Mongolia)*	IGM 100/3004 Adult	Partial skeleton - *Nesbitt et al. 2011* At only two years old, it was already an elderly adult.	Length: **60 cm** Hip height: **23 cm**	355 g
9	*Xixianykus zhangi* *eUC, EA (China)*	XMDFEC V0011 Subadult	Partial skeleton - *Xu et al. 2010* One of the few alvarezsaurs with gastralia.	Length: **65 cm** Hip height: **25 cm**	440 g
10	*Linhenykus monodactylus* *IUC, EA (China)*	IVPP V17608 Subadult	Partial skeleton - *Xu et al. 2011* Its hand was more atrophied than other alvarezsaurids.	Length: **65 cm** Hip height: **25 cm**	440 g
11	*Ceratonykus oculatus* *IUC, EA (Mongolia)*	MPC 100/24 Undetermined age	Skull and partial skeleton - *Alifanov & Barsbold 2009* The measurements of several bones were not published.	Length: **75 cm** Hip height: **30 cm**	760 g
12	*"Ornithomimus" minutus* *IUC, WNA (Colorado - USA)*	YPM 1049 Adult	Incomplete metatarsus - *Marsh 1892* The specimen was lost.	Length: **1.1 m** Hip height: **44 cm**	2.3 kg
13	*Shuvuuia deserti* *IUC, EA (Mongolia)*	IGM 100/975 Undetermined age	Skull and partial skeleton - *Chiappe et al. 1998* The specimen IGM 100/1276 is of the same size.	Length: **1.15 m** Hip height: **46 cm**	2.7 kg
14	*Albertonykus borealis* *IUC, WNA (western Canada)*	RTMP 2000.45.98 Adult	Skull and partial skeleton - *Longrich & Currie 2009* North America's smallest terrestrial theropod.	Length: **1.2 m** Hip height: **48 cm**	3 kg
15	*Mononykus olecranus* *IUC, EA (Mongolia)*	IGM N107/6 Adult	Skull and partial skeleton - *Perle et al. 1993* Was named "Mononychus," but the name was already in use.	Length: **1.25 m** Hip height: **50 cm**	3.5 kg
16	*Heptasteornis andrewsi (n.d.)* *IUC, EE (Romania)*	BMNH A1528 Undetermined age	Incomplete tibiotarsus - *Lambrecht 1929* It was considered other kind of theropod. *Martin 1997; Naish & Dyke 2004*	Length: **2.05 m** Hip height: **80 cm**	15 kg
17	*Kol ghuvia* *IUC, EA (Mongolia)*	IGM 100/2011 Undetermined age	Partial skeleton - *Turner, Nesbitt & Norell 2009* Oviraptorosaur ornithomimus? *Agnolin et al. 2012*	Length: **2.4 m** Hip height: **95 cm**	24 kg

1:150

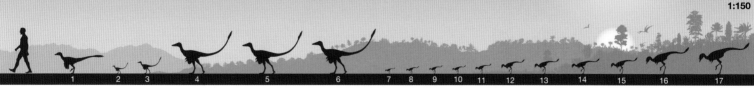

1 2 3 4 5 6 7 8 9 10 11 12 13 14 15 16 17

THERIZINOSAURIA

Therizinosaurs similar to *Falcarius* - Relatively light bodies, omnivores or herbivores.

N.º	Genus and species	Largest specimen	Material and data	Size	Mass
1	*Jianchangosaurus yixianensis* ILC, EA (China)	41HIII-0308A Juvenile	Skull, partial skeleton, and filaments - *Pu et al. 2013* The second therizinosaur with filaments.	Length: 2.2 m Hip height: 50 cm	21 kg
2	*Eshanosaurus deguchiianus* LJ, EA (China)	V11579 Undetermined age	Incomplete jaw and teeth - *Xu et al. 2001* Therizinosaur or sauropod? *Barrett 2009; Mortimer**	Length: 2.65 m Hip height: 60 cm	37 kg
3	*Thecocoelurus daviesi* (n.d.) ILC, WE (England)	BMNH R181 Undetermined age	Cervical vertebra - *Seeley 1888* Ornithomimosaur? *Allain et al. 2014*	Length: 3.2 m Hip height: 75 cm	70 kg
4	*Falcarius utahensis* ILC, WNA (Utah - USA)	UMNH VP 14526 Adult?	Jaw and teeth - *Kirkland et al. 2005* One of North America's most ancient.	Length: 4 m Hip height: 95 cm	135 kg

Therizinosaurs similar to *Beipiaosaurus* - Robust bodies, omnivores and/or herbivores.

N.º	Genus and species	Largest specimen	Material and data	Size	Mass
5	*Inosaurus tedreftensis* (n.d.) ILC, NAf (Niger)	MNNHN Undetermined age	Vertebrae and incomplete tibia - *Lapparent 1960* May have been a therizinosaur or another type of theropod.	Length: 1.2 m Hip height: 43 cm	13 kg
6	*Beipiaosaurus inexpectus* eLC, EA (China)	IVPP 11559 Subadult	Skull, partial skeleton, and filaments - *Xu et al. 1999* The tail had a pygostyle.	Length: 1.8 m Hip height: 65 cm	43 kg
7	*Neimongosaurus yangi* IUC, EA (China)	LH V0001 Undetermined age	Skull and partial skeleton - *Seeley 1888* Its neck was proportionally longer.	Length: 2.7 m Hip height: 90 cm	115 kg
8	*Martharaptor greenriverensis* ILC, WNA (Utah - USA)	UMNH VP 21400 Juvenile	Partial skeleton - *Senter et al. 2012* Oviraptosaur? *Cau**	Length: 2.6 m Hip height: 1 m	122 kg
9	*Erliansaurus bellamanus* IUC, EA (China)	LH V0002 Undetermined age	Partial skeleton - *Xu et al. 2002* Contemporary of *Neimongosaurus*	Length: 2.85 m Hip height: 1.05 m	165 kg
10	cf. *Neimongosaurus* IUC, WA (Kazakhstan)	N 601/12457 Adult	Femur - *Averianov 2007* It was considered a tyrannosaurid. *Nesov 1995*	Length: 3.5 m Hip height: 1.3 m	315 kg
11	*Alxasaurus elesitaiensis* ILC, EA (China)	IVPP 88402a Undetermined age	Skull and partial skeleton - *Russell & Dong 1994* Alxa is the region's name.	Length: 3.8 m Hip height: 1.35 m	400 kg
12	"*Nanshiungosaurus*" *bohlini* ILC, EA (China)	IVPP V 11116 Undetermined age	Jaw - *Dong & Yu 1997* Does not belong to the genus "*Nanshiungosaurus*"	Length: 5 m Hip height: 1.8 m	900 kg
13	*Enigmosaurus mongoliensis* eUC, EA (Mongolia)	IGM 100/84 Adult	Cervical vertebra - *Barsbold & Perle 1980 in Barsbold 1983* Therizinosaur or therizinosaurid? *Zanno 2010*	Length: 5 m Hip height: 1.8 m	900 kg
14	"*Chilantaisaurus*" *zheziangensis* eUC, EA (China)	ZhM V.001 Undetermined age	Partial frontal limb - *Dong 1979* It was 77% the size of PIN 551-483. *Mortimer**	Length: 5.5 m Hip height: 2 m	1.2 t
15	"*Tiantaiosaurus sifengensis*" (n.n.) ILC, EA (China)	Uncatalogued Undetermined age	Partial frontal limb - *Dong, Chen & Jiang 2007* The original document is not official. *Qian 2007, 2011*	Length: 5.5 m Hip height: 2 m	1.2 t

Therizinosaurs similar to *Nothronychus* - Robust bodies, omnivores and/or herbivores.

N.º	Genus and species	Largest specimen	Material and data	Size	Mass
16	*Erlikosaurus andrewsi* IUC, EA (China)	IGM 100/111 Undetermined age	Skull and partial skeleton - *Perle in Barsbold and Perle 1980* The best-preserved therizinosaur skull.	Length: 3.55 m Hip height: 1.25 m	280 kg
17	*Nothronychus mckinleyi* eUC, WNA (New Mexico - USA)	AzMNH P-2117 Undetermined age	Skull and partial skeleton - *Smith 2015* It is the type species of the genus.	Length: 4.85 m Hip height: 1.7 m	700 kg
18	*Nanshiungosaurus brevispinus* IUC, EA (China)	IVPP V4731 Undetermined age	Partial skeleton - *Dong 1979* It has a very strange pelvis.	Length: 4.9 m Hip height: 1.7 m	730 kg
19	*Nothronychus graffami* eUC, WNA (Utah - USA)	UMNH VP 16420 Adult	Partial skeleton - *Zanno et al. 2009* One of the most complete materials.	Length: 5 m Hip height: 1.75 m	785 kg
20	*Suzhousaurus megatheroides* ILC, EA (China)	FRDC-GSJB-2004-001 Undetermined age	Partial skeleton - *Zanno et al. 2009* Could it have been from the late Upper Cretaceous? *Currie & Eberth 1993*	Length: 5.8 m Hip height: 2.1 m	1.4 t
21	*Segnosaurus galbinensis* IUC EA (Mongolia)	IGM 100/82 Undetermined age	Partial skeleton - *Perle 1979* Could it have been from the late Upper Cretaceous? *Currie & Eberth 1993*	Length: 6.9 m Hip height: 2.4 m	2 t
22	*Therizinosaurus cheloniformis* IUC EA (Mongolia)	IGM 100/15 Adult	Coracoid scapula and arms - *Barsbold 1976* Its enormous nails were thought to be part of a giant turtle's shell. *Maleev 1954*	Length: 9 m Hip height: 3.1 m	4.5 t

1:300

OVIRAPTOSAURIA

Oviraptorosaurs similar to *Protarchaeopteryx* - Light bodies, omnivores or herbivores.

N.º	Genus and species	Largest specimen	Material and data	Size	Mass
1	*Protarchaeopteryx robusta* ILC, EA (China)	NGMC 2125 Undetermined age	Skull, partial skeleton, and feathers - *Ji & Ji 1997* Was thought to be close to *Archaeopteryx*	Length: 75 cm Hip height: 40 cm	1.8 kg
2	*Ningyuansaurus wangi* ILC, EA (China)	Uncatalogued Undetermined age	Skull, partial skeleton, and feathers - *Ji, Lu, Wei & Wang 2012* It may not be an oviraptorosaur. *Mortimer**	Length: 80 cm Hip height: 43 cm	2.3 kg

#	Taxon / Locality	Specimen / Age	Description	Size	Weight
3	*Incisivosaurus gauthieri* ILC, EA (China)	IVPP V13326 Undetermined age	Skull and incomplete cervical vertebra - *Xu et al. 2002* It had more teeth than *Protarchaeopteryx*.	Length: **1 m** Hip height: **55 cm**	4.6 kg

Oviraptorosaurs similar to *Caudipteryx* - Light bodies, omnivores or herbivores.

#	Taxon / Locality	Specimen / Age	Description	Size	Weight
4	cf. *Similicaudipteryx yixianensis* ILC, EA (China)	STM4-1 Juvenile	Partial skeleton - *Xu et al. 2010* More recent than *Similicaudipteryx yixianensis*	Length: **27 cm** Hip height: **10 cm**	105 g
5	*Caudipteryx zhoui* ILC, EA (China)	NGMC 97-9-A Undetermined age	Skull, partial skeleton, and feathers - *Ji et al. 1998* IVPP V11819 is similar. *Erickson et al. 2009*	Length: **78 cm** Hip height: **44 cm**	2.2 kg
6	*Caudipteryx dongi* ILC, EA (China)	IVPP V 12344 Undetermined age	Vetebrae and incomplete tibia - *Zhou & Wang 2000* *Caudipteryx zhoui*? *Mortimer**	Length: **80 cm** Hip height: **45 cm**	2.3 kg
7	*Similicaudipteryx yixianensis* ILC, EA (China)	IVPP V12556 Adult	Partial skeleton and feathers - *He et al. 2008* It has some characteristics similar to the caenagnathids.	Length: **1.15 m** Hip height: **65 cm**	6.7 kg
8	*Luoyanggia liudianensis* eUC, EA (China)	41HIII-00011 Undetermined age	Dentario - *Lu, Xu, Jiang, Jia, Li, Yuan, Zhang & Ji 2009* The most recent caudipterid.	Length: **1.2 m** Hip height: **70 cm**	8.5 kg

Oviraptorosaurs similar to *Avimimus* - Relatively light bodies, omnivores and/or herbivores.

#	Taxon / Locality	Specimen / Age	Description	Size	Weight
9	*Avimimus portentosus* IUC, EA (Mongolia)	PIN 3907/5 Undetermined age	Pelvis parcial - *Kurzanov 1987* Was one of the first to have its feathers recognized indirectly.	Length: **1.5 m** Hip height: **77 cm**	15 kg

1:100

Oviraptorosaurs similar to *Gigantoraptor* - Light bodies, omnivores or herbivores.

#	Taxon / Locality	Specimen / Age	Description	Size	Weight
1	*Microvenator celer* ILC, WNA (Montana - USA)	AMNH 3041 Juvenile	Skull and partial skeleton - *Ostrom 1970* A large tooth YPM 5366 was considered to belong to it, but it's *Deinonychus antirrhophus*	Length: **90 cm** Hip height: **35 cm**	3 kg
2	*Caenagnathasia martinsoni* eUC, WA (Uzbekistan)	ZIN PO 4603 Undetermined age	Cervical vertebra - *Sues & Averianov 2015* *Kuzholia mengi*? *Mortimer**	Length: **1.1 m** Hip height: **45 cm**	5.5 kg
3	*Leptorhynchos gaddisi* IUC, WNA (Texas - USA)	TMM 45920-1 Undetermined age	Incomplete jaw - *Longrich et al. 2013.* The genus name is taken by a plant. *Lessing 1832*	Length: **1.75 m** Hip height: **70 cm**	19 kg
4	*Kuszholia mengi* eUC, WA (Uzbekistan)	ZIN PO Undetermined age	Sacral vertebra - *Nesov 1992* Was thought to be a bird.	Length: **1.95 m** Hip height: **77 cm**	26 kg
5	*Leptorhynchus elegans* IUC, WNA (western Canada, Montana - USA)	RTMP 2001.12.12 Undetermined age	Incomplete metatarsus - *Parks 1933* Might be "*Elmisaurus*" elegans. *Funston & Currie 2014; Mortimer**	Length: **2.1 m** Hip height: **82 cm**	32 kg
6	*Elmisaurus rarus* eUC, EA (Mongolia)	ZPAL MgD-I Undetermined age	Metacarpals and phalanges - *Osmólska 1981* The genus that represents elmisaurs.	Length: **2.2 m** Hip height: **93 cm**	40 kg
7	cf. *Elmisaurus elegans* IUC, WNA (Montana - USA)	NS.31996.114H Undetermined age	Incomplete metatarsus - *Buckley 2002* *Elmisaurus elegans*?	Length: **2.4 m** Hip height: **95 cm**	50 kg
8	*Gigantoraptor erlianensis* IUC, EA (Mongolia)	LH V0011 Subadult	Jaw and partial skeleton - *Xu et al. 2007* It is enormous compared with other oviraptorosaurs.	Length: **8 m** Hip height: **3.2 m**	2 t
9	cf. *Gigantoraptor* IUC, EA (Mongolia)	MPC-D 107/17 Adult	Incomplete jaw - *Xu et al. 2007* It is thought to be a specimen of *Gigantoraptor erlianensis*	Length: **8.9 m** Hip height: **3.5 m**	2.7 t

1:300

Oviraptorosaurs similar to *Anzu* - Light bodies, omnivores and/or herbivores.

#	Taxon / Locality	Specimen / Age	Description	Size	Weight
1	*Ojoraptorsaurus boerei* IUC, WNA (western Canada)	SMP VP-1458 Undetermined age	Incomplete pubis - *Sullivan et al. 2011* It comes from the Ojo Alamo Formation.	Length: **2.5 m** Hip height: **1 m**	38 kg
2	*Chirostenotes pergracilis* IUC, WNA (western Canada)	RTMP 79.14.499 Undetermined age	Nail phalanx - *Currie & Russell 1988* Was originally thought to be the remains of a large Mesozoic bird. *Sternberg 1940*	Length: **2.6 m** Hip height: **1.05 m**	40 kg
3	*Hagryphus giganteus* IUC, WNA (Utah - USA)	UMNH VP 12765 Undetermined age	Partial skeleton - *Zanno & Sampson 2005* It has shorter and more robust metacarpals.	Length: **2.9 m** Hip height: **1.2 m**	110 kg
4	cf. *Chirostenotes pergracilis* IUC, WNA (western Canada)	CMN 957 Undetermined age	Metatarsus - *Russell 1984* More recent than *Chirostenotes pergracilis*	Length: **3.1 m** Hip height: **1.25 m**	130 kg
5	*Epichirostenotes curriei* IUC, WNA (western Canada)	ROM 43250 Adult	Skull and partial skeleton - *Sullivan et al. 2011* *Chirostenotes pergracilis*? *Mortimer**	Length: **3.2 m** Hip height: **1.3 m**	145 kg
6	*Anzu wyliei* IUC, WNA (North Dakota, South Dakota, Montana - USA)	MRF 319 Undetermined age	Frontal bone - *Lamanna, Sues, Schachner & Lyson 2014* The second-largest known oviraptorosaur.	Length: **3.8 m** Hip height: **1.55 m**	240 kg

Indeterminate oviraptorosaurs - Light bodies, omnivores or herbivores.

#	Taxon / Locality	Specimen / Age	Description	Size	Weight
7	*Yulong mini* IUC, EA (China)	HGM 41HIII-0107 Juvenile 1 year-old	Skull and partial skeleton - *Lu et al. 2013* It was found with smaller specimens.	Length: **50 cm** Hip height: **21 cm**	525 g
8	*Nomingia gobiensis* IUC, EA (China)	IGM 100/119 Undetermined age	Partial skeleton - *Barsbold et al. 2000* Incorrectly listed as GIN 940824. *Mortimer**	Length: **2.1 m** Hip height: **85 cm**	32 kg

Oviraptorosaurs similar to *Citipati* - Light bodies, long and thin arms, omnivores and/or herbivores.

#	Taxon / Locality	Specimen / Age	Description	Size	Weight
9	*Banji long* IUC, EA (China)	IVPP V16896 Juvenile	Skull and mandible - *Xu & Han 2010* It had unique striations in the nasal crest.	Length: **70 cm** Hip height: **30 cm**	1.25 kg
10	cf. *Oviraptor philoceratops* IUC, EA (Mongolia)	IVPP V9608 Undetermined age	Partial skeleton - *Dong & Currie 1996* *Oviraptor philoceratops* or *Citipati osmolskae*?	Length: **1.7 m** Hip height: **70 cm**	17 kg

11	*Wulatelong gobiensis* IUC, EA (China)	IVPP V18409 Adult	Skull and partial skeleton - *Xu et al. 2013* It was a member of this subfamily. *Cau*	Length: 1.85 m Hip height: 75 cm	21 kg
12	*Oviraptor philoceratops* IUC, EA (Mongolia)	AMNH 6517 Undetermined age	Skull and partial skeleton - *Osborn 1924* Another attributed skull was of *Citipati*.	Length: 1.9 m Hip height: 75 cm	22 kg
13	*Rinchenia mongoliensis* IUC, EA (Mongolia)	GI 100/32A Undetermined age	Frontal bone - *Osmólska, Currie & Barsbold 2004* The genus "Rinchenia" was created by Barsbold in 1997.	Length: 2.2 m Hip height: 90 cm	36 kg
14	*Citipati sp. nov.* IUC, EA (Mongolia)	IGM 100/42 Undetermined age	Skull and partial skeleton - *Barsbold 1981* Contemporary of *Citipati* osmolskae.	Length: 2.25 m Hip height: 90 cm	38 kg
15	*Huanansaurus ganzhouensis* IUC, EA (China)	HGM41HIII-0443 Undetermined age	Skull and partial skeleton - *Lu et al. 2015* Coexisted with five other genera of oviraptorosaurs.	Length: 2.9 m Hip height: 1.05 m	60 kg
16	*Nankangia jiangxiensis* IUC, EA (China)	GMNH F10003 Undetermined age	Partial skeleton - *Lu et al. 2013* It seems too large to be *Wulatelong* as suggested.	Length: 2.75 m Hip height: 1.1 m	69 kg
17	*Citipati osmolskae* IUC, EA (Mongolia)	IGM 100/1004 Undetermined age	Partial skeleton - *Webster 1996* The largest specimen is known as "Big Auntie."	Length: 2.9 m Hip height: 1.15 m	83 kg

Oviraptorosaurs similar to *Nemegtomaia* - Light bodies, relatively short and thick arms, omnivores and/or herbivores.

18	*Khaan mckennai* IUC, EA (Mongolia)	IGM 100/973 Undetermined age	Skull and partial skeleton - *Clark et al. 2001* Was considered cf. *Ingenia*. *Dashzeveg et al. 1995*	Length: 1.25 m Hip height: 49 cm	9.5 kg
19	*Machairasaurus leptonychus* IUC, EA (China)	IVPP V15980 Undetermined age	Partial skeleton - *Longrich et al. 2010* Its claws are very elongated.	Length: 1.3 m Hip height: 50 cm	10 kg
20	*Nemegtomaia barsboldi* IUC, EA (Mongolia)	IGM 100/2112 Undetermined age	Frontal bone - *Lu et al. 2005* The original name was already in use. *Lu et al. 2004*	Length: 1.45 m Hip height: 55 cm	15 kg
21	*Jiangxisaurus ganzhouensis* IUC, EA (China)	HGM41HIII0421 Subadult	Partial skeleton - *Bell et al. 2015* Similar to *Heyuannia*?	Length: 1.55 m Hip height: 58 cm	18 kg
22	*"Ajancingenia yanshini"* IUC, EA (Mongolia)	IGM 100/30 Undetermined age	Skull and partial skeleton - *Barsbold 1981* *Ingenia yanshini* is its synonym.	Length: 1.6 m Hip height: 60 cm	20 kg
23	*Heyuannia huangi* IUC, EA (China)	HYMV1-1 Undetermined age	Skull and partial skeleton - *Lu 2002* Associated eggs are known. *Qiu &Huang 2001*	Length: 1.6 m Hip height: 60 cm	20 kg
24	*Shanyangosaurus niupanggouensis (n.d.)* IUC, EA (China)	NWUV 1111 Undetermined age	Partial skeleton - *Xue et al. 1996* Similar to *Ajacingenia*? *Mortimer**	Length: 1.6 m Hip height: 60 cm	20 kg
25	*Ganzhousaurus nankangensis* IUC, EA (China)	SDM 20090302 Undetermined age	Jaw and partial skeleton - *Wang & Xu 2012* Oviraptosaurid?	Length: 1.65 m Hip height: 63 cm	23 kg
26	*Shixinggia oblita* IUC, EA (China)	BPV-112 Undetermined age	Partial skeleton - *Lu & Zhang 2005* Oviraptorid? *Maryanska et al. 2002*	Length: 1.8 m Hip height: 69 cm	30 kg
27	*Conchoraptor gracilis* IUC, EA (Mongolia)	IGM 100/21 Undetermined age	Jaw - *Barsbold 1976* The piece predates the genus *Conchoraptor* by ten years. *Barsbold 1986*	Length: 1.85 m Hip height: 70 cm	32 kg

1:150

1:20

PARAVES

Paravians similar to *Epidexipteryx* - Gliders, light bodies, small game hunters or insectivores.

N.º	Genus and species	Largest specimen	Material and data	Size	Mass
1	*Epidendrosaurus ningchengensis* MJ, EA (China)	IVPP V12653 Juvenile	Skull, partial skeleton, and filaments - *Zhang et al. 2002* It had a long tail unlike *Epidexipteryx*.	Length: 13 cm Hip height: 5 cm	6.7 g
2	*Scansoriopteryx heilmanni* ILC, EA (China)	CAGS02-IG-gausa-1/DM 607 - Juvenile	Skull, partial skeleton, and filaments - *Czerkas & Yuan 2002* More recent than *Epidendrosaurus*. *Wang et al. 2005*	Length: 13 cm Hip height: 5 cm	6.8 g
3	*Epidexipteryx hui* MJ, EA (China)	IVPP V15471 Subadult	Skull, partial skeleton, and filaments - *Zhang et al. 2008* Its teeth were forward-facing.	Length: 25 cm Hip height: 14 cm	220 g
4	*Yi qi* MJ, EA (China)	STM 31-2 Adult	Skull, partial skeleton, and filaments - *Xu et al. 2015* It has a unique "syiform" bone in its hand.	Length: 33 cm Hip height: 19 cm	520 g

	Indeterminate paravians - Light bodies, small game hunters.				
1	Pedopenna daohugouensis MJ, EA (China)	IVPP V12721 Undetermined age	Partial frontal limb and filaments - *Xu et al. 2005* Basal paravian or dromaeosaurid?	Length: **68 cm** Hip height: **25 cm**	590 g
2	Yixianosaurus longimanus eLC, EA (China)	IVPP V12638 Adult?	Fragments and filaments. *Xu & Wang 2003* Dromaeosaurid or troodontid? *Dececchi et al. 2012*	Length: **76 cm** Hip height: **28 cm**	815 g
3	Bradycneme draculae (n.d.) IUC, EE (Romania)	BMNH A1588 Undetermined age	Incomplete tibiotarsus - *Lambrecht 1929* Dromaeosaurid or troodontid? *Paul 1988; Osmólska & Barsbold 1990; Le Loeuff et al. 1992*	Length: **1.4 m** Hip height: **52 cm**	5.1 kg
	Paravians similar to *Mahakala* - Light bodies, small game hunters.				
4	Mahakala omnogovae IUC, EA (Mongolia)	IGM 100/1033 Undetermined age	Skull and partial skeleton - *Turner et al. 2007* It was found with other smaller specimens	Length: **77 cm** Hip height: **25 cm**	450 g
	Paravians similar to *Unenlagia* - Light bodies, piscivores or small game hunters.				
5	Pamparaptor micros eUC, SSA (Argentina)	MUCPv-1163 Undetermined age	Skull and partial skeleton - *Porfiri et al. 2011* Probable unenlagian.	Length: **1.15 m** Hip height: **35 cm**	2.6 kg
6	Buitreraptor gonzalezorum eUC, SSA (Argentina)	MPCA 245 A1 Adult?	Tooth - *Gianechini et al. 2011* The piece is larger than that of the subadult holotype MPCA 245. *Cerda & Gianechini 2015*	Length: **1.5 m** Hip height: **45 cm**	5.5 kg
7	Neuquenraptor argentinus eUC, SSA (Argentina)	MCF PVPH 77 Undetermined age	Skull and partial skeleton - *Novas & Pol 2005* *Unenlagia*? *Makovicky et al. 2005*	Length: **2.7 m** Hip height: **82 cm**	33 kg
8	Unenlagia comahuensis eUC, SSA (Argentina)	MCF PVPH 78 Undetermined age	Partial skeleton - *Novas &Puerta 1997* First recognized unenlagian.	Length: **3.4 m** Hip height: **1.05 m**	67 kg
9	Unenlagia paynemii eUC, SSA (Argentina)	MUCPv-343 Undetermined age	Nail phalanx - *Calvo, Porfiri & Kellner 2004* Contemporary of *Unenlagia comahuensis*	Length: **3.5 m** Hip height: **1.1 m**	75 kg
10	Austroraptor cabazai IUC, SSA (Argentina)	MML-195 Adult?	Skull and partial skeleton - *Novas et al. 2009* Troodontid? *Cau**	Length: **6.2 m** Hip height: **1.5 m**	340 kg
	Paravians similar to *Richardoestesia* - Light bodies, small game hunters.				
11	Richardoestesia cf. gilmorei MJ, WE (England)	GLRCM G.50823 Undetermined age	Tooth - *Metcalf & Walker 1994* The most ancient; indistinguishable from *R. gilmorei*	Length: **38 cm** Hip height: **10 cm**	100 g
12	Richardoestesia cf. isosceles UJ, WE (Portugal)	IPFUB GUI D 118/155 Undetermined age	Tooth - *Zinke 1998* The most ancient *R.* cf. *isosceles*	Length: **45 cm** Hip height: **12.5 cm**	138 g
13	Richardoestesia cf. isosceles IUC, WNA (Texas - USA)	LSUMG Undetermined age	Tooth - *Sankey et al. 2005* More recent than SMU 73779.	Length: **52 cm** Hip height: **14 cm**	200 g
14	Richardoestesia cf. gilmorei IUC, WNA (Texas - USA)	LSUMG V-6237 Undetermined age	Incomplete tooth - *Sankey 2001* Probable *Richardoestesia gilmorei*	Length: **57 cm** Hip height: **15.5 cm**	270 g
15	Richardoestesia cf. gilmorei ILC, WNA (Utah - USA)	Uncatalogued Undetermined age	Tooth - *Garrison et al. 2007* The second most ancient from North America.	Length: **62 cm** Hip height: **17 cm**	340 g
16	Richardoestesia cf. isosceles eUC, WNA (Texas - USA)	SMU 73779 Undetermined age	Tooth - *Lee 1995* New species.	Length: **64 cm** Hip height: **17.5 cm**	380 g
17	Richardoestesia cf. gilmorei UJ, WE (Portugal)	ML 939 Undetermined age	Tooth - *Hendrickx & Mateus 2014* Similar to *R. gilmorei*	Length: **64 cm** Hip height: **17.5 cm**	380 g
18	Richardoestesia cf. isosceles ILC, WE (Spain)	IPFUB Uña Th Undetermined age	Tooth - *Rauhut & Zinke 1995* Different serration. *Mortimer**	Length: **74 cm** Hip height: **20 cm**	740 g
19	Richardoestesia cf. isosceles eUC, WNA (New Mexico - USA)	NMMNH P-32753 Undetermined age	Tooth - *Williamson & Brusate 2014* New species.	Length: **84 cm** Hip height: **23 cm**	1 kg
20	Richardoestesia cf. gilmorei IUC, WE (France)	VIC 17 Undetermined age	Tooth - *Buffetaut, Marandat & Sige 1986* Very similar to *R. gilmorei*	Length: **90 cm** Hip height: **24.5 cm**	1.3 kg
21	cf. Richardoestesia gilmorei IUC, WNA (Wyoming - USA)	AMNH5545.2 Undetermined age	Tooth - *Longrich 2008* The most recent *R. gilmorei*.	Length: **1.03 m** Hip height: **28 cm**	1.9 kg
22	Richardoestesia cf. gilmorei IUC, WNA (western Canada)	RTMP Undetermined age	Tooth - *Baszio 1997* Possibly a new species.	Length: **1.1 m** Hip height: **30 cm**	2 kg
23	Richardoestesia cf. isosceles eUC, WNA (Montana, South Dakota - USA)	UCMP186840 Undetermined age	Tooth - *Longrich 2008* Another of similar size is UCMP213975.	Length: **1.13 m** Hip height: **30.5 cm**	2.3 kg
24	cf. Richardoestesia gilmorei IUC, WNA (western Canada)	Uncatalogued Undetermined age	Tooth - *Ryan & Russell 2001* Contemporary of *Hesperonychus elizabethae*.	Length: **1.2 m** Hip height: **33 cm**	2.9 kg
25	cf. Richardoestesia gilmorei eUC, WNA (Utah - USA)	UMNH VP 24115 Undetermined age	Tooth - *Eaton et al. 2014* More ancient than *Richardoestesia gilmorei*.	Length: **1.29 m** Hip height: **35 cm**	3.7 kg
26	Richardoestesia cf. isosceles eUC, WNA (Wyoming - USA)	AMNH5382.2 Undetermined age	Tooth - *Longrich 2008* The most recent *R. isosceles*.	Length: **1.47 m** Hip height: **40 cm**	5.5 kg
27	Richardoestesia isosceles IUC, WNA (western Canada)	Uncatalogued Undetermined age	Tooth - *Farlow et al 1991* Very similar to the crocodile *Doratodon* teeth. *Company et al. 2005*	Length: **1.48 m** Hip height: **40 cm**	5.7 kg
28	cf. Richardoestesia isosceles IUC, WNA (Utah - USA)	UALVP48343 Undetermined age	Tooth - *Larson 2008* Between the Santonian and Campanian.	Length: **1.5 m** Hip height: **41 cm**	6 kg
29	Richardoestesia cf. isosceles IUC, WNA (western Canada)	nmckno23b Undetermined age	Tooth - *Farlow et al. 1991* Significantly unlike *R. gilmorei*	Length: **1.5 m** Hip height: **41 cm**	6 kg
30	cf. Richardoestesia gilmorei IUC, WNA (western Canada)	LSUMG 489:6237 Undetermined age	Tooth - *Baszio 1997* More recent than *R. gilmorei*.	Length: **1.6 m** Hip height: **43 cm**	7 kg
31	cf. Richardoestesia gilmorei IUC, WNA (Montana - USA)	UCMP123543 Undetermined age	Tooth - *Sankey 2008* One of the largest teeth of the species.	Length: **1.8 m** Hip height: **49 cm**	10 kg
32	Richardoestesia asiatica eUC, WA (Uzbekistan)	N 460/12457 Undetermined age	Tooth - *Nesov 1995* Synonym of *Richardoestesia*. *Ryan 1997; Sues & Averianov 2013*	Length: **1.95 m** Hip height: **53 cm**	13 kg
33	Richardoestesia gilmorei IUC, WNA (western Canada)	P83452 Undetermined age	Dentary and teeth - *Farlow et al. 1991* The CMN 343 specimen in an incomplete jaw with teeth. *Gilmore 1924*	Length: **3.3 m** Hip height: **90 cm**	63 kg
	Paravians similar to *Microraptor* - Light bodies, small game hunters.				
34	Pneumatoraptor fodori IUC, EE (Hungary)	MTM PAL 2011.18 Adult	Tibiotarsus - *Osi & Buffetaut 2011* It is larger than the adult holotype MTM V.2008.38.1.	Length: **67 cm** Hip height: **18 cm**	510 g
35	Shanag ashile ILC, EA (Mongolia)	IGM 100/1119 Undetermined age	Incomplete skull and teeth - *Turner et al. 2007* There are teeth in other locations. *Kurochkin 1988*	Length: **84 cm** Hip height: **23 cm**	1 kg

N.º	Genus and species	Largest specimen	Material and data	Size	Mass
36	Microraptor zhaoianus ILC, EA (China)	LVH 0026 Adult	Skull, partial skeleton, and feathers - *Lu et al. 2013* The three *Microraptor* species may be synonymous. *Senter et al. 2004; Pei et al. 2014*	Length: **95 cm** Hip height: **26 cm**	1.45 kg
37	Graciliraptor lujiatunensis ILC, EA (China)	IVPP V13474 Undetermined age	Jaw and partial skeleton - *Xu & Wang 2004* *Microraptor?* *Turner et al. 2012*	Length: **1.05 m** Hip height: **28 cm**	2 kg
38	Sinornithosaurus millenii ILC, EA (China)	IVPP V12811 Undetermined age	Skull, partial skeleton, and filaments - *Xu et al. 1999* Along with *Microraptor*, it is the best-known microraptor.	Length: **1.18 m** Hip height: **32 cm**	2.9 kg
39	Changyuraptor yangi ILC, EA (China)	HG B016 Adult	Skull, partial skeleton, and feathers - *Han et al. 2014* Its caudal feathers were very long.	Length: **1.22 m** Hip height: **33 cm**	3.2 kg
40	Hesperonychus elizabethae IUC, WNA (western Canada)	RTMP 1983.67.7 Undetermined age	Foot phalanx - *Longrich & Currie 2009* It is one of the smallest non-avian theropods of North America.	Length: **1.27 m** Hip height: **35 cm**	3.6 kg
41	Zhenyuanlong suni ILC, EA (China)	JPM-0008 Undetermined age	Skull, partial skeleton, and feathers - *Lu & Brusatte 2015* With short arms, much like *Tianyuraptor ostromi*	Length: **1.55 m** Hip height: **42 cm**	6.4 kg
42	Tianyuraptor ostromi ILC, EA (China)	STM1-3 Subadult	Skull, partial skeleton, and feathers - *Zheng et al. 2010* Microraptorinae?	Length: **1.6 m** Hip height: **43 cm**	7.1 kg
43	"Julieraptor" (n.n.) IUC, WNA (western Canada)	ROM Undetermined age	Skull and partial skeleton Unpublished, similar to *Bambiraptor* or *Saurornithoides*?	Length: **1.6 m** Hip height: **43 cm**	7.1 kg
44	Bambiraptor feinbergi eUC, WNA (Montana - USA)	FIP 002-136 Adult	Long bones and vertebrae - *Burnham 2004* Saurornitholestine or *Microraptor*?	Length: **1.8 m** Hip height: **49 cm**	10 kg
45	Palaeopteryx thompsoni UJ, WNA (Colorado - USA)	BYU 2022 Undetermined age	Radius - *Jensen 1981; Jensen & Padian 1989* Similar to *Microraptor. Mortimer**	Length: **1.9 m** Hip height: **51 cm**	11.5 kg

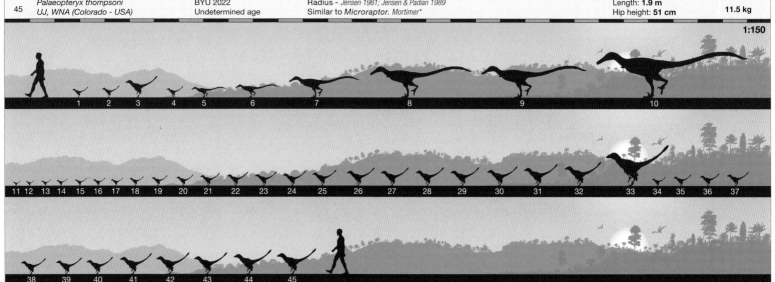

EUDROMAEOSAURIA

Indeterminate eudromaeosaurids - Relatively robust bodies, medium or large game hunters.

N.º	Genus and species	Largest specimen	Material and data	Size	Mass
1	Archaeornithoides deinosauriscus IUC, EA (Mongolia)	ZPAL MgD-II/29 Juvenile	Tooth - *Elzanowski & Wellnhofer 1992* Dromaeosaurid or troodontid. *Chiappe et al. 1996*	Length: **50 cm** Hip height: **9.6 cm**	90 g
2	"Dromaeosaurus" gracilis ILC, ENA (Maryland - USA)	USNM 8176 Undetermined age	Tooth - *Lull 1911* Attributed material; the holotype is the nail phalanx USNM 4973. *Marsh 1888*	Length: **1.2 m** Hip height: **38 cm**	7 kg
3	Ornithodesmus cluniculus ILC, WE (England)	BMNH R187 Undetermined age	Sacrum - *Seeley 1887* Dromaeosaurid? *Naish 2011*	Length: **1.25 m** Hip height: **40 cm**	8 kg
4	"Kitadanisaurus" (n.n.) ILC, EA (Japan)	FPDM-V98081115 Undetermined age	Humerus - *Currie & Azuma 2006* There is an undescribed skeleton.	Length: **1.5 m** Hip height: **48 cm**	13 kg
5	"Megalosaurus" lonzeensis IUC, WE (Belgium)	ALMNH 2001.1 Undetermined age	Nail phalanx - *Dollo 1903* Noasaurid or dromaeosaurid? *Mortimer**	Length: **2.5 m** Hip height: **85 cm**	60 kg

Eudromeosaurids similar to *Saurornitholestes* - Relatively robust bodies, medium or big game hunters.

6	Saurornitholestes sullivani IUC, WNA (New Mexico - USA)	SMP VP-1901 Undetermined age	Tooth - *Sullivan 2006* Were assigned to "*Saurornitholestes*" robustus.	Length: **2.4 m** Hip height: **65 cm**	31 kg
7	Saurornitholestes langstoni IUC, WNA (Canada, Montana - USA)	RTMP 95.92.16 Adult?	Tooth - *Fanti et al. 2014* There are more complete specimens. *Sues 1978; Currie et al. 1993*	Length: **3.2 m** Hip height: **85 cm**	65 kg
8	Saurornitholestes cf. langtoni IUC, WNA (Texas - USA)	LSUMG Undetermined age	Tooth - *Rowe et al. 1992; Sankey 2001* Is more ancient than *Saurornitholestes langstoni*.	Length: **3.5 m** Hip height: **95 cm**	87 kg

Eudromeosaurids similar to *Deinonychus* - Relatively robust bodies, medium or big game hunters.

9	cf. Deinonychus antirrhopus ILC, ENA (Maryland - USA)	USNM Coll. Undetermined age	Tooth- *Lipka 1998* Not guaranteed to be *Deinonychus antirrhopus*	Length: **1.45 m** Hip height: **40 cm**	6.7 kg
10	Tsaagan mangas IUC, EA (Mongolia)	IVPP V16923 Adult	Skull and partial skeleton - *Lipka 1998* Both genera are synonymous. *Turner et al. 2012*	Length: **2.15 m** Hip height: **60 cm**	23 kg
11	Deinonychus antirrhopus ILC, WNA (Montana, Oklahoma, Wyoming - USA)	YPM 5236 Adult	Coracoid - *Ostrom 1969* The specimens AMNH 3015 and MCZ 4371 are very complete.	Length: **3.55 m** Hip height: **1 m**	103 kg

Eudromeosaurids similar to *Velociraptor* - Relatively robust bodies, medium or big game hunters.

12	"Ichabodcraniosaurus" (n.n.) IUC, EA (Mongolia)	IGM 100/980 Undetermined age	Partial skeleton - *Novacek 1996* Synonym of *Velociraptor*?	Length: **1.35 m** Hip height: **37 cm**	5.2 kg
13	Luanchuanraptor henanensis IUC, EA (China)	41HIII-0100 Undetermined age	Skull and partial skeleton - *Lu et al. 2007* Velociraptorinean? *Mortimer**	Length: **1.8 m** Hip height: **49 cm**	11.8 kg

14	Velociraptor osmolskae IUC, EA (China)	IMM 99NM-BYM-3/3 Undetermined age	Incomplete jaw and lacrimal bone - *Godefroit et al. 2008* The second recognized species of the genus.	Length: **1.85 m** Hip height: **50 cm**	**13 kg**
15	Acheroraptor temertyorum IUC, WNA (Montana - USA)	ROM 63777 Undetermined age	Incomplete jaw and teeth - *Evans et al. 2013* Related teeth are of velociraptorine.	Length: **2.3 m** Hip height: **63 cm**	**25 kg**
16	Velociraptor mongoliensis IUC, EA (Mongolia)	AMNH 6518 Adult	Skull and partial skeleton - *Ostrom 1969* Lived in a desert.	Length: **2.65 m** Hip height: **72 cm**	**38 kg**
17	Adasaurus mongoliensis IUC, EA (Mongolia)	IGM 100/21 Adult	Skull and partial skeleton - *Barsbold 1983* The type specimen is an adult and quite smaller.	Length: **3.5 m** Hip height: **95 cm**	**87 kg**
18	"Koreanosaurus" (n.n.) ILC, EA (South Korea)	DGBU-78 Undetermined age	Femur - *Kim 1979* Similar to *Adasaurus* and *Velociraptor*. *Mortimer**	Length: **3.8 m** Hip height: **1.05 m**	**114 kg**

Eudromeosaurids similar to *Dromaeosaurus* - Relatively robust bodies, medium or big game hunters.

19	cf. Dromaeosaurus albertensis IUC, WNA (Alaska - USA)	AK-153-V-1 Undetermined age	Tooth - *Clemens & Nelms 1993* More recent than *Dromaeosaurus albertensis*	Length: **53 cm** Hip height: **17 cm**	**420 g**
20	cf. Zapsalis abradens IUC, WNA (western Canada)	UA 103 Undetermined age	Tooth - *Baszio 1997* More recent than *Zapsalis abradens*	Length: **1.1 cm** Hip height: **35 cm**	**3.5 kg**
21	cf. Dromaeosaurus albertensis IUC, WNA (western Canada)	UA KUA-1:106 Undetermined age	Tooth - *Baszio 1997* More recent than *Dromaeosaurus albertensis*	Length: **1.25 m** Hip height: **40 cm**	**5.3 kg**
22	cf. Zapsalis abradens IUC, WNA (Wyoming - USA)	UA 132 Undetermined age	Tooth - *Baszio 1997* More ancient than *Zapsalis abradens*	Length: **1.25 m** Hip height: **40 cm**	**5.5 kg**
23	Variraptor mechinorum IUC, WE (France)	MDE-D169 Undetermined age	Sacrum - *Le Loeuff & Buffetaut 1998* Dromaeosaurine? *Senter et al. 2004*	Length: **1.65 m** Hip height: **54 cm**	**12.8 kg**
24	Atrociraptor marshalli IUC, WNA (western Canada)	RTMP 95.166.1 Undetermined age	Partial skull and teeth - *Currie & Varricchio 2004* Initially considered possible *Saurornitholestes sp. Baszio 1997*	Length: **1.8 m** Hip height: **57 cm**	**15.5 kg**
25	Zapsalis abradens IUC, WNA (Montana - USA, (western Canada)	TMP 1982.019.0007 Undetermined age	Tooth - *Sankey et al. 2002, Larson & Currie 2013* The holotype material AMNH 3953 is a juvenile from Montan. *Cope 1876*	Length: **1.95 m** Hip height: **63 cm**	**20 kg**
26	cf. Pyroraptor olympius IUC, WE (Spain)	MCNA-14623 Undetermined age	Tooth - *Company et al. 2009* Probable *Pyroraptor olympius* adult.	Length: **2.15 m** Hip height: **68 cm**	**27 kg**
27	Pyroraptor olympius IUC, WE (France)	MNHN BO003 Undetermined age	Metatarsus - *Allain & Taquet 2000* Synonymous with *Variraptor mechinorum*? *Mortimer**	Length: **2.2 m** Hip height: **70 cm**	**29 kg**
28	Dromaeosaurus albertensis IUC, WNA (USA, Canada)	TMP 1984.36.19 Adult	Tooth - *Fanti et al. 2014* The specimen AMNH 5356 is smaller and an adult.	Length: **2.3 m** Hip height: **74 cm**	**34 kg**
29	Dromaeosauroides borholmensis eLC, WE (Denmark)	MGUH DK No. 315 Undetermined age	Tooth - *Christiansen & Bonde 2003* The first dinosaur found in Denmark.	Length: **2.6 m** Hip height: **83 cm**	**49 kg**
30	Yurgovuchia doellingi ILC, WNA (Utah - USA)	UMNH VP 20211 Adult	Thoracic vertebra - *Senter et al. 2012* Dromaeosaurine or velociraptorine? *Mortimer**	Length: **3 m** Hip height: **97 cm**	**76 kg**
31	Itemirus medullaris IUC, WA (Uzbekistan)	ZIN PH 2352/16 Adult	Tooth - *Sues & Averianov 2014* PIN 327/699 is smaller and an adult. *Kurzanov 1976*	Length: **3.1 m** Hip height: **1 m**	**85 kg**
32	Achillobator giganticus IUC, EA (Mongolia)	FR.MNUFR-15 Adult	Skull and partial skeleton - *Perle et al. 1999* The largest Asian dromeosaurid.	Length: **3.9 m** Hip height: **1.25 m**	**165 kg**
33	Dakotaraptor steini IUC, WNA (South Dakota - USA)	PBMNH.P.10.113.T Adult	Partial skeleton - *DePalma et al. 2015* There are two forms: one graceful, the other robust.	Length: **4.35 m** Hip height: **1.4m**	**220 kg**
34	Utahraptor ostrommaysorum ILC, WNA (Utah - USA)	BYU 15465 Adult	Femur - *Erickson et al. 2009* Nine specimens were mixed together (juveniles, subadults, and adults), which caused an overestimation of 11 m. *Britt et al. 2001*	Length: **4.65 m** Hip height: **1.5 m**	**280 kg**

1:150

TROODONTIDAE

Troodontids similar to *Anchiornis* - Light bodies, arboreal, small game hunters and/or insectivores.

N.º	Genus and species	Largest specimen	Material and data	Size	Mass
1	*Eosinopteryx brevipenna* MJ, EA (China)	YFGP-T5197 Subadult	Skull, partial skeleton, and feathers - *Godefroit et al. 2013a* Basal paravian or troodontid? *Godefroit et al. 2013b*	Length: 30 cm Hip height: 15 cm	100 g
2	*Aurornis xui* MJ, EA (China)	YFGP-T5198 Adult	Skull, partial skeleton, and feathers - *Godefroit et al. 2013* Related with *Anchiornis huxleyi*.	Length: 40 cm Hip height: 15 cm	260 g
3	*Xiaotingia zhengi* MJ, EA (China)	STM 27-2 Adult	Skull, partial skeleton, and feathers - *Xu et al. 2011* Basal troodontids looked a lot like primitive avialaen.	Length: 50 cm Hip height: 20 cm	530 g
4	*Anchiornis huxleyi* MJ, EA (China)	STM 0-8 Adult	Femur - *Zheng et al. 2014* The first Mesozoic dinosaur whose color we know. *Li et al 2010*	Length: 50 cm Hip height: 20 cm	570 g

Troodontids similar to *Jinfengopteryx* - Light bodies, omnivores or herbivores.

N.º	Genus and species	Largest specimen	Material and data	Size	Mass
5	*Jinfengopteryx elegans* ILC, EA (China)	CAGS-IG-04-0801 Undetermined age	Skull, partial skeleton, and feathers - *Ji et al. 2005* May have been phytophagous. *Zanno & Markoviky 2010*	Length: 55 cm Hip height: 21 cm	370 g

1:30

Troodontids similar to *Sinovenator* - Light bodies, small game hunters.

N.º	Genus and species	Largest specimen	Material and data	Size	Mass
1	*Hulsanpes perlei* IUC, EA (Mongolia)	ZPAL MgD-I/173 Juvenile	Incomplete metatarsus and phalanges - *Osmólska 1982* Basal troodontid.	Length: 35 cm Hip height: 14 cm	135 g
2	*Mei long* ILC, EA (China)	IVPP V12733 Subadult	Skull and partial skeleton - *Xu & Norell 2004* Its name means "deep sleep."	Length: 50 cm Hip height: 21 cm	430 g
3	cf. *Paronychodon lacustris* IUC, WNA (New Mexico - USA)	NMMNH P-33479 Undetermined age	Tooth - *Williamson & Brusate 2014* Eudromeosauria, *Zapsalis*?	Length: 53 ccm Hip height: 22 cm	510 g
4	*Euronychodon portucalensis* (n.d.) IUC, WE (Portugal)	CEPUNL TV 18 Undetermined age	Tooth - *Antunes & Sigogneau-Russell 1992* Synonym of *Paronychodon*? *Rauhut 2002*	Length: 54 cm Hip height: 22.5 cm	545 g
5	*Sinovenator changii* ILC, EA (China)	IVPP V12615 Adult	Skull and partial skeleton - *Xu et al. 2002* Mistakenly known as "Sinovenator changiae."	Length: 73 cm Hip height: 30 cm	1.3 kg
6	*Paronychodon* cf. *lacustris* IUC, WNA (western Canada)	UA MR-48 Undetermined age	Tooth - *Baszio 1997* More recent than *Paronychodon lacustris*	Length: 75 cm Hip height: 32 cm	1.45 kg
7	*Paronychodon caperatus* (n.d.) IUC, WNA (Wyoming - USA)	UCM 38288 Undetermined age	Tooth - *Baszio 1997* *Paronychodon* were thought either to be teeth with pathologies or to belong to mammals.	Length: 87 cm Hip height: 36.5 cm	2.2 kg
8	*Sinusonasus magnodens* ILC, EA (China)	IVPP V 11527 Adult	Skull and partial skeleton - *Xu & Wang 2004* Mistakenly mentioned as "Sinocerasaurus." *Xu & Norell 2006*	Length: 87 cm Hip height: 37 cm	2.3 kg
9	*Yaverlandia bitholus* ILC, WE (England)	MIWG 1530 Undetermined age	Partial skull - *Galton 1971* It is a troodontid. *Sullivan 2006; Naish 2011*	Length: 87 cm Hip height: 37 cm	2.3 kg
10	*Elopteryx nopcsai* (n.d.) eUC, EE (Romania)	BMNH A1234 Adult	Incomplete femur - *Andrews 1913; Lambrecht 1929* Troodontid or alvarezaurid? *Naish & Dyke 2004; Kessler, Grigorescu & Csiki 2005*	Length: 93 cm Hip height: 39 cm	2.7 kg
11	cf. *Sinovenator changii* ILC, EA (China)	IVPP Undetermined age	Jaw and partial skeleton - *White 2009* Too large to be *Sinovenator changii*	Length: 1.3 m Hip height: 54 cm	7 kg
12	*Talos sampsoni* IUC, WNA (Utah - USA)	UMNH VP 19479 Subadult	Partial skeleton - *Zanno et al. 2011* Its claw was damaged by some accident while it was alive.	Length: 1.4 m Hip height: 58 cm	9 kg
13	*Urbacodon itemirensis* eUC WA (Uzbekistan)	ZIN PH 944/16 Undetermined age	Dentary and teeth - *Averianov & Sues 2007* Larger remains exist, but they are more recent.	Length: 1.45 m Hip height: 62 cm	11 kg
14	*Paronychodon lacustris* (n.d.) IUC, WNA (Montana, New Mexico - USA, western Canada)	AMNH 8522 Undetermined age	Tooth - *Sahni 1972* Several teeth assigned to this species may be assigned to others in future revisions.	Length: 1.6 m Hip height: 67 cm	13 kg
15	*Boreonykus certekorum* IUC, WNA (western Canada)	TMP 1989.055.0047 Undetermined age	Incomplete frontal bone - *Ryan & Russell 2001; Bell & Currie 2015* Does not belong to a dromeosaurid. *Cau*; Mortimer*	Length: 1.65 m Hip height: 68 cm	13.5 kg
16	*Paronychodon asiaticus* eUC, WA (Uzbekistan)	ZIN PH Undetermined age	Tooths - *Nesov 1995* "Plesiosaurodon" is an informal name mentioned by Nessov 1985. *Sues & Averianov 2013*	Length: 1.7 m Hip height: 70 cm	16 kg

Troodontids similar to *Sinornithoides* - Light bodies, small game hunters.

N.º	Genus and species	Largest specimen	Material and data	Size	Mass
17	*Xixiasaurus henanensis* IUC, EA (China)	41HIII-0201 Undetermined age	Skull and partial skeleton - *Lu et al. 2010* Close to *Byronosaurus*?	Length: 1.05 m Hip height: 37 cm	2.3 kg
18	*Koparion douglassi* UJ, WNA (Utah - USA)	DINO 3353 Undetermined age	Tooth - *Chure 1994* The only piece is just 2 mm long.	Length: 1.05 m Hip height: 37 cm	2.4 kg
19	*Sinornithoides youngi* ILC, EA (China)	IVPP V9612 Undetermined age	Skull and partial skeleton - *Russell & Dong 1993* Was published in 1994.	Length: 1.1 m Hip height: 38 cm	2.5 kg
20	*Byronosaurus jaffei* IUC, EA (Mongolia)	IGM 100/983 Undetermined age	Skull and partial skeleton - *Norell et al. 2000* The adult specimen IGM 100/1003 is based on a tooth. *Varricchio & Barta 2015*	Length: 1.55 m Hip height: 55 cm	7.2 kg
21	*Geminiraptor suarezarum* ILC, WNA (Utah - USA)	CEUM 73719 Undetermined age	Maxilar - *Senter, Kirkland, Bird & Bartlett 2010* The name refers to the twin Suárez doctors.	Length: 2 m Hip height: 70 cm	16 kg

Troodontids similar to *Saurornithoides* - Large serrated teeth, light bodies, small game hunters.

N.º	Genus and species	Largest specimen	Material and data	Size	Mass
22	*Philovenator curriei* IUC, EA (China)	IVPP V 10597 Adult	Partial skeleton - *Currie & Peng 1994; Xu et al. 2012* Was considered a juvenile *Saurornithoides*.	Length: 75 cm Hip height: 23 cm	980 g
23	*Gobivenator mongoliensis* IUC, EA (Mongolia)	IGM 100/86 Undetermined age	Skull and partial skeleton - *Tsuihiji et al. 2014* One of the most complete troodontids that has been found.	Length: 1.65 m Hip height: 51 cm	10.5 kg

| 24 | "Saurornitholestes" robustus
IUC, WNA (New Mexico - USA) | NMMNH P-68396
Undetermined age | Tooth - *Sullivan 2006*
This specimen is larger than SMP VP-1955. | Length: 1.9 m
Hip height: 60 cm | 17 kg |
| 25 | Borogovia gracilicrus
IUC, EA (Mongolia) | ZPAL MgD-I/174
Undetermined age | Partial frontal limb - *Osmólska 1987*
Zanabazar junior? *Mortimer** | Length: 2 m
Hip height: 62 cm | 18 kg |
| 26 | Saurornithoides mongoliensis
IUC, EA (Mongolia) | IGM 100/1083
Undetermined age | Teeth - *Norell & Hwang 2004*
The specimen AMNH 6516 is smaller. *Osborn 1924* | Length: 2 m
Hip height: 64 cm | 19.5 kg |
| 27 | Linhevenator tani
IUC, EA (China) | LH V0021
Adult | Skull and partial skeleton - *Xu et al. 2011*
Short and robust arms. | Length: 2.05 m
Hip height: 65 cm | 21 kg |
| 28 | Zanabazar junior
IUC, EA (Mongolia) | IGM 100/1
Adult | Skull and partial skeleton - *Barsbold 1974*
"Mongolodon" is a name used in a thesis. *Franzosa 2004* | Length: 2.15 m
Hip height: 67 cm | 24 kg |
| 29 | Tochisaurus nemegtensis
IUC, EA (Mongolia) | PIN 551-224
Undetermined age | Teeth - *Kurzanov & Osmólska 1991*
Contemporary of *Borogovia* and *Zanabazar* | Length: 2.35 m
Hip height: 72 cm | 31 kg |
| 30 | Troodon bakkeri
IUC, WNA (Montana, Wyoming - USA) | UCM 38445
Undetermined age | Tooth - *Carpenter 1982*
Pectinodon is a synonym of *Troodon*. *Olshevsky 1991* | Length: 2.4 m
Hip height: 73 cm | 32 kg |
| 31 | Troodon inequalis
IUC, WNA (western Canada) | CMN 12433
Undetermined age | Ulna - *Russell 1969*
Troodon formosus? | Length: 3 m
Hip height: 92 cm | 64 kg |
| 32 | cf. Troodon formosus
IUC, WNA (Mexico) | IGM
Undetermined age | Tooth - *Romo de Vivar 2011*
Indeterminate genus. *Ramírez-Velasco et al. 2015* | Length: 3.15 m
Hip height: 97 cm | 76 kg |
| 33 | Troodon formosus
IUC, WNA (Montana - USA) | MOR 553-7.24.8.64
Adult | Tibia - *Varricchio 1993*
MOR 748 is a 12-year-old adult that is 4.1 m long and weighs 170 kg. | Length: 4.3 m
Hip height: 1.35 m | 195 kg |
| 34 | Troodon cf. formosus
IUC, EA (eastern Russia) | ZIN PH 1/28
Undetermined age | Incomplete tooth - *Nesov & Golovneva 1990*
Asia's largest troodontid. | Length: 4.4 m
Hip height: 1.4 m | 208 kg |

1:150

AVIALAE

Avialans similar to *Archaeopteryx* - Gliders, light bodies, arboreal, small game hunters.

N.º	Genus and species	Largest specimen	Material and data	Size	Mass
1	Archaeopteryx lithographica
UJ, WE (Germany)	"Solnhofen specimen"
BMMS 500
Subadult	Skull, partial skeleton, and feathers - *Wellnhofer 1988*
Not all authors agree that it is an avialan; some say it is a basal paravian. *Xu et al. 2011*	Length: 53 cm
Hip height: 20 cm	420 g

Probable avialans similar to *Rahonavis* - Gliders, relatively light bodies, small game hunters.

| 2 | Rahonavis ostromi
IUC, SAf (Madagascar) | UA 8656
Undetermined age | Partial skeleton - *Forster et al. 1998*
Unenlagian or avialan? *Makovicky et al. 2005; Xu et al. 2008* | Length: 80 cm
Hip height: 25 cm | 950 g |

Avialans similar to *Jeholornis* - Gliders, relatively light bodies, arboreal, omnivores or herbivores.

| 3 | Dalianraptor cuhe
ILC, EA (China) | D2139
Undetermined age | Skull, partial skeleton, and feathers - *Gao & Liu 2005*
Probable chimera. *O'Connor et al. 2012* | Length: 55 cm
Hip height: 13 cm | 200 g |
| 4 | Shenzhouraptor sinensis
ILC, EA (China) | LPM 0193
Undetermined age | Partial skeleton and feathers - *Zhou & Zhang 2002*
Synonym of *Jeholornis prima*? | Length: 62 cm
Hip height: 14.5 cm | 290 g |
| 5 | "Jeholornis" palmapenis
ILC, EA (China) | SDM 20090109
Undetermined age | Skull, partial skeleton, and feathers - *O'Connor et al. 2012*
Shenzhouraptor sinensis? *Mortimer** | Length: 65 cm
Hip height: 16 cm | 350 g |
| 6 | "Jeholornis" curvipes
ILC, EA (China) | YFGP-yb2
Undetermined age | Skull and partial skeleton - *Lefevre, et al. 2014*
Jixiangornis orientalis? *Mortimer** | Length: 85 cm
Hip height: 21 cm | 780 g |
| 7 | Shenzhouraptor sinensis
ILC, EA (China) | IVPP V13274
Adult | Partial skeleton - *Zhou & Zhang 2002*
Synonym of *Shenzhouraptor sinensis*? | Length: 90 cm
Hip height: 23 cm | 850 g |
| 8 | Jixiangornis orientalis
ILC, EA (China) | STM2-51
Adult | Incomplete vertebra - *Zheng et al. 2013*
Female specimen. | Length: 1 m
Hip height: 25 cm | 1.2 kg |

Avialans similar to *Yandangornis* - Gliders, relatively light bodies, omnivores or herbivores.

| 9 | Yandangornis longicaudus
IUC, EA (China) | M1326
Undetermined age | Skull and partial skeleton - *Cai & Zhao 1999*
Adapted to terrestrial life. | Length: 60 cm
Hip height: 27 cm | 1 kg |

Avialans similar to *Balaur* - Relatively robust bodies, two falciform claws on each foot, omnivores or carnivores.

| 10 | Bauxitornis mindszentyae
IUC, EE (Hungary) | MTM V 2009.38.1
Undetermined age | Incomplete tarsometatarsus - *Dyke & Osi 2010*
Similar to *Balaur*, or is it enantiornithean? *Cau** | Length: 1.9 m
Hip height: 55 cm | 16 kg |

| 11 | Balaur bondoc
IUC, EE (Romania) | FGGUB R.1581
Adult | Incomplete ulna - *Csiki et al 2010*
Avialae? *Brusatte et al. 2013; Cau et al. 2015* | Length: **1.9**
Hip height: **55 cm** | **16 kg** |

Avialans similar to *Sapeornis* - Gliders, relatively light bodies, arboreal, omnivores or herbivores.

| 12 | Sapeornis chaoyangensis
ILC, EA (China)) | IVPP V12698
Subadult? | Partial skeleton - *Zhou & Zhang 2002*
All relatives of *Sapeornis chaoyangensis* are probably synonymous. *Pu et al. 2013* | Length: **40 cm**
Hip height: **20 cm** | **880 g** |

Avialans similar to *Confuciusornis* - Fliers, relatively light bodies, arboreal, omnivores or herbivores.

13	Zhongornis haoae ILC, EA (China)	D2455/6 Juvenile	Skull, partial skeleton, and feathers - *Gao et al. 2008* Scansoriopterygid? *O'Connor & Sullivan 2014*	Length: **8 cm** Hip height: **4 cm**	**6 g**
14	Changchengornis hengdaoziensis eLC, EA (China)	GMV 2129a/b Adult	Skull, partial skeleton, and feathers - *Ji, Chiappe & Ji 1999* More basal than *Confuciusornis*	Length: **17.5 cm** Hip height: **9 cm**	**60 g**
15	Confuciusornis dui ILC, EA (China)	IVPP V11553 Adult	Skull, partial skeleton, and feathers - *Hou et al. 1999* Contemporary of *Confuciusornis santus*	Length: **18 cm** Hip height: **9.5 cm**	**70 g**
16	Eoconfuciusornis zhengi eLC, EA (China)	IVPP V11977 Subadult	Skull and partial skeleton - *Zhang et al. 2008* The most ancient confuciusornithid.	Length: **18.5 cm** Hip height: **9.5 cm**	**70 g**
17	Jinzhouornis zhangjiyingia ILC, EA (China)	IVPP V12352 Undetermined age	Skull and partial skeleton - *Hou et al. 1999* *Confuciusornis santus?* *Mortimer**	Length: **22 cm** Hip height: **11 cm**	**120 g**
18	"Confuciusornis" jianchangensis ILC, EA (China)	PMOL-AM00114 Undetermined age	Skull and partial skeleton - *Li et al. 2010* May have been a confuciusornithid. *Cau* *	Length: **22 cm** Hip height: **11 cm**	**120 g**
19	cf. Confuciusornis sanctus ILC, EA (China)	IVPP V13313 Undetermined age	Skull and partial skeleton - *Dalsatt et al. 2006* More recent than *Confuciusornis sanctus*	Length: **24.5 cm** Hip height: **12 cm**	**200 g**
20	Confuciusornis chuonzhous ILC, EA (China)	IVPP V10919 Undetermined age	Incomplete frontal limb - *Hou 1997* Indeterminate identity.	Length: **26 cm** Hip height: **13 cm**	**200 g**
21	"Proornis coreae" (n.n.) ILC, EA (North Korea)	Uncatalogued Undetermined age	Skull, partial skeleton, and feathers - *Lim 1993 in Pak & Kim 1996* Close to *Confuciusornis*. *Li & Gao 2007*	Length: **27.5 cm** Hip height: **14 cm**	**240 g**
22	Confuciusornis feducciai ILC, EA (China)	D2454 Adult	Skull, partial skeleton, and feathers - *Zhang et al. 2009* Similar characteristics to ornithothoracines	Length: **30.5 cm** Hip height: **15 cm**	**325 g**
23	Confuciusornis sanctus ILC, EA (China)	BSP 1999 I 15 Adult	Radius - *Marugán-Lobón et al. 2011* There are hundreds of specimens, many of them complete. The largest female was 30 cm long and weighed 330 g.	Length: **31 cm** Hip height: **15.5 cm**	**340 g**

1:50

ENANTIORNITHES

Enantiornitheans similar to *Protopteryx* - Fliers, light bodies, small game hunters or piscivores.

N.º	Genus and species	Largest specimen	Material and data	Size	Mass
1	Dalingheornis liweii ILC, EA (China)	CNU VB2005001 Juvenile	Skull and partial skeleton - *Zhang et al. 2006* The only enantiornithine with zigodactyl feet.	Length: **5.7 cm** Hip height: **2.8 cm**	**2.7 g**
2	Iberomesornis romerali ILC, WE (Spain)	LH-22 Adult	Partial skeleton and feathers - *Sanz & Bonaparte 1992* It had a very developed pygostyle.	Length: **8.7 cm** Hip height: **4.3 cm**	**9.5 g**
3	Protopteryx fengningensis ILC, EA (China)	IVPP V11665 Juvenile?	Skull, partial skeleton, and feathers - *Zhang & Zhou 2000* The specimen IVPP V11844 is of similar size.	Length: **11.4 cm** Hip height: **5.6 cm**	**21 g**
4	Jibeinia luanhera ILC, EA (China)	IVPP Juvenile	Skull and partial skeleton - *Hou 2000* Previously illustrated in Hou 1997	Length: **11.5 cm** Hip height: **5.7 cm**	**21 g**
5	Paraprotopteryx gracilis ILC, EA (China)	STM V001 Subadult	Skull, partial skeleton, and feathers - *Zheng, Zhang & Hou 2007* The skull and two flight feathers were artificially placed. *O'Connor et al. 2012*	Length: **11.5 cm** Hip height: **5.7 cm**	**21 g**
6	Eopengornis martini eLC, EA (China)	STM24-1 Subadult	Skull, partial skeleton, and feathers - *Wang et al. 2014* Second-oldest known enantiornithine.	Length: **14 cm** Hip height: **7 cm**	**39 g**
7	Otogornis genghisi ILC, EA (Mongolia)	IVPP V9607 Undetermined age	Partial skeleton - *Hou 1994* Enantiornithean or Ornithuran? *Kurochkin 1999*	Length: **15 cm** Hip height: **7 cm**	**42 g**
8	Wyleyia valdensis (n.d.) eLC, WE (England)	BMNH A3658 Undetermined age	Incomplete humerus - *Harrison & Walker 1973* Basal enantiornithean or indeterminate maniraptoran? *Naish 2011*	Length: **15.5 cm** Hip height: **7.5 cm**	**52 g**
9	Parapengornis eurycaudatus ILC, EA (China)	VPP V18687 Juvenile	Skull and partial skeleton - *Hu et al. 2015* It would be adapted to climb vertically.	Length: **18.8 cm** Hip height: **10.2 cm**	**126 g**
10	Pengornis houi ILC, EA (China)	IVPP V15336 Adult	Skull and partial skeleton - *Zhou et al. 2008* "Pengornithidae" is doubtful. *Wang et al. 2014*	Length: **23 cm** Hip height: **12.3 cm**	**210 g**
11	Evgenavis nobilis ILC, EA (China)	ZIN PH 1/154 Undetermined age	Incomplete tarsometatarsus - *O'Connor et al. 2014* It is not a confuciusornithid. *Cau**	Length: **33 cm** Hip height: **16 cm**	**515 g**
12	Elsornis keni IUC, EA (Mongolia)	MPD-b 100/201 Undetermined age	Partial skeleton - *Chiappe et al. 2007* Despite being more recent, it is one of the most basal.	Length: **35.9 cm** Hip height: **17.5 cm**	**660 g**
13	Flexomornis howei IUC, WNA (Texas - USA)	DMNH 18137 Undetermined age	Partial skeleton - *Tykoski & Fiorillo 2010* Probably close to *Elsornis*	Length: **50.7 cm** Hip height: **25 cm**	**1.85 kg**

Enantiornitheans similar to *Longipteryx* - Fliers, light bodies, waders.

#	Species / Locality	Specimen / Age	Description	Size	Mass
14	*Boluochia zhengi* ILC, EA (China)	IVPP V9770 Undetermined age	Skull and partial skeleton - *Zhou 1995* *Longipteryx? Mortimer*	Length: 15 cm Hip height: 7 cm	34 g
15	*Camptodontus yangi* ILC, EA (China)	SG2005-B1 Undetermined age	Skull and partial skeleton - *O'Connor & Chiappe 2011* *Boluochia zheng? Martyniuk 2012*	Length: 17 cm Hip height: 8 cm	51 g
16	*Longipteryx chaoyangensis* ILC, EA (China)	DNHM D2889 Undetermined age	Skull, partial skeleton, and feathers - *Wang et al. 2015* IVPP V12325 is of similar size.	Length: 18 cm Hip height: 8.5 cm	57 g
17	*Shengjingornis yangi* ILC, EA (China)	PMOL AB00179 Undetermined age	Skull and partial skeleton - *Li et al. 2012* Longipterygidae?	Length: 21.5 cm Hip height: 10 cm	100 g

Enantiornitheans similar to *Rapaxavis* - Fliers, light bodies, piscivores.

#	Species / Locality	Specimen / Age	Description	Size	Mass
18	*Shanweiniao cooperorum* ILC, EA (China)	DMNH D1878 Adult	Skull, partial skeleton, and feathers - *O'Connor et al. 2009* It is the smallest basal erantiornithine.	Length: 9 cm Hip height: 4 cm	10.5 g
19	*Noguerornis gonzalezi* ILC, WE (Spain)	LP.1702 Undetermined age	Partial skeleton - *Lacasa-Ruiz 1986* Similar to *Longirostravis Cau*	Length: 10 cm Hip height: 4.5 cm	13 g
20	*Rapaxavis pani* ILC, EA (China)	DMNH D2522 Subadult	Skull and partial skeleton - *Morschhauser et al. 2009* Similar to *Longirostravis*	Length: 10 cm Hip height: 4.5 cm	14 g
21	*Longirostravis hani* ILC, EA (China)	IVPP V11309 Undetermined age	Skull and partial skeleton - *Hou et al. 2004* Long, conical snout like *Rapaxavis* and *Shanweiniao*	Length: 12.5 cm Hip height: 5.5 cm	28 g

Enantiornitheans similar to *Cathayornis* - Fliers, light bodies, small game hunters.

#	Species / Locality	Specimen / Age	Description	Size	Mass
22	*Cathayornis yandica* ILC, EA (China)	IVPP V 9769 Undetermined age	Skull and partial skeleton - *Zhou, Jin & Zhang 1992* The only sure cathayornithid species. *Wang & Liu 2015*	Length: 12 cm Hip height: 5.3 cm	24 g

Enantiornitheans similar to *Bohaiornis* - Fliers, light bodies, small game hunters, and tough.

#	Species / Locality	Specimen / Age	Description	Size	Mass
23	*Longusunguis kurochkini* ILC, EA (China)	IVPP V17964 Subadult	Skull and partial skeleton - *Wang et al. 2014* It posses several unique traits compared to other bohaiornithids.	Length: 18 cm Hip height: 9 cm	88 g
24	*Parabohaiornis martini* ILC, EA (China)	IVPP V18690 Subadult	Partial skeleton - *Wang et al. 2014* The holotype specimen IVPP V18691 is the smallest.	Length: 18.5 cm Hip height: 9.5 cm	100 g
25	*Sulcavis geeorum* ILC, EA (China)	BMNHC Ph-000805 Undetermined age	Skull, partial skeleton, and feathers - *O'Connor et al. 2013* Its teeth are ornamented.	Length: 21 cm Hip height: 10.5 cm	135 g
26	*Shenqiornis mengi* ILC, EA (China)	DNHM D2950-51 Subadult	Skull, partial skeleton, and feathers - *Wang et al. 2010* "Dalianornis" is a mistakenly published name. *O'Connor et al. 2011*	Length: 21 cm Hip height: 10.5 cm	135 g
27	*Bohaiornis guoi* ILC, EA (China)	IVPP V17963 Adult	Skull, partial skeleton, and feathers - *Li et al. 2014* Gastroliths were found in the specimen.	Length: 21.5 cm Hip height: 11 cm	147 g
28	*Zhouornis hani* ILC, EA (China)	CNUVB-0903 Subadult	Skull and partial skeleton - *Zhang et al. 2013* The largest bohaiornithid.	Length: 22.5 cm Hip height: 11.5 cm	168 g

1:20

Enantiornitheans similar to *Eoenantiornis* - Fliers, light bodies, small game hunters.

#	Species / Locality	Specimen / Age	Description	Size	Mass
1	*Gobipipus reshetovi* IUC, EA (Mongolia)	PIN 4492-3 Embryo	Partial skeleton - *Kurochkin et al. 2013* The specimen PIN 4492-4 is smaller: 4.7 cm and 1.4 g.	Length: 4.7 cm Hip height: 2.3 cm	2 g
2	*Cratoavis cearensis* ILC, NSA (northern Brazil)	UFRJ-DG 031 Av Subadult?	Skull, partial skeleton, and feathers - *Carvalho et al. 2015a* The smallest Mesozoic bird. *Carvalho et al. 2015b*	Length: 6.6 cm Hip height: 2.9 cm	4 g
3	*Parvavis chuxiongensis* eUC, EA (China)	IVPP V18586 Subadult	Partial skeleton and feathers - *Wang, Zhou & Xu 2014* The smallest of Asia.	Length: 7.5 cm Hip height: 3.3 cm	6 g
4	*Liaoxiornis delicatus* ILC, EA (China)	GMV 2156 - NIGP 130723 Juvenile	Skull and partial skeleton - *Hou & Chen 1999; Ji & Ji 1999* Both are counterparts of the same specimen.	Length: 8.3 cm Hip height: 3.6 cm	8 g
5	*Eocathayornis walkeri* ILC, EA (China)	IVPP V10916 Undetermined age	Skull and partial skeleton - *Zhou 2002* Appears as *Cathayornis* in Martin & Zhou 1997.	Length: 10.4 cm Hip height: 4.4 cm	14.5 g
6	*"Cathayornis" aberransis* ILC, EA (China)	IVPP V12353 Undetermined age	Skull and partial skeleton - *Hou et al. 2002* It does not belong to *Cathayornis* genus. *Wang & Liu 2015*	Length: 10.4 cm Hip height: 4.5 cm	16 g
7	*Huoshanornis huji* ILC, EA (China)	D2126 Undetermined age	Skull and partial skeleton - *Wang et al. 2010* Could have flown with great maneuverability.	Length: 10.8 cm Hip height: 4.7 cm	17.5 g
8	*Gracilornis jiufotangensis* ILC, EA (China)	PMOL-AB00170 Undetermined age	Skull and partial skeleton - *Li & Hou 2011* Its sternum was proportionally small.	Length: 11 cm Hip height: 4.8 cm	19 g
9	*Alexornis antecedens* IUC, WNA (Mexico)	LACM 32213 Undetermined age	Partial skeleton - *Brodkorb 1976* North America's smallest bird.	Length: 11.4 cm Hip height: 5 cm	22 g
10	*Intiornis inexpectatus* IUC, SSA (Argentina)	MAS-P/2 1 Undetermined age	Incomplete frontal limb - *Novas et al. 2010* Related to *Soroavisaurus*	Length: 11.9 cm Hip height: 5.2 cm	24 g
11	*Sinornis santensis* ILC, EA (China)	IVPP V9769 Undetermined age	Skull and partial skeleton - *Sereno & Rao 1992* *Cathayornis? O'Connor & Dyke 2010*	Length: 12 cm Hip height: 5.2 cm	24 g
12	*Dapingfangornis sentisorhinus* ILC, EA (China)	LPM 00039 Undetermined age	Skull, partial skeleton, and feathers - *Li et al. 2006* It had a unique process in the nostrils.	Length: 12 cm Hip height: 5.2 cm	24 g

#	Species / Age	Specimen	Description	Measurements	Weight
13	*Kizylkumavis cretacea* eUC, WA (Uzbekistan)	TsNIGRI 51/11915 Undetermined age	Incomplete humerus - *Nesov 1984* Western Asia's smallest bird.	Length: **12.5 cm** Hip height: **5.4 cm**	25 g
14	*Pterygornis dapingfangensiss* ILC, EA (China)	IVPP V20729 Undetermined age	Partial skeleton - *Wang et al. 2015* It mistakenly appears as *Dispersusia*.	Length: **12.5 cm** Hip height: **5.4 cm**	25 g
15	*Concornis lacustris* ILC, WE (Spain)	LH-2814 Undetermined age	Partial skeleton - *Sanz & Buscalioni 1992* Similar to *Sinornis santensis*	Length: **12.7 cm** Hip height: **5.5 cm**	26 g
16	*Qiliania graffini* ILC, EA (China)	GSGM-F00003 Adult	Skull and partial skeleton - *Ji et al. 2011* Similar to *Confuciusornis*. Mortimer*	Length: **13 cm** Hip height: **5.6 cm**	29 g
17	*Hebeiornis fengningensis* ILC, EA (China)	NIGP 130722 Subadult	Partial skeleton - *Yan en Xu et al. 1999* It had proportionally short wings.	Length: **13.3 cm** Hip height: **5.7 cm**	30 g
18	*Holbotia ponomarenkoi* ILC, EA (Mongolia)	PIN 3147-200 Subadult	Partial skeleton - *Zelenkov & Averianov 2015* Invalid name in Kurochkin 1991 (Kurochkin 1982)	Length: **13.5 cm** Hip height: **5.9 cm**	31 g
19	*Eoalulavis hoyasi* ILC, WE (Spain)	LH-13500 Adult	Partial skeleton and feathers - *Sanz et al. 1996* Its wingspan was ca. 17 cm.	Length: **14 cm** Hip height: **6 cm**	36 g
20	*Liaoningornis longidigitus* ILC, EA (China)	IVPP V11303 Undetermined age	Partial skeleton - *Hou 1996* It was considered an ornithuromorph bird.	Length: **14 cm** Hip height: **6 cm**	36 g
21	*Feitianius paradisi* ILC, EA (China)	GSGM-05-CM-004 Adult	Partial skeleton - *O'Connor et al. 2015* Its tail was more complex than on other enantiornithean birds.	Length: **14.1 cm** Hip height: **6.2 cm**	39 g
22	*Cuspirostrisornis houi* ILC, EA (China)	IVPP V10897 Undetermined age	Skull and partial skeleton - *Hou 1997* Its beak is similar to those of sandpipers (*Motacilla*).	Length: **14.2 cm** Hip height: **6.2 cm**	40 g
23	*Platanavis nana* eUC, WA (Uzbekistan)	ZIN PO 4601 Undetermined age	Incomplete synsacrum - *Nesov 1992* Synonym of other contemporary genera?	Length: **14.3 cm** Hip height: **6.3 cm**	40 g
24	*Alethoalaornis agitornis* ILC, EA (China)	LPM 00038 Subadult?	Skull and partial skeleton - *Li et al. 2007* There are several referred specimens.	Length: **14.7 cm** Hip height: **6.4 cm**	44 g
25	*Largirostrisornis sexdentoris* ILC, EA (China)	IVPP V10531 Undetermined age	Skull and partial skeleton - *Hou 1997* It had six teeth in each premaxilla.	Length: **14.8 cm** Hip height: **6.4 cm**	45 g
26	*Houornis caudatus* ILC, EA (China)	IVPP V10904 Undetermined age	Incomplete frontal limb - *Hou 1997* Used to be *Cathayornis caudatus*. Wang & Liu 2015	Length: **15 cm** Hip height: **6.6 cm**	48 g
27	*Eoenantiornis buhleri* ILC, EA (China)	IVPP V11537 Undetermined age	Skull and partial skeleton - *Hou et al. 1999* Phylogenetically an intermediate genus.	Length: **15 cm** Hip height: **6.6 cm**	48 g
28	*"Cathayornis" chabuensis* ILC, EA (China)	BMNHC-Ph000110 Subadult	Partial skeleton - *Li et al. 2008* Uncertain if it is a *Cathayornis*. Wang & Liu 2015	Length: **15.6 cm** Hip height: **6.8 cm**	53 g
29	*Sazavis prisca* eUC, WA (Uzbekistan)	ZIN PO 3472 Undetermined age	Incomplete tibiotarsus - *Nesov en Nesov & Jarkov 1989* Synonymous of other contemporary genera?	Length: **15.6 cm** Hip height: **6.8 cm**	53 g
30	*Grabauornis lingyuanensis* ILC, EA (China)	IVPP V14595 Adult	Skull and partial skeleton - *Dalsatt, Ericson & Zhou 2014* May have been a very maneuverable bird.	Length: **16.2 cm** Hip height: **7.1 cm**	63 g
31	*Explorornis nessovi* eUC, WA (Uzbekistan)	ZIN PO 4819 Undetermined age	Incomplete coracoid - *Panteleev 1998* Synonymous of other contemporary genera?	Length: **16.4 cm** Hip height: **7.1 cm**	63 g
32	*Dunhuangia cuii* ILC, EA (China)	CAGS-IG-05-CM-030 Undetermined age	Partial skeleton - *Wang et al. 2015* Similar to *Fortunguavis xiaotaizicus*	Length: **17.1 cm** Hip height: **7.5 cm**	70 g
33	*Catenoleimus anachoretus* eUC, WA (Uzbekistan)	ZIN PO 4606 Undetermined age	Incomplete coracoid - *Panteleev 1998* Synonymous of other contemporary genera?	Length: **18.7 cm** Hip height: **8.2 cm**	91 g
34	*Fortunguavis xiaotaizicus* ILC, EA (China)	IVPP V18631 Adult	Skull and partial skeleton - *Wang et al. 2014* Its nails were quite curved.	Length: **21.3 cm** Hip height: **9.3 cm**	140 g
35	*Incolornis silvae* eUC, WA (Uzbekistan)	ZIN PO 4604 Undetermined age	Incomplete coracoid - *You et al. 2005* Synonymous of other contemporary genera?	Length: **21.8 cm** Hip height: **9.5 cm**	144 g
36	*Yuanjiawaornis viriosus* ILC, EA (China)	PMOL-AB00032 Undetermined age	Incomplete tarsometatarsus - *Dyke & Osi 2010* The largest of the Lower Cretaceous.	Length: **22.3 cm** Hip height: **9.8 cm**	155 g
37	*Lenesornis maltshevskyi* eUC, WA (Uzbekistan)	ZIN PO 3434 Undetermined age	Incomplete synsacrum - *Nesov 1986* Considered to be of the genus *Ichthyornis*. Kurochkin 1996	Length: **22.6 cm** Hip height: **9.9 cm**	160 g
38	*"Ichthyornis" minusculus* eUC, WA (Uzbekistan)	ZIN PO 3941 Undetermined age	Thoracic vertebra - *Nesov 1990* Not of the genus *Ichthyornis*.	Length: **22.9 cm** Hip height: **10 cm**	166 g
39	*Gobipteryx minuta* IUC, EA (Mongolia)	ZPAL MgR-I/12 Subadult?	Skull and vertebrae - *Elzanowski 1974* Embryos are known as small as 7.5 cm and 4 g.	Length: **23.4 cm** Hip height: **10.2 cm**	175 g
40	*Mystiornis cyrili* ILC, WA (western Russia)	PM TSU 16/5-45 Undetermined age	Tarsometatarsus - *Kurochkin et al. 2011* May have been a diving bird.	Length: **24.6 cm** Hip height: **10.8 cm**	206 g
41	*Abavornis bonaparti* eUC, WA (Uzbekistan)	TsNIGRI 56/11915 Undetermined age	Incomplete coracoid - *Panteleev 1998* Synonymous of other contemporary genera?	Length: **24.7 cm** Hip height: **10.8 cm**	209 g
42	*Enantiophoenix electrophyla* IUC, ME (Lebanon)	MSNM V3882 Undetermined age	Partial skeleton and feathers - *Cau & Arduini 2008* The only one found in the Middle East.	Length: **26.3 cm** Hip height: **11.5 cm**	250 g
43	*Nanantius eos* ILC, OC (Australia)	QM F12992 Undetermined age	Incomplete tibiotarsus - *Molnar 1986* QM F16811 was found in an ichthyosaur. Kurochkin & Molnar 1997	Length: **29.1 cm** Hip height: **12.7 cm**	343 g
44	*Avisaurus gloriae* IUC, WNA (Montana - USA)	MOR 553E/6.19.91.64 Undetermined age	Tarsometatarsus - *Varricchio & Chiappe 1995* More ancient than *Avisaurus archibaldi*	Length: **29.1 cm** Hip height: **12.7 cm**	343 g
45	*Explorornis walkeri* eUC, WA (Uzbekistan)	ZIN PO 4825 Undetermined age	Incomplete coracoid - *Nesov & Panteleev 1993* Largest specimen of its time.	Length: **29.4 cm** Hip height: **12.8 cm**	350 g
46	*Xiangornis shenmi* IUC, EA (China)	PMOL-AB00245 Undetermined age	Partial skeleton - *Hu et al. 2012* It is one of the largest of the Lower Cretaceous.	Length: **30.1 cm** Hip height: **13.1 cm**	375 g
47	*Halimornis thompsoni* IUC, ENA (Alabama - USA)	UAMNH PV996.1.1 Undetermined age	Vertebrae and incomplete scapula - *Chiappe et al. 2002* Lived in marine environments.	Length: **32.8 cm** Hip height: **14.3 cm**	490 g
48	*Incolornis martini* eUC, WA (Uzbekistan)	ZIN PO 4609 Undetermined age	Incomplete coracoid - *Nesov & Panteleev 1993* Synonymous of other contemporary genera?	Length: **33.8 cm** Hip height: **14.8 cm**	535 g
49	*Martinavis minor* IUC, SSA (Argentina)	PVL-4046 Undetermined age	Incomplete humerus - *Walker & Dyke 2009* The smallest species attributed to *Martinavis*	Length: **34.1 cm** Hip height: **14.9 cm**	550 g
50	*Martinavis whetstonei* IUC, SSA (Argentina)	PVL-4028 Undetermined age	Incomplete humerus - *Walker & Dyke 2009* The least complete *Martinavis*	Length: **35.5 cm** Hip height: **15.4 cm**	620 g

N.º	Genus and species	Largest specimen	Material and data	Size	Mass
51	*Martinavis cruzyensis* IUC, WE (France)	ACAP-M 1957 Undetermined age	Humerus - *Walker & Dyke 2010* *Martinavis* type specimen.	Length: **42.7 cm** Hip height: **18.7 cm**	**1.1 kg**
52	*Martinavis saltariensis* IUC, SSA (Argentina)	PVL-4025 Undetermined age	Humerus - *Walker & Dyke 2010* Second largest species of the genus.	Length: **44.2 cm** Hip height: **19.3 cm**	**1.2 kg**
53	*Soroavisaurus australis* IUC, SSA (Argentina)	PVL-4048 Undetermined age	Tibiotarsus - *Chiappe 1993* The largest specimen that is not in doubt.	Length: **48.6 cm** Hip height: **21.1 cm**	**1.55 kg**
54	*Martinavis vincei* IUC, SSA (Argentina)	PVL-4054 Undetermined age	Humerus - *Buffetaut 2010* The specimen PVL-4059 is the same size.	Length: **51 cm** Hip height: **22.3 cm**	**1.8 kg**
55	*Gurilynia nessovi* IUC, EA (Mongolia)	PIN 4499-12 Undetermined age	Incomplete humerus - *Kurochkin 1999* Mongolia's largest bird.	Length: **53 cm** Hip height: **23.2 cm**	**2.1 kg**
56	cf. *Avisaurus archibaldi* IUC, WNA (Utah - USA)	RAM 14306 Undetermined age	Incomplete coracoid - *Farke & Patel 2012* Larger than the holotype UCMP 117600. *Hutchison 1993*	Length: **72 cm** Hip height: **31.5 cm**	**5.1 kg**
57	"*Coelurosaurus*" (n.n.) IUC, SSA (Argentina)	MLP CS 1478 Undetermined age	Nail phalanx - *Huene 1929* Might have been a bird, but it is difficult to confirm. *Mortimer**	Length: **76.4 cm** Hip height: **33.5 cm**	**6.25 kg**
58	*Enantiornis leali* IUC, SSA (Argentina)	PVL-4267 Undetermined age	Ulna - *Walker 1981* The specimen PVL-4039 is the same size.	Length: **78.5 cm** Hip height: **34 cm**	**6.75 kg**
59	cf. *Soroavisaurus australis* IUC, SSA (Argentina)	PVL-4033 Undetermined age	Tibiotarsus - *Walker 1981; Chiappe 1993* Specimen of *Soroavisaurus australis*? *Walker & Dyke 2009*	Length: **80 cm** Hip height: **35 cm**	**7.25 kg**

1:20

1:100

Enantiornithes similar to *Neuquenornis* - Fliers, light bodies, long-legged, small game hunters.

N.º	Genus and species	Largest specimen	Material and data	Size	Mass
1	*Longchengornis sanyanensis* ILC, EA (China)	IVPP V10530 Undetermined age	Skull and partial skeleton - *Hou 1997* The oldest long-legged enantiornithean bird.	Length: **11 cm** Hip height: **7.5 cm**	**20 g**
2	*Neuquenornis volans* IUC, SSA (Argentina)	MUCPv-142 Undetermined age	Skull and partial skeleton - *Chiappe & Calvo 1994* South America's most complete.	Length: **23.5 cm** Hip height: **17 cm**	**205 g**
3	*Elbretornis bonapartei* IUC, SSA (Argentina)	PVL-4022 Undetermined age	Partial skeleton - *Walker & Dyke 2009* *Lectavis* or *Yungavolucris*?	Length: **33 cm** Hip height: **24 cm**	**570 g**
4	*Lectavis bretincola* IUC, SSA (Argentina)	PVL-4021 Undetermined age	Tibiotarsus and incomplete metatarsus - *Chiappe 1993* It had very long legs.	Length: **41 cm** Hip height: **30 cm**	**1.15 kg**

Enantiornithes similar to *Yungavolucris* - Fliers, light bodies, very robust legs.

N.º	Genus and species	Largest specimen	Material and data	Size	Mass
5	*Yungavolucris brevipedalis* IUC, SSA (Argentina)	PVL-4052 Undetermined age	Incomplete tarsometatarsus - *Chiappe 1993* It had very robust legs.	Length: **50 cm** Hip height: **25 cm**	**1.75 kg**

1:40

ORNITHUROMORPHA

Ornithuromorphs similar to *Archaeorhynchus* - Fliers, light bodies, semi-aquatic, long-legged, small game hunters and/or piscivores.

N.º	Genus and species	Largest specimen	Material and data	Size	Mass
1	*Archaeorhynchus spathula* ILC, EA (China)	IVPP V14287 Subadult	Skull, partial skeleton, and feathers - *Zhou & Zhang 2006* There was a previous description. *Zhou & Zhang 2005*	Length: **20 cm** Hip height: **15.5 cm**	**275 g**
2	*Jianchangornis microdonta* ILC, EA (China)	IVPP V16708 Subadult	Skull, partial skeleton, and feathers - *Zhou et al. 2009* It had all of its teeth.	Length: **33 cm** Hip height: **26 cm**	**1.2 kg**

Ornithuromorphs similar to *Xinghaiornis* - Fliers, elongated face, light bodies, small game hunters and/or piscivores.

3	Juehuaornis zhangi ILC, EA (China)	SJG 00001A Subadult	Partial skeleton - *Wang et al. 2015* Synonym of *Xinghaiornis*?	Length: **17 cm** Hip height: **7 cm**	**80 g**
4	Xinghaiornis lini ILC, EA (China)	XHPM 1121 Adult	Partial skeleton and feathers - *Wang et al. 2013* Exceptionally long head.	Length: **24 cm** Hip height: **10 cm**	**225 g**

Indeterminate ornithuromorphs similar to *Hollanda* - Fliers, light bodies, sprinters, small game hunters.

5	Hollanda luceria IUC, EA (Mongolia)	MPC-b100/202 Undetermined age	Incomplete frontal limb - *Bell et al. 2010* Very long metatarsi. *Mortimer**	Length: **55 cm** Hip height: **35 cm**	**3 kg**

Indeterminate ornithuromorphs - Fliers, light bodies, semi-aquatic, long-legged, small game hunters and/or piscivores.

6	Cerebavis cenomanica eUC, WA (western Russia)	PIN 5028/2 Undetermined age	Incomplete skull - *Kurochkin et al. 2006* Was considered a "brain." *Walsh et al. 2015*	Length: **18 cm** Hip height: **13.5 cm**	**180 g**
7	Jiuquanornis niui ILC, EA (China)	GSGM-05-CM-021 Undetermined age	Furcula and sternum - *Wang et al. 2013* One of the most incomplete ornithuromorph named.	Length: **21 cm** Hip height: **16 cm**	**300 g**
8	Schizooura liia ILC, EA (China)	IVPP V16861 Undetermined age	Skull and partial skeleton - *Zhou et al. 2012* Its tail was designed for display.	Length: **24 cm** Hip height: **19 cm**	**500 g**
9	Chaoyangia beishanensis ILC, EA (China)	IVPP V9934 Undetermined age	Partial skeleton - *Hou & Zhang 1993* *Chaoyangia* was a plant, so the name was changed to *Chaoyangicarpus*. *Duan 1998*	Length: **24 cm** Hip height: **19 cm**	**500 g**
10	Zhongjianornis yangi ILC, EA (China)	IVPP V15900 Undetermined age	Skull and partial skeleton - *Zhou, Zhang & Li 2010* Enantiornithean bird or basal Avialae? *O'Connor et al. 2013*	Length: **26 m** Hip height: **20 cm**	**600 g**
11	Piscivoravis lii ILC, EA (China)	IVPP V17078 Subadult	Skull, partial skeleton, and feathers - *Zhou et al. 2013* Fish bones were found inside it.	Length: **30 cm** Hip height: **24 cm**	**960 g**
12	Vorona berivotrensis IUC, SAf (Madagascar)	FMNH PA 717 Undetermined age	Incomplete femur - *Forster et al. 1996* Enantiornithean bird? *Mortimer**	Length: **51 cm** Hip height: **40 cm**	**4.5 kg**

1:40

Ornithuromorphs similar to *Patagopteryx* - Relatively heavy bodies, small game hunters or omnivores.

1	Alamitornis minutus IUC, SSA (Argentina)	MACN PV RN 1108 Undetermined age	Incomplete tibiotarsus - *Agnolin & Martinelli 2009* MACN PV RN 1110 a juvenile 5.5 cm long.	Length: **8.5 cm** Hip height: **5.5 cm**	**20 g**
2	Patagopteryx deferrariisi IUC, SSA (Argentina)	MACN-N-03 Undetermined age	Partial skeleton - *Alvarenga & Bonaparte 1992* The first known Mesozoic land bird.	Length: **40 cm** Hip height: **26 cm**	**2 kg**
3	Gargantuavis philoinos IUC, WE (France)	MDE-A08 Undetermined age	Skull and partial skeleton - *Wang et al. 2015* IVPP V12325 is of similar size.	Length: **1.8 m** Hip height: **1.3 m**	**120 kg**

1:100

Ornithuromorphs similar to *Hongshanornis* - Fliers, with or without teeth, light bodies, semi-aquatic, long-legged, small game hunters and/or piscivores.

1	Hongshanornis longicresta ILC, EA (China)	IVPP V14533 Adult	Skull, partial skeleton and feathers - *Zhou & Zhang 2005* DNHM D2945/6 is similar. *O'Connor et al. 2010*	Length: **12 cm** Hip height: **4 cm**	**58 g**
2	Tianyuornis cheni ILC, EA (China)	STM7-53 Subadult	Jaw and partial skeleton - *Zheng et al. 2014* The first with teeth.	Length: **12.8 cm** Hip height: **4.3 cm**	**71 g**
3	Archaeornithura meemannae ILC, EA (China)	STM7-145 Subadult	Partial skeleton - *Wang et al. 2015* One of the oldest ornithuromorph birds.	Length: **13 cm** Hip height: **4.4 cm**	**74 g**
4	Parahongshanornis chaoyangensis eLC, EA (China)	PMOL.AB00161 Undetermined age	Partial skeleton - *Li et al. 2011* Hongshanornithid?	Length: **13 m** Hip height: **4.4 cm**	**75 g**
5	Longicrusavis houi eLC, EA (China)	PKUP V1069 Undetermined age	Skull and partial skeleton - *O'Connor et al. 2010* Hongshanornithid?	Length: **13.5 cm** Hip height: **4.5 cm**	**78 g**

Ornithuromorphs similar to *Yanornis* - Fliers, semi-aquatic, light bodies, piscivores.

6	Songlingornis linghensis ILC, EA (China)	IVPP V10913 Undetermined age	Skull and partial skeleton - *Hou 1997* Songlingornithidae predates Yanornithidae.	Length: **18 cm** Hip height: **11 cm**	**170 g**
7	Yixianornis grabaui ILC, EA (China)	IVPP V12631 Undetermined age	Skull, partial skeleton, and feathers - *Zhou & Zhang 2001* Was a diurnal bird. *Schmitz & Motani 2011*	Length: **19 cm** Hip height: **11.5 cm**	**200 g**
8	Aberratiodontus wui ILC, EA (China)	LHV0001a/b Undetermined age	Skull and partial skeleton - *Gong, Hou & Wang 2004* *Yanornis martini*? *Zhou et al. 2008*	Length: **25 cm** Hip height: **15 cm**	**450 g**
9	Yanornis guozhangi ILC, EA (China)	XHPM 1205 Undetermined age	Skull and partial skeleton - *Wang et al. 2013* More ancient than *Yanornis martini*	Length: **25 cm** Hip height: **15 cm**	**450 g**
10	Yanornis martini ILC, EA (China)	IVPP V12444 Undetermined age	Skull, partial skeleton, and feathers - *Sloan 1999* The specimen IVPP V13358 is eating a fish. *Zhou et al. 2004*	Length: **30 cm** Hip height: **18 cm**	**775 g**

Ornithuromorphs similar to *Gansus* - Fliers, semi-aquatic or terrestrial, light bodies, small game hunters or omnivores.

11	Guildavis tener IUC, ENA (Kansas - USA)	YPM 1760 Undetermined age	Incomplete synsacrum - *Marsh 1880* More basal than *Ichthyornis*. *Clarke 2004*	Length: **14 cm** Hip height: **11.5 cm**	**100 g**
12	Changmaornis houi ILC, EA (China)	GSGM-08-CM-002 Undetermined age	Partial skeleton - *Wang, O'Connor, Li & You 2013* Similar to *Gansus*	Length: **14 cm** Hip height: **12 cm**	**107 g**
13	Yumenornis huangi ILC, EA (China)	GSGM-06-CM-013 Undetermined age	Partial skeleton - *Wang, O'Connor, Li & You 2013* Possibly close to *Iteravis*. *Mortimer**	Length: **16 cm** Hip height: **14 cm**	**155 g**

14	Zhyraornis logunovi eUC, WA (Uzbekistan)	ZIN PO 4600 Undetermined age	Incomplete synsacrum - *Nesov 1992* Smallest theropod of western Asia.	Length: **17 cm** Hip height: **14.5 cm**	**183 g**
15	Horezmavis eocretacea eUC, WA (Uzbekistan)	ZIN PO 3390 Undetermined age	Incomplete tarsometatarsus - *Nesov & Borkin 1983* Was considered a crane. *Nesov 1992*	Length: **17 cm** Hip height: **14.5 cm**	**192 g**
16	Iteravis huchzermeyeri ILC, EA (China)	IVPP V18958 Subadult	Skull and partial skeleton - *Zhou, et al. 2014*	Length: **18 cm** Hip height: **15 cm**	**205 g**
17	Gansus zheni ILC, EA (China)	BMNHC-Ph1342 Adult	Skull and partial skeleton - *Liu et al. 2014* *Iteravis huchzermeyeri?* Mortimer*	Length: **18.5 cm** Hip height: **16 cm**	**230 g**
18	Gansus yumenensis ILC, EA (China)	CAGS-IG-04-CM-002 Undetermined age	Skull and partial skeleton - *You et al. 2006* A footprint IVPP V15083. *Li et al. 2011*	Length: **19 cm** Hip height: **16 cm**	**240 g**
19	Zhyraornis kashkarovi eUC, WA (Uzbekistan)	TsNIGRI 43/11915 Undetermined age	Incomplete synsacrum - *Nesov 1984* Considered a enantiornithean. *Kurochkin 1995*	Length: **20 cm** Hip height: **16.5 cm**	**260 g**
20	cf. "Zhyraornis kashkarovi" eUC, WA (Uzbekistan)	TsNIGRI 43/11915 Undetermined age	Thoracic vertebra - *Nesov & Borkin 1983* Could have been any other bird. *Nesov 1992*	Length: **28 cm** Hip height: **24 cm**	**850 g**

1:40

Ornithuromorphs similar to *Ichthyornis* - Fliers, light bodies, piscivores.

1	cf. Ichthyornis antecessor eUC, WNA (Texas - USA)	ET 4396 Undetermined age	Incomplete carpometacarpus - *Parris & Echols 1992* Older than TMM 31051-24.	Length: **19 cm** Hip height: **8 cm**	**57 g**
2	Ichthyornis cf. anceps eUC, WNA (western Canada)	UA 18456 Undetermined age	Skull and partial skeleton - *Lucas & Sullivan 1982* More recent than SMNH P2077.111.	Length: **21.3 cm** Hip height: **9 cm**	**83 g**
3	Ichthyornis cf. anceps eUC, WNA (Texas - USA)	TMM 31051-24 Undetermined age	Skull and partial skeleton - *Parris & Echols 1992* Too ancient to be *I. anceps*	Length: **27.5 cm** Hip height: **11.6 cm**	**163 g**
4	Ichthyornis anceps eUC, WNA (Kansas - USA, Mexico)	YPM 1739 Adult	Carpometacarpus - *Marsh 1880* The first name is *Ichthyornis anceps*, but *I. dispar* is used more frequently. MUZ-689A was reported in Mexico; it was 24 cm long and weighed 118 g. *Porras-Múzquiz & Chaterjee 2014*	Length: **31 cm** Hip height: **13 cm**	**260 g**

Ornithuromorphs similar to *Ambiortus* - Fliers, light bodies, small game hunters or seed and fruit pickers.

5	Apsaravis ukhaana IUC, EA (Mongolia)	IGM 100/1017 Undetermined age	Skull and partial skeleton - *Norell & Clarke 2001* Lived in very arid environments.	Length: **20.5 cm** Hip height: **12 cm**	**265 g**
6	Ambiortus dementjevi ILC, EA (Mongolia)	PIN 3790-271/272 Undetermined age	Partial skeleton and feathers - *Kurochkin 1982* It is very ancient, despite being a derived genus.	Length: **28 cm** Hip height: **17 cm**	**700 g**
7	Palintropus retusus IUC, WNA (Wyoming - USA)	YPM 513 Undetermined age	Incomplete coracoid - *Marsh 1892* Considered a galliform. *Hope 2002*	Length: **31 cm** Hip height: **19 cm**	**900 g**

1:100

HESPERORNITHIFORMES

Hesperornithiformes similar to *Pasquiaornis* - Short and heavy bodies, aquatic, piscivores.

N.º	Genus and species	Largest specimen	Material and data	Size	Mass
1	Pasquiaornis hardiei eUC, WNA (western Canada)	SMNH P2077.60 Undetermined age	Femur - *Tokaryk, Cumbaa & Storer 1997* Similar to *Enaliornis*	Length: **57 cm** Hip height: -	**1.55 kg**
2	Potamornis skutchi eUC, WNA (Wyoming - USA)	UCMP 73103 Undetermined age	Quadrate bone - *Elzanowski et al. 2001* The specimen UCMP 117605 is also attributed to it.	Length: **64 cm** Hip height: -	**2.2 kg**
3	Enaliornis sedgwicki ILC, WE (England)	SMC B55315 Adult	Incomplete tibiotarsus - *Seeley 1869* There are numerous adult specimens.	Length: **67 cm** Hip height: -	**2.3 kg**
4	"Enaliornis" seeleyi ILC, WE (England)	SMC B55317 Adult	Incomplete tibiotarsus - *Seeley 1869* Probably a different genus. Mortimer*	Length: **76 cm** Hip height: -	**3.5 kg**
5	Enaliornis barretti ILC, WE (England)	SMC B55331 Undetermined age	Incomplete tarsometatarsus - *Galton & Martin 2002* The type specimen of the genus.	Length: **86 cm** Hip height: -	**5.1 kg**
6	"Pasquiaornis" tankei eUC, WNA (western Canada)	SMNH P2077.63 Undetermined age	Incomplete tarsometatarsus - *Tokaryk et al. 1997* Possibly a different genus.	Length: **90 cm** Hip height: -	**5.7 kg**

Hesperornithiformes similar to *Baptornis* - Long and heavy bodies, aquatic, piscivores.

7	"Hesperornis" macdonaldi IUC, WNA (South Dakota - USA)	LACM 9728 Adult	Femur - *Martin & Lim 2002* More basal than *Hesperornis*	Length: **55 cm** Hip height: -	**1.55 kg**
8	Judinornis nogontsavensis IUC, EA (Mongolia)	ZIN PO 3389 Undetermined age	Incomplete thoracic vertebra - *Nesov & Borkin 1983* Originally thought to be close to gulls.	Length: **58 cm** Hip height: -	**1.85 kg**
9	Parascaniornis stensioei IUC, WE (Sweden)	RM PZ R1261 Undetermined age	Incomplete tarsometatarsus - *Rees & Lindgren 2005* The type specimen is the thoracic vertebra MGUH 1908.214. *Lambrecht 1933*	Length: **71 cm** Hip height: -	**2.8 kg**
10	Baptornis advenus IUC, ENA (Kansas - USA)	KUVP 2290 Adult?	Partial skeleton - *Lucas 1903* Other smaller specimens were adults.	Length: **92 cm** Hip height: -	**7.2 kg**

Hesperornithiformes similar to *Brodavis* - Long and heavy bodies, aquatic, piscivores.

| 11 | Brodavis baileyi
IUC, WNA (South Dakota - USA) | USNM 50665
Undetermined age | Incomplete tarsometatarsus - *Martin et al. 2012*
The piece is longer than that of *B. americanus* | Length: **62 cm**
Hip height: - | **1.55 kg** |

12	*Brodavis americanus* IUC, WNA (western Canada)	RSM P2315.1 Undetermined age	Incomplete tarsometatarsus - *Martin et al. 2012* Intermediate between *B. baileyi* and *B. varneri*.	Length: **92 cm** Hip height: -	**5 kg**
13	*Brodavis mongoliensis* IUC, EA (Mongolia)	PIN 4491-8 Undetermined age	Incomplete tarsometatarsus - *Kurochkin 2000* Interpreted to be of the genus *Brodavis*. *Martin et al. 2012*	Length: **96 cm** Hip height: -	**5.7 kg**
14	*Brodavis varneri* IUC, WNA (South Dakota - USA)	SDSM 68430 Adult	Partial skeleton - *Martin & Cordes-Person 2007* Considered to belong to the genus *Baptornis*.	Length: **1.15 m** Hip height: -	**9.2 kg**

Hesperornithiformes similar to *Hesperornis* - Long and heavy bodies, aquatic, piscivores.

15	*Hesperornis altus* IUC, WNA (Montana - USA)	YPM 515 Adult	Incomplete tibiotarsus - *Marsh 1893* Difficult to identify. *Mortimer**	Length: **80 cm** Hip height: -	**3.8 kg**
16	*Hesperornis mengeli* IUC, WNA (western Canada)	BO 780106 Undetermined age	Incomplete tarsometatarsus - *Martin & Lim 2002* May not be of the genus *Hesperornis*.	Length: **88 cm** Hip height: -	**5 kg**
17	*Fumicollis hoffmani* IUC, ENA (Kansas - USA)	UNSM 20030 Undetermined age	Partial skeleton - *Martin & Tate 1976* Considered *Baptornis advenus*. *Bell & Chiappe 2015*	Length: **95 cm** Hip height: -	**6.4 kg**
18	*Parahesperornis alexi* eUC, ENA (Kansas - USA)	KUVP 2287 Subadult	Skull, partial skeleton, and feathers - *Martin 1984* It had impressions of scales and feathers. *Williston 1898*	Length: **1.05 m** Hip height: -	**8.4 kg**
19	*Hesperornis bairdi* IUC, WNA (South Dakota - USA)	PU 17208A Undetermined age	Partial skeleton - *Martin & Lim 2002* Contemporary of "*Hesperornis*" *macdonaldi*.	Length: **1.05 m** Hip height: -	**8.6 kg**
20	*Canadaga arctica* IUC, WNA (western Canada)	NMC 41050 Undetermined age	Partial skeleton - *Hou 1999* Northernmost Mesozoic theropod.	Length: **1.15 m** Hip height: -	**11.5 kg**
21	cf. *Coniornis altus* IUC, WNA (South Dakota - USA)	YPM PU 17208D? Undetermined age	Partial frontal limb - *Bell et al. 2015* Not described.	Length: **1.15 m** Hip height: -	**11.7 kg**
22	*Asiahesperornis bazhanovi* IUC, WA (Kazakhstan)	IZASK 5/287/86a Undetermined age	Incomplete tarsometatarsus - *Nesov & Prizemlin 1991* IZASK 1/KM 97 is of similar size. *Malakhov & Ustinov 1998*	Length: **1.25 m** Hip height: -	**14.8 kg**
23	*Hesperornis gracilis* IUC, ENA (Kansas - USA)	YPM 1478 Undetermined age	Incomplete tarsometatarsus and phalanx - *Marsh 1880* Its metatarsus is more elongated than that of *Hesperornis regalis*.	Length: **1.3 m** Hip height: -	**17.5 kg**
24	*Hesperornis crassipes* IUC, ENA (Kansas - USA)	YPM 1474 Undetermined age	Jaw and partial skeleton - *Marsh 1876* Contemporary of, but unlike, *Hesperornis regalis*	Length: **1.35 m** Hip height: -	**19.4 kg**
25	*Hesperornis regalis* IUC, ENA, WNA (Kansas, South Dakota - USA)	YPM 1476 Undetermined age	Partial skeleton - *Marsh 1880* The specimen YPM 1206 is the most complete Hesperornithes. *Marsh 1875*	Length: **1.4 m** Hip height: -	**20.4 kg**
26	*Hesperornis chowi* IUC, WNA (South Dakota - USA)	PU 17208 Undetermined age	Tarsometatarsus - *Martin & Lim 2002* Unlike and more recent than *Hesperornis regalis*.	Length: **1.4 m** Hip height: -	**20.6 kg**
27	*Hesperornis* cf. *regalis* IUC, WNA (western Canada)	UA 9716 Undetermined age	Tarsometatarsus - *Fox 1974* More northern and recent than *Hesperornis regalis*.	Length: **1.48 m** Hip height: -	**25.6 kg**
28	*Hesperornis rossicus* IUC, WA (western Russia)	ZIN PO 5463 Adult?	Incomplete tarsometatarsus - *Nesov & Yarkov 1993* Other smaller specimens are subadults.	Length: **1.6 m** Hip height: -	**30 kg**
29	*Hesperornis rossicus* IUC, WE (Sweden)	SGU 3442 Ve02 Adult?	Thoracic vertebra - *Rees & Lindgren 2005* The largest specimen is from Sweden and is identical in size to the one from Russia.	Length: **1.6 m** Hip height: -	**30 kg**

1:100

1 2 3 4 5 6 7 8 9 10 11 12 13 14 15 16

17 18 19 20 21 22 23 24 25 26 27 28 29

NEORNITHES

Indeterminate Neornithes - Fliers, light bodies, small game hunters or omnivores.

N.°	Genus and species	Largest specimen	Material and data	Size	Mass
1	*Apatornis celer* IUC, ENA (Kansas - USA)	YPM 1451 Undetermined age	Incomplete synsacrum - *Marsh 1873* Was considered a *Ichthyornis* species.	Length: **13 cm** Hip height: **12 cm**	**100 g**
2	*Iaceornis marshi* IUC, ENA (Kansas - USA)	YPM 1734 Undetermined age	Partial skeleton - *Clarke 2004* Similar to *Apatornis celer*	Length: **14 cm** Hip height: **13.5 cm**	**145 g**
3	*Gallornis straeleni* IUC, WE (France)	RBINS Undetermined age	Humerus and incomplete femur - *Lambrecht 1931* Its age is in doubt.	Length: **27.5 cm** Hip height: **23.5 cm**	**775 g**

Neornithes similar to *Limenavis* - Fliers, light bodies, small game hunters or omnivores.

4	*Limenavis patagonica* IUC, SSA (Argentina)	PVL 4731 Undetermined age	Partial frontal limb - *Clarke & Chiappe 2001* Similar to *Lithornis*? *Mortimer**	Length: **20 cm** Hip height: **15.5 cm**	**600 g**

Neornithes similar to *Austinornis* - Fliers, light bodies, small game hunters or omnivores.

5	*Austinornis lentus* IUC, WNA (Texas - USA)	YPM 1796 Undetermined age	Incomplete tarsometatarsus - *Marsh 1877* Was considered a *Ichthyornis*. *Clarke 2004*	Length: **30 cm** Hip height: **18 cm**	**370 g**

Neornithes similar to *Vegavis* - Light bodies, small game hunters and/or filter feeders.

6	*Vegavis iaai* IUC, AN (Antarctica)	MLP 93-I-3-1 Adult	Partial skeleton - *Clarke et al. 2005*	Length: **46 cm** Hip height: **27 cm**	**340 g**
7	*Teviornis gobiensis* IUC, EA (Mongolia)	PIN 4499-1 Undetermined age	Partial frontal limb - *Kurochkin et al. 2002* Its name was composed by Victor Tereschenko.	Length: **58 cm** Hip height: **33 cm**	**680 g**

#	Species / Location	Specimen / Age	Notes	Size	Weight
8	cf. *Cimolopteryx rara* *IUC, WNA (Wyoming - USA)*	UCMP 53964 Undetermined age	Carpometacarpus and scapula - *Brodkorb 1963* Was considered a *Cimolopteryx rara* specimen but was a duck.	Length: **1.15 m** Hip height: **67 cm**	5.2 kg
9	"Styginetta lofgreni" (n.n.) *IUC, WNA (Montana - USA)*	RAM 6707 (V94078) Undetermined age	Incomplete coracoid - *Stidham 2001* Described in a thesis. *Stidham 2009*	Length: **1.2 m** Hip height: **70 cm**	6 kg

Neornithes similar to *Polarornis* - Fliers, aquatic, relatively light bodies, piscivores.

#	Species / Location	Specimen / Age	Notes	Size	Weight
10	*Neogaeornis wetzeli* *IUC, SSA (Chile)*	GPMK 123 Undetermined age	Tarsometatarsus - *Lambrecht 1929* There is a referred specimen. *Schneider 1940*	Length: **68 cm** Hip height: -	3 kg
11	*Polarornis gregorii* *IUC, AN (Antarctica)*	TTU P 9265 Undetermined age	Skull and partial skeleton - *Chatterjee 2002* Was thought to be a chimera.	Length: **75 cm** Hip height: -	4 kg

Neornithes similar to *Lonchodytes* - Fliers, light bodies, piscivores.

#	Species / Location	Specimen / Age	Notes	Size	Weight
12	*Lonchodytes estesi* *IUC, WNA (Texas - USA)*	UCMP 53954 Undetermined age	Incomplete tarsometatarsus - *Brodkorb 1963* Gaviiform or Procellariiform? *Hope 2002*	Length: **32 cm** Hip height: -	610 g

Neornithes similar to *Volgavis* - Fliers, light bodies, piscivores.

#	Species / Location	Specimen / Age	Notes	Size	Weight
13	*Torotix clemensi* *IUC, WNA (Wyoming - USA)*	UCMP 53958 Undetermined age	Incomplete humerus - *Brodkorb 1963* Pelecaniform? *Hope 2002*	Length: **32 cm** Hip height: **12 cm**	140 g
14	*Volgavis marina* *IUC, EE (European Russia)*	ZIN PO 3638 Undetermined age	Partial skull - *Nesov & Jarkov 1989* Very similar to Fregatidae. *Hope 2002*	Length: **60 cm** Hip height: **24 cm**	1 kg

1:100

Neornithes similar to *Cimolopteryx* - Fliers, light bodies, small game hunters or piscivores.

#	Species / Location	Specimen / Age	Notes	Size	Weight
1	*Lamarqueavis australis* *IUC, SSA (Argentina)*	MML 207 Undetermined age	Incomplete coracoid - *Agnolin 2010* South America's smallest flying bird.	Total height: **9.5 cm** Hip height: **7 cm**	18 g
2	*Lamarqueavis minima* *IUC, WNA (Wyoming - USA)*	UCMP 53976 Undetermined age	Incomplete coracoid - *Brodkorb 1963* Previously known as *Cimolopteryx minima*	Total height: **11.5 cm** Hip height: **8 cm**	31 g
3	*Lamarqueavis petra* *IUC, WNA (Wyoming - USA)*	AMNH 21911 Undetermined age	Incomplete coracoid - *Hope 2002* Very similar to *Cimolopteryx rara*	Total height: **15.3 cm** Hip height: **11 cm**	73 g
4	"*Lonchodytes*" *pterygius* (n.d.). *IUC, WNA (Wyoming - USA)*	UCMP 53961 Undetermined age	Incomplete carpometacarpus - *Brodkorb 1963* Close to Charadriiforme. *Hope 2002*	Total height: **16.2 cm** Hip height: **11.8 cm**	88 g
5	*Cimolopteryx rara* *IUC, WNA (Wyoming - USA)*	YPM 1805 Undetermined age	Incomplete coracoid - *Marsh 1892* The name informally appears in Marsh 1889.	Total height: **20.4 cm** Hip height: **14.8 cm**	175 g
6	*Ceramornis major* *IUC, WNA (Wyoming - USA)*	UCMP 53959 Undetermined age	Incomplete coracoid - *Brodkorb 1963* Its name means "bird molded in clay."	Total height: **26.7 cm** Hip height: **19 cm**	390 g
7	*Cimolopteryx maxima* *IUC, WNA (Wyoming - USA)*	UCMP 53973 Undetermined age	Incomplete coracoid - *Brodkorb 1963* Largest species of *Cimolopteryx*	Total height: **30 cm** Hip height: **22 cm**	550 g
8	cf. *Cimolopteryx maxima* *IUC, WNA (Wyoming - USA)*	UCMP 53957 Undetermined age	Incomplete coracoid - *Brodkorb 1963* Known as "Ornithurine F."	Total height: **32 cm** Hip height: **23 cm**	660 g
9	*Graculavus augustus* *IUC, WNA (Wyoming - USA)*	AMNH 25223 Undetermined age	Incomplete humerus - *Hope 1999* Discovered in 1985. *Mortimer**	Total height: **34.5 cm** Hip height: **25 cm**	840 g

1:30

Glossary

Acetabulum
The cavity or socket formed by the ilium, the ischium and the pubis that articulates with the femur or thigh bone.

Aetosaur
Herbivorous armored archisauromorphs that lived in the late Triassic period.

Air sacs
Organs connected to the lungs of some dinosaurs, including birds.

Allometric growth
Growth of an organism in which different parts develop at different rates, causing its appearance to change as it grows.

Amniote
A group of vertebrates capable of developing an embryo protected by a shell, several membranes and amniotic fluid. Reptiles, birds and mammals are all amniotes.

Angiosperm
A group of plants that produce flowers and fruit.

Ankylosaurids
A family of herbivorous quadruped dinosaurs characterized by their armor and their broad, low-slung bodies. Some species had a club at the end of their tails.

Apophysis
A bony protrusion that forms part of a joint or is used as a muscle anchor.

Archosauromorph
A group of vertebrates that includes archosaurs and other primitive vertebrates such as proterosuchids and trilophosaurs.

Archosaurs
A group of amniotes with several cranial foramina. Includes pterosaurs and extinct dinosaurs as well as birds and crocodylomorphs.

Arctometatarsal
A condition in which the proximal part of the middle metatarsal is compressed between two other metatarsals.

Arthropods
A group of invertebrates having an exoskeleton, legs and antennae made of jointed parts. Insects, spiders, and crustaceans are all in this group.

Asymmetrical footprint
Footprint in which the digits on the right and left halves are of different proportions.

Avulsion
Severe detachment of a ligament.

Basal
An organism or feature that is phylogenetically primitive.

Basipodium
The area of the hand or foot that includes the carpals, tarsals, metacarpals and metatarsals.

Biomechanics
The scientific field focused on the study of mechanical structures in living beings.

Biped
Animal that uses only its hind legs to move around.

Ceratopsids
A family of four-legged herbivorous dinosaurs characterized by facial horns and large frills on the head.

Cetaceans
A group of marine mammals that includes whales, dolphins and porpoises.

Clade
A group of various related organisms with a common ancestor.

Cladistics
The branch of biology that defines evolutionary relations among organisms based on common derived characteristics.

Cololite
Fossilized stomach contents.

Convergent evolution
The process in which species of different lineages evolve similar characteristics.

Cranial kinesis
The ability of an organism's skull bones to move relative to each other to facilitate eating.

Crest
A bunch of feathers found on the top of some birds' heads.

Cursorial
Animal adapted for running.

Denticles
Small pointed structures that together form the serrated edges of most theropod teeth.

Dermestids
A family of beetles consisting of species that consume foods such as carrion, excrement or pollen.

Didactyl
Having only two digits.

Dura matter
The outermost membrane enveloping the brain.

Ectaxonic
Hand or foot in which the fourth or fifth digit is the longest.

Egagropila
Balls made of bone with feathers or hair that some birds disgorge after eating their prey whole. When fossilized they are known as regurgitaliths.

Epicontinental
Describes the shallow sea overlying the continental shelf.

Epidermis
Outermost layer of the skin.

Erythrosuchids
A family of quadruped carnivorous archosaurs that existed in the Triassic.

Evolution
The process that gives rise to and transforms living beings over time.

Exostosis
An abnormal bone outgrowth.

Fossil
Any naturally-preserved evidence of life in the past.

Frontal (bone)
Bone located on the anterior dorsal part of the cranium.

Gastrolith
Stones ingested by certain animal species to aid digestion or as mineral supplements.

Gland
A group of cells that produce and expel liquid substances required by an organism to function.

Hadrosaurids
A family of phytophagous facultative bipeds commonly known as duck-billed dinosaurs.

Hemangioma
A benign tumor caused by the abnormal buildup of blood vessels.

Ichnospecies
A species based entirely on fossil footprints.

Ichnotaxon
Genus based exclusively on footprints and/or other trace fossils.

Ichthyosaur
A group of aquatic sauropods that lived in the Mesozoic Era, similar in appearance to dolphins.

Intraspecific
Biological interaction between members of the same species.

Ischial callosities
Fossil impression of the ischial zone.

Limulids
A family of very primitive marine arthropods related to spiders and scorpions. The horseshoe crab is a living example.

Lobed fingers
Fingers that are rounded or very wide.

Locomotion
The act of moving from one place to another.

Long bones
Elongated or cylindrical bones such as the femur, the phalanges and the radius.

Mammal-like reptiles
A clade of amniotes with attributes that situate them between reptiles and true mammals.

Melanosome
Organelles located inside melanocytes that contain the pigments responsible for skin, feather, and hair color.

Mesaxonic
Having hands or feet in which the middle digit is longest.

Metabolism
The series of primordial chemical reactions that enable an organism to function properly.

Mosasaurs
The flat bone that forms the nasal passage or cavity.

Neonate
A newborn.

Oligocene
An epoch of the Cenozoic Era that began 33.9 Ma and ended 23 Ma.

Ontogeny
The history of the growth and development of an organism from embryo to adult.

Ooespecies
Species based entirely on fossilized eggs.

Orionides
A group of tetanurin theropods that includes megalosauroids, allosauroids and coelurosaurs but not celofisoids or ceratosauria.

Ornithischians
Phytophagous dinosaurs with horny beaks, such as ankylosaurs, stegosaurs, ceratopses, pachycephalosaurs and ornithopods.

Orogeny
A group of mountains in a zone, also called ranges or orogenic belts.

Osteoarthritis
An illness caused by the wearing down of the joints over time.

Osteoderm
An ossified portion of skin on some animals, such as the dorsal scales on a crocodile or the armor of an armadillo.

Osteomyelitis
Inflammation of the bone and bone marrow.

Osteopetrosis
An avian disease that causes an increase in bone density.

Osteophytes
Small tissue deposits near an area of inflammation.

Ostracods
Tiny saltwater/freshwater crustaceans protected by two valves like those found on clams.

Paraphyletic
In cladistics, a group of organisms descended from a common ancestor but not including all of that ancestors' descendants.

Perinatal
Around the time of birth or hatching.

Petroglyph
Human-made images carved on stone.

Phreatomagmatics
Material resulting from an eruption caused when water and magma interact, as in geysers.

Phylogeny
Evolutionary history of a group of organisms.

Phytophagus
Animals that eat plants.

Phytosaur
A primitive archosaur from the late Triassic similar to a crocodile.

Plantigrade
An animal that walks entirely on the soles of its feet.

Pneumatic bones
Hollow bones with air chambers inside, such as bird bones.

Podotheca
Integument (tough outer layer) of skin, scales or pads that covers the feet of a bird or theropod.

Postcrania
The parts of the skeleton that include the vertebrae and long bones but not the skulls.

Prestosuchids
A family of carnivorous quadruped archosaurs that lived during the Triassic.

Prolacertiforms
A group of terrestrial, aquatic, or tree-dwelling sauropsids that lived from the Permian to the Triassic.

Proterochampsids
A family of archosauromorphs similar in appearance to crocodiles.

Protozoa
Microscopic single-celled eukaryotes such as amoebas, ciliates and apicomplexa.

Pygostyle
The final part of the vertebral column on birds and some non-avian theropods resulting from the fusion of the final caudal vertebrae.

Pyroclastics
A mixture of gas and hot volcanic material.

Quadruped
Animal that moves around on four legs.

Ratites
A group of flightless birds that includes ostriches, cassowaries, emus, ñandus, tinamous and several extinct bird species.

Rauisuchid
A family of carnivorous and plant-eating archosaurs that existed during the Triassic.

Rectrices
Tail feathers.

Regurgitalith
Fossilized egagropila (regurgitated food pellet).

Remiges
Wing feathers.

Rynchocephalic
Group of sauropsids that includes the modern-day species of tuatara (*Sphenodon*). They are also known as sphenodonts.

Saurischia
A group of dinosaurs that includes theropods and sauropodomorphs.

Sauropodomorpha
A group of herbivorous dinosaurs that includes sauropods and their ancestors.

Sauropsids
A group of vertebrates that includes reptiles and birds, but not synapsids or mammal-like reptiles.

Sexual dimorphism
A set of morphological and physiological variations that distinguish males and females of the same species.

Specific epithet
One part of the binomial that comprises a species' scientific name.

Spondylitis
An infection of the vertebra.

Styliform bone
A thin, elongated, pointed bone with a needle- or awl-like shape.

Supratemporal foramen
A space located on the posterior dorsal part of the cranium that facilitates attachment of the jaw muscles.

Taphonomy
The branch of paleontology that studies factors that influence the preservation or destruction of organic remains.

Taxonomy
The branch of biology concerned with the hierarchical classification of living things.

Tetrapods
A group of vertebrates with two pairs of limbs. It includes all vertebrates except fish.

Therapsid
The group of synapsids that gave rise to mammals.

Trace fossil (ichnofossil)
A fossil that provides direct or indirect evidence of an organism's interaction with sediments.

Trilophosaur
A group of phytophagous archosauromorphs that lived during the Triassic.

Type specimen (holotype)
The principal specimen (complete organism or part thereof) used to represent a new species when first reported in a publication.

Vertebral Arch
The upper half of a vertebra, comprising the neural arch and neural spine and more commonly known as a backbone.

Web-footed
Birds, usually aquatic, with membranes between their toes.

Taxonomic index

Bibliography

General

www.eofauna.com/book/theropoda_records_refB.pdf

Scientific webpages

Databases
Mortimer, Mickey – Theropod Database
www.theropoddatabase.com
Ford, Tracy - Paleofile
www.paleofile.com
www.dinohunter.info
Tweet, Justin – Thescelosaurus page (extinct)
Aragosaurus
www.aragosaurus.com
Stuchlik, Krzysztof - Tribute to Dinosaurs
www.dinoanimals.pl/pliki/Baza_Dinozaurow.xlsx
www.encyklopedia.dinozaury.com
Olshevsky, Jorge – Dinogenera
www.polychora.com/dinolist.html

Reconstructions
Paul, Gregory – The Science and Art of Gregory S. Paul
www.gspauldino.com
Hartman, Scott – Skeletal Drawing
www.skeletaldrawing.com

Chronology and fossils
The Paleobiology Database
www.paleobiodb.org

Paleomaps
Scotese, Christopher R. – Paleomap Project
www.scotese.com
Blakey, Ron – Colorado Plateau Geosystems, Inc.
www.cpgeosystems.com/paleomaps.html

Blogs
Headden, Jaime A. – The Bite Stuff
www.qilong.wordpress.com/
Cau, Andrea – Theropoda blog
www.theropoda.blogspot.com

Appendix

Statistics and other data

www.eofauna.com/book/theropoda_appendixB.pdf

Acknowledgments

Many friends and colleagues have assisted us on our passionate journey of making this book. First of all, we would like to thank all of those who helped us to compile scientific documentation: Augusto Haro, Tracy Ford, Ignacio Díaz Martínez, René Hernández Rivera, Sue Turner, Tony Thulborn, José Rubén Guzmán Gutiérrez, Alexander Elistratov, Jorge Aragón Palacios, Pedro de Luna, Krzysztof Rogoz, Octavio Mateus, Roman Ulasnsky, Richard Hofmann, and especially the Facebook group Wikipaleo. We also wish to thank all of the authors who answered questions that arose during the publication of this book: Gregory S. Paul, Dick Mol, Adrian Lister, Scott Hartman, Spencer Lucas, and Mauricio Antón. Particular thanks are due to Mickey Mortimer for his advice on the identification of incomplete specimens, as well as for his incredible website, Theropod Database. Thanks also to Paola Franco Echecury for her assistance with historic information. Our appreciation is also extended to everyone who helped with visits and access to different museums and collections: Marcela Milani, Nehuen Gionto, Juan D. Porfiri, and David Alejandro Vouillat. Thanks to Leví Bernardo Martínez Reza for his comments after reviewing the work. Of course, we are deeply grateful to the people who inspired us as children, instilling in us a life-long passion for dinosaurs and prehistoric life in general, as without them this book would not have been possible: David Norman, Francisco Vega Vera, and Beatriz Inés González de Suso. Lastly, we thank the entire editorial team at Larousse for its support as we brought this project to completion and for having enabled us to make our dream a reality. To anyone we may have inadvertently overlooked, we extend our sincerest apologies.